P9-BYJ-353

Compilation of E.P.A.'s Sampling and Analysis Methods

Edited by
Lawrence H. Keith

Compiled by
William Mueller
&
David L. Smith

Library of Congress Cataloging-in-Publication Data

Catalog information is available from the Library of Congress.

ISBN 0-87371-433-4

LEWIS PUBLISHERS, INC.
121 South Main Street, Chelsea, MI 48118

PRINTED IN THE UNITED STATES OF AMERICA
2 3 4 5 6 7 8 9 0

THE EDITOR

Dr. Keith is a Senior Program Manager and Principle Scientist at Radian Corporation in Austin, Texas. A pioneer in environmental sampling and analysis, method development, and handling of hazardous compounds, Dr. Keith has published over a dozen books and presented and published more than one hundred technical articles. Recent publications have involved electronic book formats and expert systems. Dr. Keith serves on numerous government, academic, publishing, and environmental committees and is a past chairman of the ACS Division of Environmental Chemistry. Prior to joining Radian Corporation in 1977, he was a research scientist with the U.S. Environmental Protection Agency.

INTRODUCTION

There is an increasing number of analytes and corresponding methods for measuring them in the environment and this often makes selection of the most appropriate ones difficult. In 1988 David L. Smith and William Mueller at EPA's Risk Reduction Engineering Laboratory (RREL) in Cincinnati, Ohio compiled summaries of many of these methods and analytes in a *D-BASE III* format so they could be searched by key words.

The objective of this database was to help EPA contractors and other scientists to rapidly and easily select the most appropriate methods of sampling and analysis for a particular situation without having to become an expert in using these methods or searching through many volumes of published EPA methods. Unfortunately, as the database grew in size the time for a search took longer and longer. Another problem was that not everyone had access to *D-BASE III*.

In talking to David, I suggested the use of a simple *"free-text"* searching program in conjunction with a text version of EPA's files. The advantage is production of compact files that can be published and widely distributed so that everyone can easily use them. Accordingly, William Mueller and David L. Smith provided an *ASCII* text dump of their *D-BASE III* files and I provided an extensive edit of them, added the searching program, wrapped all the files and programs with a menu for easy access, and wrote the tutorial and manual. In the ensuing years I was able to also help double the number of the method/analyte summaries in this electronic publication.

The result is a first-of-its kind *"electronic reference book"* on EPA's Sampling and Analytical Methods. Appropriate methods can quickly be located using any number of keywords to focus on the specific needs of users. Each summary of an analyte and method is designed to be *"self standing"*. In other words, each summary may be used by itself in a report or as a file in combination with other method/analyte summaries without loss of information.

This book is a printed version of EPA's *Sampling and Analysis Methods Database*. As such, it is designed to be a convenient reference for use with that database. Each chapter consists of the data in a file from the above reference so those summaries can readily be looked up after the searching program has located them. Although all of the method/analyte summaries can be printed using the searching program supplied with the database, there are so many of them that most people may prefer to also have a bound volume of that data. In other cases, where an IBM-compatible computer is not readily accessible, this bound volume can also serve as convenient reference source although it is not readily searchable like the electronic database.

In spite of the large number of method/analyte summaries in this publication (more than 650) not all of them are covered yet. There are still numerous methods for *"specialized"* analytes (for example, the chlorinated acids or the nitrogen- and phosphorus-containing pesticides) and for the large number of semivolatile compounds with GC/MS methods. While many of the semivolatile compounds in the latter category are covered in this publication, the reader should be aware that they are not as completely covered as the volatile (purgeable) compounds. These omitted method/analyte summaries may be added later if reader interest indicates that the work required to add them would be warranted.

I hope you will find these summaries to be a valuable addition to your technical references. At Radian we routinely use all of these methods and we'll be glad to try to answer questions you may have about any of them. If you would like information to help solve a sampling and analysis problem not covered in this volume, just write to me or give me a call.

My address and telephone number is:

Dr. Lawrence H. Keith
Radian Corporation
P.O. Box 201088
Austin, TX 78720-1088

Telephone: (512) 454-4797

If I don't know the answer to your questions, someone else at Radian surely will so together we'll get you the answer or help that you need. We have a team of technical experts that is over 1,000 strong to call on!

TABLE OF CONTENTS

Chapter 1
Chlorinated Aliphatic Volatile Organic Compounds

PRIMARY NAME: Carbon Tetrachloride

Method 502.1

TITLE: Halogenated VOCs in Water

MATRIX: Drinking water (finished or any treatment stage) and raw source water.

CAS # : 56-23-5

APPLICATION: Method covers 40 halogenated VOCs. An inert gas is bubbled through a 5 mL water sample. Purged sample components are trapped in tube of sorbent materials. When purging is complete, sorbent tube is heated and backflushed with inert gas to desorb trapped sample onto a packed GC column where it is separated.

INTERFERENCES: During analysis, major contaminant sources are volatile materials in the lab and impurities in purging gas and sorbent trap. With high and low level samples, there can be carry-over contamination. Watch for methylene chloride, it permeates through PTFE tubing.

INSTRUMENTATION: Purge and Trap GC w/halide specific detector. (Two GC columns are recommended); Column #1: 1% SP-1000 on Carbopack B; Column #2: n-octane bonded on Poracil C.

RANGE: 8.0-505 μg/L (Drinking water)

MDL: 0.003 μg/L in water

PRECISION: RSD = 7.0% @ 0.20 μg/L spike (S.L.) (17 Samples)

ACCURACY: Avg Recovery = 90% @ 0.20 μg/L spike (S.L.) (17 Samples)

SAMPLING METHOD: Use a 40-120 mL screw cap vial (pre-washed with detergent, rinsed with distilled water and oven dried at 105°C) with a PTFE-faced silicone septum. If residual chlorine is in the water add about 25 mg of ascorbic acid to each vial before sample collection. Collect bubble free samples.

STABILITY: Cool to 4°C; HCl to pH <2. M.H.T. = 14 Days.

QUALITY CONTROL: As an initial demonstration of lab accuracy and precision, analyze 4 to 7 replicates of a lab fortified blank containing analyte at 0.1-5 μg/L. Collect all samples in duplicate.

REFERENCE: Method 502.1, Volatile Halogenated Organic Compounds in Water by Purge and Trap GC, EPA 600/4-88/039.

PRIMARY NAME: Carbon Tetrachloride Method 502.2

TITLE: VOCs in H_2O by Purge and Trap capillary column GC

MATRIX: Drinking water (finished or any treatment stage) and raw source water.

CAS # : 56-23-5

APPLICATION: Method covers 60 VOCs. An inert gas is bubbled through a 5 mL sample in a purging chamber. Purged sample components are trapped in tube of sorbent materials. When purging is done, sorbent tube is heated and backflushed w/helium to desorb trapped sample onto a capillary GC column.

INTERFERENCES: During analysis, major contaminant sources are volatile materials in the lab and impurities in purging gas and sorbent trap. With high and low level samples, there can be carry-over contamination. Beware of methylene chloride, it permeates through PTFE tubing.

INSTRUMENTATION: (Capillary Column) GC, w/PID and electrolytic conductivity detectors in series. Column #1: VOCOL wide bore column; Column #2: RTX-502.2 mega bore column; Column #3: DB-62 mega bore column.

RANGE: Approximately 0.02-200 µg/L

MDL: 0.01 µg/L (reagent H_2O) (ELCD) 60 m × 0.75 mm VOCOL capillary column

PRECISION: RSD = 3.6% @ Spike of 10 µg/L (reagent H_2O) (ELCD) 60 m × 0.75 mm VOCOL capillary column

ACCURACY: Recovery = 92% @ spike of 10 µg/L (reagent H_2O) (ELCD) 60 m × 0.75 mm VOCOL capillary column

SAMPLING METHOD: Use a 40-120 mL screw cap vial (pre-washed with detergent, rinsed with distilled water and oven dried at 105°C) with a PTFE-faced silicone septum. If residual chlorine is in the water add about 25 mg of ascorbic acid to each vial before sample collection. Collect bubble free samples.

STABILITY: Cool to 4°C; HCl to pH <2.

M.H.T. = 14 Days.

QUALITY CONTROL: As an initial demonstration of lab accuracy and precision, analyze 4 to 7 replicates of a lab fortified blank containing analyte at 0.1-5 µg/L. Collect all samples in duplicate.

REFERENCE: Method 502.2, Volatile Organic Compounds in Water by Purge and Trap Capillary Column GC, EPA 600/4-88/039.

PRIMARY NAME: Carbon Tetrachloride

Method 524.1

TITLE: VOCs in H_2O by Purge and Trap GC/MS

MATRIX: Drinking water (finished or any treatment stage) and raw source water.

CAS # : 56-23-5

APPLICATION: Method covers 51 VOCs. An inert gas is bubbled through a 25 mL water sample. Purged sample components are trapped in tube of sorbent materials. When purging is complete, sorbent tube is heated and backflushed with helium to desorb trapped sample onto a packed GC column.

INTERFERENCES: During analysis, major contaminant sources are volatile

materials in the lab and impurities in purging gas and sorbent trap. With high and low level samples, there can be carry-over contamination. Watch for methylene chloride, it permeates through PTFE tubing.

INSTRUMENTATION: Gas Chromatography/Mass Spectrometry/Data System; 1% SP-1000 on Carbopack B column.

RANGE: Approximately 0.2-200 μg/L MDL: 0.3 μg/L

PRECISION: RSD = 11% @ 1.0 μg/L

ACCURACY: Mean Accuracy = 88% @ 1.0 μg/L

SAMPLING METHOD: Use a 60-120 mL screw cap vial (pre-washed with detergent, rinsed with distilled water and oven dried at 105°C) with a PTFE-faced silicone septum. If residual chlorine is in the water add about 25 mg of ascorbic acid to each vial before sample collection. Collect bubble free samples.

STABILITY: Cool to 4°C; HCl to pH <2. M.H.T. = 14 Days.

QUALITY CONTROL: As an initial demonstration of lab accuracy and precision, analyze 4 to 7 replicates of a lab fortified blank containing analyte at 0.2-5 μg/L. Collect all samples in duplicate.

REFERENCE: Method 524.1, VOCs in Water by Purge and Trap Gas Chromatography/Mass Spectrometry, EPA 600/4-88/039.

PRIMARY NAME: Carbon Tetrachloride Method 524.2

TITLE: VOCs in H_2O (Purge and Trap capillary column GC/MS

MATRIX: Drinking water (finished or any treatment stage) and raw source water.

CAS # : 56-23-5

APPLICATION: Method covers 60 VOCs. An inert gas is bubbled through a 25 mL water sample. Purged sample components are trapped in tube of sorbent materials. When purging is complete, sorbent tube is heated and backflushed with helium to desorb trapped sample onto a capillary GC column.

INTERFERENCES: During analysis, major contaminant sources are volatile materials in the lab and impurities in purging gas and sorbent trap. With high and low level samples, there can be carry-over contamination. Watch for methylene chloride, it permeates through PTFE tubing.

INSTRUMENTATION: (Capillary Column) Gas Chromatography/Mass Spectrometry/Data System. Column #1: VOCOL glass wide bore; Column #2: DB-624 fused silica; Column #3: DB-5 fused silica.

RANGE: Approximately 0.1-200 μg/L

MDL: 0.21 μg/L on wide bore capillary column.

PRECISION: RSD = 8.8% @ (0.5-10) μg/L (24 samples) wide bore capillary column.

ACCURACY: Recovery = 84% @ (0.5-10) μg/L (24 samples) wide bore capillary column.

SAMPLING METHOD: Use a 60-120 mL screw cap vial (pre-washed with detergent, rinsed with distilled water and oven dried at 105°C) with a PTFE-faced silicone septum. If residual chlorine is in the water add about 25 mg of ascorbic acid to each vial before sample collection. Collect bubble free samples.

STABILITY: Cool to 4°C; HCl to pH <2.

M.H.T. = 14 Days.

QUALITY CONTROL: As an initial demonstration of lab accuracy and precision, analyze 4 to 7 replicates of a lab fortified blank containing analyte at 0.2-5 μg/L. Collect all samples in duplicate.

REFERENCE: Method 524.2, VOCs in H_2O by Purge and Trap Capillary column GC/MS, EPA 600/4-88/039.

PRIMARY NAME: Carbon Tetrachloride

Method 601

TITLE: Purgeable Halocarbons

MATRIX: Wastewater

CAS # : 56-23-5

APPLICATION: Method covers 29 purgeable halocarbons. (Method 624 provides GC/MS conditions appropriate for the qualitative and quantitative

confirmation of results). Method describes conditions for a second GC column to confirm measurements made with primary column.

INTERFERENCES: Impurities in the purge gas and organic compounds outgassing from the plumbing ahead of the trap. With high and low level samples, there can be carry-over contamination. Diffusion of volatile organics through the septum seal into the sample.

INSTRUMENTATION: GC equipped with halide-specific detector. (With purge and trap unit).

RANGE: 8.0 To 500 µg/L. MDL: 0.12 µg/L.

PRECISION: 0.20X+0.39 µg/L (overall precision)

ACCURACY: 0.98C-1.04 µg/L (as recovery)

SAMPLING METHOD: 25 mL glass vial. Teflon lined septum.

STABILITY: Cool, 4°C, 0.008% sodium thiosulfate. M.H.T. = 14 Days.

QUALITY CONTROL: The laboratory must on an ongoing basis, spike at least 10% of the samples from each sample site being monitored to assess accuracy.

REFERENCE: Method 601, Federal Register Part VIII 40 CFR Part 136, Oct 26, 1984.

PRIMARY NAME: Carbon Tetrachloride Method 624

TITLE: Purgeables MATRIX: Wastewater

CAS # : 56-23-5

APPLICATION: Method covers 31 purgeable organics. An inert gas is bubbled through a 5 mL water sample in a specially designed purging chamber. Here, purgeables are transferred from aqueous to gaseous phase, passed onto a sorbent column and trapped. Trap is heated and backflushed with inert gas to desorb purgeables onto a GC column, where purgeables are separated.

INTERFERENCES: Impurities in the purge gas, organic compounds outgassing from the plumbing ahead of the trap, and solvent vapors in the laboratory. With high and low level samples, there can be carry-over contamination.

INSTRUMENTATION: GC/MS with purge and trap unit.

RANGE: 5-600 μg/L MDL: 2.8 μg/L

PRECISION: 0.11X+0.37 μg/L (overall precision)

ACCURACY: 1.10C-1.68 μg/L (as recovery)

SAMPLING METHOD: 25 mL glass vial. Teflon lined septum.

STABILITY: Cool, 4°C, 0.008% sodium thiosulfate. M.H.T. = 14 Days.

QUALITY CONTROL: The laboratory must on an ongoing basis, spike at least 5% of the samples from each sample site being monitored to assess accuracy.

REFERENCE: Method 624, Federal Register Part VIII 40 CFR Part 136, Oct 26, 1984.

PRIMARY NAME: Carbon tetrachloride Method 8010

TITLE: Halogenated Volatile Organics

MATRIX: groundwater, soils, sludges water miscible liquid wastes, and non-water miscible wastes.

CAS # : 56-23-5

APPLICATION: This method is used for the analysis of 39 halogenated VOCs. Samples are analyzed using direct injection or purge and trap methods. Groundwater must be analyzed by the purge and trap method. The method provides an optional GC column which is used for analyte confirmation and that may help resolve analytes from interferences.

INTERFERENCES: There can be carry-over contamination with high and low level samples. Impurities may come from the purge and trap apparatus, organic compounds outgassing from the plumbing ahead of trap, diffusion of VOCs through the sample bottle septum during shipping or storage, or from solvent vapors in the lab.

INSTRUMENTATION: GC capable of on-column injections or purge and trap sample introduction and a halogen specific detector. Column 1: 8 foot by 0.1 inch 1% SP-1000 on Carbopack-B. Column 2: 6 foot by 0.1 inch bonded n-octane on Porasil-C.

RANGE: 8 to 500 µg/L (reagent water)

MDL: 0.12 µg/L (reagent water)

PRACTICAL QUANTITATION LIMIT FACTORS FOR MULTIPLYING TIMES MDL VALUE

Matrix	Multiplication Factor
Groundwater	10
Low-level soil	10
Water miscible liquid waste	500
High-level soil and sludge	1250
Non-water miscible waste	1250

PRECISION: 0.20X + 0.39 µg/L (overall precision)

ACCURACY: 0.98C - 1.04 µg/L (as recovery)

SAMPLING METHOD: For water and liquid samples; use glass 40 mL vials with Teflon lined septum caps and collect two vials per sample location with no headspace. For solids and concentrated waste samples; use widemouth glass bottles with Teflon liners.

STABILITY: For concentrated wastes, soils, sediments or sludges; cool to 4°C. For liquids; add 4 drops of concentrated hydrochloric acid and cool to 4°C. M.H.T. = 14 days.

QUALITY CONTROL: Analyze a reagent blank, matrix spike and matrix spike duplicate/duplicate for each analytical batch (up to 20 Samples). Demonstrate the purity of glassware and reagents by analyzing a reagent water method blank. Internal, surrogate and five concentration level calibration standards are used.

REFERENCE: Method 8010, SW-846, 3rd ed., Nov 1986.

PRIMARY NAME: Chloroethane Method 502.1

TITLE: Halogenated VOCs in Water

MATRIX: Drinking water (finished or any treatment stage) and raw source water.

CAS # : 75-00-3

APPLICATION: Method covers 40 halogenated VOCs. An inert gas is bubbled through a 5 mL water sample. Purged sample components are trapped in tube of sorbent materials. When purging is complete, sorbent tube is heated and backflushed with inert gas to desorb trapped sample onto a packed GC column where it is separated.

INTERFERENCES: During analysis, major contaminant sources are volatile materials in the lab and impurities in purging gas and sorbent trap. With high and low level samples, there can be carry-over contamination. Watch for methylene chloride, it permeates through PTFE tubing.

INSTRUMENTATION: Purge and Trap GC w/halide specific detector. (Two GC columns are recommended); Column #1: 1% SP-1000 on Carbopack B; Column #2: n-octane bonded on Poracil C.

RANGE: 8.0-505 μg/L (Drinking water) MDL: 0.008 μg/L in water

PRECISION: RSD = 18% @ 0.40 μg/L spike (S.L.) (20 Samples)

ACCURACY: Avg Recovery = 93% @ 0.40 μg/L spike (S.L.) (20 Samples)

SAMPLING METHOD: Use a 40-120 mL screw cap vial (pre-washed with detergent, rinsed with distilled water and oven dried at 105°C) with a PTFE-faced silicone septum. If residual chlorine is in the water add about 25 mg of ascorbic acid to each vial before sample collection. Collect bubble free samples.

STABILITY: Cool to 4°C; HCl to pH <2. M.H.T. = 14 Days.

QUALITY CONTROL: As an initial demonstration of lab accuracy and precision, analyze 4 to 7 replicates of a lab fortified blank containing analyte at 0.1-5 μg/L. Collect all samples in duplicate.

REFERENCE: Method 502.1, Volatile Halogenated Organic Compounds in Water by Purge and Trap GC, EPA 600/4-88/039.

PRIMARY NAME: Chloroethane

Method 502.2

TITLE: VOCs in H_2O by Purge and Trap

MATRIX: Drinking water (finished or capillary column GC any treatment stage) and raw source water.

CAS # : 75-00-3

APPLICATION: Method covers 60 VOCs. An inert gas is bubbled through a 5 mL sample in a purging chamber. Purged sample components are trapped in tube of sorbent materials. When purging is done, sorbent tube is heated and backflushed w/helium to desorb trapped sample onto a capillary GC column.

INTERFERENCES: During analysis, major contaminant sources are volatile materials in the lab and impurities in purging gas and sorbent trap. With high and low level samples, there can be carry-over contamination. Beware of methylene chloride, it permeates through PTFE tubing.

INSTRUMENTATION: (Capillary Column) GC, w/PID and electrolytic conductivity detectors in series. Column #1: VOCOL wide bore column; Column #2: RTX-502.2 mega bore column; Column #3: DB-62 mega bore column.

RANGE: Approximately 0.02-200 µg/L

MDL: 0.1 µg/L (reagent H_2O) (ELCD) 60 m × 0.75 mm VOCOL capillary column

PRECISION: RSD = 3.9% @ Spike of 10 µg/L (reagent H_2O) (ELCD). 60 m × 0.75 mm VOCOL capillary column

ACCURACY: Recovery = 96% @ spike of 10 µg/L (reagent H_2O) (ELCD) 60 m × 0.75 mm VOCOL capillary column

SAMPLING METHOD: Use a 40-120 mL screw cap vial (pre-washed with detergent, rinsed with distilled water and oven dried at 105°C) with a PTFE-faced silicone septum. If residual chlorine is in the water add about 25 mg of ascorbic acid to each vial before sample collection. Collect bubble free samples.

STABILITY: Cool to 4°C; HCl to pH <2.

M.H.T. = 14 Days.

QUALITY CONTROL: As an initial demonstration of lab accuracy and pre-

cision, analyze 4 to 7 replicates of a lab fortified blank containing analyte at 0.1-5 µg/L. Collect all samples in duplicate.

REFERENCE: Method 502.2, Volatile Organic Compounds in Water by Purge and Trap Capillary Column GC, EPA 600/4-88/039.

PRIMARY NAME: Chloroethane

Method 524.1

TITLE: VOCs in H_2O by Purge and Trap GC/MS

MATRIX: Drinking water (finished or any treatment stage) and raw source water.

CAS # : 75-00-3

APPLICATION: Method covers 51 VOCs. An inert gas is bubbled through a 25 mL water sample. Purged sample components are trapped in tube of sorbent materials. When purging is complete, sorbent tube is heated and backflushed with helium to desorb trapped sample onto a packed GC column.

INTERFERENCES: During analysis, major contaminant sources are volatile materials in the lab and impurities in purging gas and sorbent trap. With high and low level samples, there can be carry-over contamination. Watch for methylene chloride, it permeates through PTFE tubing.

INSTRUMENTATION: Gas Chromatography/Mass Spectrometry/Data System; 1% SP-1000 on Carbopack B column.

RANGE: Approximately 0.2-200 µg/L

MDL: not listed.

PRECISION: not listed.

ACCURACY: not listed.

SAMPLING METHOD: Use a 60-120 mL screw cap vial (pre-washed with detergent, rinsed with distilled water and oven dried at 105°C) with a PTFE-faced silicone septum. If residual chlorine is in the water add about 25 mg of ascorbic acid to each vial before sample collection. Collect bubble free samples.

STABILITY: Cool to 4°C; HCl to pH <2.

M.H.T. = 14 Days.

QUALITY CONTROL: As an initial demonstration of lab accuracy and precision, analyze 4 to 7 replicates of a lab fortified blank containing analyte at 0.2-5 µg/L. Collect all samples in duplicate.

REFERENCE: Method 524.1, VOCs in Water by Purge and Trap Gas Chromatography/Mass Spectrometry, EPA 600/4-88/039.

PRIMARY NAME: Chloroethane

Method 524.2

TITLE: VOCs in H_2O (Purge and Trap capillary column GC/MS

MATRIX: Drinking water (finished or any treatment stage) and raw source water.

CAS # : 75-00-3

APPLICATION: Method covers 60 VOCs. An inert gas is bubbled through a 25 mL water sample. Purged sample components are trapped in tube of sorbent materials. When purging is complete, sorbent tube is heated and backflushed with helium to desorb trapped sample onto a capillary GC column.

INTERFERENCES: During analysis, major contaminant sources are volatile materials in the lab and impurities in purging gas and sorbent trap. With high and low level samples, there can be carry-over contamination. Watch for methylene chloride, it permeates through PTFE tubing.

INSTRUMENTATION: (capillary column) Gas Chromatography/Mass Spectrometry/Data System. Column #1: VOCOL glass wide bore; Column #2: DB-624 fused silica; Column #3: DB-5 fused silica.

RANGE: Approximately 0.1-200 µg/L MDL: 0.10 µg/L on wide bore capillary column.

PRECISION: RSD = 9.0% @ (0.5-10) µg/L (24 samples) wide bore capillary column.

ACCURACY: Recovery = 89% @ (0.5-10) µg/L (24 samples) wide bore capillary column.

SAMPLING METHOD: Use a 60-120 mL screw cap vial (pre-washed with detergent, rinsed with distilled water and oven dried at 105°C) with a PTFE-faced silicone septum. If residual chlorine is in the water add about 25 mg of

ascorbic acid to each vial before sample collection. Collect bubble free samples.

STABILITY: Cool to 4°C; HCl to pH <2. M.H.T. = 14 Days.

QUALITY CONTROL: As an initial demonstration of lab accuracy and precision, analyze 4 to 7 replicates of a lab fortified blank containing analyte at 0.2-5 μg/L. Collect all samples in duplicate.

REFERENCE: Method 524.2, VOCs in H_2O by Purge and Trap Capillary column GC/MS, EPA 600/4-88/039.

PRIMARY NAME: Chloroethane Method 601

TITLE: Purgeable Halocarbons MATRIX: Wastewater

CAS # : 75-00-3

APPLICATION: Method covers 29 purgeable halocarbons. (Method 624 provides GC/MS conditions appropriate for the qualitative and quantitative confirmation of results). Method describes conditions for a second GC column to confirm measurements made with primary column.

INTERFERENCES: Impurities in the purge gas and organic compounds outgassing from the plumbing ahead of the trap. With high and low level samples, there can be carry-over contamination. Diffusion of volatile organics through the septum seal into the sample.

INSTRUMENTATION: GC equipped with halide-specific detector. (With purge and trap unit).

RANGE: 8.0 To 500 μg/L. MDL: 0.52 μg/L.

PRECISION: 0.17X+0.63 μg/L (overall precision)

ACCURACY: 0.99C-1.53 μg/L (as recovery)

SAMPLING METHOD: 25 mL glass vial. Teflon lined septum.

STABILITY: Cool, 4°C, 0.008% sodium thiosulfate. M.H.T. = 14 Days.

QUALITY CONTROL: The laboratory must on an ongoing basis, spike at

least 10% of the samples from each sample site being monitored to assess accuracy.

REFERENCE: Method 601, Federal Register Part VIII 40 CFR Part 136, Oct 26, 1984.

PRIMARY NAME: Chloroethane Method 624

TITLE: Purgeables MATRIX: Wastewater

CAS # : 75-00-3

APPLICATION: Method covers 31 purgeable organics. An inert gas is bubbled through a 5 mL water sample in a specially designed purging chamber. Here, purgeables are transferred from aqueous to gaseous phase, passed onto a sorbent column and trapped. Trap is heated and backflushed with inert gas to desorb purgeables onto a GC column, where purgeables are separated.

INTERFERENCES: Impurities in the purge gas, organic compounds outgassing from the plumbing ahead of the trap, and solvent vapors in the laboratory. With high and low level samples, there can be carry-over contamination.

INSTRUMENTATION: GC/MS with purge and trap unit.

RANGE: 5-600 μg/L MDL: not determined

PRECISION: 0.29X+1.75 μg/L (overall precision)

ACCURACY: 1.18C+0.81 μg/L (as recovery)

SAMPLING METHOD: 25 mL glass vial. Teflon lined septum.

STABILITY: Cool, 4°C, 0.008% sodium thiosulfate. M.H.T. = 14 Days.

QUALITY CONTROL: The laboratory must on an ongoing basis, spike at least 5% of the samples from each sample site being monitored to assess accuracy.

REFERENCE: Method 624, Federal Register Part VIII 40 CFR Part 136, Oct 26, 1984.

PRIMARY NAME: Chloroethane

Method 8010

TITLE: Halogenated Volatile Organics

MATRIX: groundwater, soils, sludges water miscible liquid wastes, and non-water miscible wastes.

CAS # : 75-00-3

APPLICATION: This method is used for the analysis of 39 halogenated VOCs. Samples are analyzed using direct injection or purge and trap methods. Groundwater must be analyzed by the purge and trap method. The method provides an optional GC column which is used for analyte confirmation and that may help resolve analytes from interferences.

INTERFERENCES: There can be carry-over contamination with high and low level samples. Impurities may come from the purge and trap apparatus, organic compounds outgassing from the plumbing ahead of trap, diffusion of VOCs through the sample bottle septum during shipping or storage, or from solvent vapors in the lab.

INSTRUMENTATION: GC capable of on-column injections or purge and trap sample introduction and a halogen specific detector. Column 1: 8 foot by 0.1 inch 1% SP-1000 on Carbopack-B. Column 2: 6 foot by 0.1 inch bonded n-octane on Porasil-C.

RANGE: 8 to 500 μg/L (reagent water)

MDL: 0.52 μg/L (reagent water)

PRACTICAL QUANTITATION LIMIT FACTORS FOR MULTIPLYING TIMES MDL VALUE

Matrix	Multiplication Factor
Groundwater	10
Low-level soil	10
Water miscible liquid waste	500
High-level soil and sludge	1250
Non-water miscible waste	1250

PRECISION: 0.17X + 0.63 μg/L (overall precision)

ACCURACY: 0.99C - 1.53 μg/L (as recovery)

SAMPLING METHOD: For water and liquid samples; use glass 40 mL vials with Teflon lined septum caps and collect two vials per sample location with no headspace. For solids and concentrated waste samples; use widemouth glass bottles with Teflon liners.

STABILITY: For concentrated wastes, soils, sediments or sludges; cool to 4°C. For liquids; add 4 drops of concentrated hydrochloric acid and cool to 4°C. M.H.T. = 14 days.

QUALITY CONTROL: Analyze a reagent blank, matrix spike and matrix spike duplicate/duplicate for each analytical batch (up to 20 Samples). Demonstrate the purity of glassware and reagents by analyzing a reagent water method blank. Internal, surrogate and five concentration level calibration standards are used.

REFERENCE: Method 8010, SW-846, 3rd ed., Nov 1986.

PRIMARY NAME: Chloroform Method 502.1

TITLE: Halogenated VOCs in Water

MATRIX: Drinking water (finished or any treatment stage) and raw source water.

CAS # : 67-66-3

APPLICATION: Method covers 40 halogenated VOCs. An inert gas is bubbled through a 5 mL water sample. Purged sample components are trapped in tube of sorbent materials. When purging is complete, sorbent tube is heated and backflushed with inert gas to desorb trapped sample onto a packed GC column where it is separated.

INTERFERENCES: During analysis, major contaminant sources are volatile materials in the lab and impurities in purging gas and sorbent trap. With high and low level samples, there can be carry-over contamination. Watch for methylene chloride, it permeates through PTFE tubing.

INSTRUMENTATION: Purge and Trap GC w/halide specific detector. (Two GC columns are recommended); Column #1: 1% SP-1000 on Carbopack B; Column #2: n-octane bonded on Poracil C.

RANGE: 8.0-505 µg/L (Drinking water) MDL: 0.002 µg/L in water

PRECISION: SD = 0.09X + 6.21 (Overall precision) (x in μg/L) (SA)

ACCURACY: mean Recovery = (0.90C + 3.44) (C in μg/L) (SA)

SAMPLING METHOD: Use a 40-120 mL screw cap vial (pre-washed with detergent, rinsed with distilled water and oven dried at 105°C) with a PTFE-faced silicone septum. If residual chlorine is in the water add about 25 mg of ascorbic acid to each vial before sample collection. Collect bubble free samples.

STABILITY: Cool to 4°C; HCl to pH <2. M.H.T. = 14 Days.

QUALITY CONTROL: As an initial demonstration of lab accuracy and precision, analyze 4 to 7 replicates of a lab fortified blank containing analyte at 0.1-5 μg/L. Collect all samples in duplicate.

REFERENCE: Method 502.1, Volatile Halogenated Organic Compounds in Water by Purge and Trap GC, EPA 600/4-88/039.

PRIMARY NAME: Chloroform Method 502.2

TITLE: VOCs in H_2O by Purge and Trap capillary column GC

MATRIX: Drinking water (finished or any treatment stage) and raw source water.

CAS # : 67-66-3

APPLICATION: Method covers 60 VOCs. An inert gas is bubbled through a 5 mL sample in a purging chamber. Purged sample components are trapped in tube of sorbent materials. When purging is done, sorbent tube is heated and backflushed w/helium to desorb trapped sample onto a capillary GC column.

INTERFERENCES: During analysis, major contaminant sources are volatile materials in the lab and impurities in purging gas and sorbent trap. With high and low level samples, there can be carry-over contamination. Beware of methylene chloride, it permeates through PTFE tubing.

INSTRUMENTATION: (capillary column) GC, w/PID and electrolytic conductivity detectors in series. Column #1: VOCOL wide bore column; Column #2: RTX-502.2 mega bore column; Column #3: DB-62 mega bore column.

RANGE: Approximately 0.02-200 µg/L

MDL: 0.02 µg/L (reagent H_2O) (ELCD) 60 m × 0.75 mm VOCOL capillary column

PRECISION: RSD = 2.5% @ Spike of 10 µg/L (reagent H_2O) (ELCD). 60 m × 0.75 mm VOCOL capillary column

ACCURACY: Recovery = 98% @ spike of 10 µg/L (reagent H_2O) (ELCD) 60 m × 0.75 mm VOCOL capillary column

SAMPLING METHOD: Use a 40-120 mL screw cap vial (pre-washed with detergent, rinsed with distilled water and oven dried at 105°C) with a PTFE-faced silicone septum. If residual chlorine is in the water add about 25 mg of ascorbic acid to each vial before sample collection. Collect bubble free samples.

STABILITY: Cool to 4°C; HCl to pH <2.

M.H.T. = 14 Days.

QUALITY CONTROL: As an initial demonstration of lab accuracy and precision, analyze 4 to 7 replicates of a lab fortified blank containing analyte at 0.1-5 µg/L. Collect all samples in duplicate.

REFERENCE: Method 502.2, Volatile Organic Compounds in Water by Purge and Trap Capillary Column GC, EPA 600/4-88/039.

PRIMARY NAME: Chloroform

Method 524.1

TITLE: VOCs in H_2O by Purge and Trap GC/MS

MATRIX: Drinking water (finished or any treatment stage) and raw source water.

CAS # : 67-66-3

APPLICATION: Method covers 51 VOCs. An inert gas is bubbled through a 25 mL water sample. Purged sample components are trapped in tube of sorbent materials. When purging is complete,sorbent tube is heated and backflushed with helium to desorb trapped sample onto a packed GC column.

INTERFERENCES: During analysis, major contaminant sources are volatile materials in the lab and impurities in purging gas and sorbent trap. With

high and low level samples, there can be carry-over contamination. Watch for methylene chloride, it permeates through PTFE tubing.

INSTRUMENTATION: Gas Chromatography/Mass Spectrometry/Data System; 1% SP-1000 on Carbopack B column.

RANGE: Approximately 0.2-200 μg/L MDL: 0.2 μg/L

PRECISION: RSD = 8.3% @ 1.0 μg/L

ACCURACY: Mean Accuracy = 103% @ 1.0 μg/L

SAMPLING METHOD: Use a 60-120 mL screw cap vial (pre-washed with detergent, rinsed with distilled water and oven dried at 105°C) with a PTFE-faced silicone septum. If residual chlorine is in the water add about 25 mg of ascorbic acid to each vial before sample collection. Collect bubble free samples.

STABILITY: Cool to 4°C; HCl to pH <2. M.H.T. = 14 Days.

QUALITY CONTROL: As an initial demonstration of lab accuracy and precision, analyze 4 to 7 replicates of a lab fortified blank containing analyte at 0.2-5 μg/L. Collect all samples in duplicate.

REFERENCE: Method 524.1, VOCs in Water by Purge and Trap Gas Chromatography/Mass Spectrometry, EPA 600/4-88/039.

PRIMARY NAME: Chloroform Method 524.2

TITLE: VOCs in H_2O (Purge and Trap capillary column GC/MS

MATRIX: Drinking water (finished or any treatment stage) and raw source water.

CAS # : 67-66-3

APPLICATION: Method covers 60 VOCs. An inert gas is bubbled through a 25 mL water sample. Purged sample components are trapped in tube of sorbent materials. When purging is complete,sorbent tube is heated and backflushed with helium to desorb trapped sample onto a capillary GC column.

INTERFERENCES: During analysis, major contaminant sources are volatile materials in the lab and impurities in purging gas and sorbent trap. With high and low level samples, there can be carry-over contamination. Watch for methylene chloride, it permeates through PTFE tubing.

INSTRUMENTATION: (capillary column) Gas Chromatography/Mass Spectrometry/Data System. Column #1: VOCOL glass wide bore; Column #2: DB-624 fused silica; Column #3: DB-5 fused silica.

RANGE: Approximately 0.1-200 µg/L

MDL: 0.03 µg/L on wide bore capillary column.

PRECISION: RSD = 6.1% @ (0.5-10) µg/L (24 samples) wide bore capillary column.

ACCURACY: Recovery = 90% @ (0.5-10) µg/L (24 samples) wide bore capillary column.

SAMPLING METHOD: Use a 60-120 mL screw cap vial (pre-washed with detergent, rinsed with distilled water and oven dried at 105°C) with a PTFE-faced silicone septum. If residual chlorine is in the water add about 25 mg of ascorbic acid to each vial before sample collection. Collect bubble free samples.

STABILITY: Cool to 4°C; HCl to pH <2.

M.H.T. = 14 Days.

QUALITY CONTROL: As an initial demonstration of lab accuracy and precision, analyze 4 to 7 replicates of a lab fortified blank containing analyte at 0.2-5 µg/L. Collect all samples in duplicate.

REFERENCE: Method 524.2, VOCs in H_2O by Purge and Trap Capillary column GC/MS, EPA 600/4-88/039.

PRIMARY NAME: Chloroform

Method 601

TITLE: Purgeable Halocarbons

MATRIX: Wastewater

CAS # : 67-66-3

APPLICATION: Method covers 29 purgeable halocarbons. (Method 624 provides GC/MS conditions appropriate for the qualitative and quantitative

confirmation of results). Method describes conditions for a second GC column to confirm measurements made with primary column.

INTERFERENCES: Impurities in the purge gas and organic compounds outgassing from the plumbing ahead of the trap. With high and low level samples, there can be carry-over contamination. Diffusion of volatile organics through the septum seal into the sample.

INSTRUMENTATION: GC equipped with halide-specific detector. (With purge and trap unit).

RANGE: 8.0 To 500 µg/L. MDL: 0.05 µg/L.

PRECISION: 0.19X-0.02 µg/L (overall precision)

ACCURACY: 0.93C-0.39 µg/L (as recovery)

SAMPLING METHOD: 25 mL glass vial. Teflon lined septum.

STABILITY: Cool, 4°C, 0.008% sodium thiosulfate. M.H.T. = 14 Days.

QUALITY CONTROL: The laboratory must on an ongoing basis, spike at least 10% of the samples from each sample site being monitored to assess accuracy.

REFERENCE: Method 601, Federal Register Part VIII 40 CFR Part 136, Oct 26, 1984.

PRIMARY NAME: Chloroform Method 624

TITLE: Purgeables MATRIX: Wastewater

CAS # : 67-66-3

APPLICATION: Method covers 31 purgeable organics. An inert gas is bubbled through a 5 mL water sample in a specially designed purging chamber. Here, purgeables are transferred from aqueous to gaseous phase, passed onto a sorbent column and trapped. Trap is heated and backflushed with inert gas to desorb purgeables onto a GC column, where purgeables are separated.

INTERFERENCES: Impurities in the purge gas, organic compounds outgassing from the plumbing ahead of the trap, and solvent vapors in the laboratory. With high and low level samples, there can be carry-over contamination.

INSTRUMENTATION: GC/MS with purge and trap unit.

RANGE: 5-600 μg/L MDL: 1.6 μg/L

PRECISION: 0.18X+0.16 μg/L (overall precision)

ACCURACY: 0.93C+0.33 μg/L (as recovery)

SAMPLING METHOD: 25 mL glass vial. Teflon lined septum.

STABILITY: Cool, 4°C, 0.008% sodium thiosulfate. M.H.T. = 14 Days.

QUALITY CONTROL: The laboratory must on an ongoing basis, spike at least 5% of the samples from each sample site being monitored to assess accuracy.

REFERENCE: Method 624, Federal Register Part VIII 40 CFR Part 136, Oct 26, 1984.

PRIMARY NAME: Chloroform Method 8010

TITLE: Halogenated Volatile Organics

MATRIX: groundwater, soils, sludges water miscible liquid wastes, and non-water miscible wastes.

CAS # : 67-66-3

APPLICATION: This method is used for the analysis of 39 halogenated VOCs. Samples are analyzed using direct injection or purge and trap methods. Groundwater must be analyzed by the purge and trap method. The method provides an optional GC column which is used for analyte confirmation and that may help resolve analytes from interferences.

INTERFERENCES: There can be carry-over contamination with high and low level samples. Impurities may come from the purge and trap apparatus,

organic compounds outgassing from the plumbing ahead of trap, diffusion of VOCs through the sample bottle septum during shipping or storage, or from solvent vapors in the lab.

INSTRUMENTATION: GC capable of on-column injections or purge and trap sample introduction and a halogen specific detector. Column 1: 8 foot by 0.1 inch 1% SP-1000 on Carbopack-B. Column 2: 6 foot by 0.1 inch bonded n-octane on Porasil-C.

RANGE: 8 to 500 µg/L (reagent water)

MDL: 0.05 µg/L (reagent water)

PRACTICAL QUANTITATION LIMIT FACTORS FOR MULTIPLYING TIMES MDL VALUE

Matrix	Multiplication Factor
Groundwater	10
Low-level soil	10
Water miscible liquid waste	500
High-level soil and sludge	1250
Non-water miscible waste	1250

PRECISION: 0.19X - 0.02 µg/L (overall precision)

ACCURACY: 0.93C - 0.39 µg/L (as recovery)

SAMPLING METHOD: For water and liquid samples; use glass 40 mL vials with Teflon lined septum caps and collect two vials per sample location with no headspace. For solids and concentrated waste samples; use widemouth glass bottles with Teflon liners.

STABILITY: For concentrated wastes, soils, sediments or sludges; cool to 4°C. For liquids; add 4 drops of concentrated hydrochloric acid and cool to 4°C. M.H.T. = 14 days.

QUALITY CONTROL: Analyze a reagent blank, matrix spike and matrix spike duplicate/duplicate for each analytical batch (up to 20 Samples). Demonstrate the purity of glassware and reagents by analyzing a reagent water method blank. Internal, surrogate and five concentration level calibration standards are used.

REFERENCE: Method 8010, SW-846, 3rd ed., Nov 1986.

PRIMARY NAME: Chloroform Method 1624,B

TITLE: Volatile Organic Compounds MATRIX: Wastewater

CAS # : 67-66-3

APPLICATION: Method covers determination of 33 volatile organic compounds by isotope dilution GC/MS. Stable isotopically labelled analogs are added to 5 mL water sample and purged with inert gas. VOCs are transferred from aqueous to gaseous phase, passed onto sorbent column and trapped. The trap is backflushed and heated rapidly to desorb compounds into a GC.

INTERFERENCES: Impurities in the purge gas and organic compds outgassing from the plumbing ahead of the trap. With high and low level samples, there can be carry-over contamination. Diffusion of volatile organics through septum seal into the sample.

INSTRUMENTATION: GC/MS with purge and trap unit.

RANGE: Not listed MDL: 10μg (with no interferences)

PRECISION: Not listed

ACCURACY: Not listed

SAMPLING METHOD: 25 mL glass vial. Teflon lined septum.

STABILITY: Cool, 4°C, 0.008% sodium thiosulfate M.H.T. = 14 Days.

QUALITY CONTROL: The laboratory shall, on an ongoing basis, demonstrate through the analysis of the aqueous performance std that analysis system is in control.

REFERENCE: Method 1624,B, Federal Register Part VIII 40 CFR Part 136, Oct 26, 1984.

PRIMARY NAME: 2-Chloroethyl vinyl ether Method 601

TITLE: Purgeable Halocarbons MATRIX: Wastewater

CAS # : 110-75-8

APPLICATION: Method covers 29 purgeable halocarbons. (Method 624 provides GC/MS conditions appropriate for the qualitative and quantitative confirmation of results). Method describes conditions for a second GC column to confirm measurements made with primary column.

INTERFERENCES: Impurities in the purge gas and organic compounds outgassing from the plumbing ahead of the trap. With high and low level samples, there can be carry-over contamination. Diffusion of volatile organics through the septum seal into the sample.

INSTRUMENTATION: GC equipped with halide-specific detector. (With purge and trap unit).

RANGE: 8.0 To 500 μg/L. MDL: 0.13 μg/L.

PRECISION: 0.35X μg/L (overall precision)

ACCURACY: 1.00C μg/L (as recovery)

SAMPLING METHOD: 25 mL glass vial. Teflon lined septum.

STABILITY: Cool, 4°C, 0.008% sodium thiosulfate. M.H.T. = 14 Days.

QUALITY CONTROL: The laboratory must on an ongoing basis, spike at least 10% of the samples from each sample site being monitored to assess accuracy.

REFERENCE: Method 601, Federal Register Part VIII 40 CFR Part 136, Oct 26, 1984.

PRIMARY NAME: 2-Chloroethyl vinyl ether Method 624

TITLE: Purgeables MATRIX: Wastewater

CAS # : 110-75-8

APPLICATION: Method covers 31 purgeable organics. An inert gas is bubbled through a 5 mL water sample in a specially designed purging chamber. Here, purgeables are transferred from aqueous to gaseous phase,

passed onto a sorbent column and trapped. Trap is heated and backflushed with inert gas to desorb purgeables onto a GC column, where purgeables are separated.

INTERFERENCES: Impurities in the purge gas, organic compounds outgassing from the plumbing ahead of the trap, and solvent vapors in the laboratory. With high and low level samples, there can be carry-over contamination.

INSTRUMENTATION: GC/MS with purge and trap unit.

RANGE: 5-600 µg/L MDL: not determined

PRECISION: 0.84X µg/L (overall precision)

ACCURACY: 1.00C µg/L (as recovery)

SAMPLING METHOD: 25 mL glass vial. Teflon lined septum.

STABILITY: Cool, 4°C, 0.008% sodium thiosulfate. M.H.T. = 14 Days.

QUALITY CONTROL: The laboratory must on an ongoing basis, spike at least 5% of the samples from each sample site being monitored to assess accuracy.

REFERENCE: Method 624, Federal Register Part VIII 40 CFR Part 136, Oct 26, 1984.

PRIMARY NAME: 2-Chloroethyl vinyl ether Method 8010

TITLE: Halogenated Volatile Organics

MATRIX: groundwater, soils, sludges water miscible liquid wastes, and non-water miscible wastes.

CAS # : 110-75-8

APPLICATION: This method is used for the analysis of 39 halogenated VOCs. Samples are analyzed using direct injection or purge and trap methods. Groundwater must be analyzed by the purge and trap method. The method provides an optional GC column which is used for analyte confirmation and that may help resolve analytes from interferences.

INTERFERENCES: There can be carry-over contamination with high and low level samples. Impurities may come from the purge and trap apparatus, organic compounds outgassing from the plumbing ahead of trap, diffusion of VOCs through the sample bottle septum during shipping or storage, or from solvent vapors in the lab.

INSTRUMENTATION: GC capable of on-column injections or purge and trap sample introduction and a halogen specific detector. Column 1: 8 foot by 0.1 inch 1% SP-1000 on Carbopack-B. Column 2: 6 foot by 0.1 inch bonded n-octane on Porasil-C.

RANGE: 8 to 500 μg/L (reagent water)

MDL: 0.13 μg/L (reagent water)

PRACTICAL QUANTITATION LIMIT FACTORS FOR MULTIPLYING TIMES MDL VALUE

Matrix	Multiplication Factor
Groundwater	10
Low-level soil	10
Water miscible liquid waste	500
High-level soil and sludge	1250
Non-water miscible waste	1250

PRECISION: 0.35X μg/L (overall precision; estimate)

ACCURACY: 1.00C μg/L (as recovery; estimate)

SAMPLING METHOD: For water and liquid samples; use glass 40 mL vials with Teflon lined septum caps and collect two vials per sample location with no headspace. For solids and concentrated waste samples; use widemouth glass bottles with Teflon liners.

STABILITY: For concentrated wastes, soils, sediments or sludges; cool to 4°C. For liquids; add 4 drops of concentrated hydrochloric acid and cool to 4°C. M.H.T. = 14 days.

QUALITY CONTROL: Analyze a reagent blank, matrix spike and matrix spike duplicate/duplicate for each analytical batch (up to 20 Samples). Demonstrate the purity of glassware and reagents by analyzing a reagent water method blank. Internal, surrogate and five concentration level calibration standards are used.

REFERENCE: Method 8010, SW-846, 3rd ed., Nov 1986.

PRIMARY NAME: Chloromethane | Method 502.1

TITLE: Halogenated VOCs in Water

MATRIX: Drinking water (finished or any treatment stage) and raw source water.

CAS # : 74-87-3

APPLICATION: Method covers 40 halogenated VOCs. An inert gas is bubbled through a 5 mL water sample. Purged sample components are trapped in tube of sorbent materials.When purging is complete, sorbent tube is heated and backflushed with inert gas to desorb trapped sample onto a packed GC column where it is separated.

INTERFERENCES: During analysis, major contaminant sources are volatile materials in the lab and impurities in purging gas and sorbent trap. With high and low level samples, there can be carry-over contamination. Watch for methylene chloride, it permeates through PTFE tubing.

INSTRUMENTATION: Purge and Trap GC w/halide specific detector. (Two GC columns are recommended); Column #1: 1% SP-1000 on Carbopack B; Column #2: n-octane bonded on Poracil C.

RANGE: 8.0-505 µg/L (Drinking water) MDL: 0.01 µg/L in water

PRECISION: RSD = 8.5% @ 0.40 µg/L spike (S.L.) (16 Samples)

ACCURACY: Avg Recovery = 93% @ 0.40 µg/L spike (S.L.) (16 Samples)

SAMPLING METHOD: Use a 40-120 mL screw cap vial (pre-washed with detergent, rinsed with distilled water and oven dried at 105°C) with a PTFE-faced silicone septum. If residual chlorine is in the water add about 25 mg of ascorbic acid to each vial before sample collection. Collect bubble free samples.

STABILITY: Cool to 4°C; HCl to pH <2. M.H.T. = 14 Days.

QUALITY CONTROL: As an initial demonstration of lab accuracy and precision, analyze 4 to 7 replicates of a lab fortified blank containing analyte at 0.1-5 µg/L. Collect all samples in duplicate.

REFERENCE: Method 502.1, Volatile Halogenated Organic Compounds in Water by Purge and Trap GC, EPA 600/4-88/039.

PRIMARY NAME: Chloromethane

Method 502.2

TITLE: VOCs in H_2O by Purge and Trap capillary column GC

MATRIX: Drinking water (finished or any treatment stage) and raw source water.

CAS # : 74-87-3

APPLICATION: Method covers 60 VOCs. An inert gas is bubbled through a 5 mL sample in a purging chamber. Purged sample components are trapped in tube of sorbent materials. When purging is done, sorbent tube is heated and backflushed w/helium to desorb trapped sample onto a capillary GC column.

INTERFERENCES: During analysis, major contaminant sources are volatile materials in the lab and impurities in purging gas and sorbent trap. With high and low level samples, there can be carry-over contamination. Beware of methylene chloride, it permeates through PTFE tubing.

INSTRUMENTATION: (capillary column) GC, w/PID and electrolytic conductivity detectors in series. Column #1: VOCOL wide bore column; Column #2: RTX-502.2 mega bore column; Column #3: DB-62 mega bore column.

RANGE: Approximately 0.02-200 μg/L

MDL: 0.03 μg/L (reagent H_2O) (ELCD) 60 m × 0.75 mm VOCOL capillary column

PRECISION: RSD = 9.2% @ Spike of 10 μg/L (reagent H_2O) (ELCD). 60 m × 0.75 mm VOCOL capillary column

ACCURACY: Recovery = 96% @ spike of 10 μg/L (reagent H_2O) (ELCD) 60 m × 0.75 mm VOCOL capillary column

SAMPLING METHOD: Use a 40-120 mL screw cap vial (pre-washed with detergent, rinsed with distilled water and oven dried at 105°C) with a PTFE-faced silicone septum. If residual chlorine is in the water add about 25 mg of ascorbic acid to each vial before sample collection. Collect bubble free samples.

STABILITY: Cool to 4°C; HCl to pH <2.

M.H.T. = 14 Days.

QUALITY CONTROL: As an initial demonstration of lab accuracy and precision, analyze 4 to 7 replicates of a lab fortified blank containing analyte at 0.1-5 μg/L. Collect all samples in duplicate.

REFERENCE: Method 502.2, Volatile Organic Compounds in Water by Purge and Trap Capillary Column GC, EPA 600/4-88/039.

PRIMARY NAME: Chloromethane

Method 524.1

TITLE: VOCs in H_2O by Purge and Trap GC/MS

MATRIX: Drinking water (finished or any treatment stage) and raw source water.

CAS # : 74-87-3

APPLICATION: Method covers 51 VOCs. An inert gas is bubbled through a 25 mL water sample. Purged sample components are trapped in tube of sorbent materials. When purging is complete, sorbent tube is heated and backflushed with helium to desorb trapped sample onto a packed GC column.

INTERFERENCES: During analysis, major contaminant sources are volatile materials in the lab and impurities in purging gas and sorbent trap. With high and low level samples, there can be carry-over contamination. Watch for methylene chloride, it permeates through PTFE tubing.

INSTRUMENTATION: Gas Chromatography/Mass Spectrometry/Data System; 1% SP-1000 on Carbopack B column.

RANGE: Approximately 0.2-200 µg/L

MDL: not listed.

PRECISION: not listed.

ACCURACY: not listed.

SAMPLING METHOD: Use a 60-120 mL screw cap vial (pre-washed with detergent, rinsed with distilled water and oven dried at 105°C) with a PTFE-faced silicone septum. If residual chlorine is in the water add about 25 mg of ascorbic acid to each vial before sample collection. Collect bubble free samples.

STABILITY: Cool to 4°C; HCl to pH <2.

M.H.T. = 14 Days.

QUALITY CONTROL: As an initial demonstration of lab accuracy and precision, analyze 4 to 7 replicates of a lab fortified blank containing analyte at 0.2-5 µg/L. Collect all samples in duplicate.

REFERENCE: Method 524.1, VOCs in Water by Purge and Trap Gas Chromatography/Mass Spectrometry, EPA 600/4-88/039.

PRIMARY NAME: Chloromethane

Method 524.2

TITLE: VOCs in H_2O Purge and Trap capillary column GC/MS

MATRIX: Drinking water (finished or any treatment stage) and raw source water.

CAS # : 74-87-3

APPLICATION: Method covers 60 VOCs. An inert gas is bubbled through a 25 mL water sample. Purged sample components are trapped in tube of sorbent materials. When purging is complete, sorbent tube is heated and backflushed with helium to desorb trapped sample onto a capillary GC column.

INTERFERENCES: During analysis, major contaminant sources are volatile materials in the lab and impurities in purging gas and sorbent trap. With high and low level samples, there can be carry-over contamination. Watch for methylene chloride, it permeates through PTFE tubing.

INSTRUMENTATION: (capillary column) Gas Chromatography/Mass Spectrometry/Data System. Column #1: VOCOL glass wide bore; Column #2: DB-624 fused silica; Column #3: DB-5 fused silica.

RANGE: Approximately 0.1-200 µg/L

MDL: 0.13 µg/L on wide bore capillary column.

PRECISION: RSD = 8.9% @ (0.5-10) µg/L (23 samples) wide bore capillary column.

ACCURACY: Recovery = 93% @ (0.5-10) µg/L (23 samples) wide bore capillary column.

SAMPLING METHOD: Use a 60-120 mL screw cap vial (pre-washed with detergent, rinsed with distilled water and oven dried at 105°C) with a PTFE-faced silicone septum. If residual chlorine is in the water add about 25 mg of ascorbic acid to each vial before sample collection. Collect bubble free samples.

STABILITY: Cool to 4°C; HCl to pH <2.

M.H.T. = 14 Days.

QUALITY CONTROL: As an initial demonstration of lab accuracy and precision, analyze 4 to 7 replicates of a lab fortified blank containing analyte at 0.2-5 µg/L. Collect all samples in duplicate.

REFERENCE: Method 524.2, VOCs in H_2O by Purge and Trap Capillary column GC/MS, EPA 600/4-88/039.

PRIMARY NAME: Chloromethane Method 601

TITLE: Purgeable Halocarbons MATRIX: Wastewater

CAS # : 74-87-3

APPLICATION: Method covers 29 purgeable halocarbons. (Method 624 provides GC/MS conditions appropriate for the qualitative and quantitative confirmation of results). Method describes conditions for a second GC column to confirm measurements made with primary column.

INTERFERENCES: Impurities in the purge gas and organic compounds outgassing from the plumbing ahead of the trap. With high and low level samples, there can be carry-over contamination. Diffusion of volatile organics through the septum seal into the sample.

INSTRUMENTATION: GC equipped with halide-specific detector. (With purge and trap unit).

RANGE: 8.0 To 500 mg/L. MDL: 0.08 mg/L.

PRECISION: 0.52X+1.31 mg/L (overall precision)

ACCURACY: 0.77C+0.18 mg/L (as recovery)

SAMPLING METHOD: 25 mL glass vial. Teflon lined septum.

STABILITY: Cool, 4°C, 0.008% sodium thiosulfate. M.H.T. = 14 Days.

QUALITY CONTROL: The laboratory must on an ongoing basis, spike at least 10% of the samples from each sample site being monitored to assess accuracy.

REFERENCE: Method 601, Federal Register Part VIII 40 CFR Part 136, Oct 26, 1984.

PRIMARY NAME: Chloromethane Method 624

TITLE: Purgeables MATRIX: Wastewater

CAS # : 74-87-3

APPLICATION: Method covers 31 purgeable organics. An inert gas is bubbled through a 5 mL water sample in a specially designed purging chamber. Here, purgeables are transferred from aqueous to gaseous phase, passed onto a sorbent column and trapped. Trap is heated and backflushed with inert gas to desorb purgeables onto a GC column, where purgeables are separated.

INTERFERENCES: Impurities in the purge gas, organic compounds outgassing from the plumbing ahead of the trap, and solvent vapors in the laboratory. With high and low level samples, there can be carry-over contamination.

INSTRUMENTATION: GC/MS with purge and trap unit.

RANGE: 5-600 μg/L MDL: not determined.

PRECISION: 0.58X+0.43 μg/L (overall precision)

ACCURACY: 1.03C-1.81 μg/L (as recovery)

SAMPLING METHOD: 25 mL glass vial. Teflon lined septum.

STABILITY: Cool, 4°C, 0.008% sodium thiosulfate. M.H.T. = 14 Days.

QUALITY CONTROL: The laboratory must on an ongoing basis, spike at least 5% of the samples from each sample site being monitored to assess accuracy.

REFERENCE: Method 624, Federal Register Part VIII 40 CFR Part 136, Oct 26, 1984.

PRIMARY NAME: Chloromethane

Method 8010

TITLE: Halogenated Volatile Organics

MATRIX: groundwater, soils, sludges water miscible liquid wastes, and non-water miscible wastes.

CAS # : 74-87-3

APPLICATION: This method is used for the analysis of 39 halogenated VOCs. Samples are analyzed using direct injection or purge and trap methods. Groundwater must be analyzed by the purge and trap method. The method provides an optional GC column which is used for analyte confirmation and that may help resolve analytes from interferences.

INTERFERENCES: There can be carry-over contamination with high and low level samples. Impurities may come from the purge and trap apparatus, organic compounds outgassing from the plumbing ahead of trap, diffusion of VOCs through the sample bottle septum during shipping or storage, or from solvent vapors in the lab.

INSTRUMENTATION: GC capable of on-column injections or purge and trap sample introduction and a halogen specific detector. Column 1: 8 foot by 0.1 inch 1% SP-1000 on Carbopack-B. Column 2: 6 foot by 0.1 inch bonded n-octane on Porasil-C.

RANGE: 8 to 500 μg/L (reagent water)

MDL: 0.08 μg/L (reagent water)

PRACTICAL QUANTITATION LIMIT FACTORS FOR MULTIPLYING TIMES MDL VALUE

Matrix	Multiplication Factor
Groundwater	10
Low-level soil	10
Water miscible liquid waste	500
High-level soil and sludge	1250
Non-water miscible waste	1250

PRECISION: 0.52X + 1.31 μg/L (overall precision)

ACCURACY: 0.77C + 0.18 μg/L (as recovery)

SAMPLING METHOD: For water and liquid samples; use glass 40 mL vials with Teflon lined septum caps and collect two vials per sample location with no headspace. For solids and concentrated waste samples; use widemouth glass bottles with Teflon liners.

STABILITY: For concentrated wastes, soils, sediments or sludges; cool to 4°C. For liquids; add 4 drops of concentrated hydrochloric acid and cool to 4°C. M.H.T. = 14 days.

QUALITY CONTROL: Analyze a reagent blank, matrix spike and matrix spike duplicate/duplicate for each analytical batch (up to 20 Samples). Demonstrate the purity of glassware and reagents by analyzing a reagent water method blank. Internal, surrogate and five concentration level calibration standards are used.

REFERENCE: Method 8010, SW-846, 3rd ed., Nov 1986.

PRIMARY NAME: 1,1-Dichloroethane

Method 502.1

TITLE: Halogenated VOCs in Water

MATRIX: Drinking water (finished or any treatment stage) and raw source water.

CAS # : 75-34-3

APPLICATION: Method covers 40 halogenated VOCs. An inert gas is bubbled through a 5 mL water sample. Purged sample components are trapped in tube of sorbent materials. When purging is complete, sorbent tube is heated and backflushed with inert gas to desorb trapped sample onto a packed GC column where it is separated.

INTERFERENCES: During analysis, major contaminant sources are volatile materials in the lab and impurities in purging gas and sorbent trap. With high and low level samples, there can be carry-over contamination. Watch for methylene chloride, it permeates through PTFE tubing.

INSTRUMENTATION: Purge and Trap GC w/halide specific detector. (Two GC columns are recommended); Column #1: 1% SP-1000 on Carbopack B; Column #2: n-octane bonded on Poracil C.

RANGE: 8.0-505 µg/L (Drinking water)

MDL: 0.003 µg/L in water

PRECISION: RSD = 6.0% @ 0.20 µg/L spike (S.L.) (17 Samples)

ACCURACY: Avg Recovery = 95% @ 0.20 µg/L spike (S.L.) (17 Samples)

SAMPLING METHOD: Use a 40-120 mL screw cap vial (pre-washed with detergent, rinsed with distilled water and oven dried at 105°C) with a PTFE-faced silicone septum. If residual chlorine is in the water add about 25 mg of ascorbic acid to each vial before sample collection. Collect bubble free samples.

STABILITY: Cool to 4°C; HCl to pH <2. M.H.T. = 14 Days.

QUALITY CONTROL: As an initial demonstration of lab accuracy and precision, analyze 4 to 7 replicates of a lab fortified blank containing analyte at 0.1-5 µg/L. Collect all samples in duplicate.

REFERENCE: Method 502.1, Volatile Halogenated Organic Compounds in Water by Purge and Trap GC, EPA 600/4-88/039.

PRIMARY NAME: 1,1-Dichloroethane Method 502.2

TITLE: VOCs in H_2O by purge and trap capillary column GC

MATRIX: Drinking water (finished or any treatment stage) and raw source water.

CAS # : 75-34-3

APPLICATION: Method covers 60 VOCs. An inert gas is bubbled through a 5 mL sample in a purging chamber. Purged sample components are trapped in tube of sorbent materials. When purging is done, sorbent tube is heated and backflushed w/helium to desorb trapped sample onto a capillary GC column.

INTERFERENCES: During analysis, major contaminant sources are volatile materials in the lab and impurities in purging gas and sorbent trap. With high and low level samples, there can be carry-over contamination. Beware of methylene chloride, it permeates through PTFE tubing.

INSTRUMENTATION: (Capillary Column) GC, w/PID and electrolytic conductivity detectors in series. Column #1: VOCOL wide bore column; Column #2: RTX-502.2 mega bore column; Column #3: DB-62 mega bore column.

RANGE: Approximately 0.02-200 µg/L

MDL: 0.07 µg/L (reagent H_2O) (ELCD) 60 m × 0.75 mm VOCOL capillary column

PRECISION: RSD = 5.7% @ Spike of 10 µg/L (reagent H_2O) (ELCD) 60 m × 0.75 mm VOCOL capillary column

ACCURACY: Recovery = 100% @ spike of 10 µg/L (reagent H_2O) (ELCD) 60 m × 0.75 mm VOCOL capillary column

SAMPLING METHOD: Use a 40-120 mL screw cap vial (pre-washed with detergent, rinsed with distilled water and oven dried at 105°C) with a PTFE-faced silicone septum. If residual chlorine is in the water add about 25 mg of ascorbic acid to each vial before sample collection. Collect bubble free samples.

STABILITY: Cool to 4°C; HCl to pH <2. M.H.T. = 14 Days.

QUALITY CONTROL: As an initial demonstration of lab accuracy and precision, analyze 4 to 7 replicates of a lab fortified blank containing analyte at 0.1-5 µg/L. Collect all samples in duplicate.

REFERENCE: Method 502.2, Volatile Organic Compounds in Water by Purge and Trap Capillary Column GC, EPA 600/4-88/039.

PRIMARY NAME: 1,1-Dichloroethane

Method 524.1

TITLE: VOCs in H_2O by purge and trap GC/MS

MATRIX: Drinking water (finished or any treatment stage) and raw source water.

CAS # : 75-34-3

APPLICATION: Method covers 51 VOCs. An inert gas is bubbled through a 25 mL water sample. Purged sample components are trapped in tube of sorbent materials. When purging is complete, sorbent tube is heated and backflushed with helium to desorb trapped sample onto a packed GC column.

INTERFERENCES: During analysis, major contaminant sources are volatile materials in the lab and impurities in purging gas and sorbent trap. With

high and low level samples, there can be carry-over contamination. Watch for methylene chloride, it permeates through PTFE tubing.

INSTRUMENTATION: Gas Chromatography/Mass Spectrometry/Data System; 1% SP-1000 on Carbopack B column.

RANGE: Approximately 0.2-200 µg/L MDL: 0.2 µg/L

PRECISION: RSD = 5.7% @ 1.0 µg/L

ACCURACY: Mean Accuracy = 105% @ 1.0 µg/L

SAMPLING METHOD: Use a 60-120 mL screw cap vial (pre-washed with detergent, rinsed with distilled water and oven dried at 105°C) with a PTFE-faced silicone septum. If residual chlorine is in the water add about 25 mg of ascorbic acid to each vial before sample collection. Collect bubble free samples.

STABILITY: Cool to 4°C; HCl to pH <2. M.H.T. = 14 Days.

QUALITY CONTROL: As an initial demonstration of lab accuracy and precision, analyze 4 to 7 replicates of a lab fortified blank containing analyte at 0.2-5 µg/L. Collect all samples in duplicate.

REFERENCE: Method 524.1, VOCs in Water by Purge and Trap Gas Chromatography/Mass Spectrometry, EPA 600/4-88/039.

PRIMARY NAME: 1,1-Dichloroethane Method 524.2

TITLE: VOCs in H_2O (Purge and Trap capillary column GC/MS

MATRIX: Drinking water (finished or) any treatment stage) and raw source water.

CAS # : 75-34-3

APPLICATION: Method covers 60 VOCs. An inert gas is bubbled through a 25 mL water sample. Purged sample components are trapped in tube of sorbent materials. When purging is complete, sorbent tube is heated and backflushed with helium to desorb trapped sample onto a capillary GC column.

INTERFERENCES: During analysis, major contaminant sources are volatile materials in the lab and impurities in purging gas and sorbent trap. With high and low level samples, there can be carry-over contamination. Watch for methylene chloride, it permeates through PTFE tubing.

INSTRUMENTATION: (Capillary Column) Gas Chromatography/Mass Spectrometry/Data System. Column #1: VOCOL glass wide bore; Column #2: DB-624 fused silica; Column #3: DB-5 fused silica.

RANGE: Approximately 0.1-200 μg/L

MDL: 0.04 μg/L on wide bore capillary column.

PRECISION: RSD = 5.3% @ (0.5-10) μg/L (24 samples) wide bore capillary column.

ACCURACY: Recovery = 96% @ (0.5-10) μg/L (24 samples) wide bore capillary column.

SAMPLING METHOD: Use a 60-120 mL screw cap vial (pre-washed with detergent, rinsed with distilled water and oven dried at 105°C) with a PTFE-faced silicone septum. If residual chlorine is in the water add about 25 mg of ascorbic acid to each vial before sample collection. Collect bubble free samples.

STABILITY: Cool to 4°C; HCl to pH <2.

M.H.T. = 14 Days.

QUALITY CONTROL: As an initial demonstration of lab accuracy and precision, analyze 4 to 7 replicates of a lab fortified blank containing analyte at 0.2-5 μg/L. Collect all samples in duplicate.

REFERENCE: Method 524.2, VOCs in H_2O by Purge and Trap Capillary column GC/MS, EPA 600/4-88/039.

PRIMARY NAME: 1,1-Dichloroethane

Method 601

TITLE: Purgeable Halocarbons

MATRIX: Wastewater

CAS # : 75-34-3

APPLICATION: Method covers 29 purgeable halocarbons. (Method 624 provides GC/MS conditions appropriate for the qualitative and quantitative confirmation of results). Method describes conditions for a second GC column to confirm measurements made with primary column.

INTERFERENCES: Impurities in the purge gas and organic compounds outgassing from the plumbing ahead of the trap. With high and low level samples, there can be carry-over contamination. Diffusion of volatile organics through the septum seal into the sample.

INSTRUMENTATION: GC equipped with halide-specific detector. (With purge and trap unit).

RANGE: 8.0 To 500 μg/L. MDL: 0.07 μg/L.

PRECISION: 0.14X+0.94 μg/L (overall precision)

ACCURACY: 0.95C-1.08 μg/L (as recovery)

SAMPLING METHOD: 25 mL glass vial. Teflon lined septum.

STABILITY: Cool, 4°C, 0.008% sodium thiosulfate. M.H.T. = 14 Days.

QUALITY CONTROL: The laboratory must on an ongoing basis, spike at least 10% of the samples from each sample site being monitored to assess accuracy.

REFERENCE: Method 601, Federal Register Part VIII 40 CFR Part 136, Oct 26, 1984.

PRIMARY NAME: 1,1-Dichloroethane Method 624

TITLE: Purgeables MATRIX: Wastewater

CAS # : 75-34-3

APPLICATION: Method covers 31 purgeable organics. An inert gas is bubbled through a 5 mL water sample in a specially designed purging chamber. Here, purgeables are transferred from aqueous to gaseous phase, passed onto a sorbent column and trapped. Trap is heated and backflushed with inert gas to desorb purgeables onto a GC column, where purgeables are separated.

INTERFERENCES: Impurities in the purge gas, organic compounds outgassing from the plumbing ahead of the trap, and solvent vapors in the laboratory. With high and low level samples, there can be carry-over contamination.

INSTRUMENTATION: GC/MS with purge and trap unit.

RANGE: 5-600 μg/L MDL: 4.7 μg/L

PRECISION: 0.16X+0.47 μg/L (overall precision)

ACCURACY: 1.05C+0.36 μg/L (as recovery)

SAMPLING METHOD: 25 mL glass vial. Teflon lined septum.

STABILITY: Cool, 4°C, 0.008% sodium thiosulfate. M.H.T. = 14 Days.

QUALITY CONTROL: The laboratory must on an ongoing basis, spike at least 5% of the samples from each sample site being monitored to assess accuracy.

REFERENCE: Method 624, Federal Register Part VIII 40 CFR Part 136, Oct 26, 1984.

PRIMARY NAME: 1,1-Dichloroethane Method 8010

TITLE: Halogenated Volatile Organics

MATRIX: groundwater, soils, sludges water miscible liquid wastes, and non-water miscible wastes.

CAS # : 75-34-3

APPLICATION: This method is used for the analysis of 39 halogenated VOCs. Samples are analyzed using direct injection or purge and trap methods. Groundwater must be analyzed by the purge and trap method. The method provides an optional GC column which is used for analyte confirmation and that may help resolve analytes from interferences.

INTERFERENCES: There can be carry-over contamination with high and low level samples. Impurities may come from the purge and trap apparatus,

organic compounds outgassing from the plumbing ahead of trap, diffusion of VOCs through the sample bottle septum during shipping or storage, or from solvent vapors in the lab.

INSTRUMENTATION: GC capable of on-column injections or purge and trap sample introduction and a halogen specific detector. Column 1: 8 foot by 0.1 inch 1% SP-1000 on Carbopack-B. Column 2: 6 foot by 0.1 inch bonded n-octane on Porasil-C.

RANGE: 8 to 500 µg/L (reagent water)

MDL: 0.07 µg/L (reagent water)

PRACTICAL QUANTITATION LIMIT FACTORS FOR MULTIPLYING TIMES MDL VALUE

Matrix	Multiplication Factor
Groundwater	10
Low-level soil	10
Water miscible liquid waste	500
High-level soil and sludge	1250
Non-water miscible waste	1250

PRECISION: 0.14X + 0.94 µg/L (overall precision)

ACCURACY: 0.95C - 1.08 µg/L (as recovery)

SAMPLING METHOD: For water and liquid samples; use glass 40 mL vials with Teflon lined septum caps and collect two vials per sample location with no headspace. For solids and concentrated waste samples; use widemouth glass bottles with Teflon liners.

STABILITY: For concentrated wastes, soils, sediments or sludges; cool to 4°C. For liquids; add 4 drops of concentrated hydrochloric acid and cool to 4°C. M.H.T. = 14 days.

QUALITY CONTROL: Analyze a reagent blank, matrix spike and matrix spike duplicate/duplicate for each analytical batch (up to 20 Samples). Demonstrate the purity of glassware and reagents by analyzing a reagent water method blank. Internal, surrogate and five concentration level calibration standards are used.

REFERENCE: Method 8010, SW-846, 3rd ed., Nov 1986.

PRIMARY NAME: 1,2-Dichloroethane

Method 502.1

TITLE: Halogenated VOCs in Water

MATRIX: Drinking water (finished or any treatment stage) and raw source water.

CAS # : 107-06-2

APPLICATION: Method covers 40 halogenated VOCs. An inert gas is bubbled through a 5 mL water sample. Purged sample components are trapped in tube of sorbent materials. When purging is complete, sorbent tube is heated and backflushed with inert gas to desorb trapped sample onto a packed GC column where it is separated.

INTERFERENCES: During analysis, major contaminant sources are volatile materials in the lab and impurities in purging gas and sorbent trap. With high and low level samples, there can be carry-over contamination. Watch for methylene chloride, it permeates through PTFE tubing.

INSTRUMENTATION: Purge and Trap GC w/Halide Specific Detector. (Two GC columns are recommended); Column #1: 1% SP-1000 on Carbopack B; Column #2: n-octane bonded on Poracil C.

RANGE: 8.0-505 μg/L (Drinking water)

MDL: 0.002 μg/L in water

PRECISION: RSD = 7.0% @ 0.20 μg/L spike (S.L.) (17 Samples)

ACCURACY: Avg Recovery = 110% @ 0.20 μg/L spike (S.L.) (17 Samples)

SAMPLING METHOD: Use a 40-120 mL screw cap vial (pre-washed with detergent, rinsed with distilled water and oven dried at 105°C) with a PTFE-faced silicone septum. If residual chlorine is in the water add about 25 mg of ascorbic acid to each vial before sample collection. Collect bubble free samples.

STABILITY: Cool to 4°C; HCl to pH <2.

M.H.T. = 14 Days.

QUALITY CONTROL: As an initial demonstration of lab accuracy and precision, analyze 4 to 7 replicates of a lab fortified blank containing analyte at 0.1-5 μg/L. Collect all samples in duplicate.

REFERENCE: Method 502.1, Volatile Halogenated Organic Compounds in Water by Purge and Trap GC, EPA 600/4-88/039.

PRIMARY NAME: 1,2-Dichloroethane Method 502.2

TITLE: VOCs in H_2O by Purge and Trap capillary column GC

MATRIX: Drinking water (finished or any treatment stage) and raw source water.

CAS # : 107-06-2

APPLICATION: Method covers 60 VOCs. An inert gas is bubbled through a 5 mL sample in a purging chamber. Purged sample components are trapped in tube of sorbent materials. When purging is done, sorbent tube is heated and backflushed w/helium to desorb trapped sample onto a capillary GC column.

INTERFERENCES: During analysis, major contaminant sources are volatile materials in the lab and impurities in purging gas and sorbent trap. With high and low level samples, there can be carry-over contamination. Beware of methylene chloride, it permeates through PTFE tubing.

INSTRUMENTATION: (Capillary Column) GC, w/PID and electrolytic conductivity detectors in series. Column #1: VOCOL wide bore column; Column #2: RTX-502.2 mega bore column; Column #3: DB-62 mega bore column.

RANGE: Approximately 0.02-200 µg/L

MDL: 0.03 µg/L (reagent H_2O) (ELCD) 60 m × 0.75 mm VOCOL capillary column

PRECISION: RSD = 3.8% @ Spike of 10 µg/L (reagent H_2O) (ELCD) 60 m × 0.75 mm VOCOL capillary column

ACCURACY: Recovery = 100% @ spike of 10 µg/L (reagent H_2O) (ELCD) 60 m × 0.75 mm VOCOL capillary column

SAMPLING METHOD: Use a 40-120 mL screw cap vial (pre-washed with detergent, rinsed with distilled water and oven dried at 105°C) with a PTFE-faced silicone septum. If residual chlorine is in the water add about 25 mg of ascorbic acid to each vial before sample collection. Collect bubble free samples.

STABILITY: Cool to 4°C; HCl to pH <2. M.H.T. = 14 Days.

QUALITY CONTROL: As an initial demonstration of lab accuracy and precision, analyze 4 to 7 replicates of a lab fortified blank containing analyte at 0.1-5 µg/L. Collect all samples in duplicate.

REFERENCE: Method 502.2, Volatile Organic Compounds in Water by Purge and Trap Capillary Column GC, EPA 600/4-88/039.

PRIMARY NAME: 1,2-Dichloroethane

Method 524.1

TITLE: VOCs in H_2O by Purge and Trap GC/MS

MATRIX: Drinking water (finished or any treatment stage) and raw source water.

CAS # : 107-06-2

APPLICATION: Method covers 51 VOCs. An inert gas is bubbled through a 25 mL water sample. Purged sample components are trapped in tube of sorbent materials. When purging is complete, sorbent tube is heated and backflushed with helium to desorb trapped sample onto a packed GC column.

INTERFERENCES: During analysis, major contaminant sources are volatile materials in the lab and impurities in purging gas and sorbent trap. With high and low level samples, there can be carry-over contamination.Watch for methylene chloride, it permeates through PTFE tubing.

INSTRUMENTATION: Gas Chromatography/Mass Spectrometry/Data System; 1% SP-1000 on Carbopack B column.

RANGE: Approximately 0.2-200 µg/L

MDL: 0.2 µg/L

PRECISION: RSD = 7.9% @ 1.0 µg/L

ACCURACY: Mean Accuracy = 97% @ 1.0 µg/L

SAMPLING METHOD: Use a 60-120 mL screw cap vial (pre-washed with detergent, rinsed with distilled water and oven dried at 105°C) with a PTFE-faced silicone septum. If residual chlorine is in the water add about 25 mg of ascorbic acid to each vial before sample collection. Collect bubble free samples.

STABILITY: Cool to 4°C; HCl to pH <2.

M.H.T. = 14 Days.

QUALITY CONTROL: As an initial demonstration of lab accuracy and precision, analyze 4 to 7 replicates of a lab fortified blank containing analyte at 0.2-5 µg/L. Collect all samples in duplicate.

REFERENCE: Method 524.1, VOCs in Water by Purge and Trap Gas Chromatography/Mass Spectrometry, EPA 600/4-88/039.

PRIMARY NAME: 1,2-Dichloroethane

Method 524.2

TITLE: VOCs in H_2O (Purge and Trap capillary column GC/MS

MATRIX: Drinking water (finished or any treatment stage) and raw source water.

CAS # : 107-06-2

APPLICATION: Method covers 60 VOCs. An inert gas is bubbled through a 25 mL water sample. Purged sample components are trapped in tube of sorbent materials. When purging is complete, sorbent tube is heated and backflushed with helium to desorb trapped sample onto a capillary GC column.

INTERFERENCES: During analysis, major contaminant sources are volatile materials in the lab and impurities in purging gas and sorbent trap. With high and low level samples, there can be carry-over contamination.Watch for methylene chloride, it permeates through PTFE tubing.

INSTRUMENTATION: (capillary column) gas chromatography/mass spectrometry/data system. Column #1: VOCOL glass wide bore; Column #2: DB-624 fused silica; Column #3: DB-5 fused silica.

RANGE: Approximately 0.1-200 µg/L

MDL: 0.06 µg/L on wide bore capillary column.

PRECISION: RSD = 5.4% @ (0.1-10) µg/L (31 samples) wide bore capillary column.

ACCURACY: Recovery = 95% @ (0.1-10) µg/L (31 samples) wide bore capillary column.

SAMPLING METHOD: Use a 60-120 mL screw cap vial (pre-washed with detergent, rinsed with distilled water and oven dried at 105°C) with a PTFE-faced silicone septum. If residual chlorine is in the water add about 25 mg of ascorbic acid to each vial before sample collection. Collect bubble free samples.

STABILITY: Cool to 4°C; HCl to pH <2. M.H.T. = 14 Days.

QUALITY CONTROL: As an initial demonstration of lab accuracy and precision, analyze 4 to 7 replicates of a lab fortified blank containing analyte at 0.2-5 μg/L. Collect all samples in duplicate.

REFERENCE: Method 524.2, VOCs in H_2O by Purge and Trap Capillary column GC/MS, EPA 600/4-88/039.

PRIMARY NAME: 1,2-Dichloroethane Method 601

TITLE: Purgeable Halocarbons MATRIX: Wastewater

CAS # : 107-06-2

APPLICATION: Method covers 29 purgeable halocarbons. (Method 624 provides GC/MS conditions appropriate for the qualitative and quantitative confirmation of results). Method describes conditions for a second GC column to confirm measurements made with primary column.

INTERFERENCES: Impurities in the purge gas and organic compounds outgassing from the plumbing ahead of the trap. With high and low level samples, there can be carry-over contamination. Diffusion of volatile organics through the septum seal into the sample.

INSTRUMENTATION: GC equipped with halide-specific detector. (With purge and trap unit).

RANGE: 8.0 To 500 μg/L. MDL: 0.03 μg/L.

PRECISION: 0.15X+0.94 μg/L (overall precision)

ACCURACY: 1.04C-1.06 μg/L (as recovery)

SAMPLING METHOD: 25 mL glass vial. Teflon lined septum.

STABILITY: Cool, 4°C, 0.008% sodium thiosulfate. M.H.T. = 14 Days.

QUALITY CONTROL: The laboratory must on an ongoing basis, spike at least 10% of the samples from each sample site being monitored to assess accuracy.

REFERENCE: Method 601, Federal Register Part VIII 40 CFR Part 136, Oct 26, 1984.

PRIMARY NAME: 1,2-Dichloroethane Method 624

TITLE: Purgeables MATRIX: Wastewater

CAS # : 107-06-2

APPLICATION: Method covers 31 purgeable organics. An inert gas is bubbled through a 5 mL water sample in a specially designed purging chamber. Here, purgeables are transferred from aqueous to gaseous phase, passed onto a sorbent column and trapped. Trap is heated and backflushed with inert gas to desorb purgeables onto a GC column, where purgeables are separated.

INTERFERENCES: Impurities in the purge gas, organic compounds outgassing from the plumbing ahead of the trap, and solvent vapors in the laboratory. With high and low level samples, there can be carry-over contamination.

INSTRUMENTATION: GC/MS with purge and trap unit.

RANGE: 5-600 μg/L MDL: 2.8 μg/L

PRECISION: 0.21X-0.38 μg/L (overall precision)

ACCURACY: 1.02C+0.45 μg/L (as recovery)

SAMPLING METHOD: 25 mL glass vial. Teflon lined septum.

STABILITY: Cool, 4°C, 0.008% sodium thiosulfate. M.H.T. = 14 Days.

QUALITY CONTROL: The laboratory must on an ongoing basis, spike at least 5% of the samples from each sample site being monitored to assess accuracy.

REFERENCE: Method 624, Federal Register Part VIII 40 CFR Part 136, Oct 26, 1984.

PRIMARY NAME: 1,2-Dichloroethane

Method 8010

TITLE: Halogenated Volatile Organics

MATRIX: groundwater, soils, sludges water miscible liquid wastes, and non-water miscible wastes.

CAS # : 107-06-2

APPLICATION: This method is used for the analysis of 39 halogenated VOCs. Samples are analyzed using direct injection or purge and trap methods. Groundwater must be analyzed by the purge and trap method. The method provides an optional GC column which is used for analyte confirmation and that may help resolve analytes from interferences.

INTERFERENCES: There can be carry-over contamination with high and low level samples. Impurities may come from the purge and trap apparatus, organic compounds outgassing from the plumbing ahead of trap, diffusion of VOCs through the sample bottle septum during shipping or storage, or from solvent vapors in the lab.

INSTRUMENTATION: GC capable of on-column injections or purge and trap sample introduction and a halogen specific detector. Column 1: 8 foot by 0.1 inch 1% SP-1000 on Carbopack-B. Column 2: 6 foot by 0.1 inch bonded n-octane on Porasil-C.

RANGE: 8 to 500 µg/L (reagent water)

MDL: 0.03 µg/L (reagent water)

PRACTICAL QUANTITATION LIMIT FACTORS FOR MULTIPLYING TIMES MDL VALUE

Matrix	Multiplication Factor
Groundwater	10
Low-level soil	10
Water miscible liquid waste	500
High-level soil and sludge	1250
Non-water miscible waste	1250

PRECISION: 0.15X + 0.94 µg/L (overall precision)

ACCURACY: 1.04C - 1.06 μg/L (as recovery)

SAMPLING METHOD: For water and liquid samples; use glass 40 mL vials with Teflon lined septum caps and collect two vials per sample location with no headspace. For solids and concentrated waste samples; use widemouth glass bottles with Teflon liners.

STABILITY: For concentrated wastes, soils, sediments or sludges; cool to 4°C. For liquids; add 4 drops of concentrated hydrochloric acid and cool to 4°C. M.H.T. = 14 days.

QUALITY CONTROL: Analyze a reagent blank, matrix spike and matrix spike duplicate/duplicate for each analytical batch (up to 20 Samples). Demonstrate the purity of glassware and reagents by analyzing a reagent water method blank. Internal, surrogate and five concentration level calibration standards are used.

REFERENCE: Method 8010, SW-846, 3rd ed., Nov 1986.

PRIMARY NAME: 1,1-Dichloroethylene Method 502.1

TITLE: Halogenated VOCs in Water

MATRIX: Drinking water (finished or any treatment stage) and raw source water.

CAS # : 75-35-4

APPLICATION: Method covers 40 halogenated VOCs. An inert gas is bubbled through a 5 mL water sample. Purged sample components are trapped in tube of sorbent materials. When purging is complete, sorbent tube is heated and backflushed with inert gas to desorb trapped sample onto a packed GC column where it is separated.

INTERFERENCES: During analysis, major contaminant sources are volatile materials in the lab and impurities in purging gas and sorbent trap. With high and low level samples, there can be carry-over contamination. Watch for methylene chloride, it permeates through PTFE tubing.

INSTRUMENTATION: Purge and Trap GC w/halide specific detector. (Two GC columns are recommended); Column #1: 1% SP-1000 on Carbopack B; Column #2: n-octane bonded on Poracil C.

RANGE: 8.0-505 μg/L (Drinking water) MDL: 0.003 μg/L in water

PRECISION: RSD = 9.3% @ 0.40 μg/L spike (S.L.) (18 Samples)

ACCURACY: Avg Recovery = 88% @ 0.40 μg/L spike (S.L.) (18 Samples)

SAMPLING METHOD: Use a 40-120 mL screw cap vial (pre-washed with detergent, rinsed with distilled water and oven dried at 105°C) with a PTFE-faced silicone septum. If residual chlorine is in the water add about 25 mg of ascorbic acid to each vial before sample collection. Collect bubble free samples.

STABILITY: Cool to 4°C; HCl to pH <2. M.H.T. = 14 Days.

QUALITY CONTROL: As an initial demonstration of lab accuracy and precision, analyze 4 to 7 replicates of a lab fortified blank containing analyte at 0.1-5 μg/L. Collect all samples in duplicate.

REFERENCE: Method 502.1, Volatile Halogenated Organic Compounds in Water by Purge and Trap GC, EPA 600/4-88/039.

PRIMARY NAME: 1,1-Dichloroethylene Method 502.2

TITLE: VOCs in H_2O by Purge and Trap capillary column GC

MATRIX: Drinking water (finished or any treatment stage) and raw source water.

CAS # : 75-35-4

APPLICATION: Method covers 60 VOCs. An inert gas is bubbled through a 5 mL sample in a purging chamber. Purged sample components are trapped in tube of sorbent materials. When purging is done, sorbent tube is heated and backflushed w/helium to desorb trapped sample onto a capillary GC column.

INTERFERENCES: During analysis, major contaminant sources are volatile materials in the lab and impurities in purging gas and sorbent trap. With high and low level samples, there can be carry-over contamination. Beware of methylene chloride, it permeates through PTFE tubing.

INSTRUMENTATION: (Capillary Column) GC, w/PID and electrolytic conductivity detectors in series. Column #1: VOCOL wide bore column; Column #2: RTX-502.2 mega bore column; Column #3: DB-62 mega bore column.

RANGE: Approximately 0.02-200 μg/L

MDL: 0.07 μg/L (reagent H_2O) (ELCD) 60 m × 0.75 mm VOCOL capillary column

PRECISION: RSD = 2.8% @ Spike of 10 μg/L (reagent H_2O) (PID)

ACCURACY: Recovery = 103% @ spike of 10 μg/L (reagent H_2O) (PID)

SAMPLING METHOD: Use a 40-120 mL screw cap vial (pre-washed with detergent, rinsed with distilled water and oven dried at 105°C) with a PTFE-faced silicone septum. If residual chlorine is in the water add about 25 mg of ascorbic acid to each vial before sample collection. Collect bubble free samples.

STABILITY: Cool to 4°C; HCl to pH <2.

M.H.T. = 14 Days.

QUALITY CONTROL: As an initial demonstration of lab accuracy and precision, analyze 4 to 7 replicates of a lab fortified blank containing analyte at 0.1-5 μg/L. Collect all samples in duplicate.

REFERENCE: Method 502.2, Volatile Organic Compounds in Water by Purge and Trap Capillary Column GC, EPA 600/4-88/039.

PRIMARY NAME: 1,1-Dichloroethylene

Method 524.1

TITLE: VOCs in H_2O by Purge and Trap GC/MS

MATRIX: Drinking water (finished or any treatment stage) and raw source water.

CAS # : 75-35-4

APPLICATION: Method covers 51 VOCs. An inert gas is bubbled through a 25 mL water sample. Purged sample components are trapped in tube of sorbent materials. When purging is complete, sorbent tube is heated and backflushed with helium to desorb trapped sample onto a packed GC column.

INTERFERENCES: During analysis, major contaminant sources are volatile

materials in the lab and impurities in purging gas and sorbent trap. With high and low level samples, there can be carry-over contamination. Watch for methylene chloride, it permeates through PTFE tubing.

INSTRUMENTATION: Gas Chromatography/Mass Spectrometry/Data System; 1% SP-1000 on Carbopack B column.

RANGE: Approximately 0.2-200 μg/L MDL: 0.2 μg/L

PRECISION: RSD = 6.1% @ 1.0 μg/L

ACCURACY: Mean Accuracy = 109% @ 1.0 μg/L

SAMPLING METHOD: Use a 60-120 mL screw cap vial (pre-washed with detergent, rinsed with distilled water and oven dried at 105°C) with a PTFE-faced silicone septum. If residual chlorine is in the water add about 25 mg of ascorbic acid to each vial before sample collection. Collect bubble free samples.

STABILITY: Cool to 4°C; HCl to pH <2. M.H.T. = 14 Days.

QUALITY CONTROL: As an initial demonstration of lab accuracy and precision, analyze 4 to 7 replicates of a lab fortified blank containing analyte at 0.2-5 μg/L. Collect all samples in duplicate.

REFERENCE: Method 524.1, VOCs in Water by Purge and Trap Gas chromatography/mass spectrometry, EPA 600/4-88/039.

PRIMARY NAME: 1,1-Dichloroethylene Method 524.2

TITLE: VOCs in H_2O (purge and trap capillary column GC/MS

MATRIX: Drinking water (finished or) any treatment stage) and raw source water.

CAS # : 75-35-4

APPLICATION: Method covers 60 VOCs. An inert gas is bubbled through a 25 mL water sample. Purged sample components are trapped in tube of sorbent materials. When purging is complete, sorbent tube is heated and backflushed with helium to desorb trapped sample onto a capillary GC column.

INTERFERENCES: During analysis, major contaminant sources are volatile materials in the lab and impurities in purging gas and sorbent trap. With high and low level samples, there can be carry-over contamination. Watch for methylene chloride, it permeates through PTFE tubing.

INSTRUMENTATION: (Capillary Column) Gas Chromatography/Mass Spectrometry/Data System. Column #1: VOCOL glass wide bore; Column #2: DB-624 fused silica; Column #3: DB-5 fused silica.

RANGE: Approximately 0.1-200 µg/L

MDL: 0.12 µg/L on wide bore capillary column.

PRECISION: RSD = 6.7% @ (0.1-10) µg/L (34 samples) wide bore capillary column.

ACCURACY: Recovery = 94% @ (0.1-10) µg/L (34 samples) wide bore capillary column.

SAMPLING METHOD: Use a 60-120 mL screw cap vial (pre-washed with detergent, rinsed with distilled water and oven dried at 105°C) with a PTFE-faced silicone septum. If residual chlorine is in the water add about 25 mg of ascorbic acid to each vial before sample collection. Collect bubble free samples.

STABILITY: Cool to 4°C; HCl to pH <2.

M.H.T. = 14 Days.

QUALITY CONTROL: As an initial demonstration of lab accuracy and precision, analyze 4 to 7 replicates of a lab fortified blank containing analyte at 0.2-5 µg/L. Collect all samples in duplicate.

REFERENCE: Method 524.2, VOCs in H_2O by Purge and Trap Capillary column GC/MS, EPA 600/4-88/039.

PRIMARY NAME: 1,1-Dichloroethylene

Method 601

TITLE: Purgeable Halocarbons

MATRIX: Wastewater

CAS # : 75-35-4

APPLICATION: Method covers 29 purgeable halocarbons. (Method 624 provides GC/MS conditions appropriate for the qualitative and quantitative

confirmation of results). Method describes conditions for a second GC column to confirm measurements made with primary column.

INTERFERENCES: Impurities in the purge gas and organic compounds outgassing from the plumbing ahead of the trap. With high and low level samples, there can be carry-over contamination. Diffusion of volatile organics through the septum seal into the sample.

INSTRUMENTATION: GC equipped with halide-specific detector. (With purge and trap unit).

RANGE: 8.0 To 500 µg/L. MDL: 0.13 µg/L.

PRECISION: 0.29X-0.40 µg/L (overall precision)

ACCURACY: 0.98C-0.87 µg/L (as recovery)

SAMPLING METHOD: 25 mL glass vial. Teflon lined septum.

STABILITY: Cool, 4°C, 0.008% sodium thiosulfate. M.H.T. = 14 Days.

QUALITY CONTROL: The laboratory must on an ongoing basis, spike at least 10% of the samples from each sample site being monitored to assess accuracy.

REFERENCE: Method 601, Federal Register Part VIII 40 CFR Part 136, Oct 26, 1984.

PRIMARY NAME: 1,1-Dichloroethylene Method 624

TITLE: Purgeables MATRIX: Wastewater

CAS # : 75-35-4

APPLICATION: Method covers 31 purgeable organics. An inert gas is bubbled through a 5 mL water sample in a specially designed purging chamber. Here, purgeables are transferred from aqueous to gaseous phase, passed onto a sorbent column and trapped. Trap is heated and backflushed with inert gas to desorb purgeables onto a GC column, where purgeables are separated.

INTERFERENCES: Impurities in the purge gas, organic compounds outgassing from the plumbing ahead of the trap, and solvent vapors in the laboratory. With high and low level samples, there can be carry-over contamination.

INSTRUMENTATION: GC/MS with purge and trap unit.

RANGE: 5-600 μg/L MDL: 2.8 μg/L

PRECISION: 0.43X-0.22 μg/L (overall precision)

ACCURACY: 1.12C+0.61 μg/L (as recovery)

SAMPLING METHOD: 25 mL glass vial. Teflon lined septum.

STABILITY: Cool, 4°C, 0.008% sodium thiosulfate. M.H.T. = 14 Days.

QUALITY CONTROL: The laboratory must on an ongoing basis, spike at least 5% of the samples from each sample site being monitored to assess accuracy.

REFERENCE: Method 624, Federal Register Part VIII 40 CFR Part 136, Oct 26, 1984.

PRIMARY NAME: 1,1-Dichloroethylene Method 8010

TITLE: Halogenated Volatile Organics

MATRIX: groundwater, soils, sludges water miscible liquid wastes, and non-water miscible wastes.

CAS # : 75-35-4

APPLICATION: This method is used for the analysis of 39 halogenated VOCs. Samples are analyzed using direct injection or purge and trap methods. Groundwater must be analyzed by the purge and trap method. The method provides an optional GC column which is used for analyte confirmation and that may help resolve analytes from interferences.

INTERFERENCES: There can be carry-over contamination with high and low level samples. Impurities may come from the purge and trap apparatus,

organic compounds outgassing from the plumbing ahead of trap, diffusion of VOCs through the sample bottle septum during shipping or storage, or from solvent vapors in the lab.

INSTRUMENTATION: GC capable of on-column injections or purge and trap sample introduction and a halogen specific detector. Column 1: 8 foot by 0.1 inch 1% SP-1000 on Carbopack-B. Column 2: 6 foot by 0.1 inch bonded n-octane on Porasil-C.

RANGE: 8 to 500 μg/L (reagent water)

MDL: 0.13 μg/L (reagent water)

PRACTICAL QUANTITATION LIMIT FACTORS FOR MULTIPLYING TIMES MDL VALUE

Matrix	Multiplication Factor
Groundwater	10
Low-level soil	10
Water miscible liquid waste	500
High-level soil and sludge	1250
Non-water miscible waste	1250

PRECISION: 0.29X - 0.04 μg/L (overall precision)

ACCURACY: 0.98C - 0.87 μg/L (as recovery)

SAMPLING METHOD: For water and liquid samples; use glass 40 mL vials with Teflon lined septum caps and collect two vials per sample location with no headspace. For solids and concentrated waste samples; use widemouth glass bottles with Teflon liners.

STABILITY: For concentrated wastes, soils, sediments or sludges; cool to 4°C. For liquids; add 4 drops of concentrated hydrochloric acid and cool to 4°C. M.H.T. = 14 days.

QUALITY CONTROL: Analyze a reagent blank, matrix spike and matrix spike duplicate/duplicate for each analytical batch (up to 20 Samples). Demonstrate the purity of glassware and reagents by analyzing a reagent water method blank. Internal, surrogate and five concentration level calibration standards are used.

REFERENCE: Method 8010, SW-846, 3rd ed., Nov 1986.

PRIMARY NAME: cis-1,2-Dichloroethylene Method 502.1

TITLE: Halogenated VOCs in Water

MATRIX: Drinking water (finished or any treatment stage) and raw source water.

CAS # : 156-59-2

APPLICATION: Method covers 40 halogenated VOCs. An inert gas is bubbled through a 5 mL water sample. Purged sample components are trapped in tube of sorbent materials. When purging is complete, sorbent tube is heated and backflushed with inert gas to desorb trapped sample onto a packed GC column where it is separated.

INTERFERENCES: During analysis, major contaminant sources are volatile materials in the lab and impurities in purging gas and sorbent trap. With high and low level samples, there can be carry-over contamination. Watch for methylene chloride, it permeates through PTFE tubing.

INSTRUMENTATION: Purge and Trap GC w/halide specific detector. (Two GC columns are recommended); Column #1: 1% SP-1000 on Carbopack B; Column #2: n-octane bonded on Poracil C.

RANGE: 8.0-505 μg/L (Drinking water) MDL: 0.002 μg/L in water

PRECISION: RSD = 7.0% @ 0.40 μg/L spike (S.L.) (20 Samples)

ACCURACY: Avg Recovery = 88% @ 0.40 μg/L spike (S.L.) (20 Samples)

SAMPLING METHOD: Use a 40-120 mL screw cap vial (pre-washed with detergent, rinsed with distilled water and oven dried at 105°C) with a PTFE-faced silicone septum. If residual chlorine is in the water add about 25 mg of ascorbic acid to each vial before sample collection. Collect bubble free samples.

STABILITY: Cool to 4°C; HCl to pH <2. M.H.T. = 14 Days.

QUALITY CONTROL: As an initial demonstration of lab accuracy and precision, analyze 4 to 7 replicates of a lab fortified blank containing analyte at 0.1-5 μg/L. Collect all samples in duplicate.

REFERENCE: Method 502.1, Volatile Halogenated Organic Compounds in Water by Purge and Trap GC, EPA 600/4-88/039.

PRIMARY NAME: cis-1,2-Dichloroethylene

Method 502.2

TITLE: VOCs in H_2O by Purge and Trap capillary column GC

MATRIX: Drinking water (finished or any treatment stage) and raw source water.

CAS # : 156-59-2

APPLICATION: Method covers 60 VOCs. An inert gas is bubbled through a 5 mL sample in a purging chamber. Purged sample components are trapped in tube of sorbent materials. When purging is done, sorbent tube is heated and backflushed w/helium to desorb trapped sample onto a capillary GC column.

INTERFERENCES: During analysis, major contaminant sources are volatile materials in the lab and impurities in purging gas and sorbent trap. With high and low level samples, there can be carry-over contamination. Beware of methylene chloride, it permeates through PTFE tubing.

INSTRUMENTATION: (capillary column) GC, w/PID and electrolytic conductivity detectors in series. Column #1: VOCOL wide bore column; Column #2: RTX-502.2 mega bore column; Column #3: DB-62 mega bore column.

RANGE: Approximately 0.02-200 µg/L

MDL: 0.01 µg/L (reagent H_2O) (ELCD) 60 m × 0.75 mm VOCOL capillary column

PRECISION: RSD = 3.3% @ Spike of 10 µg/L (reagent H_2O) (ELCD). 60 m × 0.75 mm VOCOL capillary column

ACCURACY: Recovery = 105% @ spike of 10 µg/L (reagent H_2O) (ELCD) 60 m × 0.75 mm VOCOL capillary column

SAMPLING METHOD: Use a 40-120 mL screw cap vial (pre-washed with detergent, rinsed with distilled water and oven dried at 105°C) with a PTFE-faced silicone septum. If residual chlorine is in the water add about 25 mg of ascorbic acid to each vial before sample collection. Collect bubble free samples.

STABILITY: Cool to 4°C; HCl to pH <2.

M.H.T. = 14 Days.

QUALITY CONTROL: As an initial demonstration of lab accuracy and precision, analyze 4 to 7 replicates of a lab fortified blank containing analyte at 0.1-5 µg/L. Collect all samples in duplicate.

REFERENCE: Method 502.2, Volatile Organic Compounds in Water by Purge and Trap Capillary Column GC, EPA 600/4-88/039.

PRIMARY NAME: cis-1,2-Dichloroethylene Method 524.1

TITLE: VOCs in H_2O by Purge and Trap GC/MS

MATRIX: Drinking water (finished or any treatment stage) and raw source water.

CAS # : 156-59-2

APPLICATION: Method covers 51 VOCs. An inert gas is bubbled through a 25 mL water sample. Purged sample components are trapped in tube of sorbent materials. When purging is complete, sorbent tube is heated and backflushed with helium to desorb trapped sample onto a packed GC column.

INTERFERENCES: During analysis, major contaminant sources are volatile materials in the lab and impurities in purging gas and sorbent trap. With high and low level samples, there can be carry-over contamination. Watch for methylene chloride, it permeates through PTFE tubing.

INSTRUMENTATION: Gas Chromatography/Mass Spectometry/Data System; 1% SP-1000 on Carbopack B column.

RANGE: Approximately 0.2-200 µg/L MDL: not listed.

PRECISION: not listed.

ACCURACY: not listed.

SAMPLING METHOD: Use a 60-120 mL screw cap vial (pre-washed with detergent, rinsed with distilled water and oven dried at 105°C) with a PTFE-faced silicone septum. If residual chlorine is in the water add about 25 mg of ascorbic acid to each vial before sample collection. Collect bubble free samples.

STABILITY: Cool to 4°C; HCl to pH <2. M.H.T. = 14 Days.

QUALITY CONTROL: As an initial demonstration of lab accuracy and pre-

cision, analyze 4 to 7 replicates of a lab fortified blank containing analyte at 0.2-5 μg/L. Collect all samples in duplicate.

REFERENCE: Method 524.1, VOCs in Water by Purge and Trap Gas Chromatography/Mass Spectometry, EPA 600/4-88/039.

PRIMARY NAME: cis-1,2-Dichloroethylene

Method 524.2

TITLE: VOCs in H_2O (Purge and Trap capillary column GC/MS

MATRIX: Drinking water (finished or) any treatment stage) and raw source water.

CAS # : 156-59-2

APPLICATION: Method covers 60 VOCs. An inert gas is bubbled through a 25 mL water sample. Purged sample components are trapped in tube of sorbent materials. When purging is complete, sorbent tube is heated and backflushed with helium to desorb trapped sample onto a capillary GC column.

INTERFERENCES: During analysis, major contaminant sources are volatile materials in the lab and impurities in purging gas and sorbent trap. With high and low level samples, there can be carry-over contamination. Watch for methylene chloride, it permeates through PTFE tubing.

INSTRUMENTATION: (capillary column) Gas Chromatography/Mass Spectometry/Data System. Column #1: VOCOL glass wide bore; Column #2: DB-624 fused silica; Column #3: DB-5 fused silica.

RANGE: Approximately 0.1-200 μg/L

MDL: 0.12 μg/L on wide bore capillary column.

PRECISION: RSD = 6.7% @ (0.5-10) μg/L (18 samples) wide bore capillary column.

ACCURACY: Recovery = 101% @ (0.5-10) μg/L (18 samples) wide bore capillary column.

SAMPLING METHOD: Use a 60-120 mL screw cap vial (pre-washed with detergent, rinsed with distilled water and oven dried at 105°C) with a PTFE-

faced silicone septum. If residual chlorine is in the water add about 25 mg of ascorbic acid to each vial before sample collection. Collect bubble free samples.

STABILITY: Cool to 4°C; HCl to pH <2. M.H.T. = 14 Days.

QUALITY CONTROL: As an initial demonstration of lab accuracy and precision, analyze 4 to 7 replicates of a lab fortified blank containing analyte at 0.2-5 μg/L. Collect all samples in duplicate.

REFERENCE: Method 524.2, VOCs in H_2O by Purge and Trap Capillary column GC/MS, EPA 600/4-88/039.

PRIMARY NAME: trans-1,2-Dichloroethylene Method 502.1

TITLE: Halogenated VOCs in Water

MATRIX: Drinking water (finished or any treatment stage) and raw source water.

CAS # : 156-60-5

APPLICATION: Method covers 40 halogenated VOCs. An inert gas is bubbled through a 5 mL water sample. Purged sample components are trapped in tube of sorbent materials. When purging is complete, sorbent tube is heated and backflushed with inert gas to desorb trapped sample onto a packed GC column where it is separated.

INTERFERENCES: During analysis, major contaminant sources are volatile materials in the lab and impurities in purging gas and sorbent trap. With high and low level samples, there can be carry-over contamination. Watch for methylene chloride, it permeates through PTFE tubing.

INSTRUMENTATION: Purge and Trap GC w/halide specific detector. (Two GC columns are recommended); Column #1: 1% SP-1000 on Carbopack B; Column #2: n-octane bonded on Poracil C.

RANGE: 8.0-505 μg/L (Drinking water) MDL: 0.002 μg/L in water

PRECISION: RSD = 7.0% @ 0.40 μg/L spike (S.L.) (20 Samples)

ACCURACY: Avg Recovery = 88% @ 0.40 µg/L spike (S.L.) (20 Samples)

SAMPLING METHOD: Use a 40-120 mL screw cap vial (pre-washed with detergent, rinsed with distilled water and oven dried at 105°C) with a PTFE-faced silicone septum. If residual chlorine is in the water add about 25 mg of ascorbic acid to each vial before sample collection. Collect bubble free samples.

STABILITY: Cool to 4°C; HCl to pH <2. M.H.T. = 14 Days.

QUALITY CONTROL: As an initial demonstration of lab accuracy and precision, analyze 4 to 7 replicates of a lab fortified blank containing analyte at 0.1-5 µg/L. Collect all samples in duplicate.

REFERENCE: Method 502.1, Volatile Halogenated Organic Compounds in Water by Purge and Trap GC, EPA 600/4-88/039.

PRIMARY NAME: trans-1,2-Dichloroethylene Method 502.2

TITLE: VOCs in H_2O by Purge and Trap capillary column GC

MATRIX: Drinking water (finished or any treatment stage) and raw source water.

CAS # : 156-60-5

APPLICATION: Method covers 60 VOCs. An inert gas is bubbled through a 5 mL sample in a purging chamber. Purged sample components are trapped in tube of sorbent materials. When purging is done, sorbent tube is heated and backflushed w/helium to desorb trapped sample onto a capillary GC column.

INTERFERENCES: During analysis, major contaminant sources are volatile materials in the lab and impurities in purging gas and sorbent trap. With high and low level samples, there can be carry-over contamination. Beware of methylene chloride, it permeates through PTFE tubing.

INSTRUMENTATION: (capillary column) GC, w/PID and electrolytic conductivity detectors in series. Column #1: VOCOL wide bore column; Column #2: RTX-502.2 mega bore column; Column #3: DB-62 mega bore column.

RANGE: Approximately 0.02-200 µg/L

MDL: 0.06 µg/L (reagent H_2O) (PID) 60 m × 0.75 mm VOCOL capillary column

PRECISION: RSD = 3.7% @ Spike of 10 µg/L (reagent H_2O) (ELCD). 60 m × 0.75 mm VOCOL capillary column

ACCURACY: Recovery = 99% @ spike of 10 µg/L (reagent H_2O) (ELCD) 60 m × 0.75 mm VOCOL capillary column

SAMPLING METHOD: Use a 40-120 mL screw cap vial (pre-washed with detergent, rinsed with distilled water and oven dried at 105°C) with a PTFE-faced silicone septum. If residual chlorine is in the water add about 25 mg of ascorbic acid to each vial before sample collection. Collect bubble free samples.

STABILITY: Cool to 4°C; HCl to pH <2. M.H.T. = 14 Days.

QUALITY CONTROL: As an initial demonstration of lab accuracy and precision, analyze 4 to 7 replicates of a lab fortified blank containing analyte at 0.1-5 µg/L. Collect all samples in duplicate.

REFERENCE: Method 502.2, Volatile Organic Compounds in Water by Purge and Trap Capillary Column GC, EPA 600/4-88/039.

PRIMARY NAME: trans-1,2-Dichloroethylene Method 524.1

TITLE: VOCs in H_2O by Purge and Trap GC/MS

MATRIX: Drinking water (finished or any treatment stage) and raw source water.

CAS # : 156-60-5

APPLICATION: Method covers 51 VOCs. An inert gas is bubbled through a 25 mL water sample. Purged sample components are trapped in tube of sorbent materials. When purging is complete, sorbent tube is heated and backflushed with helium to desorb trapped sample onto a packed GC column.

INTERFERENCES: During analysis, major contaminant sources are volatile materials in the lab and impurities in purging gas and sorbent trap. With

high and low level samples, there can be carry-over contamination. Watch for methylene chloride, it permeates through PTFE tubing.

INSTRUMENTATION: Gas Chromatography/Mass Spectometry/Data System; 1% SP-1000 on Carbopack B column.

RANGE: Approximately 0.2-200 µg/L MDL: 0.2 µg/L

PRECISION: RSD = 6.7% @ 1.0 µg/L

ACCURACY: Mean Accuracy = 98% @ 1.0 µg/L

SAMPLING METHOD: Use a 60-120 mL screw cap vial (pre-washed with detergent, rinsed with distilled water and oven dried at 105°C) with a PTFE-faced silicone septum. If residual chlorine is in the water add about 25 mg of ascorbic acid to each vial before sample collection. Collect bubble free samples.

STABILITY: Cool to 4°C; HCl to pH <2. M.H.T. = 14 Days.

QUALITY CONTROL: As an initial demonstration of lab accuracy and precision, analyze 4 to 7 replicates of a lab fortified blank containing analyte at 0.2-5 µg/L. Collect all samples in duplicate.

REFERENCE: Method 524.1, VOCs in Water by Purge and Trap Gas Chromatography/Mass Spectometry, EPA 600/4-88/039.

PRIMARY NAME: trans-1,2-Dichloroethylene Method 524.2

TITLE: VOCs in H_2O (Purge and Trap capillary column GC/MS

MATRIX: Drinking water (finished or any treatment stage) and raw source water.

CAS # : 156-60-5

APPLICATION: Method covers 60 VOCs. An inert gas is bubbled through a 25 mL water sample. Purged sample components are trapped in tube of sorbent materials. When purging is complete, sorbent tube is heated and backflushed with helium to desorb trapped sample onto a capillary GC column.

INTERFERENCES: During analysis, major contaminant sources are volatile materials in the lab and impurities in purging gas and sorbent trap. With high and low level samples, there can be carry-over contamination. Watch for methylene chloride, it permeates through PTFE tubing.

INSTRUMENTATION: (capillary column) Gas Chromatography/Mass Spectometry/Data System. Column #1: VOCOL glass wide bore; Column #2: DB-624 fused silica; Column #3: DB-5 fused silica.

RANGE: Approximately 0.1-200 μg/L

MDL: 0.06 μg/L on wide bore capillary column.

PRECISION: RSD = 5.6% @ (0.1-10) μg/L (30 samples) wide bore capillary column.

ACCURACY: Recovery = 93% @ (0.1-10) μg/L (30 samples) wide bore capillary column.

SAMPLING METHOD: Use a 60-120 mL screw cap vial (pre-washed with detergent, rinsed with distilled water and oven dried at 105°C) with a PTFE-faced silicone septum. If residual chlorine is in the water add about 25 mg of ascorbic acid to each vial before sample collection. Collect bubble free samples.

STABILITY: Cool to 4°C; HCl to pH <2.

M.H.T. = 14 Days.

QUALITY CONTROL: As an initial demonstration of lab accuracy and precision, analyze 4 to 7 replicates of a lab fortified blank containing analyte at 0.2-5 μg/L. Collect all samples in duplicate.

REFERENCE: Method 524.2, VOCs in H_2O by Purge and Trap Capillary column GC/MS, EPA 600/4-88/039.

PRIMARY NAME: trans-1,2-Dichloroethylene

Method 601

TITLE: Purgeable Halocarbons

MATRIX: Wastewater

CAS # : 156-60-5

APPLICATION: Method covers 29 purgeable halocarbons. (Method 624 provides GC/MS conditions appropriate for the qualitative and quantitative confirmation of results). Method describes conditions for a second GC column to confirm measurements made with primary column.

INTERFERENCES: Impurities in the purge gas and organic compounds outgassing from the plumbing ahead of the trap. With high and low level samples, there can be carry-over contamination. Diffusion of volatile organics through the septum seal into the sample.

INSTRUMENTATION: GC equipped with halide-specific detector. (With purge and trap unit).

RANGE: 8.0 To 500 µg/L. MDL: 0.10 µg/L.

PRECISION: 0.17X+1.46 µg/L (overall precision)

ACCURACY: 0.97C-0.16 µg/L (as recovery)

SAMPLING METHOD: 25 mL glass vial. Teflon lined septum.

STABILITY: Cool, 4°C, 0.008% sodium thiosulfate. M.H.T. = 14 Days.

QUALITY CONTROL: The laboratory must on an ongoing basis, spike at least 10% of the samples from each sample site being monitored to assess accuracy.

REFERENCE: Method 601, Federal Register Part VIII 40 CFR Part 136, Oct 26, 1984.

PRIMARY NAME: trans-1,2-Dichloroethylene Method 624

TITLE: Purgeables MATRIX: Wastewater

CAS # : 156-60-5

APPLICATION: Method covers 31 purgeable organics. An inert gas is bubbled through a 5 mL water sample in a specially designed purging chamber. Here, purgeables are transferred from aqueous to gaseous phase,

passed onto a sorbent column and trapped. Trap is heated and backflushed with inert gas to desorb purgeables onto a GC column, where purgeables are separated.

INTERFERENCES: Impurities in the purge gas, organic compounds outgassing from the plumbing ahead of the trap, and solvent vapors in the laboratory. With high and low level samples, there can be carry-over contamination.

INSTRUMENTATION: GC/MS with purge and trap unit.

RANGE: 5-600 μg/L MDL: 1.6 μg/L

PRECISION: 0.19X+0.17 μg/L (overall precision)

ACCURACY: 1.05C+0.03 μg/L (as recovery)

SAMPLING METHOD: 25 mL glass vial. Teflon lined septum.

STABILITY: Cool, 4°C, 0.008% sodium thiosulfate. M.H.T. = 14 Days.

QUALITY CONTROL: The laboratory must on an ongoing basis, spike at least 5% of the samples from each sample site being monitored to assess accuracy.

REFERENCE: Method 624, Federal Register Part VIII 40 CFR Part 136, Oct 26, 1984.

PRIMARY NAME: trans-1,2-Dichloroethylene Method 8010

TITLE: Halogenated Volatile Organics

MATRIX: groundwater, soils, sludges water miscible liquid wastes, and non-water miscible wastes.

CAS # : 156-60-5

APPLICATION: This method is used for the analysis of 39 halogenated VOCs. Samples are analyzed using direct injection or purge and trap methods. Groundwater must be analyzed by the purge and trap method. The method provides an optional GC column which is used for analyte confirmation and that may help resolve analytes from interferences.

INTERFERENCES: There can be carry-over contamination with high and low level samples. Impurities may come from the purge and trap apparatus, organic compounds outgassing from the plumbing ahead of trap, diffusion of VOCs through the sample bottle septum during shipping or storage, or from solvent vapors in the lab.

INSTRUMENTATION: GC capable of on-column injections or purge and trap sample introduction and a halogen specific detector. Column 1: 8 foot by 0.1 inch 1% SP-1000 on Carbopack-B. Column 2: 6 foot by 0.1 inch bonded n-octane on Porasil-C.

RANGE: 8 to 500 µg/L (reagent water)

MDL: 0.10 µg/L (reagent water)

PRACTICAL QUANTITATION LIMIT FACTORS FOR MULTIPLYING TIMES MDL VALUE

Matrix	Multiplication Factor
Groundwater	10
Low-level soil	10
Water miscible liquid waste	500
High-level soil and sludge	1250
Non-water miscible waste	1250

PRECISION: 0.17X + 1.46 µg/L (overall precision)

ACCURACY: 0.97C - 0.16 µg/L (as recovery)

SAMPLING METHOD: For water and liquid samples; use glass 40 mL vials with Teflon lined septum caps and collect two vials per sample location with no headspace. For solids and concentrated waste samples; use widemouth glass bottles with Teflon liners.

STABILITY: For concentrated wastes, soils, sediments or sludges; cool to 4°C. For liquids; add 4 drops of concentrated hydrochloric acid and cool to 4°C. M.H.T. = 14 days.

QUALITY CONTROL: Analyze a reagent blank, matrix spike and matrix spike duplicate/duplicate for each analytical batch (up to 20 Samples). Demonstrate the purity of glassware and reagents by analyzing a reagent water method blank. Internal, surrogate and five concentration level calibration standards are used.

REFERENCE: Method 8010, SW-846, 3rd ed., Nov 1986.

PRIMARY NAME: 1,2-Dichloropropane Method 502.1

TITLE: Halogenated VOCs in Water

MATRIX: Drinking water (finished or any treatment stage) and raw source water.

CAS # : 78-87-5

APPLICATION: Method covers 40 halogenated VOCs. An inert gas is bubbled through a 5 mL water sample. Purged sample components are trapped in tube of sorbent materials. When purging is complete, sorbent tube is heated and backflushed with inert gas to desorb trapped sample onto a packed GC column where it is separated.

INTERFERENCES: During analysis, major contaminant sources are volatile materials in the lab and impurities in purging gas and sorbent trap. With high and low level samples, there can be carry-over contamination. Watch for methylene chloride, it permeates through PTFE tubing.

INSTRUMENTATION: Purge and Trap GC w/halide specific detector. (Two GC columns are recommended); Column #1: 1% SP-1000 on Carbopack B; Column #2: n-octane bonded on Poracil C.

RANGE: 8.0-505 μg/L (Drinking water) MDL: not determined.

PRECISION: RSD = 3.5% @ 0.40 μg/L spike (S.L.) (20 Samples)

ACCURACY: Avg Recovery = 95% @ 0.40 μg/L spike (S.L.) (20 Samples)

SAMPLING METHOD: Use a 40-120 mL screw cap vial (pre-washed with detergent, rinsed with distilled water and oven dried at 105°C) with a PTFE-faced silicone septum. If residual chlorine is in the water add about 25 mg of

ascorbic acid to each vial before sample collection. Collect bubble free samples.

STABILITY: Cool to 4°C; HCl to pH <2. M.H.T. = 14 Days.

QUALITY CONTROL: As an initial demonstration of lab accuracy and precision, analyze 4 to 7 replicates of a lab fortified blank containing analyte at 0.1-5 μg/L. Collect all samples in duplicate.

REFERENCE: Method 502.1, Volatile Halogenated Organic Compounds in Water by Purge and Trap GC, EPA 600/4-88/039.

PRIMARY NAME: 1,2-Dichloropropane Method 502.2

TITLE: VOCs in H_2O by Purge and Trap capillary column GC

MATRIX: Drinking water (finished or any treatment stage) and raw source water.

CAS # : 78-87-5

APPLICATION: Method covers 60 VOCs. An inert gas is bubbled through a 5 mL sample in a purging chamber. Purged sample components are trapped in tube of sorbent materials. When purging is done,sorbent tube is heated and backflushed w/helium to desorb trapped sample onto a capillary GC column.

INTERFERENCES: During analysis, major contaminant sources are volatile materials in the lab and impurities in purging gas and sorbent trap. With high and low level samples, there can be carry-over contamination. Beware of methylene chloride, it permeates through PTFE tubing.

INSTRUMENTATION: (capillary column) GC, w/PID and electrolytic conductivity detectors in series. Column #1: VOCOL wide bore column; Column #2: RTX-502.2 mega bore column; Column #3: DB-62 mega bore column.

RANGE: Approximately 0.02-200 μg/L

MDL: 0.01 μg/L (reagent H_2O) (ELCD) 60 m × 0.75 mm VOCOL capillary column

PRECISION: RSD = 3.7% @ Spike of 10 µg/L (reagent H_2O) (ELCD) 60 m × 0.75 mm VOCOL capillary column

ACCURACY: Recovery = 103% @ spike of 10 µg/L (reagent H_2O) (ELCD) 60 m × 0.75 mm VOCOL capillary column

SAMPLING METHOD: Use a 40-120 mL screw cap vial (pre-washed with detergent, rinsed with distilled water and oven dried at 105°C) with a PTFE-faced silicone septum. If residual chlorine is in the water add about 25 mg of ascorbic acid to each vial before sample collection. Collect bubble free samples.

STABILITY: Cool to 4°C; HCl to pH <2. M.H.T. = 14 Days.

QUALITY CONTROL: As an initial demonstration of lab accuracy and precision, analyze 4 to 7 replicates of a lab fortified blank containing analyte at 0.1-5 µg/L. Collect all samples in duplicate.

REFERENCE: Method 502.2, Volatile Organic Compounds in Water by Purge and Trap Capillary Column GC, EPA 600/4-88/039.

PRIMARY NAME: 1,2-Dichloropropane Method 524.1

TITLE: VOCs in H_2O by Purge and Trap GC/MS

MATRIX: Drinking water (finished or any treatment stage) and raw source water.

CAS # : 78-87-5

APPLICATION: Method covers 51 VOCs. An inert gas is bubbled through a 25 mL water sample. Purged sample components are trapped in tube of sorbent materials. When purging is complete, sorbent tube is heated and backflushed with helium to desorb trapped sample onto a packed GC column.

INTERFERENCES: During analysis, major contaminant sources are volatile materials in the lab and impurities in purging gas and sorbent trap. With high and low level samples, there can be carry-over contamination. Watch for methylene chloride, it permeates through PTFE tubing.

INSTRUMENTATION: Gas Chromatography/Mass Spectrometry/Data System; 1% SP-1000 on Carbopack B column.

RANGE: Approximately 0.2-200 μg/L MDL: 0.2 μg/L

PRECISION: RSD = 5.9% @ 1.0 μg/L

ACCURACY: Mean Accuracy = 101% @ 1.0 μg/L

SAMPLING METHOD: Use a 60-120 mL screw cap vial (pre-washed with detergent, rinsed with distilled water and oven dried at 105°C) with a PTFE-faced silicone septum. If residual chlorine is in the water add about 25 mg of ascorbic acid to each vial before sample collection. Collect bubble free samples.

STABILITY: Cool to 4°C; HCl to pH <2. M.H.T. = 14 Days.

QUALITY CONTROL: As an initial demonstration of lab accuracy and precision, analyze 4 to 7 replicates of a lab fortified blank containing analyte at 0.2-5 μg/L. Collect all samples in duplicate.

REFERENCE: Method 524.1, VOCs in Water by Purge and Trap Gas Chromatography/Mass Spectrometry, EPA 600/4-88/039.

PRIMARY NAME: 1,2-Dichloropropane Method 524.2

TITLE: VOCs in H_2O (Purge and Trap capillary column GC/MS

MATRIX: Drinking water (finished or any treatment stage) and raw source water.

CAS # : 78-87-5

APPLICATION: Method covers 60 VOCs. An inert gas is bubbled through a 25 mL water sample. Purged sample components are trapped in tube of sorbent materials. When purging is complete, sorbent tube is heated and backflushed with helium to desorb trapped sample onto a capillary GC column.

INTERFERENCES: During analysis, major contaminant sources are volatile

materials in the lab and impurities in purging gas and sorbent trap. With high and low level samples, there can be carry-over contamination. Watch for methylene chloride, it permeates through PTFE tubing.

INSTRUMENTATION: (Capillary Column) Gas Chromatography/Mass Spectrometry/Data System. Column #1: VOCOL glass wide bore; Column #2: DB-624 fused silica; Column #3: DB-5 fused silica.

RANGE: Approximately 0.1-200 μg/L

MDL: 0.04 μg/L on wide bore capillary column.

PRECISION: RSD = 6.1% @ (0.1-10) μg/L (30 samples) wide bore capillary column.

ACCURACY: Recovery = 97% @ (0.1-10) μg/L (30 samples) wide bore capillary column.

SAMPLING METHOD: Use a 60-120 mL screw cap vial (pre-washed with detergent, rinsed with distilled water and oven dried at 105°C) with a PTFE-faced silicone septum. If residual chlorine is in the water add about 25 mg of ascorbic acid to each vial before sample collection. Collect bubble free samples.

STABILITY: Cool to 4°C; HCl to pH <2.

M.H.T. = 14 Days.

QUALITY CONTROL: As an initial demonstration of lab accuracy and precision, analyze 4 to 7 replicates of a lab fortified blank containing analyte at 0.2-5 μg/L. Collect all samples in duplicate.

REFERENCE: Method 524.2, VOCs in H_2O by Purge and Trap Capillary column GC/MS, EPA 600/4-88/039.

PRIMARY NAME: 1,2-Dichloropropane

Method 601

TITLE: Purgeable Halocarbons

MATRIX: Wastewater

CAS # : 78-87-5

APPLICATION: Method covers 29 purgeable halocarbons. (Method 624 provides GC/MS conditions appropriate for the qualitative and quantitative

confirmation of results). Method describes conditions for a second GC column to confirm measurements made with primary column.

INTERFERENCES: Impurities in the purge gas and organic compounds outgassing from the plumbing ahead of the trap. With high and low level samples, there can be carry-over contamination. Diffusion of volatile organics through the septum seal into the sample.

INSTRUMENTATION: GC equipped with halide-specific detector. (With purge and trap unit).

RANGE: 8.0 To 500 μg/L. MDL: 0.04 μg/L.

PRECISION: 0.23X μg/L (overall precision)

ACCURACY: 1.00C μg/L (as recovery)

SAMPLING METHOD: 25 mL glass vial. Teflon lined septum.

STABILITY: Cool, 4°C, 0.008% sodium thiosulfate. M.H.T. = 14 Days.

QUALITY CONTROL: The laboratory must on an ongoing basis, spike at least 10% of the samples from each sample site being monitored to assess accuracy.

REFERENCE: Method 601, Federal Register Part VIII 40 CFR Part 136, Oct 26, 1984.

PRIMARY NAME: 1,2-Dichloropropane Method 624

TITLE: Purgeables MATRIX: Wastewater

CAS # : 78-87-5

APPLICATION: Method covers 31 purgeable organics. An inert gas is bubbled through a 5 mL water sample in a specially designed purging chamber. Here, purgeables are transferred from aqueous to gaseous phase, passed onto a sorbent column and trapped. Trap is heated and backflushed with inert gas to desorb purgeables onto a GC column, where purgeables are separated.

INTERFERENCES: Impurities in the purge gas, organic compounds outgassing from the plumbing ahead of the trap, and solvent vapors in the laboratory. With high and low level samples, there can be carry-over contamination.

INSTRUMENTATION: GC/MS with purge and trap unit.

RANGE: 5-600 μg/L MDL: 6.0 μg/L

PRECISION: 0.45X μg/L (overall precision)

ACCURACY: 1.00C μg/L (as recovery)

SAMPLING METHOD: 25 mL glass vial. Teflon lined septum.

STABILITY: Cool, 4°C, 0.008% sodium thiosulfate. M.H.T. = 14 Days.

QUALITY CONTROL: The laboratory must on an ongoing basis, spike at least 5% of the samples from each sample site being monitored to assess accuracy.

REFERENCE: Method 624, Federal Register Part VIII 40 CFR Part 136, Oct 26, 1984.

PRIMARY NAME: 1,2-Dichloropropane Method 8010

TITLE: Halogenated Volatile Organics

MATRIX: groundwater, soils, sludges water miscible liquid wastes, and non-water miscible wastes.

CAS # : 78-87-5

APPLICATION: This method is used for the analysis of 39 halogenated VOCs. Samples are analyzed using direct injection or purge and trap methods. Groundwater must be analyzed by the purge and trap method. The method provides an optional GC column which is used for analyte confirmation and that may help resolve analytes from interferences.

INTERFERENCES: There can be carry-over contamination with high and low level samples. Impurities may come from the purge and trap apparatus,

organic compounds outgassing from the plumbing ahead of trap, diffusion of VOCs through the sample bottle septum during shipping or storage, or from solvent vapors in the lab.

INSTRUMENTATION: GC capable of on-column injections or purge and trap sample introduction and a halogen specific detector. Column 1: 8 foot by 0.1 inch 1% SP-1000 on Carbopack-B. Column 2: 6 foot by 0.1 inch bonded n-octane on Porasil-C.

RANGE: 8 to 500 μg/L (reagent water)

MDL: 0.04 μg/L (reagent water)

PRACTICAL QUANTITATION LIMIT FACTORS FOR MULTIPLYING TIMES MDL VALUE

Matrix	Multiplication Factor
Groundwater	10
Low-level soil	10
Water miscible liquid waste	500
High-level soil and sludge	1250
Non-water miscible waste	1250

PRECISION: 0.23X μg/L (overall precision; estimated)

ACCURACY: 1.00C μg/L (as recovery; estimated)

SAMPLING METHOD: For water and liquid samples; use glass 40 mL vials with Teflon lined septum caps and collect two vials per sample location with no headspace. For solids and concentrated waste samples; use widemouth glass bottles with Teflon liners.

STABILITY: For concentrated wastes, soils, sediments or sludges; cool to 4°C. For liquids; add 4 drops of concentrated hydrochloric acid and cool to 4°C. M.H.T. = 14 days.

QUALITY CONTROL: Analyze a reagent blank, matrix spike and matrix spike duplicate/duplicate for each analytical batch (up to 20 Samples). Demonstrate the purity of glassware and reagents by analyzing a reagent water method blank. Internal, surrogate and five concentration level calibration standards are used.

REFERENCE: Method 8010, SW-846, 3rd ed., Nov 1986.

PRIMARY NAME: 1,3-Dichloropropane — Method 502.1

TITLE: Halogenated VOCs in Water

MATRIX: Drinking water (finished or any treatment stage) and raw source water.

CAS # : 142-28-9

APPLICATION: Method covers 40 halogenated VOCs. An inert gas is bubbled through a 5 mL water sample. Purged sample components are trapped in tube of sorbent materials. When purging is complete, sorbent tube is heated and backflushed with inert gas to desorb trapped sample onto a packed GC column where it is separated.

INTERFERENCES: During analysis, major contaminant sources are volatile materials in the lab and impurities in purging gas and sorbent trap. With high and low level samples, there can be carry-over contamination. Watch for methylene chloride, it permeates through PTFE tubing.

INSTRUMENTATION: Purge and Trap GC w/halide specific detector. (Two GC columns are recommended); Column #1: 1% SP-1000 on Carbopack B; Column #2: n-octane bonded on Poracil C.

RANGE: 8.0-505 μg/L (Drinking water) — MDL: not determined.

PRECISION: RSD = 6.5% @ 0.40 μg/L spike (S.L.) (21 Samples)

ACCURACY: Avg Recovery = 98% @ 0.40 μg/L spike (S.L.) (21 Samples)

SAMPLING METHOD: Use a 40-120 mL screw cap vial (pre-washed with detergent, rinsed with distilled water and oven dried at 105°C) with a PTFE-faced silicone septum. If residual chlorine is in the water add about 25 mg of ascorbic acid to each vial before sample collection. Collect bubble free samples.

STABILITY: Cool to 4°C; HCl to pH <2. — M.H.T. = 14 Days.

QUALITY CONTROL: As an initial demonstration of lab accuracy and precision, analyze 4 to 7 replicates of a lab fortified blank containing analyte at 0.1-5 μg/L. Collect all samples in duplicate.

REFERENCE: Method 502.1, Volatile Halogenated Organic Compounds in Water by Purge and Trap GC, EPA 600/4-88/039.

PRIMARY NAME: 1,3-Dichloropropane

Method 502.2

TITLE: VOCs in H_2O by Purge and Trap capillary column GC

MATRIX: Drinking water (finished or any treatment stage) and raw source water.

CAS # : 142-28-9

APPLICATION: Method covers 60 VOCs. An inert gas is bubbled through a 5 mL sample in a purging chamber. Purged sample components are trapped in tube of sorbent materials. When purging is done, sorbent tube is heated and backflushed w/helium to desorb trapped sample onto a capillary GC column.

INTERFERENCES: During analysis, major contaminant sources are volatile materials in the lab and impurities in purging gas and sorbent trap. With high and low level samples, there can be carry-over contamination. Beware of methylene chloride, it permeates through PTFE tubing.

INSTRUMENTATION: (capillary column) GC, w/PID and electrolytic conductivity detectors in series. Column #1: VOCOL wide bore column; Column #2: RTX-502.2 mega bore column; Column #3: DB-62 mega bore column.

RANGE: Approximately 0.02-200 μg/L

MDL: 0.03 μg/L (reagent H_2O) (ELCD) 60 m × 0.75 mm VOCOL capillary column

PRECISION: RSD = 3.4% @ Spike of 10 μg/L (reagent H_2O) (ELCD) 60 m × 0.75 mm VOCOL capillary column

ACCURACY: Recovery = 100% @ spike of 10 μg/L (reagent H_2O) (ELCD) 60 m × 0.75 mm VOCOL capillary column

SAMPLING METHOD: Use a 40-120 mL screw cap vial (pre-washed with detergent, rinsed with distilled water and oven dried at 105°C) with a PTFE-

faced silicone septum. If residual chlorine is in the water add about 25 mg of ascorbic acid to each vial before sample collection. Collect bubble free samples.

STABILITY: Cool to 4°C; HCl to pH <2. M.H.T. = 14 Days.

QUALITY CONTROL: As an initial demonstration of lab accuracy and precision, analyze 4 to 7 replicates of a lab fortified blank containing analyte at 0.1-5 µg/L. Collect all samples in duplicate.

REFERENCE: Method 502.2, Volatile Organic Compounds in Water by Purge and Trap Capillary Column GC, EPA 600/4-88/039.

PRIMARY NAME: 1,3-Dichloropropane Method 524.1

TITLE: VOCs in H_2O by Purge and Trap GC/MS

MATRIX: Drinking water (finished or any treatment stage) and raw source water.

CAS # : 142-28-9

APPLICATION: Method covers 51 VOCs. An inert gas is bubbled through a 25 mL water sample. Purged sample components are trapped in tube of sorbent materials.When purging is complete, sorbent tube is heated and backflushed with helium to desorb trapped sample onto a packed GC column.

INTERFERENCES: During analysis, major contaminant sources are volatile materials in the lab and impurities in purging gas and sorbent trap. With high and low level samples, there can be carry-over contamination. Watch for methylene chloride, it permeates through PTFE tubing.

INSTRUMENTATION: Gas Chromatography/Mass Spectrometry/Data System; 1% SP-1000 on Carbopack B column.

RANGE: Approximately 0.2-200 µg/L MDL: 0.1 µg/L

PRECISION: RSD = 3.3% @ 1.0 µg/L

ACCURACY: Mean Accuracy = 100% @ 1.0 µg/L

SAMPLING METHOD: Use a 60-120 mL screw cap vial (pre-washed with detergent, rinsed with distilled water and oven dried at 105°C) with a PTFE-faced silicone septum. If residual chlorine is in the water add about 25 mg of ascorbic acid to each vial before sample collection. Collect bubble free samples.

STABILITY: Cool to 4°C; HCl to pH <2. M.H.T. = 14 Days.

QUALITY CONTROL: As an initial demonstration of lab accuracy and precision, analyze 4 to 7 replicates of a lab fortified blank containing analyte at 0.2-5 μg/L. Collect all samples in duplicate.

REFERENCE: Method 524.1, VOCs in Water by Purge and Trap Gas chromatography/mass spectrometry, EPA 600/4-88/039.

PRIMARY NAME: 1,3-Dichloropropane Method 524.2

TITLE: VOCs in H_2O Purge and Trap capillary column GC/MS

MATRIX: Drinking water (finished or any treatment stage) and raw source water.

CAS # : 142-28-9

APPLICATION: Method covers 60 VOCs. An inert gas is bubbled through a 25 mL water sample. Purged sample components are trapped in tube of sorbent materials. When purging is complete, sorbent tube is heated and backflushed with helium to desorb trapped sample onto a capillary GC column.

INTERFERENCES: During analysis, major contaminant sources are volatile materials in the lab and impurities in purging gas and sorbent trap. With high and low level samples, there can be carry-over contamination. Watch for methylene chloride, it permeates through PTFE tubing.

INSTRUMENTATION: (Capillary Column) Gas Chromatography/Mass Spectrometry/Data System. Column #1: VOCOL glass wide bore; Column #2: DB-624 fused silica; Column #3: DB-5 fused silica.

RANGE: Approximately 0.1-200 μg/L

MDL: 0.04 μg/L on wide bore capillary column.

PRECISION: RSD = 6.0% @ (0.1-10) μg/L (31 samples) wide bore capillary column.

ACCURACY: Recovery = 96% @ (0.1-10) μg/L (31 samples) wide bore capillary column.

SAMPLING METHOD: Use a 60-120 mL screw cap vial (pre-washed with detergent, rinsed with distilled water and oven dried at 105°C) with a PTFE-faced silicone septum. If residual chlorine is in the water add about 25 mg of ascorbic acid to each vial before sample collection. Collect bubble free samples.

STABILITY: Cool to 4°C; HCl to pH <2. M.H.T. = 14 Days.

QUALITY CONTROL: As an initial demonstration of lab accuracy and precision, analyze 4 to 7 replicates of a lab fortified blank containing analyte at 0.2-5 μg/L. Collect all samples in duplicate.

REFERENCE: Method 524.2, VOCs in H_2O by Purge and Trap Capillary column GC/MS, EPA 600/4-88/039.

PRIMARY NAME: 2,2-Dichloropropane Method 502.1

TITLE: Halogenated VOCs in Water

MATRIX: Drinking water (finished or any treatment stage) and raw source water.

CAS # : 590-20-7

APPLICATION: Method covers 40 halogenated VOCs. An inert gas is bubbled through a 5 mL water sample. Purged sample components are trapped in tube of sorbent materials.When purging is complete, sorbent tube is heated and backflushed with inert gas to desorb trapped sample onto a packed GC column where it is separated.

INTERFERENCES: During analysis, major contaminant sources are volatile materials in the lab and impurities in purging gas and sorbent trap. With high and low level samples, there can be carry-over contamination. Watch for methylene chloride, it permeates through PTFE tubing.

INSTRUMENTATION: Purge and Trap GC w/halide specific detector. (Two

GC columns are recommended); Column #1: 1% SP-1000 on Carbopack B; Column #2: n-octane bonded on Poracil C.

RANGE: 8.0-505 μg/L (Drinking water) MDL: not listed.

PRECISION: not listed (single laboratory or single analyst)

ACCURACY: not listed (single laboratory or single analyst)

SAMPLING METHOD: Use a 40-120 mL screw cap vial (pre-washed with detergent, rinsed with distilled water and oven dried at 105°C) with a PTFE-faced silicone septum. If residual chlorine is in the water add about 25 mg of ascorbic acid to each vial before sample collection. Collect bubble free samples.

STABILITY: Cool to 4°C; HCl to pH <2. M.H.T. = 14 Days.

QUALITY CONTROL: As an initial demonstration of lab accuracy and precision, analyze 4 to 7 replicates of a lab fortified blank containing analyte at 0.1-5 μg/L. Collect all samples in duplicate.

REFERENCE: Method 502.1, Volatile Halogenated Organic Compounds in Water by Purge and Trap GC, EPA 600/4-88/039.

PRIMARY NAME: 2,2-Dichloropropane Method 502.2

TITLE: VOCs in H_2O by Purge and Trap capillary column GC

MATRIX: Drinking water (finished or any treatment stage) and raw source water.

CAS # : 590-20-7

APPLICATION: Method covers 60 VOCs. An inert gas is bubbled through a 5 mL sample in a purging chamber. Purged sample components are trapped in tube of sorbent materials. When purging is done, sorbent tube is heated and backflushed w/helium to desorb trapped sample onto a capillary GC column.

INTERFERENCES: During analysis, major contaminant sources are volatile materials in the lab and impurities in purging gas and sorbent trap. With high and low level samples, there can be carry-over contamination. Beware of methylene chloride, it permeates through PTFE tubing.

INSTRUMENTATION: (capillary column) GC, w/PID and electrolytic conductivity detectors in series. Column #1: VOCOL wide bore column; Column #2: RTX-502.2 mega bore column; Column #3: DB-62 mega bore column.

RANGE: Approximately 0.02-200 µg/L

MDL: 0.05 µg/L (reagent H_2O) (ELCD) 60 m × 0.75 mm VOCOL capillary column

PRECISION: RSD = 3.4% @ Spike of 10 µg/L (reagent H_2O) (ELCD) 60 m × 0.75 mm VOCOL capillary column

ACCURACY: Recovery = 105% @ spike of 10 µg/L (reagent H_2O) (ELCD) 60 m × 0.75 mm VOCOL capillary column

SAMPLING METHOD: Use a 40-120 mL screw cap vial (pre-washed with detergent, rinsed with distilled water and oven dried at 105°C) with a PTFE-faced silicone septum. If residual chlorine is in the water add about 25 mg of ascorbic acid to each vial before sample collection. Collect bubble free samples.

STABILITY: Cool to 4°C; HCl to pH <2.

M.H.T. = 14 Days.

QUALITY CONTROL: As an initial demonstration of lab accuracy and precision, analyze 4 to 7 replicates of a lab fortified blank containing analyte at 0.1-5 µg/L. Collect all samples in duplicate.

REFERENCE: Method 502.2, Volatile Organic Compounds in Water by Purge and Trap Capillary Column GC, EPA 600/4-88/039.

PRIMARY NAME: 2,2-Dichloropropane

Method 524.1

TITLE: VOCs in H_2O by Purge and Trap GC/MS

MATRIX: Drinking water (finished or any treatment stage) and raw source water.

CAS # : 590-20-7

APPLICATION: Method covers 51 VOCs. An inert gas is bubbled through a 25 mL water sample. Purged sample components are trapped in tube of sorbent materials. When purging is complete, sorbent tube is heated and backflushed with helium to desorb trapped sample onto a packed GC column.

INTERFERENCES: During analysis, major contaminant sources are volatile materials in the lab and impurities in purging gas and sorbent trap. With high and low level samples, there can be carry-over contamination. Watch for methylene chloride, it permeates through PTFE tubing.

INSTRUMENTATION: Gas Chromatography/Mass Spectrometry/Data System; 1% SP-1000 on Carbopack B column.

RANGE: Approximately 0.2-200 μg/L MDL: not determined.

PRECISION: not listed.

ACCURACY: not listed.

SAMPLING METHOD: Use a 60-120 mL screw cap vial (pre-washed with detergent, rinsed with distilled water and oven dried at 105°C) with a PTFE-faced silicone septum. If residual chlorine is in the water add about 25 mg of ascorbic acid to each vial before sample collection. Collect bubble free samples.

STABILITY: Cool to 4°C; HCl to pH <2. M.H.T. = 14 Days.

QUALITY CONTROL: As an initial demonstration of lab accuracy and precision, analyze 4 to 7 replicates of a lab fortified blank containing analyte at 0.2-5 μg/L. Collect all samples in duplicate.

REFERENCE: Method 524.1, VOCs in Water by Purge and Trap Gas Chromatography/Mass Spectrometry, EPA 600/4-88/039.

PRIMARY NAME: 2,2-Dichloropropane Method 524.2

TITLE: VOCs in H_2O Purge and Trap capillary column GC/MS

MATRIX: Drinking water (finished or any treatment stage) and raw source water.

CAS # : 590-20-7

APPLICATION: Method covers 60 VOCs. An inert gas is bubbled through a 25 mL water sample. Purged sample components are trapped in tube of sorbent materials. When purging is complete, sorbent tube is heated and backflushed with helium to desorb trapped sample onto a capillary GC column.

INTERFERENCES: During analysis, major contaminant sources are volatile materials in the lab and impurities in purging gas and sorbent trap. With high and low level samples, there can be carry-over contamination. Watch for methylene chloride, it permeates through PTFE tubing.

INSTRUMENTATION: (Capillary Column) Gas Chromatography/Mass Spectrometry/Data System. Column #1: VOCOL glass wide bore; Column #2: DB-624 fused silica; Column #3: DB-5 fused silica.

RANGE: Approximately 0.1-200 μg/L

MDL: 0.35 μg/L on wide bore capillary column.

PRECISION: RSD = 16.9% @ (0.5-10) μg/L (12 samples) wide bore capillary column.

ACCURACY: Recovery = 86% @ (0.5-10) μg/L (12 samples) wide bore capillary column.

SAMPLING METHOD: Use a 60-120 mL screw cap vial (pre-washed with detergent, rinsed with distilled water and oven dried at 105°C) with a PTFE-faced silicone septum. If residual chlorine is in the water add about 25 mg of ascorbic acid to each vial before sample collection. Collect bubble free samples.

STABILITY: Cool to 4°C; HCl to pH <2.

M.H.T. = 14 Days.

QUALITY CONTROL: As an initial demonstration of lab accuracy and precision, analyze 4 to 7 replicates of a lab fortified blank containing analyte at 0.2-5 μg/L. Collect all samples in duplicate.

REFERENCE: Method 524.2, VOCs in H_2O by Purge and Trap Capillary column GC/MS, EPA 600/4-88/039.

PRIMARY NAME: 1,1-Dichloropropylene Method 502.1

TITLE: Halogenated VOCs in Water

MATRIX: Drinking water (finished or any treatment stage) and raw source water.

CAS # : 563-58-6

APPLICATION: Method covers 40 halogenated VOCs. An inert gas is bubbled through a 5 mL water sample. Purged sample components are trapped in tube of sorbent materials. When purging is complete, sorbent tube is heated and backflushed with inert gas to desorb trapped sample onto a packed GC column where it is separated.

INTERFERENCES: During analysis, major contaminant sources are volatile materials in the lab and impurities in purging gas and sorbent trap. With high and low level samples,there can be carry-over contamination. Watch for methylene chloride, it permeates through PTFE tubing.

INSTRUMENTATION: Purge and Trap GC w/halide specific detector. (Two GC columns are recommended); Column #1: 1% SP-1000 on Carbopack B; Column #2: n-octane bonded on Poracil C.

RANGE: 8.0-505 μg/L (Drinking water) MDL: not determined.

PRECISION: RSD = 9.3% @ 0.40 μg/L spike (S.L.) (18 Samples)

ACCURACY: Avg Recovery = 88% @ 0.40 μg/L spike (S.L.) (18 Samples)

SAMPLING METHOD: Use a 40-120 mL screw cap vial (pre-washed with detergent, rinsed with distilled water and oven dried at 105°C) with a PTFE-faced silicone septum. If residual chlorine is in the water add about 25 mg of ascorbic acid to each vial before sample collection. Collect bubble free samples.

STABILITY: Cool to 4°C; HCl to pH <2. M.H.T. = 14 Days.

QUALITY CONTROL: As an initial demonstration of lab accuracy and precision, analyze 4 to 7 replicates of a lab fortified blank containing analyte at 0.1-5 μg/L. Collect all samples in duplicate.

REFERENCE: Method 502.1, Volatile Halogenated Organic Compounds in Water by Purge and Trap GC, EPA 600/4-88/039.

PRIMARY NAME: 1,1-Dichloropropylene Method 502.2

TITLE: VOCs in H_2O by Purge and Trap capillary column GC

MATRIX: Drinking water (finished or any treatment stage) and raw source water.

CAS # : 563-58-6

APPLICATION: Method covers 60 VOCs. An inert gas is bubbled through a 5 mL sample in a purging chamber. Purged sample components are trapped in tube of sorbent materials. When purging is done, sorbent tube is heated and backflushed w/helium to desorb trapped sample onto a capillary GC column.

INTERFERENCES: During analysis, major contaminant sources are volatile materials in the lab and impurities in purging gas and sorbent trap. With high and low level samples, there can be carry-over contamination. Beware of methylene chloride, it permeates through PTFE tubing.

INSTRUMENTATION: (Capillary Column) GC, w/PID and electrolytic conductivity detectors in series. Column #1: VOCOL wide bore column; Column #2: RTX-502.2 mega bore column; Column #3: DB-62 mega bore column.

RANGE: Approximately 0.02-200 μg/L

MDL: 0.02 μg/L (reagent H_2O) (ELCD) 60 m × 0.75 mm VOCOL capillary column

PRECISION: RSD = 3.3% of 10 μg/L (reagent H_2O) (ELCD) 60 m × 0.75 mm VOCOL capillary column

ACCURACY: Recovery = 103% @ spike of 10 μg/L (reagent H_2O) (ELCD) 60 m × 0.75 mm VOCOL capillary column

SAMPLING METHOD: Use a 40-120 mL screw cap vial (pre-washed with detergent, rinsed with distilled water and oven dried at 105°C) with a PTFE-faced silicone septum. If residual chlorine is in the water add about 25 mg of ascorbic acid to each vial before sample collection. Collect bubble free samples.

STABILITY: Cool to 4°C; HCl to pH <2. M.H.T. = 14 Days.

QUALITY CONTROL: As an initial demonstration of lab accuracy and

precision, analyze 4 to 7 replicates of a lab fortified blank containing analyte at 0.1-5 μg/L. Collect all samples in duplicate.

REFERENCE: Method 502.2, Volatile Organic Compounds in Water by Purge and Trap Capillary Column GC, EPA 600/4-88/039.

PRIMARY NAME: 1,1-Dichloropropylene Method 524.1

TITLE: VOCs in H_2O by Purge and Trap GC/MS

MATRIX: Drinking water (finished or any treatment stage) and raw source water.

CAS # : 563-58-6

APPLICATION: Method covers 51 VOCs. An inert gas is bubbled through a 25 mL water sample. Purged sample components are trapped in tube of sorbent materials. When purging is complete,sorbent tube is heated and backflushed with helium to desorb trapped sample onto a packed GC column.

INTERFERENCES: During analysis, major contaminant sources are volatile materials in the lab and impurities in purging gas and sorbent trap. With high and low level samples, there can be carry-over contamination. Watch for methylene chloride, it permeates through PTFE tubing.

INSTRUMENTATION: Gas Chromatography/Mass Spectrometry/Data System; 1% SP-1000 on Carbopack B column.

RANGE: Approximately 0.2-200 μg/L MDL: not determined.

PRECISION: not listed.

ACCURACY: not listed.

SAMPLING METHOD: Use a 60-120 mL screw cap vial (pre-washed with detergent, rinsed with distilled water and oven dried at 105°C) with a PTFE-faced silicone septum. If residual chlorine is in the water add about 25 mg of ascorbic acid to each vial before sample collection. Collect bubble free samples.

STABILITY: Cool to 4°C; HCl to pH <2. M.H.T. = 14 Days.

QUALITY CONTROL: As an initial demonstration of lab accuracy and precision, analyze 4 to 7 replicates of a lab fortified blank containing analyte at 0.2-5 µg/L. Collect all samples in duplicate.

REFERENCE: Method 524.1, VOCs in Water by Purge and Trap Gas Chromatography/Mass Spectrometry, EPA 600/4-88/039.

PRIMARY NAME: 1,1-Dichloropropylene

Method 524.2

TITLE: VOCs in H_2O Purge and Trap capillary column GC/MS

MATRIX: Drinking water (finished or) any treatment stage) and raw source water.

CAS # : 563-58-6

APPLICATION: Method covers 60 VOCs. An inert gas is bubbled through a 25 mL water sample. Purged sample components are trapped in tube of sorbent materials. When purging is complete, sorbent tube is heated and backflushed with helium to desorb trapped sample onto a capillary GC column.

INTERFERENCES: During analysis, major contaminant sources are volatile materials in the lab and impurities in purging gas and sorbent trap. With high and low level samples, there can be carry-over contamination. Watch for methylene chloride, it permeates through PTFE tubing.

INSTRUMENTATION: (Capillary Column) Gas Chromatography/Mass Spectrometry/Data System. Column #1: VOCOL glass wide bore; Column #2: DB-624 fused silica; Column #3: DB-5 fused silica.

RANGE: Approximately 0.1-200 µg/L

MDL: 0.10 µg/L on wide bore capillary column.

PRECISION: RSD = 8.9% @ (0.5-10) µg/L (18 samples) wide bore capillary column.

ACCURACY: Recovery = 98% @ (0.5-10) µg/L (18 samples) wide bore capillary column.

SAMPLING METHOD: Use a 60-120 mL screw cap vial (pre-washed with detergent, rinsed with distilled water and oven dried at 105°C) with a PTFE-

faced silicone septum. If residual chlorine is in the water add about 25 mg of ascorbic acid to each vial before sample collection. Collect bubble free samples.

STABILITY: Cool to 4°C; HCl to pH <2. M.H.T. = 14 Days.

QUALITY CONTROL: As an initial demonstration of lab accuracy and precision, analyze 4 to 7 replicates of a lab fortified blank containing analyte at 0.2-5 µg/L. Collect all samples in duplicate.

REFERENCE: Method 524.2, VOCs in H_2O by Purge and Trap Capillary column GC/MS, EPA 600/4-88/039.

PRIMARY NAME: cis-1,3-Dichloropropylene Method 502.1

TITLE: Halogenated VOCs in Water

MATRIX: Drinking water (finished or any treatment stage) and raw source water.

CAS # : 10061-01-5

APPLICATION: Method covers 40 halogenated VOCs. An inert gas is bubbled through a 5 mL water sample. Purged sample components are trapped in tube of sorbent materials. When purging is complete, sorbent tube is heated and backflushed with inert gas to desorb trapped sample onto a packed GC column where it is separated.

INTERFERENCES: During analysis, major contaminant sources are volatile materials in the lab and impurities in purging gas and sorbent trap. With high and low level samples, there can be carry-over contamination. Watch for methylene chloride, it permeates through PTFE tubing.

INSTRUMENTATION: Purge and Trap GC w/halide specific detector. (Two GC columns are recommended); Column #1: 1% SP-1000 on Carbopack B; Column #2: n-octane bonded on Poracil C.

RANGE: 8.0-505 µg/L (Drinking water) MDL: not listed

PRECISION: Not listed

ACCURACY: Not listed

SAMPLING METHOD: Use a 40-120 mL screw cap vial (pre-washed with detergent, rinsed with distilled water and oven dried at 105°C) with a PTFE-faced silicone septum. If residual chlorine is in the water add about 25 mg of ascorbic acid to each vial before sample collection. Collect bubble free samples.

STABILITY: Cool to 4°C; HCl to pH <2. M.H.T. = 14 Days.

QUALITY CONTROL: As an initial demonstration of lab accuracy and precision, analyze 4 to 7 replicates of a lab fortified blank containing analyte at 0.1-5 μg/L. Collect all samples in duplicate.

REFERENCE: Method 502.1, Volatile Halogenated Organic Compounds in Water by Purge and Trap GC, EPA 600/4-88/039.

PRIMARY NAME: cis-1,3-Dichloropropylene Method 502.2

TITLE: VOCs in H_2O by Purge and Trap capillary column GC

MATRIX: Drinking water (finished or any treatment stage) and raw source water.

CAS # : 10061-01-5

APPLICATION: Method covers 60 VOCs. An inert gas is bubbled through a 5 mL sample in a purging chamber. Purged sample components are trapped in tube of sorbent materials. When purging is done, sorbent tube is heated and backflushed w/helium to desorb trapped sample onto a capillary GC column.

INTERFERENCES: During analysis, major contaminant sources are volatile materials in the lab and impurities in purging gas and sorbent trap. With high and low level samples, there can be carry-over contamination. Beware of methylene chloride, it permeates through PTFE tubing.

INSTRUMENTATION: (capillary column) GC, w/PID and electrolytic conductivity detectors in series. Column #1: VOCOL wide bore column; Column #2: RTX-502.2 mega bore column; Column #3: DB-62 mega bore column.

RANGE: Approximately 0.02-200 μg/L

MDL: 0.08 μg/L (reagent H_2O) (ELCD) 105 mm × 0.53 mm RTX-502.2 column

PRECISION: RSD = 2.0% @ Spike of 10 µg/L (reagent H_2O) (ELCD) 105 mm × 0.53 mm RTX-502.2 column

ACCURACY: Recovery = 98% @ spike of 10 µg/L (reagent H_2O) (ELCD) 105 mm × 0.53 mm RTX-502.2 column

SAMPLING METHOD: Use a 40-120 mL screw cap vial (pre-washed with detergent, rinsed with distilled water and oven dried at 105°C) with a PTFE-faced silicone septum. If residual chlorine is in the water add about 25 mg of ascorbic acid to each vial before sample collection. Collect bubble free samples.

STABILITY: Cool to 4°C; HCl to pH <2. M.H.T. = 14 Days.

QUALITY CONTROL: As an initial demonstration of lab accuracy and precision, analyze 4 to 7 replicates of a lab fortified blank containing analyte at 0.1-5 µg/L. Collect all samples in duplicate.

REFERENCE: Method 502.2, Volatile Organic Compounds in Water by Purge and Trap Capillary Column GC, EPA 600/4-88/039.

PRIMARY NAME: cis-1,3-Dichloropropylene Method 524.1

TITLE: VOCs in H_2O by Purge and Trap GC/MS

MATRIX: Drinking water (finished or any treatment stage) and raw source water.

CAS # : 10061-01-5

APPLICATION: Method covers 51 VOCs. An inert gas is bubbled through a 25 mL water sample. Purged sample components are trapped in tube of sorbent materials.When purging is complete, sorbent tube is heated and backflushed with helium to desorb trapped sample onto a packed GC column.

INTERFERENCES: During analysis, major contaminant sources are volatile materials in the lab and impurities in purging gas and sorbent trap. With high and low level samples, there can be carry-over contamination. Watch for methylene chloride, it permeates through PTFE tubing.

INSTRUMENTATION: Gas Chromatography/Mass Spectrometry/Data System; 1% SP-1000 on Carbopack B column.

RANGE: Approximately 0.2-200 µg/L MDL: Not listed

PRECISION: Not listed

ACCURACY: Not listed

SAMPLING METHOD: Use a 60-120 mL screw cap vial (pre-washed with detergent, rinsed with distilled water and oven dried at 105°C) with a PTFE-faced silicone septum. If residual chlorine is in the water add about 25 mg of ascorbic acid to each vial before sample collection. Collect bubble free samples.

STABILITY: Cool to 4°C; HCl to pH <2. M.H.T. = 14 Days.

QUALITY CONTROL: As an initial demonstration of lab accuracy and precision, analyze 4 to 7 replicates of a lab fortified blank containing analyte at 0.2-5 µg/L. Collect all samples in duplicate.

REFERENCE: Method 524.1, VOCs in Water by Purge and Trap Gas chromatography/mass spectrometry, EPA 600/4-88/039.

PRIMARY NAME: cis-1,3-Dichloropropylene Method 601

TITLE: Purgeable Halocarbons MATRIX: Wastewater

CAS # : 10061-01-5

APPLICATION: Method covers 29 purgeable halocarbons. (Method 624 provides GC/MS conditions appropriate for the qualitative and quantitative confirmation of results). Method describes conditions for a second GC column to confirm measurements made with primary column.

INTERFERENCES: Impurities in the purge gas and organic compounds outgassing from the plumbing ahead of the trap. With high and low level samples, there can be carry-over contamination. Diffusion of volatile organics through the septum seal into the sample.

INSTRUMENTATION: GC equipped with halide-specific detector. (With purge and trap unit).

RANGE: 8.0 To 500 µg/L. MDL: 0.34 µg/L.

PRECISION: 0.32X μg/L (overall precision)

ACCURACY: 1.00C μg/L (as recovery)

SAMPLING METHOD: 25 mL glass vial. Teflon lined septum.

STABILITY: Cool, 4°C, 0.008% sodium thiosulfate. M.H.T. = 14 Days.

QUALITY CONTROL: The laboratory must on an ongoing basis, spike at least 10% of the samples from each sample site being monitored to assess accuracy.

REFERENCE: Method 601, Federal Register Part VIII 40 CFR Part 136, Oct 26, 1984.

PRIMARY NAME: cis-1,3-Dichloropropylene Method 624

TITLE: Purgeables MATRIX: Wastewater

CAS # : 10061-01-5

APPLICATION: Method covers 31 purgeable organics. An inert gas is bubbled through a 5 mL water sample in a specially designed purging chamber. Here, purgeables are transferred from aqueous to gaseous phase, passed onto a sorbent column and trapped. Trap is heated and backflushed with inert gas to desorb purgeables onto a GC column, where purgeables are separated.

INTERFERENCES: Impurities in the purge gas, organic compounds outgassing from the plumbing ahead of the trap, and solvent vapors in the laboratory. With high and low level samples, there can be carry-over contamination.

INSTRUMENTATION: GC/MS with purge and trap unit.

RANGE: 5-600 μg/L MDL: 5.0 μg/L

PRECISION: 0.52X μg/L (overall precision)

ACCURACY: 1.00C μg/L (as recovery)

SAMPLING METHOD: 25 mL glass vial. Teflon lined septum.

STABILITY: Cool, 4°C, 0.008% sodium thiosulfate. M.H.T. = 14 Days.

QUALITY CONTROL: The laboratory must on an ongoing basis, spike at least 5% of the samples from each sample site being monitored to assess accuracy.

REFERENCE: Method 624, Federal Register Part VIII 40 CFR Part 136, Oct 26, 1984.

PRIMARY NAME: cis-1,3-Dichloropropylene Method 8010

TITLE: Halogenated Volatile Organics

MATRIX: groundwater, soils, sludges water miscible liquid wastes, and non-water miscible wastes.

CAS # : 10061-01-5

APPLICATION: This method is used for the analysis of 39 halogenated VOCs. Samples are analyzed using direct injection or purge and trap methods. Groundwater must be analyzed by the purge and trap method. The method provides an optional GC column which is used for analyte confirmation and that may help resolve analytes from interferences.

INTERFERENCES: There can be carry-over contamination with high and low level samples. Impurities may come from the purge and trap apparatus, organic compounds outgassing from the plumbing ahead of trap, diffusion of VOCs through the sample bottle septum during shipping or storage, or from solvent vapors in the lab.

INSTRUMENTATION: GC capable of on-column injections or purge and trap sample introduction and a halogen specific detector. Column 1: 8 foot by 0.1 inch 1% SP-1000 on Carbopack-B. Column 2: 6 foot by 0.1 inch bonded n-octane on Porasil-C.

RANGE: 8 to 500 μg/L (reagent water) MDL: Not determined

PRACTICAL QUANTITATION LIMIT FACTORS FOR MULTIPLYING TIMES MDL VALUE

Matrix	Multiplication Factor
Groundwater	10
Low-level soil	10
Water miscible liquid waste	500
High-level soil and sludge	1250
Non-water miscible waste	1250

PRECISION: 0.32X μg/L (overall precision; estimated)

ACCURACY: 1.00C μg/L (as recovery; estimated)

SAMPLING METHOD: For water and liquid samples; use glass 40 mL vials with Teflon lined septum caps and collect two vials per sample location with no headspace. For solids and concentrated waste samples; use widemouth glass bottles with Teflon liners.

STABILITY: For concentrated wastes, soils, sediments or sludges; cool to 4°C. For liquids; add 4 drops of concentrated hydrochloric acid and cool to 4°C. M.H.T. = 14 days.

QUALITY CONTROL: Analyze a reagent blank, matrix spike and matrix spike duplicate/duplicate for each analytical batch (up to 20 Samples). Demonstrate the purity of glassware and reagents by analyzing a reagent water method blank. Internal, surrogate and five concentration level calibration standards are used.

REFERENCE: Method 8010, SW-846, 3rd ed., Nov 1986.

PRIMARY NAME: trans-1,3-Dichloropropylene Method 502.1

TITLE: Halogenated VOCs in Water

MATRIX: Drinking water (finished or any treatment stage) and raw source water.

CAS # : 10061-02-6

APPLICATION: Method covers 40 halogenated VOCs. An inert gas is bubbled through a 5 mL water sample. Purged sample components are trapped in tube of sorbent materials. When purging is complete, sorbent tube is heated and backflushed with inert gas to desorb trapped sample onto a packed GC column where it is separated.

INTERFERENCES: During analysis, major contaminant sources are volatile materials in the lab and impurities in purging gas and sorbent trap. With high and low level samples, there can be carry-over contamination. Watch for methylene chloride, it permeates through PTFE tubing.

INSTRUMENTATION: Purge and Trap GC w/halide specific detector. (Two GC columns are recommended); Column #1: 1% SP-1000 on Carbopack B; Column #2: n-octane bonded on Poracil C.

RANGE: 8.0-505 µg/L (Drinking water) MDL: not listed

PRECISION: Not listed

ACCURACY: Not listed

SAMPLING METHOD: Use a 40-120 mL screw cap vial (pre-washed with detergent, rinsed with distilled water and oven dried at 105°C) with a PTFE-faced silicone septum. If residual chlorine is in the water add about 25 mg of ascorbic acid to each vial before sample collection. Collect bubble free samples.

STABILITY: Cool to 4°C; HCl to pH <2. M.H.T. = 14 Days.

QUALITY CONTROL: As an initial demonstration of lab accuracy and precision, analyze 4 to 7 replicates of a lab fortified blank containing analyte at 0.1-5 µg/L. Collect all samples in duplicate.

REFERENCE: Method 502.1, Volatile Halogenated Organic Compounds in Water by Purge and Trap GC, EPA 600/4-88/039.

PRIMARY NAME: trans-1,3-Dichloropropylene Method 502.2

TITLE: VOCs in H_2O by Purge and Trap capillary column GC

MATRIX: Drinking water (finished or any treatment stage) and raw source water.

CAS # : 10061-02-6

APPLICATION: Method covers 60 VOCs. An inert gas is bubbled through a 5 mL sample in a purging chamber. Purged sample components are trapped in tube of sorbent materials. When purging is done, sorbent tube is heated and backflushed w/helium to desorb trapped sample onto a capillary GC column.

INTERFERENCES: During analysis, major contaminant sources are volatile materials in the lab and impurities in purging gas and sorbent trap. With high and low level samples, there can be carry-over contamination. Beware of methylene chloride, it permeates through PTFE tubing.

INSTRUMENTATION: (capillary column) GC, w/PID and electrolytic conductivity detectors in series. Column #1: VOCOL wide bore column; Column #2: RTX-502.2 mega bore column; Column #3: DB-62 mega bore column.

RANGE: Approximately 0.02-200 µg/L

MDL: 0.10 µg/L (reagent H_2O) (ELCD) 105 mm × 0.53 mm RTX-502.2 column

PRECISION: RSD = 1.4% @ Spike of 10 µg/L (reagent H_2O) (ELCD) 105 mm × 0.53 mm RTX-502.2 column

ACCURACY: Recovery = 97% @ spike of 10 µg/L (reagent H_2O) (ELCD) 105 mm × 0.53 mm RTX-502.2 column

SAMPLING METHOD: Use a 40-120 mL screw cap vial (pre-washed with detergent, rinsed with distilled water and oven dried at 105°C) with a PTFE-faced silicone septum. If residual chlorine is in the water add about 25 mg of ascorbic acid to each vial before sample collection. Collect bubble free samples.

STABILITY: Cool to 4°C; HCl to pH <2.

M.H.T. = 14 Days.

QUALITY CONTROL: As an initial demonstration of lab accuracy and precision, analyze 4 to 7 replicates of a lab fortified blank containing analyte at 0.1-5 µg/L. Collect all samples in duplicate.

REFERENCE: Method 502.2, Volatile Organic Compounds in Water by Purge and Trap Capillary Column GC, EPA 600/4-88/039.

PRIMARY NAME: trans-1,3-Dichloropropylene Method 524.1

TITLE: VOCs in H_2O by Purge and Trap GC/MS

MATRIX: Drinking water (finished or any treatment stage) and raw source water.

CAS # : 10061-02-6

APPLICATION: Method covers 51 VOCs. An inert gas is bubbled through a 25 mL water sample. Purged sample components are trapped in tube of sorbent materials.When purging is complete, sorbent tube is heated and backflushed with helium to desorb trapped sample onto a packed GC column.

INTERFERENCES: During analysis, major contaminant sources are volatile materials in the lab and impurities in purging gas and sorbent trap. With high and low level samples, there can be carry-over contamination. Watch for methylene chloride, it permeates through PTFE tubing.

INSTRUMENTATION: Gas Chromatography/Mass Spectrometry/Data System; 1% SP-1000 on Carbopack B column.

RANGE: Approximately 0.2-200 µg/L MDL: Not listed

PRECISION: Not listed

ACCURACY: Not listed

SAMPLING METHOD: Use a 60-120 mL screw cap vial (pre-washed with detergent, rinsed with distilled water and oven dried at 105°C) with a PTFE-faced silicone septum. If residual chlorine is in the water add about 25 mg of ascorbic acid to each vial before sample collection. Collect bubble free samples.

STABILITY: Cool to 4°C; HCl to pH <2. M.H.T. = 14 Days.

QUALITY CONTROL: As an initial demonstration of lab accuracy and precision, analyze 4 to 7 replicates of a lab fortified blank containing analyte at 0.2-5 µg/L. Collect all samples in duplicate.

REFERENCE: Method 524.1, VOCs in Water by Purge and Trap Gas chromatography/mass spectrometry, EPA 600/4-88/039.

PRIMARY NAME: trans-1,3-Dichloropropylene Method 601

TITLE: Purgeable Halocarbons MATRIX: Wastewater

CAS # : 10061-02-6

APPLICATION: Method covers 29 purgeable halocarbons. (Method 624 provides GC/MS conditions appropriate for the qualitative and quantitative confirmation of results). Method describes conditions for a second GC column to confirm measurements made with primary column.

INTERFERENCES: Impurities in the purge gas and organic compounds outgassing from the plumbing ahead of the trap. With high and low level samples, there can be carry-over contamination. Diffusion of volatile organics through the septum seal into the sample.

INSTRUMENTATION: GC equipped with halide-specific detector. (With purge and trap unit).

RANGE: 8.0 To 500 μg/L. MDL: 0.20 μg/L.

PRECISION: 0.32X μg/L (overall precision)

ACCURACY: 1.00C μg/L (as recovery)

SAMPLING METHOD: 25 mL glass vial. Teflon lined septum.

STABILITY: Cool, 4°C, 0.008% sodium thiosulfate. M.H.T. = 14 Days.

QUALITY CONTROL: The laboratory must on an ongoing basis, spike at least 10% of the samples from each sample site being monitored to assess accuracy.

REFERENCE: Method 601, Federal Register Part VIII 40 CFR Part 136, Oct 26, 1984.

PRIMARY NAME: trans-1,3-Dichloropropylene Method 624

TITLE: Purgeables MATRIX: Wastewater

CAS # : 10061-02-6

APPLICATION: Method covers 31 purgeable organics. An inert gas is bubbled through a 5 mL water sample in a specially designed purging chamber. Here, purgeables are transferred from aqueous to gaseous phase, passed onto a sorbent column and trapped. Trap is heated and backflushed with inert gas to desorb purgeables onto a GC column, where purgeables are separated.

INTERFERENCES: Impurities in the purge gas, organic compounds outgassing from the plumbing ahead of the trap, and solvent vapors in the laboratory. With high and low level samples, there can be carry-over contamination.

INSTRUMENTATION: GC/MS with purge and trap unit.

RANGE: 5-600 μg/L MDL: not determined

PRECISION: 0.34X μg/L (overall precision)

ACCURACY: 1.00C μg/L (as recovery)

SAMPLING METHOD: 25 mL glass vial. Teflon lined septum.

STABILITY: Cool, 4°C, 0.008% sodium thiosulfate. M.H.T. = 14 Days.

QUALITY CONTROL: The laboratory must on an ongoing basis, spike at least 5% of the samples from each sample site being monitored to assess accuracy.

REFERENCE: Method 624, Federal Register Part VIII 40 CFR Part 136, Oct 26, 1984.

PRIMARY NAME: trans-1,3-Dichloropropylene Method 8010

TITLE: Halogenated Volatile Organics

MATRIX: groundwater, soils, sludges water miscible liquid wastes, and non-water miscible wastes.

CAS # : 10061-02-6

APPLICATION: This method is used for the analysis of 39 halogenated VOCs. Samples are analyzed using direct injection or purge and trap methods. Groundwater must be analyzed by the purge and trap method. The method provides an optional GC column which is used for analyte confirmation and that may help resolve analytes from interferences.

INTERFERENCES: There can be carry-over contamination with high and low level samples. Impurities may come from the purge and trap apparatus, organic compounds outgassing from the plumbing ahead of trap, diffusion of VOCs through the sample bottle septum during shipping or storage, or from solvent vapors in the lab.

INSTRUMENTATION: GC capable of on-column injections or purge and trap sample introduction and a halogen specific detector. Column 1: 8 foot by 0.1 inch 1% SP-1000 on Carbopack-B. Column 2: 6 foot by 0.1 inch bonded n-octane on Porasil-C.

RANGE: 8 to 500 µg/L (reagent water)

MDL: 0.34 µg/L (reagent water)

PRACTICAL QUANTITATION LIMIT FACTORS FOR MULTIPLYING TIMES MDL VALUE

MATRIX	MULTIPLICATION FACTOR
Groundwater	10
Low-level soil	10
Water miscible liquid waste	500
High-level soil and sludge	1250
Non-water miscible waste	1250

PRECISION: 0.32X µg/L (overall precision; estimated)

ACCURACY: 1.00C µg/L (as recovery; estimated)

SAMPLING METHOD: For water and liquid samples; use glass 40 mL vials with Teflon lined septum caps and collect two vials per sample location with no headspace. For solids and concentrated waste samples; use widemouth glass bottles with Teflon liners.

STABILITY: For concentrated wastes, soils, sediments or sludges; cool to 4°C. For liquids; add 4 drops of concentrated hydrochloric acid and cool to 4°C. M.H.T. = 14 days.

QUALITY CONTROL: Analyze a reagent blank, matrix spike and matrix spike duplicate/duplicate for each analytical batch (up to 20 Samples). Demonstrate the purity of glassware and reagents by analyzing a reagent water method blank. Internal, surrogate and five concentration level calibration standards are used.

REFERENCE: Method 8010, SW-846, 3rd ed., Nov 1986.

PRIMARY NAME: Hexachlorobutadiene

Method 502.2

TITLE: VOCs in H_2O by Purge and Trap capillary column GC

MATRIX: Drinking water (finished or any treatment stage) and raw source water.

CAS # : 87-68-3

APPLICATION: Method covers 60 VOCs. An inert gas is bubbled through a 5 mL sample in a purging chamber. Purged sample components are trapped in tube of sorbent materials. When purging is done, sorbent tube is heated and backflushed w/helium to desorb trapped sample onto a capillary GC column.

INTERFERENCES: During analysis, major contaminant sources are volatile materials in the lab and impurities in purging gas and sorbent trap. With high and low level samples, there can be carry-over contamination. Beware of methylene chloride, it permeates through PTFE tubing.

INSTRUMENTATION: (capillary column) GC, w/PID and electrolytic conductivity detectors in series. Column #1: VOCOL wide bore column; Column #2: RTX-502.2 mega bore column; Column #3: DB-62 mega bore column.

RANGE: Approximately 0.02-200 µg/L

MDL: 0.02 µg/L (reagent H_2O) (ELCD) 60 m × 0.75 mm VOCOL capillary column

PRECISION: RSD = 8.3% @ Spike of 10 µg/L (reagent H_2O) (PID). 60 m × 0.75 mm VOCOL capillary column

ACCURACY: Recovery = 98% @ spike of 10 µg/L (reagent H_2O) (PID) 60 m × 0.75 mm VOCOL capillary column

SAMPLING METHOD: Use a 40-120 mL screw cap vial (pre-washed with detergent, rinsed with distilled water and oven dried at 105°C) with a PTFE-faced silicone septum. If residual chlorine is in the water add about 25 mg of ascorbic acid to each vial before sample collection. Collect bubble free samples.

STABILITY: Cool to 4°C; HCl to pH <2. M.H.T. = 14 Days.

QUALITY CONTROL: As an initial demonstration of lab accuracy and precision, analyze 4 to 7 replicates of a lab fortified blank containing analyte at 0.1-5 µg/L. Collect all samples in duplicate.

REFERENCE: Method 502.2, Volatile Organic Compounds in Water by Purge and Trap Capillary Column GC, EPA 600/4-88/039.

PRIMARY NAME: Hexachlorobutadiene Method 503.1

TITLE: Aromatic and Unsaturated VOCs in Water

MATRIX: Drinking water (finished or any treatment stage) and raw source water.

CAS # : 87-68-3

APPLICATION: Method covers 28 aromatic and unsaturated VOCs. An inert gas is bubbled through a 5 mL water sample. Purged sample components are trapped in tube of sorbent materials. When purging is complete, sorbent tube is heated and backflushed with inert gas to desorb trapped sample onto a packed GC column.

INTERFERENCES: During analysis, major contaminant sources are volatile materials in the lab and impurities in purging gas and sorbent trap. With high and low level samples, there can be carry-over contamination. Excess water causes a negative baseline deflection.

INSTRUMENTATION: Purge and Trap GC w/photoionization detector. (Two GC columns are recommended); Column #1: 5% SP-1200 and 1.75% Bentone 34 on Supelcoport; 5% 1,2,3-tris(2-cyanoethoxy)propane on Chromosorb W.

RANGE: 2.2-600 µg/L (Drinking water) MDL: 0.02 µg/L in water

PRECISION: RSD = 16.8% @ 0.50 µg/L; 10 samples

ACCURACY: Avg Recovery = 74% @ 0.50 µg/L; 10 samples

SAMPLING METHOD: Use a 40-120 mL screw cap vial (pre-washed with detergent, rinsed with distilled water and oven dried at 105°C) with a PTFE-faced silicone septum. If residual chlorine is in the water add about 25 mg of ascorbic acid to each vial before sample collection. Collect bubble free samples.

STABILITY: Cool to 4°C; HCl to pH <2. M.H.T. = 14 Days.

QUALITY CONTROL: As an initial demonstration of lab accuracy and precision, analyze 4 to 7 replicates of a lab fortified blank containing analyte at 0.1-5 µg/L. Collect all samples in duplicate.

REFERENCE: Method 503.1, Volatile Aromatic and Unsaturated Organic Compounds in H_2O by Purge and Trap GC, EPA 600/4-88/039.

PRIMARY NAME: Hexachlorobutadiene Method 524.1

TITLE: VOCs in H_2O by Purge and Trap GC/MS

MATRIX: Drinking water (finished or any treatment stage) and raw source water.

CAS # : 87-68-3

APPLICATION: Method covers 51 VOCs. An inert gas is bubbled through a 25 mL water sample. Purged sample components are trapped in tube of sorbent materials. When purging is complete, sorbent tube is heated and backflushed with helium to desorb trapped sample onto a packed GC column.

INTERFERENCES: During analysis, major contaminant sources are volatile materials in the lab and impurities in purging gas and sorbent trap. With high and low level samples, there can be carry-over contamination. Watch for methylene chloride, it permeates through PTFE tubing.

INSTRUMENTATION: Gas Chromatography/Mass Spectrometry/Data System; 1% SP-1000 on Carbopack B column.

RANGE: Approximately 0.2-200 µg/L MDL: not determined.

PRECISION: not listed.

ACCURACY: not listed.

SAMPLING METHOD: Use a 60-120 mL screw cap vial (pre-washed with detergent, rinsed with distilled water and oven dried at 105°C) with a PTFE-faced silicone septum. If residual chlorine is in the water add about 25 mg of ascorbic acid to each vial before sample collection. Collect bubble free samples.

STABILITY: Cool to 4°C; HCl to pH <2. M.H.T. = 14 Days.

QUALITY CONTROL: As an initial demonstration of lab accuracy and precision, analyze 4 to 7 replicates of a lab fortified blank containing analyte at 0.2-5 µg/L. Collect all samples in duplicate.

REFERENCE: Method 524.1, VOCs in Water by Purge and Trap Gas Chromatography/Mass Spectrometry, EPA 600/4-88/039.

PRIMARY NAME: Hexachlorobutadiene Method 524.2

TITLE: VOCs in H_2O (Purge and Trap capillary column GC/MS

MATRIX: Drinking water (finished or any treatment stage) and raw source water.

CAS # : 87-68-3

APPLICATION: Method covers 60 VOCs. An inert gas is bubbled through a 25 mL water sample. Purged sample components are trapped in tube of sorbent materials. When purging is complete, sorbent tube is heated and backflushed with helium to desorb trapped sample onto a capillary GC column.

INTERFERENCES: During analysis, major contaminant sources are volatile materials in the lab and impurities in purging gas and sorbent trap. With high and low level samples, there can be carry-over contamination. Watch for methylene chloride, it permeates through PTFE tubing.

INSTRUMENTATION: (capillary column) Gas Chromatography/Mass Spectrometry/Data System. Column #1: VOCOL glass wide bore; Column #2: DB-624 fused silica; Column #3: DB-5 fused silica.

RANGE: Approximately 0.1-200 µg/L

MDL: 0.11 µg/L on wide bore capillary column.

PRECISION: RSD = 6.8% @ (0.5-10) µg/L (18 samples) wide bore capillary column.

ACCURACY: Recovery = 100% @ (0.5-10) µg/L (18 samples) wide bore capillary column.

SAMPLING METHOD: Use a 60-120 mL screw cap vial (pre-washed with detergent, rinsed with distilled water and oven dried at 105°C) with a PTFE-faced silicone septum. If residual chlorine is in the water add about 25 mg of ascorbic acid to each vial before sample collection. Collect bubble free samples.

STABILITY: Cool to 4°C; HCl to pH <2.

M.H.T. = 14 Days.

QUALITY CONTROL: As an initial demonstration of lab accuracy and precision, analyze 4 to 7 replicates of a lab fortified blank containing analyte at 0.2-5 µg/L. Collect all samples in duplicate.

REFERENCE: Method 524.2, VOCs in H_2O by Purge and Trap Capillary column GC/MS, EPA 600/4-88/039.

PRIMARY NAME: Methylene chloride

Method 502.1

TITLE: Halogenated VOCs in Water

MATRIX: Drinking water (finished or any treatment stage) and raw source water.

CAS # : 75-09-2

APPLICATION: Method covers 40 halogenated VOCs. An inert gas is bubbled through a 5 mL water sample. Purged sample components are trapped in tube of sorbent materials. When purging is complete, sorbent tube is heated and backflushed with inert gas to desorb trapped sample onto a packed GC column where it is separated.

INTERFERENCES: During analysis, major contaminant sources are volatile materials in the lab and impurities in purging gas and sorbent trap. With

high and low level samples, there can be carry-over contamination. Watch for methylene chloride, it permeates through PTFE tubing.

INSTRUMENTATION: Purge and Trap GC w/halide specific detector. (Two GC columns are recommended); Column #1: 1% SP-1000 on Carbopack B; Column #2: n-octane bonded on Poracil C.

RANGE: 8.0-505 µg/L (Drinking water) MDL: not determined.

PRECISION: RSD = 12.0% @ 0.20 µg/L spike (S.L.) (17 Samples)

ACCURACY: Avg Recovery = 85% @ 0.20 µg/L spike (S.L.) (17 Samples)

SAMPLING METHOD: Use a 40-120 mL screw cap vial (pre-washed with detergent, rinsed with distilled water and oven dried at 105°C) with a PTFE-faced silicone septum. If residual chlorine is in the water add about 25 mg of ascorbic acid to each vial before sample collection. Collect bubble free samples.

STABILITY: Cool to 4°C; HCl to pH <2. M.H.T. = 14 Days.

QUALITY CONTROL: As an initial demonstration of lab accuracy and precision, analyze 4 to 7 replicates of a lab fortified blank containing analyte at 0.1-5 µg/L. Collect all samples in duplicate.

REFERENCE: Method 502.1, Volatile Halogenated Organic Compounds in Water by Purge and Trap GC, EPA 600/4-88/039.

PRIMARY NAME: Methylene chloride Method 502.2

TITLE: VOCs in H_2O by Purge and Trap capillary column GC

MATRIX: Drinking water (finished or any treatment stage) and raw source water.

CAS # : 75-09-2

APPLICATION: Method covers 60 VOCs. An inert gas is bubbled through a 5 mL sample in a purging chamber. Purged sample components are trapped in tube of sorbent materials. When purging is done, sorbent tube is heated and backflushed w/helium to desorb trapped sample onto a capillary GC column.

INTERFERENCES: During analysis, major contaminant sources are volatile materials in the lab and impurities in purging gas and sorbent trap. With high and low level samples, there can be carry-over contamination. Beware of methylene chloride, it permeates through PTFE tubing.

INSTRUMENTATION: (Capillary Column) GC, w/PID and electrolytic conductivity detectors in series. Column #1: VOCOL wide bore column; Column #2: RTX-502.2 mega bore column; Column #3: DB-62 mega bore column.

RANGE: Approximately 0.02-200 μg/L

MDL: 0.02 μg/L (reagent H_2O) (ELCD) 60 m × 0.75 mm VOCOL capillary column

PRECISION: RSD = 2.9% @ Spike of 10 μg/L (reagent H_2O) (ELCD). 60 m × 0.75 mm VOCOL capillary column

ACCURACY: Recovery = 97% @ spike of 10 μg/L (reagent H_2O) (ELCD) 60 m × 0.75 mm VOCOL capillary column

SAMPLING METHOD: Use a 40-120 mL screw cap vial (pre-washed with detergent, rinsed with distilled water and oven dried at 105°C) with a PTFE-faced silicone septum. If residual chlorine is in the water add about 25 mg of ascorbic acid to each vial before sample collection. Collect bubble free samples.

STABILITY: Cool to 4°C; HCl to pH <2.

M.H.T. = 14 Days.

QUALITY CONTROL: As an initial demonstration of lab accuracy and precision, analyze 4 to 7 replicates of a lab fortified blank containing analyte at 0.1-5 μg/L. Collect all samples in duplicate.

REFERENCE: Method 502.2, Volatile Organic Compounds in Water by Purge and Trap Capillary Column GC, EPA EPA 600/4-88/039.

PRIMARY NAME: Methylene chloride

Method 524.1

TITLE: VOCs in H_2O by Purge and Trap GC/MS

MATRIX: Drinking water (finished or any treatment stage) and raw source water.

CAS # : 75-09-2

APPLICATION: Method covers 51 VOCs. An inert gas is bubbled through a 25 mL water sample. Purged sample components are trapped in tube of sorbent materials. When purging is complete, sorbent tube is heated and backflushed with helium to desorb trapped sample onto a packed GC column.

INTERFERENCES: During analysis, major contaminant sources are volatile materials in the lab and impurities in purging gas and sorbent trap. With high and low level samples, there can be carry-over contamination. Watch for methylene chloride, it permeates through PTFE tubing.

INSTRUMENTATION: Gas Chromatography/Mass Spectrometry/Data System; 1% SP-1000 on Carbopack B column.

RANGE: Approximately 0.2-200 μg/L MDL: 1 μg/L

PRECISION: RSD = 46% @ 1.0 μg/L

ACCURACY: Mean Accuracy = 99% @ 1.0 μg/L

SAMPLING METHOD: Use a 60-120 mL screw cap vial (pre-washed with detergent, rinsed with distilled water and oven dried at 105°C) with a PTFE-faced silicone septum. If residual chlorine is in the water add about 25 mg of ascorbic acid to each vial before sample collection. Collect bubble free samples.

STABILITY: Cool to 4°C; HCl to pH <2. M.H.T. = 14 Days.

QUALITY CONTROL: As an initial demonstration of lab accuracy and precision, analyze 4 to 7 replicates of a lab fortified blank containing analyte at 0.2-5 μg/L. Collect all samples in duplicate.

REFERENCE: Method 524.1, VOCs in Water by Purge and Trap Gas Chromatography/Mass Spectrometry, EPA 600/4-88/039.

PRIMARY NAME: Methylene chloride Method 524.2

TITLE: VOCs in H_2O (Purge and Trap capillary column GC/MS

MATRIX: Drinking water (finished or any treatment stage) and raw source water.

CAS # : 75-09-2

APPLICATION: Method covers 60 VOCs. An inert gas is bubbled through a 25 mL water sample. Purged sample components are trapped in tube of sorbent materials. When purging is complete, sorbent tube is heated and backflushed with helium to desorb trapped sample onto a capillary GC column.

INTERFERENCES: During analysis, major contaminant sources are volatile materials in the lab and impurities in purging gas and sorbent trap. With high and low level samples, there can be carry-over contamination. Watch for methylene chloride, it permeates through PTFE tubing.

INSTRUMENTATION: (capillary column) Gas Chromatography/Mass Spectrometry/Data System. Column #1: VOCOL glass wide bore; Column #2: DB-624 fused silica; Column #3: DB-5 fused silica.

RANGE: Approximately 0.1-200 μg/L

MDL: 0.03 μg/L on wide bore capillary column.

PRECISION: RSD = 5.3% @ (0.1-10) μg/L (30 samples) wide bore capillary column.

ACCURACY: Recovery = 95% @ (0.1-10) μg/L (30 samples) wide bore capillary column.

SAMPLING METHOD: Use a 60-120 mL screw cap vial (pre-washed with detergent, rinsed with distilled water and oven dried at 105°C) with a PTFE-faced silicone septum. If residual chlorine is in the water add about 25 mg of ascorbic acid to each vial before sample collection. Collect bubble free samples.

STABILITY: Cool to 4°C; HCl to pH <2.

M.H.T. = 14 Days.

QUALITY CONTROL: As an initial demonstration of lab accuracy and precision, analyze 4 to 7 replicates of a lab fortified blank containing analyte at 0.2-5 μg/L. Collect all samples in duplicate.

REFERENCE: Method 524.2, VOCs in H_2O by Purge and Trap Capillary column GC/MS, EPA 600/4-88/039.

PRIMARY NAME: Methylene chloride

Method 601

TITLE: Purgeable Halocarbons

MATRIX: Wastewater

CAS # : 75-09-2

APPLICATION: Method covers 29 purgeable halocarbons. (Method 624 provides GC/MS conditions appropriate for the qualitative and quantitative confirmation of results). Method describes conditions for a second GC column to confirm measurements made with primary column.

INTERFERENCES: Impurities in the purge gas and organic compounds outgassing from the plumbing ahead of the trap. With high and low level samples, there can be carry-over contamination. Diffusion of volatile organics through the septum seal into the sample.

INSTRUMENTATION: GC equipped with halide-specific detector. (With purge and trap unit).

RANGE: 8.0 To 500 μg/L. MDL: 0.25 μg/L.

PRECISION: 0.21X+1.43 μg/L (overall precision)

ACCURACY: 0.91C-0.93 μg/L (as recovery)

SAMPLING METHOD: 25 mL glass vial. Teflon lined septum.

STABILITY: Cool, 4°C, 0.008% sodium thiosulfate. M.H.T. = 14 Days.

QUALITY CONTROL: The laboratory must on an ongoing basis, spike at least 10% of the samples from each sample site being monitored to assess accuracy.

REFERENCE: Method 601, Federal Register Part VIII 40 CFR Part 136, Oct 26, 1984.

PRIMARY NAME: Methylene chloride Method 624

TITLE: Purgeables MATRIX: Wastewater

CAS # : 75-09-2

APPLICATION: Method covers 31 purgeable organics. An inert gas is bubbled through a 5 mL water sample in a specially designed purging chamber. Here, purgeables are transferred from aqueous to gaseous phase, passed onto a sorbent column and trapped. Trap is heated and backflushed

with inert gas to desorb purgeables onto a GC column, where purgeables are separated.

INTERFERENCES: Impurities in the purge gas, organic compounds outgassing from the plumbing ahead of the trap, and solvent vapors in the laboratory. With high and low level samples, there can be carry-over contamination.

INSTRUMENTATION: GC/MS with purge and trap unit.

RANGE: 5-600 μg/L MDL: 2.8 μg/L

PRECISION: 0.32X+4.00 μg/L (overall precision)

ACCURACY: 0.87C+1.88 μg/L (as recovery)

SAMPLING METHOD: 25 mL glass vial. Teflon lined septum.

STABILITY: Cool, 4°C, 0.008% sodium thiosulfate. M.H.T. = 14 Days.

QUALITY CONTROL: The laboratory must on an ongoing basis, spike at least 5% of the samples from each sample site being monitored to assess accuracy.

REFERENCE: Method 624, Federal Register Part VIII 40 CFR Part 136, Oct 26, 1984.

PRIMARY NAME: Methylene chloride Method 8010

TITLE: Halogenated Volatile Organics

MATRIX: groundwater, soils, sludges water miscible liquid wastes, and non-water miscible wastes.

CAS # : 75-09-2

APPLICATION: This method is used for the analysis of 39 halogenated VOCs. Samples are analyzed using direct injection or purge and trap methods. Groundwater must be analyzed by the purge and trap method. The method provides an optional GC column which is used for analyte confirmation and that may help resolve analytes from interferences.

INTERFERENCES: There can be carry-over contamination with high and low level samples. Impurities may come from the purge and trap apparatus, organic compounds outgassing from the plumbing ahead of trap, diffusion of VOCs through the sample bottle septum during shipping or storage, or from solvent vapors in the lab.

INSTRUMENTATION: GC capable of on-column injections or purge and trap sample introduction and a halogen specific detector. Column 1: 8 foot by 0.1 inch 1% SP-1000 on Carbopack-B. Column 2: 6 foot by 0.1 inch bonded n-octane on Porasil-C.

RANGE: 8 to 500 µg/L (reagent water) MDL: Not determined

PRACTICAL QUANTITATION LIMIT FACTORS FOR MULTIPLYING TIMES MDL VALUE

Matrix	Multiplication Factor
Groundwater	10
Low-level soil	10
Water miscible liquid waste	500
High-level soil and sludge	1250
Non-water miscible waste	1250

PRECISION: 0.21X + 1.43 µg/L (overall precision)

ACCURACY: 0.91C - 0.93 µg/L (as recovery)

SAMPLING METHOD: For water and liquid samples; use glass 40 mL vials with Teflon lined septum caps and collect two vials per sample location with no headspace. For solids and concentrated waste samples; use widemouth glass bottles with Teflon liners.

STABILITY: For concentrated wastes, soils, sediments or sludges; cool to 4°C. For liquids; add 4 drops of concentrated hydrochloric acid and cool to 4°C. M.H.T. = 14 days.

QUALITY CONTROL: Analyze a reagent blank, matrix spike and matrix spike duplicate/duplicate for each analytical batch (up to 20 Samples). Demonstrate the purity of glassware and reagents by analyzing a reagent water method blank. Internal, surrogate and five concentration level calibration standards are used.

REFERENCE: Method 8010, SW-846, 3rd ed., Nov 1986.

PRIMARY NAME: 1,1,1,2-Tetrachloroethane Method 502.1

TITLE: Halogenated VOCs in Water

MATRIX: Drinking water (finished or any treatment stage) and raw source water.

CAS # : 630-20-6

APPLICATION: Method covers 40 halogenated VOCs. An inert gas is bubbled through a 5 mL water sample. Purged sample components are trapped in tube of sorbent materials. When purging is complete, sorbent tube is heated and backflushed with inert gas to desorb trapped sample onto a packed GC column where it is separated.

INTERFERENCES: During analysis, major contaminant sources are volatile materials in the lab and impurities in purging gas and sorbent trap. With high and low level samples, there can be carry-over contamination. Watch for methylene chloride, it permeates through PTFE tubing.

INSTRUMENTATION: Purge and trap GC w/halide specific detector. (Two GC columns are recommended); Column #1: 1% SP-1000 on Carbopack B; Column #2: n-octane bonded on Poracil C.

RANGE: 8.0-505 μg/L (Drinking water) MDL: not determined.

PRECISION: RSD = 8.0% @ 0.40 μg/L spike (S.L.) (20 Samples)

ACCURACY: Avg Recovery = 93% @ 0.40 μg/L spike (S.L.) (20 Samples)

SAMPLING METHOD: Use a 40-120 mL screw cap vial (pre-washed with detergent, rinsed with distilled water and oven dried at 105°C) with a PTFE-faced silicone septum. If residual chlorine is in the water add about 25 mg of ascorbic acid to each vial before sample collection. Collect bubble free samples.

STABILITY: Cool to 4°C; HCl to pH <2. M.H.T. = 14 Days.

QUALITY CONTROL: As an initial demonstration of lab accuracy and precision, analyze 4 to 7 replicates of a lab fortified blank containing analyte at 0.1-5 μg/L. Collect all samples in duplicate.

REFERENCE: Method 502.1, Volatile Halogenated Organic Compounds in Water by Purge and Trap GC, EPA 600/4-88/039.

PRIMARY NAME: 1,1,1,2-Tetrachloroethane Method 502.2

TITLE: VOCs in H_2O by purge and trap (capillary column GC)

MATRIX: Drinking water (finished or any treatment stage) and raw source water.

CAS # : 630-20-6

APPLICATION: Method covers 60 VOCs. An inert gas is bubbled through a 5 mL sample in a purging chamber. Purged sample components are trapped in tube of sorbent materials. When purging is done, sorbent tube is heated and backflushed w/helium to desorb trapped sample onto a capillary GC column.

INTERFERENCES: During analysis, major contaminant sources are volatile materials in the lab and impurities in purging gas and sorbent trap. With high and low level samples, there can be carry-over contamination. Beware of methylene chloride, it permeates through PTFE tubing.

INSTRUMENTATION: (capillary column) GC, w/PID and electrolytic conductivity detectors in series. Column #1: VOCOL wide bore column; Column #2: RTX-502.2 mega bore column; Column #3: DB-62 mega bore column.

RANGE: Approximately 0.02-200 µg/L

MDL: 0.01 µg/L (reagent H_2O) (ELCD) 60 m × 0.75 mm VOCOL capillary column

PRECISION: RSD = 2.3% @ Spike of 10 µg/L (reagent H_2O) (ELCD) 60 m × 0.75 mm VOCOL capillary column

ACCURACY: Recovery = 99% @ spike of 10 µg/L (reagent H_2O) (ELCD) 60 m × 0.75 mm VOCOL capillary column

SAMPLING METHOD: Use a 40-120 mL screw cap vial (pre-washed with detergent, rinsed with distilled water and oven dried at 105°C) with a PTFE-faced silicone septum. If residual chlorine is in the water add about 25 mg of ascorbic acid to each vial before sample collection. Collect bubble free samples.

STABILITY: Cool to 4°C; HCl to pH < 2. M.H.T. = 14 Days.

QUALITY CONTROL: As an initial demonstration of lab accuracy and

precision, analyze 4 to 7 replicates of a lab fortified blank containing analyte at 0.1-5 μg/L. Collect all samples in duplicate.

REFERENCE: Method 502.2, Volatile Organic Compounds in Water by Purge and Trap Capillary Column GC, EPA 600/4-88/039.

PRIMARY NAME: 1,1,1,2-Tetrachloroethane Method 524.1

TITLE: VOCs in H_2O by Purge and Trap GC/MS

MATRIX: Drinking water (finished or any treatment stage) and raw source water.

CAS # : 630-20-6

APPLICATION: Method covers 51 VOCs. An inert gas is bubbled through a 25 mL water sample. Purged sample components are trapped in tube of sorbent materials. When purging is complete, sorbent tube is heated and backflushed with helium to desorb trapped sample onto a packed GC column.

INTERFERENCES: During analysis, major contaminant sources are volatile materials in the lab and impurities in purging gas and sorbent trap. With high and low level samples, there can be carry-over contamination. Watch for methylene chloride, it permeates through PTFE tubing.

INSTRUMENTATION: Gas Chromatography/Mass Spectrometry/Data System; 1% SP-1000 on Carbopack B column.

RANGE: Approximately 0.2-200 μg/L (Drinking water) MDL: Not listed

PRECISION: Not listed

ACCURACY: Not listed

SAMPLING METHOD: Use a 60-120 mL screw cap vial (pre-washed with detergent, rinsed with distilled water and oven dried at 105°C) with a PTFE-faced silicone septum. If residual chlorine is in the water add about 25 mg of ascorbic acid to each vial before sample collection. Collect bubble free samples.

STABILITY: Cool to 4°C; HCl to pH < 2. M.H.T. = 14 Days.

QUALITY CONTROL: As an initial demonstration of lab accuracy and precision, analyze 4 to 7 replicates of a lab fortified blank containing analyte at 0.2-5 µg/L. Collect all samples in duplicate.

REFERENCE: Method 524.1, VOCs in Water by Purge and Trap Gas Chromatography/Mass Spectrometry, EPA 600/4-88/039.

PRIMARY NAME: 1,1,1,2-Tetrachloroethane Method 524.2

TITLE: VOCs in H_2O Purge and Trap capillary column GC/MS

MATRIX: Drinking water (finished or any treatment stage) and raw source water.

CAS # : 630-20-6

APPLICATION: Method covers 60 VOCs. An inert gas is bubbled through a 25 mL water sample. Purged sample components are trapped in tube of sorbent materials. When purging is complete, sorbent tube is heated and backflushed with helium to desorb trapped sample onto a capillary GC column.

INTERFERENCES: During analysis, major contaminant sources are volatile materials in the lab and impurities in purging gas and sorbent trap. With high and low level samples, there can be carry-over contamination. Watch for methylene chloride, it permeates through PTFE tubing.

INSTRUMENTATION: (Capillary Column) Gas Chromatography/Mass Spectrometry/Data System. Column #1: VOCOL glass wide bore; Column #2: DB-624 fused silica; Column #3: DB-5 fused silica.

RANGE: Approximately 0.1-200 µg/L

MDL: 0.05 µg/L on wide bore capillary column.

PRECISION: RDS = 6.8% @ (0.5-10) µg/L (24 samples) wide bore capillary column.

ACCURACY: Recovery = 90% @ (0.5-10) µg/L (24 samples) wide bore capillary column.

SAMPLING METHOD: Use a 60-120 mL screw cap vial (pre-washed with detergent, rinsed with distilled water and oven dried at 105°C) with a PTFE-

faced silicone septum. If residual chlorine is in the water add about 25 mg of ascorbic acid to each vial before sample collection. Collect bubble free samples.

STABILITY: Cool to 4°C; HCl to pH < 2. M.H.T. = 14 Days.

QUALITY CONTROL: As an initial demonstration of lab accuracy and precision, analyze 4 to 7 replicates of a lab fortified blank containing analyte at 0.2-5 µg/L. Collect all samples in duplicate.

REFERENCE: Method 524.2, VOCs in H_2O by Purge and Trap Capillary Column GC/MS, EPA 600/4-88/039.

PRIMARY NAME: 1,1,2,2-Tetrachloroethane Method 502.1

TITLE: Halogenated VOCs in Water

MATRIX: Drinking water (finished or any treatment stage) and raw source water.

CAS # : 79-34-5

APPLICATION: Method covers 40 halogenated VOCs. An inert gas is bubbled through a 5 mL water sample. Purged sample components are trapped in tube of sorbent materials. When purging is complete, sorbent tube is heated and backflushed with inert gas to desorb trapped sample onto a packed GC column where it is separated.

INTERFERENCES: During analysis, major contaminant sources are volatile materials in the lab and impurities in purging gas and sorbent trap. With high and low level samples, there can be carry-over contamination. Watch for methylene chloride, it permeates through PTFE tubing.

INSTRUMENTATION: Purge and trap GC w/halide specific detector. (Two GC columns are recommended); Column #1: 1% SP-1000 on Carbopack B; Column #2: n-octane bonded on Poracil C.

RANGE: 8.0-505 µg/L (Drinking water) MDL: 0.01 µg/L in water

PRECISION: RSD = 9.0% @ 0.40 µg/L spike (S.L.) (18 Samples)

ACCURACY: Avg Recovery = 95% @ 0.40 μg/L spike (S.L.) (18 Samples)

SAMPLING METHOD: Use a 40-120 mL screw cap vial (pre-washed with detergent, rinsed with distilled water and oven dried at 105°C) with a PTFE-faced silicone septum. If residual chlorine is in the water add about 25 mg of ascorbic acid to each vial before sample collection. Collect bubble free samples.

STABILITY: Cool to 4°C; HCl to pH <2. M.H.T. = 14 Days.

QUALITY CONTROL: As an initial demonstration of lab accuracy and precision, analyze 4 to 7 replicates of a lab fortified blank containing analyte at 0.1-5 μg/L. Collect all samples in duplicate.

REFERENCE: Method 502.1, Volatile Halogenated Organic Compounds in Water by Purge and Trap GC, EPA 600/4-88/039.

PRIMARY NAME: 1,1,2,2-Tetrachloroethane Method 502.2

TITLE: VOCs in H_2O by Purge and Trap capillary column GC

MATRIX: Drinking water (finished or any treatment stage) and raw source water.

CAS # : 79-34-5

APPLICATION: Method covers 60 VOCs. An inert gas is bubbled through a 5 mL sample in a purging chamber. Purged sample components are trapped in tube of sorbent materials. When purging is done, sorbent tube is heated and backflushed w/helium to desorb trapped sample onto a capillary GC column.

INTERFERENCES: During analysis, major contaminant sources are volatile materials in the lab and impurities in purging gas and sorbent trap. With high and low level samples, there can be carry-over contamination. Beware of methylene chloride,it permeates through PTFE tubing.

INSTRUMENTATION: (capillary column) GC, w/PID and electrolytic conductivity detectors in series. Column #1: VOCOL wide bore column; Column #2: RTX-502.2 mega bore column; Column #3: DB-62 mega bore column.

RANGE: Approximately 0.02-200 µg/L

MDL: 0.01 µg/L (reagent H_2O) (ELCD) 60 m × 0.75 mm VOCOL capillary column

PRECISION: RSD = 6.8% @ Spike of 10 µg/L (reagent H_2O) (ELCD) 60 m × 0.75 mm VOCOL capillary column

ACCURACY: Recovery = 99% @ spike of 10 µg/L (reagent H_2O) (ELCD) 60 m × 0.75 mm VOCOL capillary column

SAMPLING METHOD: Use a 40-120 mL screw cap vial (pre-washed with detergent, rinsed with distilled water and oven dried at 105°C) with a PTFE-faced silicone septum. If residual chlorine is in the water add about 25 mg of ascorbic acid to each vial before sample collection. Collect bubble free samples.

STABILITY: Cool to 4°C; HCl to pH <2.

M.H.T. = 14 Days.

QUALITY CONTROL: As an initial demonstration of lab accuracy and precision, analyze 4 to 7 replicates of a lab fortified blank containing analyte at 0.1-5 µg/L Collect all samples in duplicate.

REFERENCE: Method 502.2, Volatile Organic Compounds in Water by Purge and Trap Capillary Column GC, EPA 600/4-88/039.

PRIMARY NAME: 1,1,2,2-Tetrachloroethane

Method 524.1

TITLE: VOCs in H_2O by Purge and Trap GC/MS

MATRIX: Drinking water (finished or any treatment stage) and raw source water.

CAS # : 79-34-5

APPLICATION: Method covers 51 VOCs. An inert gas is bubbled through a 25 mL water sample. Purged sample components are trapped in tube of sorbent materials. When purging is complete,sorbent tube is heated and backflushed with helium to desorb trapped sample onto a packed GC column.

INTERFERENCES: During analysis, major contaminant sources are volatile materials in the lab and impurities in purging gas and sorbent trap. With

high and low level samples, there can be carry-over contamination. Watch for methylene chloride,it permeates through PTFE tubing.

INSTRUMENTATION: Gas Chromatography/Mass Spectrometry/Data System; 1% SP-1000 on Carbopack B column.

RANGE: Approximately 0.2-200 μg/L MDL: 0.4 μg/L

PRECISION: RSD = 13% @ 1.0 μg/L

ACCURACY: Mean Accuracy = 111% @ 1.0 μg/L

SAMPLING METHOD: Use a 60-120 mL screw cap vial (pre-washed with detergent, rinsed with distilled water and oven dried at 105°C) with a PTFE-faced silicone septum. If residual chlorine is in the water add about 25 mg of ascorbic acid to each vial before sample collection. Collect bubble free samples.

STABILITY: Cool to 4°C; HCl to pH <2. M.H.T. = 14 Days.

QUALITY CONTROL: As an initial demonstration of lab accuracy and precision, analyze 4 to 7 replicates of a lab fortified blank containing analyte at 0.2-5 μg/L. Collect all samples in duplicate.

REFERENCE: Method 524.1, VOCs in Water by Purge and Trap Gas chromatography/mass spectrometry, EPA 600/4-88/039.

PRIMARY NAME: 1,1,2,2-Tetrachloroethane Method 524.2

TITLE: VOCs in H_2O (purge and trap capillary column GC/MS

MATRIX: Drinking water (finished or) any treatment stage) and raw source water.

CAS # : 79-34-5

APPLICATION: Method covers 60 VOCs. An inert gas is bubbled through a 25 mL water sample. Purged sample components are trapped in tube of sorbent materials. When purging is complete, sorbent tube is heated and backflushed with helium to desorb trapped sample onto a capillary GC column.

INTERFERENCES: During analysis, major contaminant sources are volatile materials in the lab and impurities in purging gas and sorbent trap. With high and low level samples, there can be carry-over contamination. Watch for methylene chloride, it permeates through PTFE tubing.

INSTRUMENTATION: (Capillary Column) Gas Chromatography/Mass Spectrometry/Data System. Column #1: VOCOL glass wide bore; Column #2: DB-624 fused silica; Column #3: DB-5 fused silica.

RANGE: Approximately 0.1-200 μg/L

MDL: 0.04 μg/L on wide bore capillary column.

PRECISION: RDS = 6.3% @ (0.1-10) μg/L (30 samples) wide bore capillary column.

ACCURACY: Recovery = 91% @ (0.1-10) μg/L (30 samples) wide bore capillary column.

SAMPLING METHOD: Use a 60-120 mL screw cap vial (pre-washed with detergent, rinsed with distilled water and oven dried at 105°C) with a PTFE-faced silicone septum. If residual chlorine is in the water add about 25 mg of ascorbic acid to each vial before sample collection. Collect bubble free samples.

STABILITY: Cool to 4°C; HCl to pH <2.

M.H.T. = 14 Days.

QUALITY CONTROL: As an initial demonstration of lab accuracy and precision, analyze 4 to 7 replicates of a lab fortified blank containing analyte at 0.2-5 μg/L. Collect all samples in duplicate.

REFERENCE: Method 524.2, VOCs in H_2O by Purge and Trap Capillary Column GC/MS, EPA 600/4-88/039.

PRIMARY NAME: 1,1,2,2-Tetrachloroethane

Method 601

TITLE: Purgeable Halocarbons

MATRIX: Wastewater

CAS # : 79-34-5

APPLICATION: Method covers 29 purgeable halocarbons. (Method 624 provides GC/MS conditions appropriate for the qualitative and quantitative

confirmation of results). Method describes conditions for a second GC column to confirm measurements made with primary column.

INTERFERENCES: Impurities in the purge gas and organic compounds outgassing from the plumbing ahead of the trap. With high and low level samples, there can be carry-over contamination. Diffusion of volatile organics through the septum seal into the sample.

INSTRUMENTATION: GC equipped with halide-specific detector. (With purge and trap unit).

RANGE: 8.0 To 500 μg/L. MDL: 0.03 μg/L.

PRECISION: 0.23X+2.79 μg/L (overall precision)

ACCURACY: 0.95C+0.19 μg/L (as recovery)

SAMPLING METHOD: 25 mL glass vial. Teflon lined septum.

STABILITY: Cool, 4°C, 0.008% sodium thiosulfate. M.H.T. = 14 Days.

QUALITY CONTROL: The laboratory must on an ongoing basis, spike at least 10% of the samples from each sample site being monitored to assess accuracy.

REFERENCE: Method 601, Federal Register Part VIII 40 CFR Part 136, Oct 26, 1984.

PRIMARY NAME: 1,1,2,2-Tetrachloroethane Method 624

TITLE: Purgeables MATRIX: Wastewater

CAS # : 79-34-5

APPLICATION: Method covers 31 purgeable organics. An inert gas is bubbled through a 5 mL water sample in a specially designed purging chamber. Here, purgeables are transferred from aqueous to gaseous phase, passed onto a sorbent column and trapped. Trap is heated and backflushed with inert gas to desorb purgeables onto a GC column, where purgeables are separated.

INTERFERENCES: Impurities in the purge gas, organic compounds outgassing from the plumbing ahead of the trap, and solvent vapors in the laboratory. With high and low level samples, there can be carry-over contamination.

INSTRUMENTATION: GC/MS with purge and trap unit.

RANGE: 5-600 µg/L MDL: 6.9 µg/L

PRECISION: 0.20X+0.41 µg/L (overall precision)

ACCURACY: 0.93C+1.76 µg/L (as recovery)

SAMPLING METHOD: 25 mL glass vial. Teflon lined septum.

STABILITY: Cool, 4°C, 0.008% sodium thiosulfate. M.H.T. = 14 Days.

QUALITY CONTROL: The laboratory must on an ongoing basis, spike at least 5% of the samples from each sample site being monitored to assess accuracy.

REFERENCE: Method 624, Federal Register Part VIII 40 CFR Part 136, Oct 26, 1984.

PRIMARY NAME: 1,1,2,2-Tetrachloroethane Method 8010

TITLE: Halogenated Volatile Organics

MATRIX: groundwater, soils, sludges, water miscible liquid wastes, and non-water miscible wastes

CAS # : 79-34-5

APPLICATION: This method is used for the analysis of 39 halogenated VOCs. Samples are analyzed using direct injection or purge and trap methods. Groundwater must be analyzed by the purge and trap method. The method provides an optional GC column which is used for analyte confirmation and that may help resolve analytes from interferences.

INTERFERENCES: There can be carry-over contamination with high and low level samples. Impurities may come from the purge and trap apparatus,

organic compounds outgassing from the plumbing ahead of trap, diffusion of VOCs through the sample bottle septum during shipping or storage, or from solvent vapors in the lab.

INSTRUMENTATION: GC capable of on-column injections or purge and trap sample introduction and a halogen specific detector. Column 1: 8 foot by 0.1 inch 1% SP-1000 on Carbopack-B. Column 2: 6 foot by 0.1 inch bonded n-octane on Porasil-C.

RANGE: 8 to 500 μg/L (reagent water)

MDL: 0.03 μg/L (reagent water)

PRACTICAL QUANTITATION LIMIT FACTORS FOR MULTIPLYING TIMES MDL VALUE

Matrix	Multiplication Factor
Groundwater	10
Low-level soil	10
Water miscible liquid waste	500
High-level soil and sludge	1250
Non-water miscible waste	1250

PRECISION: 0.23X + 2.79 μg/L (overall precision)

ACCURACY: 0.95C + 0.19 μg/L (as recovery)

SAMPLING METHOD: For water and liquid samples; use glass 40 mL vials with Teflon lined septum caps and collect two vials per sample location with no headspace. For solids and concentrated waste samples; use widemouth glass bottles with Teflon liners.

STABILITY: For concentrated wastes, soils, sediments or sludges; cool to 4°C. For liquids; add 4 drops of concentrated hydrochloric acid and cool to 4°C. M.H.T. = 14 days.

QUALITY CONTROL: Analyze a reagent blank, matrix spike and matrix spike duplicate/duplicate for each analytical batch (up to 20 Samples). Demonstrate the purity of glassware and reagents by analyzing a reagent water method blank. Internal, surrogate and five concentration level calibration standards are used.

REFERENCE: Method 8010, SW-846, 3rd ed., Nov 1986.

PRIMARY NAME: Tetrachloroethylene Method 502.1

TITLE: Halogenated VOCs in Water

MATRIX: Drinking water (finished or any treatment stage) and raw source water.

CAS # : 127-18-4

APPLICATION: Method covers 40 halogenated VOCs. An inert gas is bubbled through a 5 mL water sample. Purged sample components are trapped in tube of sorbent materials. When purging is complete, sorbent tube is heated and backflushed with inert gas to desorb trapped sample onto a packed GC column where it is separated.

INTERFERENCES: During analysis, major contaminant sources are volatile materials in the lab and impurities in purging gas and sorbent trap. With high and low level samples, there can be carry-over contamination. Watch for methylene chloride, it permeates through PTFE tubing.

INSTRUMENTATION: Purge and Trap GC w/halide specific detector. (Two GC columns are recommended); Column #1: 1% SP-1000 on Carbopack B; Column #2: n-octane bonded on Poracil C.

RANGE: 8.0-505 µg/L (Drinking water) MDL: 0.001 µg/L in water

PRECISION: RSD = 9.5% @ 0.20 µg/L spike (S.L.) (17 Samples)

ACCURACY: Avg Recovery = 90% @ 0.20 µg/L spike (S.L.) (17 Samples)

SAMPLING METHOD: Use a 40-120 mL screw cap vial (pre-washed with detergent, rinsed with distilled water and oven dried at 105°C) with a PTFE-faced silicone septum. If residual chlorine is in the water add about 25 mg of ascorbic acid to each vial before sample collection. Collect bubble free samples.

STABILITY: Cool to 4°C; HCl to pH <2. M.H.T. = 14 Days.

QUALITY CONTROL: As an initial demonstration of lab accuracy and precision, analyze 4 to 7 replicates of a lab fortified blank containing analyte at 0.1-5 µg/L. Collect all samples in duplicate.

REFERENCE: Method 502.1, Volatile Halogenated Organic Compounds in Water by Purge and Trap GC, EPA 600/4-88/039.

PRIMARY NAME: Tetrachloroethylene

Method 502.2

TITLE: VOCs in H_2O by Purge and Trap capillary column GC

MATRIX: Drinking water (finished or any treatment stage) and raw source water.

CAS # : 127-18-4

APPLICATION: Method covers 60 VOCs. An inert gas is bubbled through a 5 mL sample in a purging chamber. Purged sample components are trapped in tube of sorbent materials. When purging is done, sorbent tube is heated and backflushed w/helium to desorb trapped sample onto a capillary GC column.

INTERFERENCES: During analysis, major contaminant sources are volatile materials in the lab and impurities in purging gas and sorbent trap. With high and low level samples, there can be carry-over contamination. Beware of methylene chloride, it permeates through PTFE tubing.

INSTRUMENTATION: (Capillary Column) GC, w/PID and electrolytic conductivity detectors in series. Column #1: VOCOL wide bore column; Column #2: RTX-502.2 mega bore column; Column #3: DB-62 mega bore column.

RANGE: Approximately 0.02-200 µg/L

MDL: 0.04 µg/L (reagent H_2O) (ELCD) 60 m × 0.75 mm VOCOL capillary column

PRECISION: RSD = 2.5% @ Spike of 10 µg/L (reagent H_2O) (PID). 60 m × 0.75 mm VOCOL capillary column

ACCURACY: Recovery = 97% @ spike of 10 µg/L (reagent H_2O) (PID) 60 m × 0.75 mm VOCOL capillary column

SAMPLING METHOD: Use a 40-120 mL screw cap vial (pre-washed with detergent, rinsed with distilled water and oven dried at 105°C) with a PTFE-faced silicone septum. If residual chlorine is in the water add about 25 mg of ascorbic acid to each vial before sample collection. Collect bubble free samples.

STABILITY: Cool to 4°C; HCl to pH <2.

M.H.T. = 14 Days.

QUALITY CONTROL: As an initial demonstration of lab accuracy and

precision, analyze 4 to 7 replicates of a lab fortified blank containing analyte at 0.1-5 μg/L. Collect all samples in duplicate.

REFERENCE: Method 502.2, Volatile Organic Compounds in Water by Purge and Trap Capillary Column GC, EPA 600/4-88/039.

PRIMARY NAME: Tetrachloroethylene Method 503.1

TITLE: Aromatic and Unsaturated VOCs in Water

MATRIX: Drinking water (finished or any treatment stage) and raw source water.

CAS # : 127-18-4

APPLICATION: Method covers 28 aromatic and unsaturated VOCs. An inert gas is bubbled through a 5 mL water sample. Purged sample components are trapped in tube of sorbent materials. When purging is complete, sorbent tube is heated and backflushed with inert gas to desorb trapped sample onto a packed GC column.

INTERFERENCES: During analysis, major contaminant sources are volatile materials in the lab and impurities in purging gas and sorbent trap. With high and low level samples, there can be carry-over contamination. Excess water causes a negative baseline deflection.

INSTRUMENTATION: Purge and Trap GC w/photoionization detector. (Two GC columns are recommended); Column #1: 5% SP-1200 and 1.75% Bentone 34 on Supelcoport; 5% 1,2,3-tris(2-cyanoethoxy)propane on Chromosorb W.

RANGE: 2.2-600 μg/L (Drinking water) MDL: 0.01 μg/L in water

PRECISION: RSD = 7.8% @ 0.50 μg/L; 19 samples

ACCURACY: Avg Recovery = 97% @ 0.50 μg/L; 19 samples

SAMPLING METHOD: Use a 40-120 mL screw cap vial (pre-washed with detergent, rinsed with distilled water and oven dried at 105°C) with a PTFE-faced silicone septum. If residual chlorine is in the water add about 25 mg of

ascorbic acid to each vial before sample collection. Collect bubble free samples.

STABILITY: Cool to 4°C; HCl to pH <2. M.H.T. = 14 Days.

QUALITY CONTROL: As an initial demonstration of lab accuracy and precision, analyze 4 to 7 replicates of a lab fortified blank containing analyte at 0.1-5 μg/L. Collect all samples in duplicate.

REFERENCE: Method 503.1, Volatile Aromatic and Unsaturated Organic Compounds in H_2O by Purge and Trap GC, EPA 600/4-88/039.

PRIMARY NAME: Tetrachloroethylene Method 524.1

TITLE: VOCs in H_2O by Purge and Trap GC/MS

MATRIX: Drinking water (finished or any treatment stage) and raw source water.

CAS # : 127-18-4

APPLICATION: Method covers 51 VOCs. An inert gas is bubbled through a 25 mL water sample. Purged sample components are trapped in tube of sorbent materials. When purging is complete, sorbent tube is heated and backflushed with helium to desorb trapped sample onto a packed GC column.

INTERFERENCES: During analysis, major contaminant sources are volatile materials in the lab and impurities in purging gas and sorbent trap. With high and low level samples, there can be carry-over contamination. Watch for methylene chloride, it permeates through PTFE tubing.

INSTRUMENTATION: Gas Chromatography/Mass Spectrometry/Data System; 1% SP-1000 on Carbopack B column.

RANGE: Approximately 0.2-200 μg/L MDL: 0.3 μg/L

PRECISION: RSD = 11% @ 1.0 μg/L

ACCURACY: Mean Accuracy = 93% @ 1.0 μg/L

SAMPLING METHOD: Use a 60-120 mL screw cap vial (pre-washed with detergent, rinsed with distilled water and oven dried at 105°C) with a PTFE-faced silicone septum. If residual chlorine is in the water add about 25 mg of ascorbic acid to each vial before sample collection. Collect bubble free samples.

STABILITY: Cool to 4°C; HCl to pH <2. M.H.T. = 14 Days.

QUALITY CONTROL: As an initial demonstration of lab accuracy and precision, analyze 4 to 7 replicates of a lab fortified blank containing analyte at 0.2-5 µg/L. Collect all samples in duplicate.

REFERENCE: Method 524.1, VOCs in Water by Purge and Trap Gas Chromatography/Mass Spectrometry, EPA 600/4-88/039.

PRIMARY NAME: Tetrachloroethylene Method 524.2

TITLE: VOCs in H_2O Purge and Trap capillary column GC/MS

MATRIX: Drinking water (finished or any treatment stage) and raw source water.

CAS # : 127-18-4

APPLICATION: Method covers 60 VOCs. An inert gas is bubbled through a 25 mL water sample. Purged sample components are trapped in tube of sorbent materials. When purging is complete, sorbent tube is heated and backflushed with helium to desorb trapped sample onto a capillary GC column.

INTERFERENCES: During analysis, major contaminant sources are volatile materials in the lab and impurities in purging gas and sorbent trap. With high and low level samples, there can be carry-over contamination. Watch for methylene chloride, it permeates through PTFE tubing.

INSTRUMENTATION: (capillary column) Gas Chromatography/Mass Spectrometry/Data System. Column #1: VOCOL glass wide bore; Column #2: DB-624 fused silica; Column #3: DB-5 fused silica.

RANGE: Approximately 0.1-200 µg/L

MDL: 0.14 µg/L on wide bore capillary column.

PRECISION: RSD = 6.8% @ (0.5-10) µg/L (24 samples) wide bore capillary column.

ACCURACY: Recovery = 89% @ (0.5-10) µg/L (24 samples) wide bore capillary column.

SAMPLING METHOD: Use a 60-120 mL screw cap vial (pre-washed with detergent, rinsed with distilled water and oven dried at 105°C) with a PTFE-faced silicone septum. If residual chlorine is in the water add about 25 mg of ascorbic acid to each vial before sample collection. Collect bubble free samples.

STABILITY: Cool to 4°C; HCl to pH <2. M.H.T. = 14 Days.

QUALITY CONTROL: As an initial demonstration of lab accuracy and precision, analyze 4 to 7 replicates of a lab fortified blank containing analyte at 0.2-5 µg/L. Collect all samples in duplicate.

REFERENCE: Method 524.2, VOCs in H_2O by Purge and Trap Capillary column GC/MS, EPA 600/4-88/039.

PRIMARY NAME: Tetrachloroethylene Method 601

TITLE: Purgeable Halocarbons MATRIX: Wastewater

CAS # : 127-18-4

APPLICATION: Method covers 29 purgeable halocarbons. (Method 624 provides GC/MS conditions appropriate for the qualitative and quantitative confirmation of results). Method describes conditions for a second GC column to confirm measurements made with primary column.

INTERFERENCES: Impurities in the purge gas and organic compounds outgassing from the plumbing ahead of the trap. With high and low level samples, there can be carry-over contamination. Diffusion of volatile organics through the septum seal into the sample.

INSTRUMENTATION: GC equipped with halide-specific detector. (With purge and trap unit).

RANGE: 8.0 To 500 µg/L. MDL: 0.03 µg/L.

PRECISION: 0.18X+2.21 µg/L (overall precision)

ACCURACY: 0.94C+0.06 µg/L (as recovery)

SAMPLING METHOD: 25 mL glass vial. Teflon lined septum.

STABILITY: Cool, 4°C, 0.008% sodium thiosulfate. M.H.T. = 14 Days.

QUALITY CONTROL: The laboratory must on an ongoing basis, spike at least 10% of the samples from each sample site being monitored to assess accuracy.

REFERENCE: Method 601, Federal Register Part VIII 40 CFR Part 136, Oct 26, 1984.

PRIMARY NAME: Tetrachloroethylene Method 624

TITLE: Purgeables MATRIX: Wastewater

CAS # : 127-18-4

APPLICATION: Method covers 31 purgeable organics. An inert gas is bubbled through a 5 mL water sample in a specially designed purging chamber. Here, purgeables are transferred from aqueous to gaseous phase, passed onto a sorbent column and trapped. Trap is heated and backflushed with inert gas to desorb purgeables onto a GC column, where purgeables are separated.

INTERFERENCES: Impurities in the purge gas, organic compounds outgassing from the plumbing ahead of the trap, and solvent vapors in the laboratory. With high and low level samples, there can be carry-over contamination.

INSTRUMENTATION: GC/MS with purge and trap unit.

RANGE: 5-600 µg/L MDL: 4.1 µg/L

PRECISION: 0.16X-0.45 µg/L (overall precision)

ACCURACY: 1.06C+0.60 μg/L (as recovery)

SAMPLING METHOD: 25 mL glass vial. Teflon lined septum.

STABILITY: Cool, 4°C, 0.008% sodium thiosulfate. M.H.T. = 14 Days.

QUALITY CONTROL: The laboratory must on an ongoing basis, spike at least 5% of the samples from each sample site being monitored to assess accuracy.

REFERENCE: Method 624, Federal Register Part VIII 40 CFR Part 136, Oct 26, 1984.

PRIMARY NAME: Tetrachloroethylene Method 8010

TITLE: Halogenated Volatile Organics

MATRIX: groundwater, soils, sludges water miscible liquid wastes, and non-water miscible wastes.

CAS # : 127-18-4

APPLICATION: This method is used for the analysis of 39 halogenated VOCs. Samples are analyzed using direct injection or purge and trap methods. Groundwater must be analyzed by the purge and trap method. The method provides an optional GC column which is used for analyte confirmation and that may help resolve analytes from interferences.

INTERFERENCES: There can be carry-over contamination with high and low level samples. Impurities may come from the purge and trap apparatus, organic compounds outgassing from the plumbing ahead of trap, diffusion of VOCs through the sample bottle septum during shipping or storage, or from solvent vapors in the lab.

INSTRUMENTATION: GC capable of on-column injections or purge and trap sample introduction and a halogen specific detector. Column 1: 8 foot by 0.1 inch 1% SP-1000 on Carbopack-B. Column 2: 6 foot by 0.1 inch bonded n-octane on Porasil-C.

RANGE: 8 to 500 μg/L (reagent water)

MDL: 0.03 μg/L (reagent water)

PRACTICAL QUANTITATION LIMIT FACTORS FOR MULTIPLYING TIMES MDL VALUE

Matrix	Multiplication Factor
Groundwater	10
Low-level soil	10
Water miscible liquid waste	500
High-level soil and sludge	1250
Non-water miscible waste	1250

PRECISION: 0.18X + 2.21 μg/L (overall precision)

ACCURACY: 0.94C + 0.06 μg/L (as recovery)

SAMPLING METHOD: For water and liquid samples; use glass 40 mL vials with Teflon lined septum caps and collect two vials per sample location with no headspace. For solids and concentrated waste samples; use widemouth glass bottles with Teflon liners.

STABILITY: For concentrated wastes, soils, sediments or sludges; cool to 4°C. For liquids; add 4 drops of concentrated hydrochloric acid and cool to 4°C. M.H.T. = 14 days.

QUALITY CONTROL: Analyze a reagent blank, matrix spike and matrix spike duplicate/duplicate for each analytical batch (up to 20 Samples). Demonstrate the purity of glassware and reagents by analyzing a reagent water method blank. Internal, surrogate and five concentration level calibration standards are used.

REFERENCE: Method 8010, SW-846, 3rd ed., Nov 1986.

PRIMARY NAME: 1,1,1-Trichloroethane

Method 502.1

TITLE: Halogenated VOCs in water

MATRIX: Drinking water (finished or any treatment stage) and raw source water.

CAS # : 71-55-6

APPLICATION: Method covers 40 halogenated VOCs. An inert gas is bubbled through a 5 mL water sample. Purged sample components are trapped in tube of sorbent materials. When purging is complete, sorbent tube is heated and backflushed with inert gas to desorb trapped sample onto a packed GC column where it is separated.

INTERFERENCES: During analysis, major contaminant sources are volatile materials in the lab and impurities in purging gas and sorbent trap. With high and low level samples, there can be carry-over contamination. Watch for methylene chloride, it permeates through PTFE tubing.

INSTRUMENTATION: Purge and Trap GC w/Halide Specific Detector. (Two GC columns are recommended); Column #1: 1% SP-1000 on Carbopack B; Column #2: n-octane bonded on Poracil C.

RANGE: 8.0-505 μg/L (Drinking water) MDL: 0.003 μg/L in water

PRECISION: RSD = 8.0% @ 0.40 μg/L spike (S.L.) (20 Samples)

ACCURACY: Avg Recovery = 93% @ 0.40 μg/L spike (S.L.) (20 Samples)

SAMPLING METHOD: Use a 40-120 mL screw cap vial (pre-washed with detergent, rinsed with distilled water and oven dried at 105°C) with a PTFE-faced silicone septum. If residual chlorine is in the water add about 25 mg of ascorbic acid to each vial before sample collection. Collect bubble free samples.

STABILITY: Cool to 4°C; HCl to pH < 2. M.H.T. = 14 Days.

QUALITY CONTROL: As an initial demonstration of lab accuracy and precision, analyze 4 to 7 replicates of a lab fortified blank containing analyte at 0.1-5 μg/L. Collect all samples in duplicate.

REFERENCE: Method 502.1, Volatile Halogenated Organic Compounds in Water by Purge and Trap GC, EPA 600/4-88/039.

PRIMARY NAME: 1,1,1-Trichloroethane Method 502.2

TITLE: VOCs in H_2O by Purge and Trap Capillary Column GC

MATRIX: Drinking water (finished or any treatment stage) and raw source water.

CAS # : 71-55-6

APPLICATION: Method covers 60 VOCs. An inert gas is bubbled through a 5 mL sample in a purging chamber. Purged sample components are trapped in tube of sorbent materials. When purging is done, sorbent tube is heated and backflushed w/helium to desorb trapped sample onto a capillary GC column.

INTERFERENCES: During analysis, major contaminant sources are volatile materials in the lab and impurities in purging gas and sorbent trap. With high and low level samples, there can be carry-over contamination. Beware of methylene chloride, it permeates through PTFE tubing.

INSTRUMENTATION: (capillary column) GC, w/PID and electrolytic conductivity detectors in series. Column #1: VOCOL wide bore column; Column #2: RTX-502.2 mega bore column; Column #3: DB-62 mega bore column.

RANGE: Approximately 0.02-200 µg/L

MDL: 0.03 µg/L (reagent H_2O) (ELCD) 60 m × 0.75 mm VOCOL capillary column

PRECISION: RSD = 3.3% @ Spike of 10 µg/L (reagent H_2O) (ELCD) 60 m × 0.75 mm VOCOL capillary column

ACCURACY: Recovery = 104% @ spike of 10 µg/L (reagent H_2O) (ELCD) 60 m × 0.75 mm VOCOL capillary column

SAMPLING METHOD: Use a 40-120 mL screw cap vial (pre-washed with detergent, rinsed with distilled water and oven dried at 105°C) with a PTFE-faced silicone septum. If residual chlorine is in the water add about 25 mg of ascorbic acid to each vial before sample collection. Collect bubble free samples.

STABILITY: Cool to 4°C; HCl to pH <2.

M.H.T. = 14 Days.

QUALITY CONTROL: As an initial demonstration of lab accuracy and precision, analyze 4 to 7 replicates of a lab fortified blank containing analyte at 0.1-5 µg/L. Collect all samples in duplicate.

REFERENCE: Method 502.2, Volatile Organic Compounds in Water by Purge and Trap Capillary Column GC, EPA 600/4-88/039.

PRIMARY NAME: 1,1,1-Trichloroethane | Method 524.1

TITLE: VOCs in H_2O by Purge and Trap GC/MS

MATRIX: Drinking water (finished or any treatment stage) and raw source water.

CAS # : 71-55-6

APPLICATION: Method covers 51 VOCs. An inert gas is bubbled through a 25 mL water sample. Purged sample components are trapped in tube of sorbent materials. When purging is complete, sorbent tube is heated and backflushed with helium to desorb trapped sample onto a packed GC column.

INTERFERENCES: During analysis, major contaminant sources are volatile materials in the lab and impurities in purging gas and sorbent trap. With high and low level samples, there can be carry-over contamination. Watch for methylene chloride, it permeates through PTFE tubing.

INSTRUMENTATION: Gas Chromatography/Mass Spectrometry/Data System; 1% SP-1000 on Carbopack B column.

RANGE: Approximately 0.2-200 µg/L MDL: 0.3 µg/L

PRECISION: RSD = 8.9% @ 1.0 µg/L

ACCURACY: Mean Accuracy = 105% @ 1.0 µg/L

SAMPLING METHOD: Use a 60-120 mL screw cap vial (pre-washed with detergent, rinsed with distilled water and oven dried at 105°C) with a PTFE-faced silicone septum. If residual chlorine is in the water add about 25 mg of ascorbic acid to each vial before sample collection. Collect bubble free samples.

STABILITY: Cool to 4°C; HCl to pH < 2. M.H.T. = 14 Days.

QUALITY CONTROL: As an initial demonstration of lab accuracy and precision, analyze 4 to 7 replicates of a lab fortified blank containing analyte at 0.2-5 µg/L. Collect all samples in duplicate.

REFERENCE: Method 524.1, VOCs in Water by Purge and Trap Gas chromatography/mass spectrometry, EPA 600/4-88/039.

PRIMARY NAME: 1,1,1-Trichloroethane Method 524.2

TITLE: VOCs in H_2O (purge and trap) capillary column GC/MS

MATRIX: Drinking water (finished or) any treatment stage) and raw source water.

CAS # : 71-55-6

APPLICATION: Method covers 60 VOCs. An inert gas is bubbled through a 25 mL water sample. Purged sample components are trapped in tube of sorbent materials. When purging is complete, sorbent tube is heated and backflushed with helium to desorb trapped sample onto a capillary GC column.

INTERFERENCES: During analysis, major contaminant sources are volatile materials in the lab and impurities in purging gas and sorbent trap. With high and low level samples, there can be carry-over contamination. Watch for methylene chloride, it permeates through PTFE tubing.

INSTRUMENTATION: (Capillary Column) Gas Chromatography/Mass Spectrometry/Data System. Column #1: VOCOL glass wide bore; Column #2: DB-624 fused silica; Column #3: DB-5 fused silica.

RANGE: Approximately 0.1-200 μg/L

MDL: 0.08 μg/L on wide bore capillary column.

PRECISION: RDS = 8.1% @ (0.5-10) μg/L (18 samples) wide bore capillary column.

ACCURACY: Recovery = 98% @ (0.5-10) μg/L (18 samples) wide bore capillary column.

SAMPLING METHOD: Use a 60-120 mL screw cap vial (pre-washed with detergent, rinsed with distilled water and oven dried at 105°C) with a PTFE-faced silicone septum. If residual chlorine is in the water add about 25 mg of ascorbic acid to each vial before sample collection. Collect bubble free samples.

STABILITY: Cool to 4°C; HCl to pH < 2. M.H.T. = 14 Days.

QUALITY CONTROL: As an initial demonstration of lab accuracy and

precision, analyze 4 to 7 replicates of a lab fortified blank containing analyte at 0.2-5 μg/L. Collect all samples in duplicate.

REFERENCE: Method 524.2, VOCs in H_2O by Purge and Trap Capillary Column GC/MS, EPA 600/4-88/039.

PRIMARY NAME: 1,1,1-Trichloroethane Method 601

TITLE: Purgeable Halocarbons MATRIX: Wastewater

CAS # : 71-55-6

APPLICATION: Method covers 29 purgeable halocarbons. (Method 624 provides GC/MS conditions appropriate for the qualitative and quantitative confirmation of results). Method describes conditions for a second GC column to confirm measurements made with primary column.

INTERFERENCES: Impurities in the purge gas and organic compounds outgassing from the plumbing ahead of the trap. With high and low level samples, there can be carry-over contamination. Diffusion of volatile organics through the septum seal into the sample.

INSTRUMENTATION: GC equipped with halide-specific detector. (With purge and trap unit).

RANGE: 8.0 To 500 μg/L. MDL: 0.03 μg/L.

PRECISION: 0.20X+0.37 μg/L (overall precision)

ACCURACY: 0.90C-0.16 μg/L (as recovery)

SAMPLING METHOD: 25 mL glass vial. Teflon lined septum.

STABILITY: Cool, 4°C, 0.008% sodium thiosulfate. M.H.T. = 14 Days.

QUALITY CONTROL: The laboratory must on an ongoing basis, spike at least 10% of the samples from each sample site being monitored to assess accuracy.

REFERENCE: Method 601, Federal Register Part VIII 40 CFR Part 136, Oct 26, 1984.

PRIMARY NAME: 1,1,1-Trichloroethane Method 624

TITLE: Purgeables MATRIX: Wastewater

CAS # : 71-55-6

APPLICATION: Method covers 31 purgeable organics. An inert gas is bubbled through a 5 mL water sample in a specially designed purging chamber. Here, purgeables are transferred from aqueous to gaseous phase, passed onto a sorbent column and trapped. Trap is heated and backflushed with inert gas to desorb purgeables onto a GC column, where purgeables are separated.

INTERFERENCES: Impurities in the purge gas, organic compounds outgassing from the plumbing ahead of the trap, and solvent vapors in the laboratory. With high and low level samples, there can be carry-over contamination.

INSTRUMENTATION: GC/MS with purge and trap unit.

RANGE: 5-600 µg/L MDL: 3.8 µg/L

PRECISION: 0.21X-0.39 µg/L (overall precision)

ACCURACY: 1.06C+0.73 µg/L (as recovery)

SAMPLING METHOD: 25 mL glass vial. Teflon lined septum.

STABILITY: Cool, 4°C, 0.008% sodium thiosulfate. M.H.T. = 14 Days.

QUALITY CONTROL: The laboratory must on an ongoing basis, spike at least 5% of the samples from each sample site being monitored to assess accuracy.

REFERENCE: Method 624, Federal Register Part VIII 40 CFR Part 136, Oct 26, 1984.

PRIMARY NAME: 1,1,1-Trichloroethane Method 8010

TITLE: Halogenated Volatile Organics

MATRIX: groundwater, soils, sludges water miscible liquid wastes, and non-water miscible wastes.

CAS # : 71-55-6

APPLICATION: This method is used for the analysis of 39 halogenated VOCs. Samples are analyzed using direct injection or purge and trap methods. Groundwater must be analyzed by the purge and trap method. The method provides an optional GC column which is used for analyte confirmation and that may help resolve analytes from interferences.

INTERFERENCES: There can be carry-over contamination with high and low level samples. Impurities may come from the purge and trap apparatus, organic compounds outgassing from the plumbing ahead of trap, diffusion of VOCs through the sample bottle septum during shipping or storage, or from solvent vapors in the lab.

INSTRUMENTATION: GC capable of on-column injections or purge and trap sample introduction and a halogen specific detector. Column 1: 8 foot by 0.1 inch 1% SP-1000 on Carbopack-B. Column 2: 6 foot by 0.1 inch bonded n-octane on Porasil-C.

RANGE: 8 to 500 μg/L (reagent water)

MDL: 0.03 μg/L (reagent water)

PRACTICAL QUANTITATION LIMIT FACTORS FOR MULTIPLYING TIMES MDL VALUE

Matrix	Multiplication Factor
Groundwater	10
Low-level soil	10
Water miscible liquid waste	500
High-level soil and sludge	1250
Non-water miscible waste	1250

PRECISION: 0.20X + 0.37 μg/L (overall precision)

ACCURACY: 0.90C - 0.16 μg/L (as recovery)

SAMPLING METHOD: For water and liquid samples; use glass 40 mL vials with Teflon lined septum caps and collect two vials per sample location with no headspace. For solids and concentrated waste samples; use widemouth glass bottles with Teflon liners.

STABILITY: For concentrated wastes, soils, sediments or sludges; cool to 4°C. For liquids; add 4 drops of concentrated hydrochloric acid and cool to 4°C. M.H.T. = 14 days.

QUALITY CONTROL: Analyze a reagent blank, matrix spike and matrix spike duplicate/duplicate for each analytical batch (up to 20 Samples). Demonstrate the purity of glassware and reagents by analyzing a reagent water method blank. Internal, surrogate and five concentration level calibration standards are used.

REFERENCE: Method 8010, SW-846, 3rd ed., Nov 1986.

PRIMARY NAME: 1,1,2-Trichloroethane Method 502.1

TITLE: Halogenated VOCs in Water

MATRIX: Drinking water (finished or any treatment stage) and raw source water.

CAS # : 79-00-5

APPLICATION: Method covers 40 halogenated VOCs. An inert gas is bubbled through a 5 mL water sample. Purged sample components are trapped in tube of sorbent materials. When purging is complete, sorbent tube is heated and backflushed with inert gas to desorb trapped sample onto a packed GC column where it is separated.

INTERFERENCES: During analysis, major contaminant sources are volatile materials in the lab and impurities in purging gas and sorbent trap. With high and low level samples, there can be carry-over contamination. Watch for methylene chloride, it permeates through PTFE tubing.

INSTRUMENTATION: Purge and Trap GC w/halide specific detector. (Two GC columns are recommended); Column #1: 1% SP-1000 on Carbopack B;

Column #2: n-octane bonded on Poracil C.

RANGE: 8.0-505 μg/L (Drinking water) MDL: 0.007 μg/L in water

PRECISION: RSD = 6.0% @ 0.40 μg/L spike (S.L.) (15 Samples)

ACCURACY: Avg Recovery = 95% @ 0.40 μg/L spike (S.L.) (15 Samples)

SAMPLING METHOD: Use a 40-120 mL screw cap vial (pre-washed with detergent, rinsed with distilled water and oven dried at 105°C) with a PTFE-faced silicone septum. If residual chlorine is in the water add about 25 mg of ascorbic acid to each vial before sample collection. Collect bubble free samples.

STABILITY: Cool to 4°C; HCl to pH <2. M.H.T. = 14 Days.

QUALITY CONTROL: As an initial demonstration of lab accuracy and precision, analyze 4 to 7 replicates of a lab fortified blank containing analyte at 0.1-5 μg/L. Collect all samples in duplicate.

REFERENCE: Method 502.1, Volatile Halogenated Organic Compounds in Water by Purge and Trap GC, EPA 600/4-88/039.

PRIMARY NAME: 1,1,2-Trichloroethane Method 502.2

TITLE: VOCs in H_2O by Purge and Trap GC

MATRIX: Drinking water (finished or any treatment stage) and raw source water.

CAS # : 79-00-5

APPLICATION: Method covers 60 VOCs. An inert gas is bubbled through a 5 mL sample in a purging chamber. Purged sample components are trapped in tube of sorbent materials. When purging is done, sorbent tube is heated and backflushed w/helium to desorb trapped sample onto a capillary GC column.

INTERFERENCES: During analysis, major contaminant sources are volatile materials in the lab and impurities in purging gas and sorbent trap. With high and low level samples, there can be carry-over contamination. Beware of methylene chloride, it permeates through PTFE tubing.

INSTRUMENTATION: (Capillary Column) GC, w/PID and electrolytic conductivity detectors in series. Column #1: VOCOL wide bore column; Column #2: RTX-502.2 mega bore column; Column #3: DB-62 mega bore column.

RANGE: Approximately 0.02-200 µg/L

MDL: 0.04 µg/L (ELCD) 105 m × 0.53 mm RTX-502.2 column

PRECISION: RSD = 5.6% @ Spike of 10 µg/L (reagent H_2O) (ELCD) 60 m × 0.75 mm VOCOL capillary column

ACCURACY: Recovery = 109% @ spike of 10 µg/L (reagent H_2O) (ELCD) 60 m × 0.75 mm VOCOL capillary column

SAMPLING METHOD: Use a 40-120 mL screw cap vial (pre-washed with detergent, rinsed with distilled water and oven dried at 105°C) with a PTFE-faced silicone septum. If residual chlorine is in the water add about 25 mg of ascorbic acid to each vial before sample collection. Collect bubble free samples.

STABILITY: Cool to 4°C; HCl to pH < 2.

M.H.T. = 14 Days.

QUALITY CONTROL: As an initial demonstration of lab accuracy and precision, analyze 4 to 7 replicates of a lab fortified blank containing analyte at 0.1-5 µg/L. Collect all samples in duplicate.

REFERENCE: Method 502.2, Volatile Organic Compounds in Water by Purge and Trap Capillary Column GC, EPA 600/4-88/039.

PRIMARY NAME: 1,1,2-Trichloroethane

Method 524.1

TITLE: VOCs in H_2O by purge and trap GC/MS

MATRIX: Drinking water (finished or any treatment stage) and raw source water.

CAS # : 79-00-5

APPLICATION: Method covers 51 VOCs. An inert gas is bubbled through a 25 mL water sample. Purged sample components are trapped in tube of sorbent materials. When purging is complete, sorbent tube is heated and backflushed with helium to desorb trapped sample onto a packed GC column.

INTERFERENCES: During analysis, major contaminant sources are volatile materials in the lab and impurities in purging gas and sorbent trap. With high and low level samples, there can be carry-over contamination. Watch for methylene chloride, it permeates through PTFE tubing.

INSTRUMENTATION: Gas Chromatography/Mass Spectrometry/Data System; 1% SP-1000 on Carbopack B column.

RANGE: Approximately 0.2-200 µg/L MDL: not listed.

PRECISION: not listed.

ACCURACY: not listed.

SAMPLING METHOD: Use a 60-120 mL screw cap vial (pre-washed with detergent, rinsed with distilled water and oven dried at 105°C) with a PTFE-faced silicone septum. If residual chlorine is in the water add about 25 mg of ascorbic acid to each vial before sample collection. Collect bubble free samples.

STABILITY: Cool to 4°C; HCl to pH <2. M.H.T. = 14 Days.

QUALITY CONTROL: As an initial demonstration of lab accuracy and precision, analyze 4 to 7 replicates of a lab fortified blank containing analyte at 0.2-5 µg/L. Collect all samples in duplicate.

REFERENCE: Method 524.1, VOCs in Water by Purge and Trap Gas Chromatography/Mass Spectrometry, EPA 600/4-88/039.

PRIMARY NAME: 1,1,2-Trichloroethane Method 524.2

TITLE: VOCs in H_2O (purge and trap) capillary column GC/MS

MATRIX: Drinking water (finished or) any treatment stage) and raw source water.

CAS # : 79-00-5

APPLICATION: Method covers 60 VOCs. An inert gas is bubbled through a 25 mL water sample. Purged sample components are trapped in tube of sorbent materials. When purging is complete, sorbent tube is heated and backflushed with helium to desorb trapped sample onto a capillary GC column.

INTERFERENCES: During analysis, major contaminant sources are volatile materials in the lab and impurities in purging gas and sorbent trap. With high and low level samples, there can be carry-over contamination. Watch for methylene chloride, it permeates through PTFE tubing.

INSTRUMENTATION: (Capillary Column) Gas Chromatography/Mass Spectrometry/Data System. Column #1: VOCOL glass wide bore; Column #2: DB-624 fused silica; Column #3: DB-5 fused silica.

RANGE: Approximately 0.1-200 µg/L

MDL: 0.10 µg/L on wide bore capillary column.

PRECISION: RDS = 7.3% @ (0.5-10) µg/L (18 samples) wide bore capillary column.

ACCURACY: Recovery = 104% @ (0.5-10) µg/L (18 samples) wide bore capillary column.

SAMPLING METHOD: Use a 60-120 mL screw cap vial (pre-washed with detergent, rinsed with distilled water and oven dried at 105°C) with a PTFE-faced silicone septum. If residual chlorine is in the water add about 25 mg of ascorbic acid to each vial before sample collection. Collect bubble free samples.

STABILITY: Cool to 4°C; HCl to pH <2.

M.H.T. = 14 Days.

QUALITY CONTROL: As an initial demonstration of lab accuracy and precision, analyze 4 to 7 replicates of a lab fortified blank containing analyte at 0.2-5 µg/L. Collect all samples in duplicate.

REFERENCE: Method 524.2, VOCs in H_2O by Purge and Trap Capillary column GC/MS, EPA 600/4-88/039.

PRIMARY NAME: 1,1,2-Trichloroethane

Method 601

TITLE: Purgeable Halocarbons

MATRIX: Wastewater

CAS # : 79-00-5

APPLICATION: Method covers 29 purgeable halocarbons. (Method 624 provides GC/MS conditions appropriate for the qualitative and quantitative confirmation of results). Method describes conditions for a second GC column to confirm measurements made with primary column.

INTERFERENCES: Impurities in the purge gas and organic compounds outgassing from the plumbing ahead of the trap. With high and low level samples, there can be carry-over contamination. Diffusion of volatile organics through the septum seal into the sample.

INSTRUMENTATION: GC equipped with halide-specific detector. (With purge and trap unit).

RANGE: 8.0 To 500 μg/L. MDL: 0.02 μg/L.

PRECISION: 0.19X+0.67 μg/L (overall precision)

ACCURACY: 0.86C+0.30 μg/L (as recovery)

SAMPLING METHOD: 25 mL glass vial. Teflon lined septum.

STABILITY: Cool, 4°C, 0.008% sodium thiosulfate. M.H.T. = 14 Days.

QUALITY CONTROL: The laboratory must on an ongoing basis, spike at least 10% of the samples from each sample site being monitored to assess accuracy.

REFERENCE: Method 601, Federal Register Part VIII 40 CFR Part 136, Oct 26, 1984.

PRIMARY NAME: 1,1,2-Trichloroethane Method 624

TITLE: Purgeables MATRIX: Wastewater

CAS # : 79-00-5

APPLICATION: Method covers 31 purgeable organics. An inert gas is bubbled through a 5 mL water sample in a specially designed purging chamber. Here, purgeables are transferred from aqueous to gaseous phase, passed onto a sorbent column and trapped. Trap is heated and backflushed

with inert gas to desorb purgeables onto a GC column, where purgeables are separated.

INTERFERENCES: Impurities in the purge gas, organic compounds outgassing from the plumbing ahead of the trap, and solvent vapors in the laboratory. With high and low level samples, there can be carry-over contamination.

INSTRUMENTATION: GC/MS with purge and trap unit.

RANGE: 5-600 μg/L MDL: 5.0 μg/L

PRECISION: 0.18X+0.00 μg/L (overall precision)

ACCURACY: 0.95C+1.71 μg/L (as recovery)

SAMPLING METHOD: 25 mL glass vial. Teflon lined septum.

STABILITY: Cool, 4°C, 0.008% sodium thiosulfate. M.H.T. = 14 Days.

QUALITY CONTROL: The laboratory must on an ongoing basis, spike at least 5% of the samples from each sample site being monitored to assess accuracy.

REFERENCE: Method 624, Federal Register Part VIII 40 CFR Part 136, Oct 26, 1984.

PRIMARY NAME: 1,1,2-Trichloroethane Method 8010

TITLE: Halogenated Volatile Organics

MATRIX: groundwater, soils, sludges water miscible liquid wastes, and non-water miscible wastes.

CAS # : 79-00-5

APPLICATION: This method is used for the analysis of 39 halogenated VOCs. Samples are analyzed using direct injection or purge and trap methods. Groundwater must be analyzed by the purge and trap method. The method provides an optional GC column which is used for analyte confirmation and that may help resolve analytes from interferences.

INTERFERENCES: There can be carry-over contamination with high and low level samples. Impurities may come from the purge and trap apparatus, organic compounds outgassing from the plumbing ahead of trap, diffusion of VOCs through the sample bottle septum during shipping or storage, or from solvent vapors in the lab.

INSTRUMENTATION: GC capable of on-column injections or purge and trap sample introduction and a halogen specific detector. Column 1: 8 foot by 0.1 inch 1% SP-1000 on Carbopack-B. Column 2: 6 foot by 0.1 inch bonded n-octane on Porasil-C.

RANGE: 8 to 500 μg/L (reagent water)

MDL: 0.02 μg/L (reagent water)

PRACTICAL QUANTITATION LIMIT FACTORS FOR MULTIPLYING TIMES MDL VALUE

Matrix	Multiplication Factor
Groundwater	10
Low-level soil	10
Water miscible liquid waste	500
High-level soil and sludge	1250
Non-water miscible waste	1250

PRECISION: 0.19X + 0.67 μg/L (overall precision)

ACCURACY: 0.86C + 0.30 μg/L (as recovery)

SAMPLING METHOD: For water and liquid samples; use glass 40 mL vials with Teflon lined septum caps and collect two vials per sample location with no headspace. For solids and concentrated waste samples; use widemouth glass bottles with Teflon liners.

STABILITY: For concentrated wastes, soils, sediments or sludges; cool to 4°C. For liquids; add 4 drops of concentrated hydrochloric acid and cool to 4°C. M.H.T. = 14 days.

QUALITY CONTROL: Analyze a reagent blank, matrix spike and matrix spike duplicate/duplicate for each analytical batch (up to 20 Samples). Demonstrate the purity of glassware and reagents by analyzing a reagent water method blank. Internal, surrogate and five concentration level calibration standards are used.

REFERENCE: Method 8010, SW-846, 3rd ed., Nov 1986.

PRIMARY NAME: Trichloroethylene

Method 502.1

TITLE: Halogenated VOCs in Water

MATRIX: Drinking water (finished or any treatment stage) and raw source water.

CAS # : 79-01-6

APPLICATION: Method covers 40 halogenated VOCs. An inert gas is bubbled through a 5 mL water sample. Purged sample components are trapped in tube of sorbent materials. When purging is complete, sorbent tube is heated and backflushed with inert gas to desorb trapped sample onto a packed GC column where it is separated.

INTERFERENCES: During analysis, major contaminant sources are volatile materials in the lab and impurities in purging gas and sorbent trap. With high and low level samples, there can be carry-over contamination. Watch for methylene chloride, it permeates through PTFE tubing.

INSTRUMENTATION: Purge and Trap GC w/halide specific detector. (Two GC columns are recommended); Column #1: 1% SP-1000 on Carbopack B; Column #2: n-octane bonded on Poracil C.

RANGE: 8.0-505 μg/L (Drinking water)

MDL: 0.001 μg/L in water

PRECISION: RSD = 6.0% @ 0.20 μg/L spike (S.L.) (17 Samples)

ACCURACY: Avg Recovery = 94% @ 0.20 μg/L spike (S.L.) (17 Samples)

SAMPLING METHOD: Use a 40-120 mL screw cap vial (pre-washed with detergent, rinsed with distilled water and oven dried at 105°C) with a PTFE-faced silicone septum. If residual chlorine is in the water add about 25 mg of ascorbic acid to each vial before sample collection. Collect bubble free samples.

STABILITY: Cool to 4°C; HCl to pH <2.

M.H.T. = 14 Days.

QUALITY CONTROL: As an initial demonstration of lab accuracy and precision, analyze 4 to 7 replicates of a lab fortified blank containing analyte at 0.1-5 μg/L. Collect all samples in duplicate.

REFERENCE: Method 502.1, Volatile Halogenated Organic Compounds in Water by Purge and Trap GC, EPA 600/4-88/039.

PRIMARY NAME: Trichloroethylene Method 502.2

TITLE: VOCs in H_2O by Purge and Trap capillary column GC

MATRIX: Drinking water (finished or any treatment stage) and raw source water.

CAS # : 79-01-6

APPLICATION: Method covers 58 VOC's. An inert gas is bubbled through a 5 mL sample in a purging chamber. Purged sample components are trapped in tube of sorbent materials. When purging is done, sorbent tube is heated and backflushed w/helium to desorb trapped sample onto a capillary GC column.

INTERFERENCES: During analysis, major contaminant sources are volatile materials in the lab and impurities in purging gas and sorbent trap. With high and low level samples, there can be carry-over contamination. Beware of methylene chloride, it permeates through PTFE tubing.

INSTRUMENTATION: (capillary column) GC, w/PID and electrolytic conductivity detectors in series. Column #1: VOCOL wide bore column; Column #2: RTX-502.2 mega bore column; Column #3: DB-62 mega bore column.

RANGE: Approximately 0.02-200 μg/L

MDL: 0.01 μg/L (reagent H_2O) (ELCD) 60 m × 0.75 mm VOCOL capillary column

PRECISION: RSD = 3.6% @ Spike of 10 μg/L (reagent H_2O) (PID). 60 m × 0.75 mm VOCOL capillary column

ACCURACY: Recovery = 96% @ spike of 10 μg/L (reagent H_2O) (PID) 60 m × 0.75 mm VOCOL capillary column

SAMPLING METHOD: Use a 40-120 mL screw cap vial (pre-washed with detergent, rinsed with distilled water and oven dried at 105°C) with a PTFE-faced silicone septum. If residual chlorine is in the water add about 25 mg of ascorbic acid to each vial before sample collection. Collect bubble free samples.

STABILITY: Cool to 4°C; HCl to pH <2. M.H.T. = 14 Days.

QUALITY CONTROL: As an initial demonstration of lab accuracy and

precision, analyze 4 to 7 replicates of a lab fortified blank containing analyte at 0.1-5 µg/L. Collect all samples in duplicate.

REFERENCE: Method 502.2, Volatile Organic Compounds in Water by Purge and Trap Capillary Column GC, EPA 600/4-88/039.

PRIMARY NAME: Trichloroethylene

Method 503.1

TITLE: Aromatic and Unsaturated VOCs in Water

MATRIX: Drinking water (finished or any treatment stage) and raw source water.

CAS # : 79-01-6

APPLICATION: Method covers 28 aromatic and unsaturated VOCs. An inert gas is bubbled through a 5 mL water sample. Purged sample components are trapped in tube of sorbent materials. When purging is complete, sorbent tube is heated and backflushed with inert gas to desorb trapped sample onto a packed GC column.

INTERFERENCES: During analysis, major contaminant sources are volatile materials in the lab and impurities in purging gas and sorbent trap. With high and low level samples, there can be carry-over contamination. Excess water causes a negative baseline deflection.

INSTRUMENTATION: Purge and Trap GC w/photoionization detector. (Two GC columns are recommended); Column #1: 5% SP-1200 and 1.75% Bentone 34 on Supelcoport; 5% 1,2,3-tris(2-cyanoethoxy)propane on Chromosorb W.

RANGE: 2.2-600 µg/L (Drinking water)

MDL: 0.01 µg/L in water

PRECISION: RSD = 6.8% @ 0.50 µg/L; 19 samples

ACCURACY: Avg Recovery = 97% @ 0.50 µg/L; 19 samples

SAMPLING METHOD: Use a 40-120 mL screw cap vial (pre-washed with detergent, rinsed with distilled water and oven dried at 105°C) with a PTFE-faced silicone septum. If residual chlorine is in the water add about 25 mg of ascorbic acid to each vial before sample collection. Collect bubble free samples.

STABILITY: Cool to 4°C; HCl to pH <2. M.H.T. = 14 Days.

QUALITY CONTROL: As an initial demonstration of lab accuracy and precision, analyze 4 to 7 replicates of a lab fortified blank containing analyte at 0.1-5 μg/L. Collect all samples in duplicate.

REFERENCE: Method 503.1, Volatile Aromatic and Unsaturated Organic Compounds in H_2O by Purge and Trap GC, EPA 600/4-88/039.

PRIMARY NAME: Trichloroethylene Method 524.1

TITLE: VOCs in H_2O by Purge and Trap GC/MS

MATRIX: Drinking water (finished or any treatment stage) and raw source water.

CAS # : 79-01-6

APPLICATION: Method covers 51 VOCs. An inert gas is bubbled through a 25 mL water sample. Purged sample components are trapped in tube of sorbent materials. When purging is complete, sorbent tube is heated and backflushed with helium to desorb trapped sample onto a packed GC column.

INTERFERENCES: During analysis, major contaminant sources are volatile materials in the lab and impurities in purging gas and sorbent trap. With high and low level samples, there can be carry-over contamination. Watch for methylene chloride, it permeates through PTFE tubing.

INSTRUMENTATION: Gas Chromatography/Mass Spectrometry/Data System; 1% SP-1000 on Carbopack B column.

RANGE: Approximately 0.2-200 μg/L MDL: 0.4 μg/L

PRECISION: RSD = 13% @ 1.0 μg/L

ACCURACY: Mean Accuracy = 90% @ 1.0 μg/L

SAMPLING METHOD: Use a 60-120 mL screw cap vial (pre-washed with detergent, rinsed with distilled water and oven dried at 105°C) with a PTFE-faced silicone septum. If residual chlorine is in the water add about 25 mg of

ascorbic acid to each vial before sample collection. Collect bubble free samples.

STABILITY: Cool to 4°C; HCl to pH <2. M.H.T. = 14 Days.

QUALITY CONTROL: As an initial demonstration of lab accuracy and precision, analyze 4 to 7 replicates of a lab fortified blank containing analyte at 0.2-5 µg/L. Collect all samples in duplicate.

REFERENCE: Method 524.1, VOCs in Water by Purge and Trap Gas Chromatography/Mass Spectrometry, EPA 600/4-88/039.

PRIMARY NAME: Trichloroethylene Method 524.2

TITLE: VOCs in H_2O (Purge and Trap) capillary column GC/MS

MATRIX: Drinking water (finished or any treatment stage) and raw source water.

CAS # : 79-01-6

APPLICATION: Method covers 60 VOCs. An inert gas is bubbled through a 25 mL water sample. Purged sample components are trapped in tube of sorbent materials. When purging is complete, sorbent tube is heated and backflushed with helium to desorb trapped sample onto a capillary GC column.

INTERFERENCES: During analysis, major contaminant sources are volatile materials in the lab and impurities in purging gas and sorbent trap. With high and low level samples, there can be carry-over contamination. Watch for methylene chloride, it permeates through PTFE tubing.

INSTRUMENTATION: (capillary column) Gas Chromatography/Mass Spectrometry/Data System. Column #1: VOCOL glass wide bore; Column #2: DB-624 fused silica; Column #3: DB-5 fused silica.

RANGE: Approximately 0.1-200 µg/L

MDL: 0.19 µg/L on wide bore capillary column.

PRECISION: RSD = 7.3% @ (0.5-10) µg/L (24 samples) wide bore capillary column.

ACCURACY: Recovery = 90% @ (0.5-10) μg/L (24 samples) wide bore capillary column.

SAMPLING METHOD: Use a 60-120 mL screw cap vial (pre-washed with detergent, rinsed with distilled water and oven dried at 105°C) with a PTFE-faced silicone septum. If residual chlorine is in the water add about 25 mg of ascorbic acid to each vial before sample collection. Collect bubble free samples.

STABILITY: Cool to 4°C; HCl to pH <2. M.H.T. = 14 Days.

QUALITY CONTROL: As an initial demonstration of lab accuracy and precision, analyze 4 to 7 replicates of a lab fortified blank containing analyte at 0.2-5 μg/L. Collect all samples in duplicate.

REFERENCE: Method 524.2, VOCs in H_2O by Purge and Trap Capillary column GC/MS, EPA 600/4-88/039.

PRIMARY NAME: Trichloroethylene Method 601

TITLE: Purgeable Halocarbons MATRIX: Wastewater

CAS # : 79-01-6

APPLICATION: Method covers 29 purgeable halocarbons. (Method 624 provides GC/MS conditions appropriate for the qualitative and quantitative confirmation of results). Method describes conditions for a second GC column to confirm measurements made with primary column.

INTERFERENCES: Impurities in the purge gas and organic compounds outgassing from the plumbing ahead of the trap. With high and low level samples, there can be carry-over contamination. Diffusion of volatile organics through the septum seal into the sample.

INSTRUMENTATION: GC equipped with halide-specific detector. (With purge and trap unit).

RANGE: 8.0 To 500 μg/L. MDL: 0.12 μg/L.

PRECISION: 0.23X+0.30 μg/L (overall precision)

ACCURACY: 0.87C+0.48 μg/L (as recovery)

SAMPLING METHOD: 25 mL glass vial. Teflon lined septum.

STABILITY: Cool, 4°C, 0.008% sodium thiosulfate. M.H.T. = 14 Days.

QUALITY CONTROL: The laboratory must on an ongoing basis, spike at least 10% of the samples from each sample site being monitored to assess accuracy.

REFERENCE: Method 601, Federal Register Part VIII 40 CFR Part 136, Oct 26, 1984.

PRIMARY NAME: Trichloroethylene Method 624

TITLE: Purgeables MATRIX: Wastewater

CAS # : 79-01-6

APPLICATION: Method covers 31 purgeable organics. An inert gas is bubbled through a 5 mL water sample in a specially designed purging chamber. Here, purgeables are transferred from aqueous to gaseous phase, passed onto a sorbent column and trapped. Trap is heated and backflushed with inert gas to desorb purgeables onto a GC column, where purgeables are separated.

INTERFERENCES: Impurities in the purge gas, organic compounds outgassing from the plumbing ahead of the trap, and solvent vapors in the laboratory. With high and low level samples, there can be carry-over contamination.

INSTRUMENTATION: GC/MS with purge and trap unit.

RANGE: 5-600 μg/L MDL: 1.9 μg/L

PRECISION: 0.12X+0.59 μg/L (overall precision)

ACCURACY: 1.04C+2.27 μg/L (as recovery)

SAMPLING METHOD: 25 mL glass vial. Teflon lined septum.

STABILITY: Cool, 4°C, 0.008% sodium thiosulfate. M.H.T. = 14 Days.

QUALITY CONTROL: The laboratory must on an ongoing basis, spike at least 5% of the samples from each sample site being monitored to assess accuracy.

REFERENCE: Method 624, Federal Register Part VIII 40 CFR Part 136, Oct 26, 1984.

PRIMARY NAME: Trichloroethylene Method 8010

TITLE: Halogenated Volatile Organics

MATRIX: groundwater, soils, sludges water miscible liquid wastes, and non-water miscible wastes.

CAS # : 79-01-6

APPLICATION: This method is used for the analysis of 39 halogenated VOCs. Samples are analyzed using direct injection or purge and trap methods. Groundwater must be analyzed by the purge and trap method. The method provides an optional GC column which is used for analyte confirmation and that may help resolve analytes from interferences.

INTERFERENCES: There can be carry-over contamination with high and low level samples. Impurities may come from the purge and trap apparatus, organic compounds outgassing from the plumbing ahead of trap, diffusion of VOCs through the sample bottle septum during shipping or storage, or from solvent vapors in the lab.

INSTRUMENTATION: GC capable of on-column injections or purge and trap sample introduction and a halogen specific detector. Column 1: 8 foot by 0.1 inch 1% SP-1000 on Carbopack-B. Column 2: 6 foot by 0.1 inch bonded n-octane on Porasil-C.

RANGE: 8 to 500 μg/L (reagent water)

MDL: 0.12 μg/L (reagent water)

PRACTICAL QUANTITATION LIMIT FACTORS FOR MULTIPLYING TIMES MDL VALUE

Matrix	Multiplication Factor
Groundwater	10
Low-level soil	10
Water miscible liquid waste	500
High-level soil and sludge	1250
Non-water miscible waste	1250

PRECISION: 0.23X + 0.30 μg/L (overall precision)

ACCURACY: 0.87C + 0.48 μg/L (as recovery)

SAMPLING METHOD: For water and liquid samples; use glass 40 mL vials with Teflon lined septum caps and collect two vials per sample location with no headspace. For solids and concentrated waste samples; use widemouth glass bottles with Teflon liners.

STABILITY: For concentrated wastes, soils, sediments or sludges; cool to 4°C. For liquids; add 4 drops of concentrated hydrochloric acid and cool to 4°C. M.H.T. = 14 days.

QUALITY CONTROL: Analyze a reagent blank, matrix spike and matrix spike duplicate/duplicate for each analytical batch (up to 20 Samples). Demonstrate the purity of glassware and reagents by analyzing a reagent water method blank. Internal, surrogate and five concentration level calibration standards are used.

REFERENCE: Method 8010, SW-846, 3rd ed., Nov 1986.

PRIMARY NAME: 1,2,3-Trichloropropane

Method 502.1

TITLE: Halogenated VOCs in Water

MATRIX: Drinking water (finished or any treatment stage) and raw source water.

CAS # : 96-18-4

APPLICATION: Method covers 40 halogenated VOCs. An inert gas is bubbled through a 5 mL water sample. Purged sample components are trapped in tube of sorbent materials. When purging is complete, sorbent tube is heated and backflushed with inert gas to desorb trapped sample onto a packed GC column where it is separated.

INTERFERENCES: During analysis, major contaminant sources are volatile materials in the lab and impurities in purging gas and sorbent trap. With high and low level samples, there can be carry-over contamination. Watch for methylene chloride, it permeates through PTFE tubing.

INSTRUMENTATION: Purge and Trap GC w/halide specific detector. (Two GC columns are recommended); Column #1: 1% SP-1000 on Carbopack B; Column #2: n-octane bonded on Poracil C.

RANGE: 8.0-505 μg/L (Drinking water)   MDL: not determined.

PRECISION: RSD = 9.5% @ 0.40 μg/L spike (S.L.) (20 Samples)

ACCURACY: Avg Recovery = 100% @ 0.40 μg/L spike (S.L.) (20 Samps)

SAMPLING METHOD: Use a 40-120 mL screw cap vial (pre-washed with detergent, rinsed with distilled water and oven dried at 105°C) with a PTFE-faced silicone septum. If residual chlorine is in the water add about 25 mg of ascorbic acid to each vial before sample collection. Collect bubble free samples.

STABILITY: Cool to 4°C; HCl to pH <2.   M.H.T. = 14 Days.

QUALITY CONTROL: As an initial demonstration of lab accuracy and precision, analyze 4 to 7 replicates of a lab fortified blank containing analyte at 0.1-5 μg/L. Collect all samples in duplicate.

REFERENCE: Method 502.1, Volatile Halogenated Organic Compounds in Water by Purge and Trap GC, EPA 600/4-88/039.

PRIMARY NAME: 1,2,3-Trichloropropane   Method 502.2

TITLE: VOCs in H_2O by Purge and Trap capillary column GC

MATRIX: Drinking water (finished or any treatment stage) and raw source water.

CAS # : 96-18-4

APPLICATION: Method covers 60 VOCs. An inert gas is bubbled through a 5 mL sample in a purging chamber. Purged sample components are trapped in tube of sorbent materials. When purging is done, sorbent tube is heated and backflushed w/helium to desorb trapped sample onto a capillary GC column.

INTERFERENCES: During analysis, major contaminant sources are volatile materials in the lab and impurities in purging gas and sorbent trap. With high and low level samples, there can be carry-over contamination. Beware of methylene chloride, it permeates through PTFE tubing.

INSTRUMENTATION: (Capillary Column) GC, w/PID and electrolytic conductivity detectors in series. Column #1: VOCOL wide bore column; Column #2: RTX-502.2 mega bore column; Column #3: DB-62 mega bore column.

RANGE: Approximately 0.02-200 μg/L

MDL: 0.4 μg/L (reagent H_2O) (ELCD) 60 m × 0.75 mm VOCOL capillary column

PRECISION: RSD = 2.3% @ Spike of 10 μg/L (reagent H_2O) (ELCD) 60 m × 0.75 mm VOCOL capillary column

ACCURACY: Recovery = 99% @ spike of 10 μg/L (reagent H_2O) (ELCD) 60 m × 0.75 mm VOCOL capillary column

SAMPLING METHOD: Use a 40-120 mL screw cap vial (pre-washed with detergent, rinsed with distilled water and oven dried at 105°C) with a PTFE-faced silicone septum. If residual chlorine is in the water add about 25 mg of ascorbic acid to each vial before sample collection. Collect bubble free samples.

STABILITY: Cool to 4°C; HCl to pH <2.

M.H.T. = 14 Days.

QUALITY CONTROL: As an initial demonstration of lab accuracy and precision, analyze 4 to 7 replicates of a lab fortified blank containing analyte at 0.1-5 μg/L. Collect all samples in duplicate.

REFERENCE: Method 502.2, Volatile Organic Compounds in Water by Purge and Trap Capillary Column GC, EPA 600/4-88/039.

PRIMARY NAME: 1,2,3-Trichloropropane   Method 524.1

TITLE: VOCs in H_2O by Purge and Trap GC/MS

MATRIX: Drinking water (finished or any treatment stage) and raw source water.

CAS # : 96-18-4

APPLICATION: Method covers 51 VOCs. An inert gas is bubbled through a 25 mL water sample. Purged sample components are trapped in tube of sorbent materials. When purging is complete, sorbent tube is heated and backflushed with helium to desorb trapped sample onto a packed GC column.

INTERFERENCES: During analysis, major contaminant sources are volatile materials in the lab and impurities in purging gas and sorbent trap. With high and low level samples, there can be carry-over contamination. Watch for methylene chloride, it permeates through PTFE tubing.

INSTRUMENTATION: Gas Chromatography/Mass Spectrometry/Data System; 1% SP-1000 on Carbopack B column.

RANGE: Approximately 0.2-200 µg/L   MDL: not listed.

PRECISION: not listed.

ACCURACY: not listed.

SAMPLING METHOD: Use a 60-120 mL screw cap vial (pre-washed with detergent, rinsed with distilled water and oven dried at 105°C) with a PTFE-faced silicone septum. If residual chlorine is in the water add about 25 mg of ascorbic acid to each vial before sample collection. Collect bubble free samples.

STABILITY: Cool to 4°C; HCl to pH <2.   M.H.T. = 14 Days.

QUALITY CONTROL: As an initial demonstration of lab accuracy and precision, analyze 4 to 7 replicates of a lab fortified blank containing analyte at 0.2-5 µg/L. Collect all samples in duplicate.

REFERENCE: Method 524.1, VOCs in Water by Purge and Trap Gas Chromatography/Mass Spectrometry, EPA 600/4-88/039.

PRIMARY NAME: 1,2,3-Trichloropropane

Method 524.2

TITLE: VOCs in H_2O (Purge and Trap) capillary column GC/MS

MATRIX: Drinking water (finished or) any treatment stage) and raw source water.

CAS # : 96-18-4

APPLICATION: Method covers 60 VOCs. An inert gas is bubbled through a 25 mL water sample. Purged sample components are trapped in tube of sorbent materials. When purging is complete, sorbent tube is heated and backflushed with helium to desorb trapped sample onto a capillary GC column.

INTERFERENCES: During analysis, major contaminant sources are volatile materials in the lab and impurities in purging gas and sorbent trap. With high and low level samples, there can be carry-over contamination. Watch for methylene chloride, it permeates through PTFE tubing.

INSTRUMENTATION: (Capillary Column) Gas Chromatography/Mass Spectrometry/Data System. Column #1: VOCOL glass wide bore; Column #2: DB-624 fused silica; Column #3: DB-5 fused silica.

RANGE: Approximately 0.1-200 µg/L

MDL: 0.32 µg/L on wide bore capillary column.

PRECISION: RSD = 14.4% @ (0.5-10) µg/L (16 samples) wide bore capillary column.

ACCURACY: Recovery = 108% @ (0.5-10) µg/L (16 samples) wide bore capillary column.

SAMPLING METHOD: Use a 60-120 mL screw cap vial (pre-washed with detergent, rinsed with distilled water and oven dried at 105°C) with a PTFE-faced silicone septum. If residual chlorine is in the water add about 25 mg of ascorbic acid to each vial before sample collection. Collect bubble free samples.

STABILITY: Cool to 4°C; HCl to pH <2.

M.H.T. = 14 Days.

QUALITY CONTROL: As an initial demonstration of lab accuracy and

precision, analyze 4 to 7 replicates of a lab fortified blank containing analyte at 0.2-5 µg/L. Collect all samples in duplicate.

REFERENCE: Method 524.2, VOCs in H_2O by Purge and Trap Capillary column GC/MS, EPA 600/4-88/039.

PRIMARY NAME: Vinyl Chloride | Method 502.1

TITLE: Halogenated VOCs in Water

MATRIX: Drinking water (finished or any treatment stage) and raw source water.

CAS # : 75-01-4

APPLICATION: Method covers 40 halogenated VOCs. An inert gas is bubbled through a 5 mL water sample. Purged sample components are trapped in tube of sorbent materials. When purging is complete, sorbent tube is heated and backflushed with inert gas to desorb trapped sample onto a packed GC column where it is separated.

INTERFERENCES: During analysis, major contaminant sources are volatile materials in the lab and impurities in purging gas and sorbent trap. With high and low level samples, there can be carry-over contamination. Watch for methylene chloride, it permeates through PTFE tubing.

INSTRUMENTATION: Purge and Trap GC w/halide specific detector. (Two GC columns are recommended); Column #1: 1% SP-1000 on Carbopack B; Column #2: n-octane bonded on Poracil C.

RANGE: 8.0-505 µg/L (Drinking water) MDL: 0.01 µg/L in water

PRECISION: RSD = 15% @ 0.20 µg/L spike (S.L.) (12 Samples)

ACCURACY: Avg Recovery = 110% @ 0.20 µg/L spike (S.L.) (12 Samples)

SAMPLING METHOD: Use a 40-120 mL screw cap vial (pre-washed with detergent, rinsed with distilled water and oven dried at 105°C) with a PTFE-faced silicone septum. If residual chlorine is in the water add about 25 mg of ascorbic acid to each vial before sample collection. Collect bubble free samples.

STABILITY: Cool to 4°C; HCl to pH <2. M.H.T. = 14 Days.

QUALITY CONTROL: As an initial demonstration of lab accuracy and precision, analyze 4 to 7 replicates of a lab fortified blank containing analyte at 0.1-5 μg/L. Collect all samples in duplicate.

REFERENCE: Method 502.1, Volatile Halogenated Organic Compounds in Water by Purge and Trap GC, EPA 600/4-88/039.

PRIMARY NAME: Vinyl Chloride Method 502.2

TITLE: VOCs in H_2O by Purge and Trap capillary column GC

MATRIX: Drinking water (finished or any treatment stage) and raw source water.

CAS # : 75-01-4

APPLICATION: Method covers 60 VOCs. An inert gas is bubbled through a 5 mL sample in a purging chamber. Purged sample components are trapped in tube of sorbent materials. When purging is done, sorbent tube is heated and backflushed w/helium to desorb trapped sample onto a capillary GC column.

INTERFERENCES: During analysis, major contaminant sources are volatile materials in the lab and impurities in purging gas and sorbent trap. With high and low level samples, there can be carry-over contamination. Beware of methylene chloride, it permeates through PTFE tubing.

INSTRUMENTATION: (capillary column) GC, w/PID and electrolytic conductivity detectors in series. Column #1: VOCOL wide bore column; Column #2: RTX-502.2 mega bore column; Column #3: DB-62 mega bore column.

RANGE: Approximately 0.02-200 μg/L

MDL: 0.04 μg/L (reagent H_2O) (ELCD) 60 m × 0.75 mm VOCOL capillary column

PRECISION: RSD = 5.9% @ Spike of 10 μg/L (reagent H_2O) (ELCD). 60 m × 0.75 mm VOCOL capillary column

ACCURACY: Recovery = 95% @ spike of 10 µg/L (reagent H_2O) (ELCD) 60 m × 0.75 mm VOCOL capillary column

SAMPLING METHOD: Use a 40-120 mL screw cap vial (pre-washed with detergent, rinsed with distilled water and oven dried at 105°C) with a PTFE-faced silicone septum. If residual chlorine is in the water add about 25 mg of ascorbic acid to each vial before sample collection. Collect bubble free samples.

STABILITY: Cool to 4°C; HCl to pH <2. M.H.T. = 14 Days.

QUALITY CONTROL: As an initial demonstration of lab accuracy and precision, analyze 4 to 7 replicates of a lab fortified blank containing analyte at 0.1-5 µg/L. Collect all samples in duplicate.

REFERENCE: Method 502.2, Volatile Organic Compounds in Water by Purge and Trap Capillary Column GC, EPA 600/4-88/039.

PRIMARY NAME: Vinyl Chloride Method 524.1

TITLE: VOCs in H_2O by Purge and Trap GC/MS

MATRIX: Drinking water (finished or any treatment stage) and raw source water.

CAS # : 75-01-4

APPLICATION: Method covers 51 VOCs. An inert gas is bubbled through a 25 mL water sample. Purged sample components are trapped in tube of sorbent materials. When purging is complete, sorbent tube is heated and backflushed with helium to desorb trapped sample onto a packed GC column.

INTERFERENCES: During analysis, major contaminant sources are volatile materials in the lab and impurities in purging gas and sorbent trap. With high and low level samples, there can be carry-over contamination. Watch for methylene chloride, it permeates through PTFE tubing.

INSTRUMENTATION: Gas Chromatography/Mass Spectometry/Data System; 1% SP-1000 on Carbopack B column.

RANGE: Approximately 0.2-200 µg/L MDL: 0.3 µg/L

PRECISION: RSD = 11% @ 1.0 µg/L

ACCURACY: Mean Accuracy = 98% @ 1.0 µg/L

SAMPLING METHOD: Use a 60-120 mL screw cap vial (pre-washed with detergent, rinsed with distilled water and oven dried at 105°C) with a PTFE-faced silicone septum. If residual chlorine is in the water add about 25 mg of ascorbic acid to each vial before sample collection. Collect bubble free samples.

STABILITY: Cool to 4°C; HCl to pH <2. M.H.T. = 14 Days.

QUALITY CONTROL: As an initial demonstration of lab accuracy and precision, analyze 4 to 7 replicates of a lab fortified blank containing analyte at 0.2-5 µg/L. Collect all samples in duplicate.

REFERENCE: Method 524.1, VOCs in Water by Purge and Trap Gas Chromatography/Mass Spectometry, EPA 600/4-88/039.

PRIMARY NAME: Vinyl Chloride Method 524.2

TITLE: VOCs in H_2O (Purge and Trap capillary column GC/MS

MATRIX: Drinking water (finished or any treatment stage) and raw source water.

CAS # : 75-01-4

APPLICATION: Method covers 60 VOCs. An inert gas is bubbled through a 25 mL water sample. Purged sample components are trapped in tube of sorbent materials. When purging is complete, sorbent tube is heated and backflushed with helium to desorb trapped sample onto a capillary GC column.

INTERFERENCES: During analysis, major contaminant sources are volatile materials in the lab and impurities in purging gas and sorbent trap. With high and low level samples, there can be carry-over contamination. Watch for methylene chloride, it permeates through PTFE tubing.

INSTRUMENTATION: (capillary column) Gas Chromatography/Mass

Spectometry/Data System. Column #1: VOCOL glass wide bore; Column #2: DB-624 fused silica; Column #3: DB-5 fused silica.

RANGE: Approximately 0.1-200 μg/L

MDL: 0.17 μg/L on wide bore capillary column.

PRECISION: RSD = 6.7% @ (0.5-10) μg/L (18 samples) wide bore capillary column.

ACCURACY: Recovery = 98% @ (0.5-10) μg/L (18 samples) wide bore capillary column.

SAMPLING METHOD: Use a 60-120 mL screw cap vial (pre-washed with detergent, rinsed with distilled water and oven dried at 105°C) with a PTFE-faced silicone septum. If residual chlorine is in the water add about 25 mg of ascorbic acid to each vial before sample collection. Collect bubble free samples.

STABILITY: Cool to 4°C; HCl to pH <2.

M.H.T. = 14 Days.

QUALITY CONTROL: As an initial demonstration of lab accuracy and precision, analyze 4 to 7 replicates of a lab fortified blank containing analyte at 0.2-5 μg/L. Collect all samples in duplicate.

REFERENCE: Method 524.2, VOCs in H_2O by Purge and Trap Capillary column GC/MS, EPA 600/4-88/039.

PRIMARY NAME: Vinyl Chloride

Method 601

TITLE: Purgeable Halocarbons

MATRIX: Wastewater

CAS # : 75-01-4

APPLICATION: Method covers 29 purgeable halocarbons. (Method 624 provides GC/MS conditions appropriate for the qualitative and quantitative confirmation of results). Method describes conditions for a second GC column to confirm measurements made with primary column.

INTERFERENCES: Impurities in the purge gas and organic compounds outgassing from the plumbing ahead of the trap. With high and low level samples, there can be carry-over contamination. Diffusion of volatile organics through the septum seal into the sample.

INSTRUMENTATION: GC equipped with halide-specific detector. (With purge and trap unit).

RANGE: 8.0 To 500 µg/L. MDL: 0.18 µg/L

PRECISION: 0.27X+0.40 µg/L (overall precision)

ACCURACY: 0.97C-0.36 µg/L (as recovery)

SAMPLING METHOD: 25 mL glass vial. Teflon lined septum.

STABILITY: Cool, 4°C, 0.008% sodium thiosulfate. M.H.T. = 14 Days.

QUALITY CONTROL: The laboratory must on an ongoing basis, spike at least 10% of the samples from each sample site being monitored to assess accuracy.

REFERENCE: Method 601, Federal Register Part VIII 40 CFR Part 136, Oct 26, 1984.

PRIMARY NAME: Vinyl Chloride Method 624

TITLE: Purgeables MATRIX: Wastewater

CAS # : 75-01-4

APPLICATION: Method covers 31 purgeable organics. An inert gas is bubbled through a 5 mL water sample in a specially designed purging chamber. Here, purgeables are transferred from aqueous to gaseous phase, passed onto a sorbent column and trapped. Trap is heated and backflushed with inert gas to desorb purgeables onto a GC column, where purgeables are separated.

INTERFERENCES: Impurities in the purge gas, organic compounds outgassing from the plumbing ahead of the trap, and solvent vapors in the laboratory. With high and low level samples, there can be carry-over contamination.

INSTRUMENTATION: GC/MS with purge and trap unit.

RANGE: 5-600 µg/L MDL: not determined

PRECISION: 0.65X µg/L (overall precision)

ACCURACY: 1.00C µg/L (as recovery)

SAMPLING METHOD: 25 mL glass vial. Teflon lined septum.

STABILITY: Cool, 4°C, 0.008% sodium thiosulfate. M.H.T. = 14 Days.

QUALITY CONTROL: The laboratory must on an ongoing basis, spike at least 5% of the samples from each sample site being monitored to assess accuracy.

REFERENCE: Method 624, Federal Register Part VIII 40 CFR Part 136, Oct 26, 1984.

PRIMARY NAME: Vinyl chloride Method 8010

TITLE: Halogenated Volatile Organics MATRIX: groundwater, soils, sludges water miscible liquid wastes, and non-water miscible wastes.

CAS # : 75-01-4

APPLICATION: This method is used for the analysis of 39 halogenated VOCs. Samples are analyzed using direct injection or purge and trap methods. Groundwater must be analyzed by the purge and trap method. The method provides an optional GC column which is used for analyte confirmation and that may help resolve analytes from interferences.

INTERFERENCES: There can be carry-over contamination with high and low level samples. Impurities may come from the purge and trap apparatus, organic compounds outgassing from the plumbing ahead of trap, diffusion of VOCs through the sample bottle septum during shipping or storage, or from solvent vapors in the lab.

INSTRUMENTATION: GC capable of on-column injections or purge and trap sample introduction and a halogen specific detector. Column 1: 8 foot by

0.1 inch 1% SP-1000 on Carbopack-B. Column 2: 6 foot by 0.1 inch bonded n-octane on Porasil-C.

RANGE: 8 to 500 µg/L (reagent water)

MDL: 0.18 µg/L (reagent water)

PRACTICAL QUANTITATION LIMIT FACTORS FOR MULTIPLYING TIMES MDL VALUE

Matrix	Multiplication Factor
Groundwater	10
Low-level soil	10
Water miscible liquid waste	500
High-level soil and sludge	1250
Non-water miscible waste	1250

PRECISION: 0.27X + 0.40 µg/L (overall precision)

ACCURACY: 0.97C - 0.36 µg/L (as recovery)

SAMPLING METHOD: For water and liquid samples; use glass 40 mL vials with Teflon lined septum caps and collect two vials per sample location with no headspace. For solids and concentrated waste samples; use widemouth glass bottles with Teflon liners.

STABILITY: For concentrated wastes, soils, sediments or sludges; cool to 4°C. For liquids; add 4 drops of concentrated hydrochloric acid and cool to 4°C. M.H.T. = 14 days.

QUALITY CONTROL: Analyze a reagent blank, matrix spike and matrix spike duplicate/duplicate for each analytical batch (up to 20 Samples). Demonstrate the purity of glassware and reagents by analyzing a reagent water method blank. Internal, surrogate and five concentration level calibration standards are used.

REFERENCE: Method 8010, SW-846, 3rd ed., Nov 1986.

Chapter 2
Other Halogenated Volatile Organic Compounds

PRIMARY NAME: Bromobenzene

Method 502.1

TITLE: Halogenated VOCs in Water

MATRIX: Drinking water (finished or any treatment stage) and raw source water.

CAS #: 108-86-1

APPLICATION: Method covers 40 halogenated VOCs. An inert gas is bubbled through a 5 mL water sample. Purged sample components are trapped in tube of sorbent materials. When purging is complete, sorbent tube is heated and backflushed with inert gas to desorb trapped sample onto a packed GC column where it is separated.

INTERFERENCES: During analysis, major contaminant sources are volatile materials in the lab and impurities in purging gas and sorbent trap. With high and low level samples, there can be carry-over contamination. Watch for methylene chloride, it permeates through PTFE tubing.

INSTRUMENTATION: Purge and Trap GC w/halide specific detector. (Two GC columns are recommended); Column #1: 1% SP-1000 on Carbopack B; Column #2: n-octane bonded on Poracil C.

RANGE: 8.0-505 μg/L (Drinking water)

MDL: not determined

PRECISION: RSD = 12% @ 0.40 μg/L spike (S.L.) (20 Samples)

ACCURACY: Avg Recovery = 93% @ 0.40 μg/L spike (S.L.) (20 Samples)

SAMPLING METHOD: Use a 40-120 mL screw cap vial (pre-washed with detergent, rinsed with distilled water and oven dried at 105°C) with a PTFE-faced silicone septum. If residual chlorine is in the water add about 25 mg of ascorbic acid to each vial before sample collection. Collect bubble free samples.

STABILITY: Cool to 4°C; HCl to pH <2. M.H.T. = 14 Days.

QUALITY CONTROL: As an initial demonstration of lab accuracy and precision, analyze 4 to 7 replicates of a lab fortified blank containing analyte at 0.1-5 μg/L. Collect all samples in duplicate.

REFERENCE: Method 502.1, Volatile Halogenated Organic Compounds in Water by Purge and Trap GC, EPA 600/4-88/039.

PRIMARY NAME: Bromobenzene Method 502.2

TITLE: VOCs in Water by Purge and Trap capillary column GC

MATRIX: Drinking water (finished or any treatment stage) and raw source water.

CAS #: 108-86-1

APPLICATION: Method covers 60 VOCs. An inert gas is bubbled through a 5 mL sample in a purging chamber. Purged sample components are trapped in tube of sorbent materials. When purging is done, sorbent tube is heated and backflushed w/helium to desorb trapped sample onto a capillary GC column.

INTERFERENCES: During analysis, major contaminant sources are volatile materials in the lab and impurities in purging gas and sorbent trap. With high and low level samples, there can be carry-over contamination. Beware of methylene chloride, it permeates through PTFE tubing.

INSTRUMENTATION: (capillary column) GC, w/PID and electrolytic conductivity detectors in series. Column #1: VOCOL wide bore column; Column #2: RTX-502.2 mega bore column; Column #3: DB-62 mega bore column.

RANGE: Approximately 0.02-200 μg/L

MDL: 0.03 μg/L (reagent C) (ELCD) 60 m × 0.75 mm VOCOL capillary column

PRECISION: RSD = 2.7% @ Spike of 10 μg/L (reagent C) (ELCD) 60 m × 0.75 mm VOCOL capillary column

ACCURACY: Recovery = 97% @ spike of 10 μg/L (reagent C) (ELCD) 60 m × 0.75 mm VOCOL capillary column

SAMPLING METHOD: Use a 40-120 mL screw cap vial (pre-washed with detergent, rinsed with distilled water and oven dried at 105°C) with a PTFE-faced silicone septum. If residual chlorine is in the water add about 25 mg of ascorbic acid to each vial before sample collection. Collect bubble free samples.

STABILITY: Cool to 4°C; HCl to pH <2.

M.H.T. = 14 Days.

QUALITY CONTROL: As an initial demonstration of lab accuracy and precision, analyze 4 to 7 replicates of a lab fortified blank containing analyte at 0.1-5 μg/L. Collect all samples in duplicate.

REFERENCE: Method 502.2, Volatile Organic Compounds in Water by Purge and Trap Capillary Column GC, EPA 600/4-88/039.

PRIMARY NAME: Bromobenzene

Method 503.1

TITLE: Aromatic and Unsaturated VOCs in Water

MATRIX: Drinking water (finished or any treatment stage) and raw source water.

CAS #: 108-86-1

APPLICATION: Method covers 28 aromatic and unsaturated VOCs. An inert gas is bubbled through a 5 mL water sample. Purged sample components are trapped in tube of sorbent materials. When purging is complete, sorbent tube is heated and backflushed with inert gas to desorb trapped sample onto a packed GC column.

INTERFERENCES: During analysis, major contaminant sources are volatile materials in the lab and impurities in purging gas and sorbent trap. With

high and low level samples, there can be carry-over contamination. Excess water causes a negative baseline deflection.

INSTRUMENTATION: Purge and Trap GC w/photoionization detector. (Two GC columns are recommended); Column #1: 5% SP-1200 and 1.75% Bentone 34 on Supelcoport; 5% 1,2,3-tris(2-cyanoethoxy)propane on Chromosorb W.

RANGE: 2.2-600 μg/L (Drinking water) MDL: 0.002 μg/L in water

PRECISION: RSD = 6.2% @ 0.50 μg/L; 19 samples

ACCURACY: Avg Recovery = 93% @ 0.50 μg/L; 19 samples

SAMPLING METHOD: Use a 40-120 mL screw cap vial (pre-washed with detergent, rinsed with distilled water and oven dried at 105°C) with a PTFE-faced silicone septum. If residual chlorine is in the water add about 25 mg of ascorbic acid to each vial before sample collection. Collect bubble free samples.

STABILITY: Cool to 4°C; HCl to pH <2. M.H.T. = 14 Days.

QUALITY CONTROL: As an initial demonstration of lab accuracy and precision, analyze 4 to 7 replicates of a lab fortified blank containing analyte at 0.1-5 μg/L. Collect all samples in duplicate.

REFERENCE: Method 503.1, Volatile aromatic and unsaturated organic compounds in C by Purge and Trap GC, EPA 600/4-88/039.

PRIMARY NAME: Bromobenzene Method 524.1

TITLE: VOCs in C by Purge and Trap GC/MS

MATRIX: Drinking water (finished or any treatment stage) and raw source water.

CAS #: 108-86-1

APPLICATION: Method covers 51 VOCs. An inert gas is bubbled through a 25 mL water sample. Purged sample components are trapped in tube of sorbent materials. When purging is complete, sorbent tube is heated and backflushed with helium to desorb trapped sample onto a packed GC column.

INTERFERENCES: During analysis, major contaminant sources are volatile materials in the lab and impurities in purging gas and sorbent trap. With high and low level samples, there can be carry-over contamination. Watch for methylene chloride, it permeates through PTFE tubing.

INSTRUMENTATION: Gas Chromatography/Mass Spectrometry/Data System; 1% SP-1000 on Carbopack B column.

RANGE: Approximately 0.2-200 μg/L MDL: 0.1 μg/L

PRECISION: RSD = 4.6% @ 1.0 μg/L

ACCURACY: Mean Accuracy = 92% @ 1.0 μg/L

SAMPLING METHOD: Use a 60-120 mL screw cap vial (pre-washed with detergent, rinsed with distilled water and oven dried at 105°C) with a PTFE-faced silicone septum. If residual chlorine is in the water add about 25 mg of ascorbic acid to each vial before sample collection. Collect bubble free samples.

STABILITY: Cool to 4°C; HCl to pH <2. M.H.T. = 14 Days.

QUALITY CONTROL: As an initial demonstration of lab accuracy and precision, analyze 4 to 7 replicates of a lab fortified blank containing analyte at 0.2-5 μg/L. Collect all samples in duplicate.

REFERENCE: Method 524.1, VOCs in Water by Purge and Trap Gas chromatography/mass spectrometry, EPA 600/4-88/039.

PRIMARY NAME: Bromobenzene Method 524.2

TITLE: VOCs in Water (Purge and Trap capillary column GC/MS

MATRIX: Drinking water (finished or any treatment stage) and raw source water.

CAS #: 108-86-1

APPLICATION: Method covers 60 VOCs. An inert gas is bubbled through a 25 mL water sample. Purged sample components are trapped in tube of sorbent materials. When purging is complete, sorbent tube is heated and backflushed with helium to desorb trapped sample onto a capillary GC column.

INTERFERENCES: During analysis, major contaminant sources are volatile materials in the lab and impurities in purging gas and sorbent trap. With high and low level samples, there can be carry-over contamination. Watch for methylene chloride, it permeates through PTFE tubing.

INSTRUMENTATION: (Capillary Column) Gas Chromatography/Mass Spectrometry/Data System. Column #1: VOCOL glass wide bore; Column #2: DB-624 fused silica; Column #3: DB-5 fused silica.

RANGE: Approximately 0.1-200 μg/L

MDL: 0.03 μg/L on wide bore capillary column.

PRECISION: RSD = 5.5% @ (0.1-10) μg/L (30 samples) wide bore capillary column.

ACCURACY: Recovery = 100% @ (0.1-10) μg/L (30 samples) wide bore capillary column.

SAMPLING METHOD: Use a 60-120 mL screw cap vial (pre-washed with detergent, rinsed with distilled water and oven dried at 105°C) with a PTFE-faced silicone septum. If residual chlorine is in the water add about 25 mg of ascorbic acid to each vial before sample collection. Collect bubble free samples.

STABILITY: Cool to 4°C; HCl to pH <2.

M.H.T. = 14 Days.

QUALITY CONTROL: As an initial demonstration of lab accuracy and precision, analyze 4 to 7 replicates of a lab fortified blank containing analyte at 0.2-5 μg/L. Collect all samples in duplicate.

REFERENCE: Method 524.2, VOCs in C by Purge and Trap Capillary column GC/MS, EPA 600/4-88/039.

PRIMARY NAME: Bromochloromethane

Method 502.1

TITLE: Halogenated VOCs in Water

MATRIX: Drinking water (finished or any treatment stage) and raw source water.

CAS #: 74-97-5

APPLICATION: Method covers 40 halogenated VOCs. An inert gas is

bubbled through a 5 mL water sample. Purged sample components are trapped in tube of sorbent materials. When purging is complete,sorbent tube is heated and backflushed with inert gas to desorb trapped sample onto a packed GC column where it is separated.

INTERFERENCES: During analysis, major contaminant sources are volatile materials in the lab and impurities in purging gas and sorbent trap. With high and low level samples, there can be carry-over contamination. Watch for methylene chloride, it permeates through PTFE tubing.

INSTRUMENTATION: Purge and Trap GC w/halide specific detector. (Two GC columns are recommended); Column #1: 1% SP-1000 on Carbopack B; Column #2: n-octane bonded on Poracil C.

RANGE: 8.0-505 µg/L (Drinking water) MDL: not determined

PRECISION: RSD = 9.5% @ 0.40 µg/L spike (S.L.) (19 Samples)

ACCURACY: Avg Recovery = 90% @ 0.40 µg/L spike (S.L.) (19 Samples)

SAMPLING METHOD: Use a 40-120 mL screw cap vial (pre-washed with detergent, rinsed with distilled water and oven dried at 105°C) with a PTFE-faced silicone septum. If residual chlorine is in the water add about 25 mg of ascorbic acid to each vial before sample collection. Collect bubble free samples.

STABILITY: Cool to 4°C; HCl to pH <2. M.H.T. = 14 Days.

QUALITY CONTROL: As an initial demonstration of lab accuracy and precision, analyze 4 to 7 replicates of a lab fortified blank containing analyte at 0.1-5 µg/L. Collect all samples in duplicate.

REFERENCE: Method 502.1, Volatile Halogenated Organic Compounds in Water by Purge and Trap GC, EPA 600/4-88/039.

PRIMARY NAME: Bromochloromethane Method 502.2

TITLE: VOCs in Water by Purge and Trap capillary column GC

MATRIX: Drinking water (finished or any treatment stage) and raw source water.

CAS #: 74-97-5

APPLICATION: Method covers 60 VOCs. An inert gas is bubbled through a 5 mL sample in a purging chamber. Purged sample components are trapped in tube of sorbent materials. When purging is done, sorbent tube is heated and backflushed w/helium to desorb trapped sample onto a capillary GC column.

INTERFERENCES: During analysis, major contaminant sources are volatile materials in the lab and impurities in purging gas and sorbent trap. With high and low level samples, there can be carry-over contamination. Beware of methylene chloride, it permeates through PTFE tubing.

INSTRUMENTATION: (capillary column) GC, w/PID and electrolytic conductivity detectors in series. Column #1: VOCOL wide bore column; Column #2: RTX-502.2 mega bore column; Column #3: DB-62 mega bore column.

RANGE: Approximately 0.02-200 μg/L

MDL: 0.01 μg/L (reagent C) (ELCD) 60 m × 0.75 mm VOCOL capillary column

PRECISION: RSD = 3.0% @ Spike of 10 μg/L (reagent C) (ELCD) 60 m × 0.75 mm VOCOL capillary column

ACCURACY: Recovery = 96% @ spike of 10 μg/L (reagent C) (ELCD) 60 m × 0.75 mm VOCOL capillary column

SAMPLING METHOD: Use a 40-120 mL screw cap vial (pre-washed with detergent, rinsed with distilled water and oven dried at 105°C) with a PTFE-faced silicone septum. If residual chlorine is in the water add about 25 mg of ascorbic acid to each vial before sample collection. Collect bubble free samples.

STABILITY: Cool to 4°C; HCl to pH <2.

M.H.T. = 14 Days.

QUALITY CONTROL: As an initial demonstration of lab accuracy and precision, analyze 4 to 7 replicates of a lab fortified blank containing analyte at 0.1-5 μg/L. Collect all samples in duplicate.

REFERENCE: Method 502.2, Volatile Organic Compounds in Water by Purge and Trap Capillary Column GC, EPA 600/4-88/039.

PRIMARY NAME: Bromochloromethane | Method 524.1

TITLE: VOCs in Water by Purge and Trap GC/MS | MATRIX: Drinking water (finished or any treatment stage) and raw source water.

CAS #: 74-97-5

APPLICATION: Method covers 51 VOCs. An inert gas is bubbled through a 25 mL water sample. Purged sample components are trapped in tube of sorbent materials. When purging is complete, sorbent tube is heated and backflushed with helium to desorb trapped sample onto a packed GC column.

INTERFERENCES: During analysis, major contaminant sources are volatile materials in the lab and impurities in purging gas and sorbent trap. With high and low level samples, there can be carry-over contamination. Watch for methylene chloride, it permeates through PTFE tubing.

INSTRUMENTATION: Gas Chromatography/Mass Spectrometry/Data System; 1% SP-1000 on Carbopack B column.

RANGE: Approximately 0.2-200 µg/L | MDL: not determined.

PRECISION: not listed.

ACCURACY: not listed.

SAMPLING METHOD: Use a 60-120 mL screw cap vial (pre-washed with detergent, rinsed with distilled water and oven dried at 105°C) with a PTFE-faced silicone septum. If residual chlorine is in the water add about 25 mg of ascorbic acid to each vial before sample collection. Collect bubble free samples.

STABILITY: Cool to 4°C; HCl to pH <2. | M.H.T. = 14 Days.

QUALITY CONTROL: As an initial demonstration of lab accuracy and precision, analyze 4 to 7 replicates of a lab fortified blank containing analyte at 0.2-5 µg/L. Collect all samples in duplicate.

REFERENCE: Method 524.1, VOCs in Water by Purge and Trap Gas chromatography/mass spectrometry, EPA 600/4-88/039.

PRIMARY NAME: Bromochloromethane

Method 524.2

TITLE: VOCs in Water (Purge and Trap capillary column GC/MS

MATRIX: Drinking water (finished or any treatment stage) and raw source water.

CAS #: 74-97-5

APPLICATION: Method covers 60 VOCs. An inert gas is bubbled through a 25 mL water sample. Purged sample components are trapped in tube of sorbent materials. When purging is complete, sorbent tube is heated and backflushed with helium to desorb trapped sample onto a capillary GC column.

INTERFERENCES: During analysis, major contaminant sources are volatile materials in the lab and impurities in purging gas and sorbent trap. With high and low level samples, there can be carry-over contamination. Watch for methylene chloride, it permeates through PTFE tubing.

INSTRUMENTATION: (Capillary Column) Gas Chromatography/Mass Spectrometry/Data System. Column #1: VOCOL glass wide bore; Column #2: DB-624 fused silica; Column #3: DB-5 fused silica.

RANGE: Approximately 0.1-200 µg/L

MDL: 0.04 µg/L on wide bore capillary column.

PRECISION: RSD = 6.4% @ (0.5-10) µg/L (24 samples) wide bore capillary column.

ACCURACY: Recovery = 90% @ (0.5-10) µg/L (24 samples) wide bore capillary column.

SAMPLING METHOD: Use a 60-120 mL screw cap vial (pre-washed with detergent, rinsed with distilled water and oven dried at 105°C) with a PTFE-faced silicone septum. If residual chlorine is in the water add about 25 mg of ascorbic acid to each vial before sample collection. Collect bubble free samples.

STABILITY: Cool to 4°C; HCl to pH <2.

M.H.T. = 14 Days.

QUALITY CONTROL: As an initial demonstration of lab accuracy and precision, analyze 4 to 7 replicates of a lab fortified blank containing analyte at 0.2-5 µg/L. Collect all samples in duplicate.

REFERENCE: Method 524.2, VOCs in C by Purge and Trap Capillary column GC/MS, EPA 600/4-88/039.

PRIMARY NAME: Bromodichloromethane Method 502.1

TITLE: Halogenated VOCs in Water

MATRIX: Drinking water (finished or any treatment stage) and raw source water.

CAS #: 75-27-4

APPLICATION: Method covers 40 halogenated VOCs. An inert gas is bubbled through a 5 mL water sample. Purged sample components are trapped in tube of sorbent materials. When purging is complete, sorbent tube is heated and backflushed with inert gas to desorb trapped sample onto a packed GC column where it is separated.

INTERFERENCES: During analysis, major contaminant sources are volatile materials in the lab and impurities in purging gas and sorbent trap. With high and low level samples, there can be carry-over contamination. Watch for methylene chloride, it permeates through PTFE tubing.

INSTRUMENTATION: Purge and Trap GC w/halide specific detector. (Two GC columns are recommended); Column #1: 1% SP-1000 on Carbopack B; Column #2: n-octane bonded on Poracil C.

RANGE: 8.0-505 µg/L (Drinking water) MDL: 0.003 µg/L in water

PRECISION: RSD = 6.5% @ 0.20 µg/L spike (S.L.) (17 Samples)

ACCURACY: Avg Recovery = 100% @ 0.20 µg/L spike (S.L.) (17 Samples)

SAMPLING METHOD: Use a 40-120 mL screw cap vial (pre-washed with detergent, rinsed with distilled water and oven dried at 105°C) with a PTFE-faced silicone septum. If residual chlorine is in the water add about 25 mg of ascorbic acid to each vial before sample collection. Collect bubble free samples.

STABILITY: Cool to 4°C; HCl to pH <2. M.H.T. = 14 Days.

QUALITY CONTROL: As an initial demonstration of lab accuracy and precision, analyze 4 to 7 replicates of a lab fortified blank containing analyte at 0.1-5 μg/L. Collect all samples in duplicate.

REFERENCE: Method 502.1, Volatile Halogenated Organic Compounds in Water by Purge and Trap GC, EPA 600/4-88/039.

PRIMARY NAME: Bromodichloromethane Method 502.2

TITLE: VOCs in Water by Purge and Trap capillary column GC

MATRIX: Drinking water (finished or any treatment stage) and raw source water.

CAS #: 75-27-4

APPLICATION: Method covers 60 VOCs. An inert gas is bubbled through a 5 mL sample in a purging chamber. Purged sample components are trapped in tube of sorbent materials. When purging is done, sorbent tube is heated and backflushed w/helium to desorb trapped sample onto a capillary GC column.

INTERFERENCES: During analysis, major contaminant sources are volatile materials in the lab and impurities in purging gas and sorbent trap. With high and low level samples, there can be carry-over contamination. Beware of methylene chloride, it permeates through PTFE tubing.

INSTRUMENTATION: (Capillary Column) GC, w/PID and electrolytic conductivity detectors in series. Column #1: VOCOL wide bore column; Column #2: RTX-502.2 mega bore column; Column #3: DB-62 mega bore column.

RANGE: Approximately 0.02-200 μg/L

MDL: 0.02 μg/L (reagent C) (ELCD) 60 m × 0.75 mm VOCOL capillary column

PRECISION: RSD = 2.9% @ Spike of 10 μg/L (reagent C) (ELCD) 60 m × 0.75 mm VOCOL capillary column

ACCURACY: Recovery = 97% @ spike of 10 μg/L (reagent C) (ELCD) 60 m × 0.75 mm VOCOL capillary column

SAMPLING METHOD: Use a 40-120 mL screw cap vial (pre-washed with detergent, rinsed with distilled water and oven dried at 105°C) with a PTFE-faced silicone septum. If residual chlorine is in the water add about 25 mg of ascorbic acid to each vial before sample collection. Collect bubble free samples.

STABILITY: Cool to 4°C; HCl to pH <2. M.H.T. = 14 Days.

QUALITY CONTROL: As an initial demonstration of lab accuracy and precision, analyze 4 to 7 replicates of a lab fortified blank containing analyte at 0.1-5 μg/L. Collect all samples in duplicate.

REFERENCE: Method 502.2, Volatile Organic Compounds in Water by Purge and Trap Capillary Column GC, EPA 600/4-88/039.

PRIMARY NAME: Bromodichloromethane Method 524.1

TITLE: VOCs in Water by Purge and Trap GC/MS

MATRIX: Drinking water (finished or any treatment stage) and raw source water.

CAS #: 75-27-4

APPLICATION: Method covers 51 VOCs. An inert gas is bubbled through a 25 mL water sample. Purged sample components are trapped in tube of sorbent materials. When purging is complete, sorbent tube is heated and backflushed with helium to desorb trapped sample onto a packed GC column.

INTERFERENCES: During analysis, major contaminant sources are volatile materials in the lab and impurities in purging gas and sorbent trap. With high and low level samples, there can be carry-over contamination. Watch for methylene chloride, it permeates through PTFE tubing.

INSTRUMENTATION: Gas Chromatography/Mass Spectrometry/Data System; 1% SP-1000 on Carbopack B column.

RANGE: Approximately 0.2-200 μg/L MDL: 0.5 μg/L

PRECISION: RSD = 17% @ 1.0 μg/L

ACCURACY: Mean Accuracy = 100% @ 1.0 μg/L

SAMPLING METHOD: Use a 60-120 mL screw cap vial (pre-washed with detergent, rinsed with distilled water and oven dried at 105°C) with a PTFE-faced silicone septum. If residual chlorine is in the water add about 25 mg of ascorbic acid to each vial before sample collection. Collect bubble free samples.

STABILITY: Cool to 4°C; HCl to pH <2. M.H.T. = 14 Days.

QUALITY CONTROL: As an initial demonstration of lab accuracy and precision, analyze 4 to 7 replicates of a lab fortified blank containing analyte at 0.2-5 μg/L. Collect all samples in duplicate.

REFERENCE: Method 524.1, VOCs in Water by Purge and Trap Gas chromatography/mass spectrometry, EPA 600/4-88/039.

PRIMARY NAME: Bromodichloromethane Method 524.2

TITLE: VOCs in Water (Purge and Trap capillary column GC/MS

MATRIX: Drinking water (finished or any treatment stage) and raw source water.

CAS #: 75-27-4

APPLICATION: Method covers 60 VOCs. An inert gas is bubbled through a 25 mL water sample. Purged sample components are trapped in tube of sorbent materials. When purging is complete, sorbent tube is heated and backflushed with helium to desorb trapped sample onto a capillary GC column.

INTERFERENCES: During analysis, major contaminant sources are volatile materials in the lab and impurities in purging gas and sorbent trap. With high and low level samples, there can be carry-over contamination. Watch for methylene chloride, it permeates through PTFE tubing.

INSTRUMENTATION: (Capillary Column) Gas Chromatography/Mass Spectrometry/Data System. Column #1: VOCOL glass wide bore; Column #2: DB-624 fused silica; Column #3: DB-5 fused silica.

RANGE: Approximately 0.1-200 μg/L

MDL: 0.08 μg/L on wide bore capillary column.

PRECISION: RSD = 6.1% @ (0.1-10) μg/L (30 samples) wide bore capillary column.

ACCURACY: Recovery = 95% @ (0.1-10) μg/L (30 samples) wide bore capillary column.

SAMPLING METHOD: Use a 60-120 mL screw cap vial (pre-washed with detergent, rinsed with distilled water and oven dried at 105°C) with a PTFE-faced silicone septum. If residual chlorine is in the water add about 25 mg of ascorbic acid to each vial before sample collection. Collect bubble free samples.

STABILITY: Cool to 4°C; HCl to pH <2.

M.H.T. = 14 Days.

QUALITY CONTROL: As an initial demonstration of lab accuracy and precision, analyze 4 to 7 replicates of a lab fortified blank containing analyte at 0.2-5 μg/L. Collect all samples in duplicate.

REFERENCE: Method 524.2, VOCs in C by Purge and Trap Capillary column GC/MS, EPA 600/4-88/039.

PRIMARY NAME: Bromodichloromethane

Method 601

TITLE: Purgeable Halocarbons

MATRIX: Wastewater

CAS #: 75-27-4

APPLICATION: Method covers 29 purgeable halocarbons. (Method 624 provides GC/MS conditions appropriate for the qualitative and quantitative confirmation of results). Method describes conditions for a second GC column to confirm measurements made with primary column.

INTERFERENCES: Impurities in the purge gas and organic compounds outgassing from the plumbing ahead of the trap. With high and low level samples, there can be carry-over contamination. Diffusion of volatile organics through the septum seal into the sample.

INSTRUMENTATION: GC equipped with halide-specific detector. (With purge and trap unit).

RANGE: 8.0 To 500 μg/L. MDL: 0.10 μg/L.

PRECISION: 0.20X+1.00 μg/L (overall precision)

ACCURACY: 1.12C-1.02 μg/L (as recovery)

SAMPLING METHOD: 25 mL glass vial. Teflon lined septum.

STABILITY: Cool, 4°C, 0.008% sodium thiosulfate. M.H.T. = 14 Days.

QUALITY CONTROL: The laboratory must on an ongoing basis, spike at least 10% of the samples from each sample site being monitored to assess accuracy.

REFERENCE: Method 601, Federal Register Part VIII 40 CFR Part 136, Oct 26, 1984.

PRIMARY NAME: Bromodichloromethane Method 624

TITLE: Purgeables MATRIX: Wastewater

CAS #: 75-27-4

APPLICATION: Method covers 31 purgeable organics. An inert gas is bubbled through a 5 mL water sample in a specially designed purging chamber. Here, purgeables are transferred from aqueous to gaseous phase, passed onto a sorbent column and trapped. Trap is heated and backflushed with inert gas to desorb purgeables onto a GC column, where purgeables are separated.

INTERFERENCES: Impurities in the purge gas, organic compounds outgassing from the plumbing ahead of the trap, and solvent vapors in the laboratory. With high and low level samples, there can be carry-over contamination.

INSTRUMENTATION: GC/MS with purge and trap unit.

RANGE: 5-600 μg/L MDL: 2.2 μg/L

PRECISION: 0.20X+1.13 μg/L (overall precision)

ACCURACY: 1.03C-1.58 μg/L (as recovery)

SAMPLING METHOD: 25 mL glass vial. Teflon lined septum.

STABILITY: Cool, 4°C, 0.008% sodium thiosulfate. M.H.T. = 14 Days.

QUALITY CONTROL: The laboratory must on an ongoing basis, spike at least 5% of the samples from each sample site being monitored to assess accuracy.

REFERENCE: Method 624, Federal Register Part VIII 40 CFR Part 136, Oct 26, 1984.

PRIMARY NAME: Bromodichloromethane Method 8010

TITLE: Halogenated Volatile Organics

MATRIX: groundwater, soils, sludges water miscible liquid wastes, and non-water miscible wastes.

CAS #: 75-27-4

APPLICATION: This method is used for the analysis of 39 halogenated VOCs. Samples are analyzed using direct injection or purge and trap methods. Groundwater must be analyzed by the purge and trap method. The method provides an optional GC column which is used for analyte confirmation and that may help resolve analytes from interferences.

INTERFERENCES: There can be carry-over contamination with high and low level samples. Impurities may come from the purge and trap apparatus, organic compounds outgassing from the plumbing ahead of trap, diffusion of VOCs through the sample bottle septum during shipping or storage, or from solvent vapors in the lab.

INSTRUMENTATION: GC capable of on-column injections or purge and trap sample introduction and a halogen specific detector. Column 1: 8 foot by

0.1 inch 1% SP-1000 on Carbopack-B. Column 2: 6 foot by 0.1 inch bonded n-octane on Porasil-C.

RANGE: 8 to 500 μg/L (reagent water)

MDL: 0.10 μg/L (reagent water)

PRACTICAL QUANTITATION LIMIT FACTORS FOR MULTIPLYING TIMES MDL VALUE

Matrix	Multiplication Factor
Groundwater	10
Low-level soil	10
Water miscible liquid waste	500
High-level soil and sludge	1250
Non-water miscible waste	1250

PRECISION: 0.20X + 1.00 μg/L (overall precision)

ACCURACY: 1.12C - 1.02 μg/L (as recovery)

SAMPLING METHOD: For water and liquid samples; use glass 40 mL vials with Teflon lined septum caps and collect two vials per sample location with no headspace. For solids and concentrated waste samples; use widemouth glass bottles with Teflon liners.

STABILITY: For concentrated wastes, soils, sediments or sludges; cool to 4°C. For liquids; add 4 drops of concentrated hydrochloric acid and cool to 4°C. M.H.T. = 14 days.

QUALITY CONTROL: Analyze a reagent blank, matrix spike and matrix spike duplicate/duplicate for each analytical batch (up to 20 Samples). Demonstrate the purity of glassware and reagents by analyzing a reagent water method blank. Internal, surrogate and five concentration level calibration standards are used.

REFERENCE: Method 8010, SW-846, 3rd ed., Nov 1986.

PRIMARY NAME: Bromoform

Method 502.1

TITLE: Halogenated VOCs in Water

MATRIX: Drinking water (finished or any treatment stage) and raw source water.

CAS #: 75-25-2

APPLICATION: Method covers 40 halogenated VOCs. An inert gas is bubbled through a 5 mL water sample. Purged sample components are trapped in tube of sorbent materials. When purging is complete, sorbent tube is heated and backflushed with inert gas to desorb trapped sample onto a packed GC column where it is separated.

INTERFERENCES: During analysis, major contaminant sources are volatile materials in the lab and impurities in purging gas and sorbent trap. With high and low level samples, there can be carry-over contamination. Watch for methylene chloride, it permeates through PTFE tubing.

INSTRUMENTATION: Purge and Trap GC w/halide specific detector. (Two GC columns are recommended); Column #1: 1% SP-1000 on Carbopack B; Column #2: n-octane bonded on Poracil C.

RANGE: 8.0-505 μg/L (Drinking water) MDL: 0.05 μg/L in water

PRECISION: RSD = 15.0% @ 0.20 μg/L spike (S.L.) (17 Samples)

ACCURACY: Avg Recovery = 95% @ 0.20 μg/L spike (S.L.) (17 Samples)

SAMPLING METHOD: Use a 40-120 mL screw cap vial (pre-washed with detergent, rinsed with distilled water and oven dried at 105°C) with a PTFE-faced silicone septum. If residual chlorine is in the water add about 25 mg of ascorbic acid to each vial before sample collection. Collect bubble free samples.

STABILITY: Cool to 4°C; HCl to pH <2. M.H.T. = 14 Days.

QUALITY CONTROL: As an initial demonstration of lab accuracy and precision, analyze 4 to 7 replicates of a lab fortified blank containing analyte at 0.1-5 μg/L. Collect all samples in duplicate.

REFERENCE: Method 502.1, Volatile Halogenated Organic Compounds in Water by Purge and Trap GC, EPA 600/4-88/039.

PRIMARY NAME: Bromoform Method 502.2

TITLE: VOCs in Water by Purge and Trap capillary column GC

MATRIX: Drinking water (finished or any treatment stage) and raw source water.

CAS #: 75-25-2

APPLICATION: Method covers 60 VOCs. An inert gas is bubbled through a 5 mL sample in a purging chamber. Purged sample components are trapped in tube of sorbent materials. When purging is done, sorbent tube is heated and backflushed w/helium to desorb trapped sample onto a capillary GC column.

INTERFERENCES: During analysis, major contaminant sources are volatile materials in the lab and impurities in purging gas and sorbent trap. With high and low level samples, there can be carry-over contamination. Beware of methylene chloride, it permeates through PTFE tubing.

INSTRUMENTATION: (Capillary Column) GC, w/PID and electrolytic conductivity detectors in series. Column #1: VOCOL wide bore column; Column #2: RTX-502.2 mega bore column; Column #3: DB-62 mega bore column.

RANGE: Approximately 0.02-200 μg/L

MDL: 1.6 μg/L (reagent C) (ELCD) 60 m × 0.75 mm VOCOL capillary column

PRECISION: RSD = 5.2% @ Spike of 10 μg/L (reagent C) (ELCD) 60 m × 0.75 mm VOCOL capillary column

ACCURACY: Recovery = 106% @ spike of 10 μg/L (reagent C) (ELCD) 60 m × 0.75 mm VOCOL capillary column

SAMPLING METHOD: Use a 40-120 mL screw cap vial (pre-washed with detergent, rinsed with distilled water and oven dried at 105°C) with a PTFE-faced silicone septum. If residual chlorine is in the water add about 25 mg of ascorbic acid to each vial before sample collection. Collect bubble free samples.

STABILITY: Cool to 4°C; HCl to pH <2.

M.H.T. = 14 Days.

QUALITY CONTROL: As an initial demonstration of lab accuracy and precision, analyze 4 to 7 replicates of a lab fortified blank containing analyte at 0.1-5 μg/L. Collect all samples in duplicate.

REFERENCE: Method 502.2, Volatile Organic Compounds in Water by Purge and Trap Capillary Column GC, EPA 600/4-88/039.

PRIMARY NAME: Bromoform Method 524.1

TITLE: VOCs in Water by Purge and Trap GC/MS

MATRIX: Drinking water (finished or any treatment stage) and raw source water.

CAS #: 75-25-2

APPLICATION: Method covers 51 VOCs. An inert gas is bubbled through a 25 mL water sample. Purged sample components are trapped in tube of sorbent materials. When purging is complete, sorbent tube is heated and backflushed with helium to desorb trapped sample onto a packed GC column.

INTERFERENCES: During analysis, major contaminant sources are volatile materials in the lab and impurities in purging gas and sorbent trap. With high and low level samples, there can be carry-over contamination.Watch for methylene chloride, it permeates through PTFE tubing.

INSTRUMENTATION: Gas Chromatography/Mass Spectrometry/Data System; 1% SP-1000 on Carbopack B column.

RANGE: Approximately 0.2-200 µg/L MDL: 0.7 µg/L

PRECISION: RSD = 9.6% @ 2.5 µg/L

ACCURACY: Mean Accuracy = 100% @ 2.5 µg/L

SAMPLING METHOD: Use a 60-120 mL screw cap vial (pre-washed with detergent, rinsed with distilled water and oven dried at 105°C) with a PTFE-faced silicone septum. If residual chlorine is in the water add about 25 mg of ascorbic acid to each vial before sample collection. Collect bubble free samples.

STABILITY: Cool to 4°C; HCl to pH <2. M.H.T. = 14 Days.

QUALITY CONTROL: As an initial demonstration of lab accuracy and precision, analyze 4 to 7 replicates of a lab fortified blank containing analyte at 0.2-5 µg/L. Collect all samples in duplicate.

REFERENCE: Method 524.1, VOCs in Water by Purge and Trap Gas Chromatography/Mass Spectrometry, EPA 600/4-88/039.

PRIMARY NAME: Bromoform Method 524.2

TITLE: VOCs in Water (Purge and Trap) capillary column GC/MS

MATRIX: Drinking water (finished or any treatment stage) and raw source water.

CAS #: 75-25-2

APPLICATION: Method covers 60 VOCs. An inert gas is bubbled through a 25 mL water sample. Purged sample components are trapped in tube of sorbent materials. When purging is complete, sorbent tube is heated and backflushed with helium to desorb trapped sample onto a capillary GC column.

INTERFERENCES: During analysis, major contaminant sources are volatile materials in the lab and impurities in purging gas and sorbent trap. With high and low level samples, there can be carry-over contamination. Watch for methylene chloride, it permeates through PTFE tubing.

INSTRUMENTATION: (Capillary Column) Gas Chromatography/Mass Spectrometry/Data System. Column #1: VOCOL glass wide bore; Column #2: DB-624 fused silica; Column #3: DB-5 fused silica.

RANGE: Approximately 0.1-200 μg/L

MDL: 0.12 μg/L on wide bore capillary column.

PRECISION: RSD = 6.3% @ (0.5-10) μg/L (18 samples) wide bore capillary column.

ACCURACY: Recovery = 101% @ (0.5-10) μg/L (18 samples) wide bore capillary column.

SAMPLING METHOD: Use a 60-120 mL screw cap vial (pre-washed with detergent, rinsed with distilled water and oven dried at 105°C) with a PTFE-faced silicone septum. If residual chlorine is in the water add about 25 mg of ascorbic acid to each vial before sample collection. Collect bubble free samples.

STABILITY: Cool to 4°C; HCl to pH <2. M.H.T. = 14 Days.

QUALITY CONTROL: As an initial demonstration of lab accuracy and precision, analyze 4 to 7 replicates of a lab fortified blank containing analyte at 0.2-5 μg/L. Collect all samples in duplicate.

REFERENCE: Method 524.2, VOCs in C by Purge and Trap Capillary column GC/MS, EPA 600/4-88/039.

PRIMARY NAME: Bromoform Method 601

TITLE: Purgeable Halocarbons MATRIX: Wastewater

CAS #: 75-25-2

APPLICATION: Method covers 29 purgeable halocarbons. (Method 624 provides GC/MS conditions appropriate for the qualitative and quantitative confirmation of results). Method describes conditions for a second GC column to confirm measurements made with primary column.

INTERFERENCES: Impurities in the purge gas and organic compounds outgassing from the plumbing ahead of the trap. With high and low level samples, there can be carry-over contamination. Diffusion of volatile organics through the septum seal into the sample.

INSTRUMENTATION: GC equipped with halide-specific detector. (With purge and trap unit).

RANGE: 8.0 To 500 μg/L. MDL: 0.20 μg/L.

PRECISION: 0.21X+2.41 μg/L (overall precision)

ACCURACY: 0.96C-2.05 μg/L (as recovery)

SAMPLING METHOD: 25 mL glass vial. Teflon lined septum.

STABILITY: Cool, 4°C, 0.008% sodium thiosulfate. M.H.T. = 14 Days.

QUALITY CONTROL: The laboratory must on an ongoing basis, spike at

least 10% of the samples from each sample site being monitored to assess accuracy.

REFERENCE: Method 601, Federal Register Part VIII 40 CFR Part 136, Oct 26, 1984.

PRIMARY NAME: Bromoform Method 624

TITLE: Purgeables MATRIX: Wastewater

CAS #: 75-25-2

APPLICATION: Method covers 31 purgeable organics. An inert gas is bubbled through a 5 mL water sample in a specially designed purging chamber. Here, purgeables are transferred from aqueous to gaseous phase, passed onto a sorbent column and trapped. Trap is heated and backflushed with inert gas to desorb purgeables onto a GC column, where purgeables are separated.

INTERFERENCES: Impurities in the purge gas, organic compounds outgassing from the plumbing ahead of the trap, and solvent vapors in the laboratory. With high and low level samples, there can be carry-over contamination.

INSTRUMENTATION: GC/MS with purge and trap unit.

RANGE: 5-600 µg/L MDL: 4.7 µg/L

PRECISION: 0.17X+1.38 µg/L (overall precision)

ACCURACY: 1.18C-2.35 µg/L (as recovery)

SAMPLING METHOD: 25 mL glass vial. Teflon lined septum.

STABILITY: Cool, 4°C, 0.008% sodium thiosulfate. M.H.T. = 14 Days.

QUALITY CONTROL: The laboratory must on an ongoing basis, spike at least 5% of the samples from each sample site being monitored to assess accuracy.

REFERENCE: Method 624, Federal Register Part VIII 40 CFR Part 136, Oct 26, 1984.

PRIMARY NAME: Bromoform

Method 8010

TITLE: Halogenated Volatile Organics

MATRIX: groundwater, soils, sludges water miscible liquid wastes, and non-water miscible wastes.

CAS #: 75-25-2

APPLICATION: This method is used for the analysis of 39 halogenated VOCs. Samples are analyzed using direct injection or purge and trap methods. groundwater must be analyzed by the purge and trap method. The method provides an optional GC column which is used for analyte confirmation and that may help resolve analytes from interferences.

INTERFERENCES: There can be carry-over contamination with high and low level samples. Impurities may come from the purge and trap apparatus, organic compounds outgassing from the plumbing ahead of trap, diffusion of VOCs through the sample bottle septum during shipping or storage, or from solvent vapors in the lab.

INSTRUMENTATION: GC capable of on-column injections or purge and trap sample introduction and a halogen specific detector. Column 1: 8 foot by 0.1 inch 1% SP-1000 on Carbopack-B. Column 2: 6 foot by 0.1 inch bonded n-octane on Porasil-C.

RANGE: 8 to 500 μg/L (reagent water)

MDL: 0.20 μg/L (reagent water)

PRACTICAL QUANTITATION LIMIT FACTORS FOR MULTIPLYING TIMES MDL VALUE

Matrix	Multiplication Factor
Groundwater	10
Low-level soil	10
Water miscible liquid waste	500
High-level soil and sludge	1250
Non-water miscible waste	1250

PRECISION: 0.21X + 2.41 μg/L (overall precision)

ACCURACY: 0.96C - 2.05 μg/L (as recovery)

SAMPLING METHOD: For water and liquid samples; use glass 40 mL vials with Teflon lined septum caps and collect two vials per sample location with no headspace. For solids and concentrated waste samples; use widemouth glass bottles with Teflon liners.

STABILITY: For concentrated wastes, soils, sediments or sludges; cool to 4°C. For liquids; add 4 drops of concentrated hydrochloric acid and cool to 4°C. M.H.T. = 14 days.

QUALITY CONTROL: Analyze a reagent blank, matrix spike and matrix spike duplicate/duplicate for each analytical batch (up to 20 Samples). Demonstrate the purity of glassware and reagents by analyzing a reagent water method blank. Internal, surrogate and five concentration level calibration standards are used.

REFERENCE: Method 8010, SW-846, 3rd ed., Nov 1986.

PRIMARY NAME: Bromomethane

Method 502.1

TITLE: Halogenated VOCs in Water

MATRIX: Drinking water (finished or any treatment stage) and raw source water.

CAS #: 74-83-9

APPLICATION: Method covers 40 halogenated VOCs. An inert gas is bubbled through a 5 mL water sample. Purged sample components are trapped in tube of sorbent materials. When purging is complete, sorbent tube is heated and backflushed with inert gas to desorb trapped sample onto a packed GC column where it is separated.

INTERFERENCES: During analysis, major contaminant sources are volatile materials in the lab and impurities in purging gas and sorbent trap. With high and low level samples, there can be carry-over contamination. Watch for methylene chloride, it permeates through PTFE tubing.

INSTRUMENTATION: Purge and Trap GC w/halide specific detector. (Two GC columns are recommended); Column #1: 1% SP-1000 on Carbopack B; Column #2: n-octane bonded on Poracil C.

RANGE: 8.0-505 μg/L (Drinking water) MDL: 0.03 μg/L in water

PRECISION: not listed (single laboratory or single analyst)

ACCURACY: not listed (single laboratory or single analyst)

SAMPLING METHOD: Use a 40-120 mL screw cap vial (pre-washed with detergent, rinsed with distilled water and oven dried at 105°C) with a PTFE-faced silicone septum. If residual chlorine is in the water add about 25 mg of ascorbic acid to each vial before sample collection. Collect bubble free samples.

STABILITY: Cool to 4°C; HCl to pH <2. M.H.T. = 14 Days.

QUALITY CONTROL: As an initial demonstration of lab accuracy and precision, analyze 4 to 7 replicates of a lab fortified blank containing analyte at 0.1-5 μg/L. Collect all samples in duplicate.

REFERENCE: Method 502.1, Volatile Halogenated Organic Compounds in Water by Purge and Trap GC, EPA 600/4-88/039.

PRIMARY NAME: Bromomethane Method 502.2

TITLE: VOCs in Water by Purge and Trap capillary column GC

MATRIX: Drinking water (finished or any treatment stage) and raw source water.

CAS #: 74-83-9

APPLICATION: Method covers 60 VOCs. An inert gas is bubbled through a 5 mL sample in a purging chamber. Purged sample components are trapped in tube of sorbent materials. When purging is done, sorbent tube is heated and backflushed w/helium to desorb trapped sample onto a capillary GC column.

INTERFERENCES: During analysis, major contaminant sources are volatile materials in the lab and impurities in purging gas and sorbent trap. With high and low level samples, there can be carry-over contamination. Beware of methylene chloride, it permeates through PTFE tubing.

INSTRUMENTATION: (capillary column) GC, w/PID and electrolytic conductivity detectors in series. Column #1: VOCOL wide bore column; Column #2: RTX-502.2 mega bore column; Column #3: DB-62 mega bore column.

RANGE: Approximately 0.02-200 µg/L

MDL: 1.1 µg/L (reagent C) (ELCD) 60 m × 0.75 mm VOCOL capillary column

PRECISION: RSD = 3.8% @ Spike of 10 µg/L (reagent C) (ELCD) 60 m × 0.75 mm VOCOL capillary column

ACCURACY: Recovery = 97% @ spike of 10 µg/L (reagent C) (ELCD) 60 m × 0.75 mm VOCOL capillary column

SAMPLING METHOD: Use a 40-120 mL screw cap vial (pre-washed with detergent, rinsed with distilled water and oven dried at 105°C) with a PTFE-faced silicone septum. If residual chlorine is in the water add about 25 mg of ascorbic acid to each vial before sample collection. Collect bubble free samples.

STABILITY: Cool to 4°C; HCl to pH <2.

M.H.T. = 14 Days.

QUALITY CONTROL: As an initial demonstration of lab accuracy and precision, analyze 4 to 7 replicates of a lab fortified blank containing analyte at 0.1-5 µg/L. Collect all samples in duplicate.

REFERENCE: Method 502.2, Volatile Organic Compounds in Water by Purge and Trap Capillary Column GC, EPA 600/4-88/039.

PRIMARY NAME: Bromomethane

Method 524.1

TITLE: VOCs in Water by Purge and Trap GC/MS

MATRIX: Drinking water (finished or any treatment stage) and raw source water.

CAS #: 74-83-9

APPLICATION: Method covers 51 VOCs. An inert gas is bubbled through a 25 mL water sample. Purged sample components are trapped in tube of sorbent materials. When purging is complete, sorbent tube is heated and backflushed with helium to desorb trapped sample onto a packed GC column.

INTERFERENCES: During analysis, major contaminant sources are volatile materials in the lab and impurities in purging gas and sorbent trap. With high and low level samples, there can be carry-over contamination. Watch for methylene chloride, it permeates through PTFE tubing.

INSTRUMENTATION: Gas Chromatography/Mass Spectrometry/Data System; 1% SP-1000 on Carbopack B column.

RANGE: Approximately 0.2-200 μg/L MDL: not listed.

PRECISION: not listed.

ACCURACY: not listed.

SAMPLING METHOD: Use a 60-120 mL screw cap vial (pre-washed with detergent, rinsed with distilled water and oven dried at 105°C) with a PTFE-faced silicone septum. If residual chlorine is in the water add about 25 mg of ascorbic acid to each vial before sample collection. Collect bubble free samples.

STABILITY: Cool to 4°C; HCl to pH <2. M.H.T. = 14 Days.

QUALITY CONTROL: As an initial demonstration of lab accuracy and precision, analyze 4 to 7 replicates of a lab fortified blank containing analyte at 0.2-5 μg/L. Collect all samples in duplicate.

REFERENCE: Method 524.1, VOCs in Water by Purge and Trap Gas Chromatography/Mass Spectrometry, EPA 600/4-88/039.

PRIMARY NAME: Bromomethane Method 524.2

TITLE: VOCs in Water (Purge and Trap capillary column GC/MS

MATRIX: Drinking water (finished or any treatment stage) and raw source water.

CAS #: 74-83-9

APPLICATION: Method covers 60 VOCs. An inert gas is bubbled through a 25 mL water sample. Purged sample components are trapped in tube of sorbent materials. When purging is complete, sorbent tube is heated and backflushed with helium to desorb trapped sample onto a capillary GC column.

INTERFERENCES: During analysis, major contaminant sources are volatile materials in the lab and impurities in purging gas and sorbent trap. With high and low level samples, there can be carry-over contamination. Watch for methylene chloride, it permeates through PTFE tubing.

INSTRUMENTATION: (capillary column) Gas Chromatography/Mass Spectrometry/Data System. Column #1: VOCOL glass wide bore; Column #2: DB-624 fused silica; Column #3: DB-5 fused silica.

RANGE: Approximately 0.1-200 μg/L

MDL: 0.11 μg/L on wide bore capillary column.

PRECISION: RSD = 8.2% @ (0.5-10) μg/L (18 samples) wide bore capillary column.

ACCURACY: Recovery = 95% @ (0.5-10) μg/L (18 samples) wide bore capillary column.

SAMPLING METHOD: Use a 60-120 mL screw cap vial (pre-washed with detergent, rinsed with distilled water and oven dried at 105°C) with a PTFE-faced silicone septum. If residual chlorine is in the water add about 25 mg of ascorbic acid to each vial before sample collection. Collect bubble free samples.

STABILITY: Cool to 4°C; HCl to pH <2.

M.H.T. = 14 Days.

QUALITY CONTROL: As an initial demonstration of lab accuracy and precision, analyze 4 to 7 replicates of a lab fortified blank containing analyte at 0.2-5 μg/L. Collect all samples in duplicate.

REFERENCE: Method 524.2, VOCs in C by Purge and Trap Capillary column GC/MS, EPA 600/4-88/039.

PRIMARY NAME: Bromomethane

Method 601

TITLE: Purgeable Halocarbons

MATRIX: Wastewater

CAS #: 74-83-9

APPLICATION: Method covers 29 purgeable halocarbons. (Method 624 provides GC/MS conditions appropriate for the qualitative and quantitative confirmation of results). Method describes conditions for a second GC column to confirm measurements made with primary column.

INTERFERENCES: Impurities in the purge gas and organic compounds outgassing from the plumbing ahead of the trap. With high and low level samples, there can be carry-over contamination. Diffusion of volatile organics through the septum seal into the sample.

INSTRUMENTATION: GC equipped with halide-specific detector. (With purge and trap unit).

RANGE: 8.0 To 500 μg/L. MDL: 1.18 μg/L.

PRECISION: 0.36X+0.94 μg/L (overall precision)

ACCURACY: 0.76C-1.27 μg/L (as recovery)

SAMPLING METHOD: 25 mL glass vial. Teflon lined septum.

STABILITY: Cool, 4°C, 0.008% sodium thiosulfate. M.H.T. = 14 Days.

QUALITY CONTROL: The laboratory must on an ongoing basis, spike at least 10% of the samples from each sample site being monitored to assess accuracy.

REFERENCE: Method 601, Federal Register Part VIII 40 CFR Part 136, Oct 26, 1984.

PRIMARY NAME: Bromomethane Method 624

TITLE: Purgeables MATRIX: Wastewater

CAS #: 74-83-9

APPLICATION: Method covers 31 purgeable organics. An inert gas is bubbled through a 5 mL water sample in a specially designed purging chamber. Here, purgeables are transferred from aqueous to gaseous phase, passed onto a sorbent column and trapped. Trap is heated and backflushed

with inert gas to desorb purgeables onto a GC column, where purgeables are separated.

INTERFERENCES: Impurities in the purge gas, organic compounds outgassing from the plumbing ahead of the trap, and solvent vapors in the laboratory. With high and low level samples, there can be carry-over contamination.

INSTRUMENTATION: GC/MS with purge and trap unit.

RANGE: 5-600 μg/L MDL: not determined

PRECISION: 0.58X μg/L (overall precision)

ACCURACY: 1.00C μg/L (as recovery)

SAMPLING METHOD: 25 mL glass vial. Teflon lined septum.

STABILITY: Cool, 4°C, 0.008% sodium thiosulfate. M.H.T. = 14 Days.

QUALITY CONTROL: The laboratory must on an ongoing basis, spike at least 5% of the samples from each sample site being monitored to assess accuracy.

REFERENCE: Method 624, Federal Register Part VIII 40 CFR Part 136, Oct 26, 1984.

PRIMARY NAME: Bromomethane Method 8010

TITLE: Halogenated Volatile Organics

MATRIX: groundwater, soils, sludges water miscible liquid wastes, and non-water miscible wastes.

CAS #: 74-83-9

APPLICATION: This method is used for the analysis of 39 halogenated VOCs. Samples are analyzed using direct injection or purge and trap methods. Groundwater must be analyzed by the purge and trap method. The method provides an optional GC column which is used for analyte confirmation and that may help resolve analytes from interferences.

INTERFERENCES: There can be carry-over contamination with high and low level samples. Impurities may come from the purge and trap apparatus, organic compounds outgassing from the plumbing ahead of trap, diffusion of VOCs through the sample bottle septum during shipping or storage, or from solvent vapors in the lab.

INSTRUMENTATION: GC capable of on-column injections or purge and trap sample introduction and a halogen specific detector. Column 1: 8 foot by 0.1 inch 1% SP-1000 on Carbopack-B. Column 2: 6 foot by 0.1 inch bonded n-octane on Porasil-C.

RANGE: 8 to 500 μg/L (reagent water) MDL: Not determined

PRACTICAL QUANTITATION LIMIT FACTORS FOR MULTIPLYING TIMES MDL VALUE

Matrix	Multiplication Factor
Groundwater	10
Low-level soil	10
Water miscible liquid waste	500
High-level soil and sludge	1250
Non-water miscible waste	1250

PRECISION: 0.36X + 0.94 μg/L (overall precision)

ACCURACY: 0.76C - 1.27 μg/L (as recovery)

SAMPLING METHOD: For water and liquid samples; use glass 40 mL vials with Teflon lined septum caps and collect two vials per sample location with no headspace. For solids and concentrated waste samples; use widemouth glass bottles with Teflon liners.

STABILITY: For concentrated wastes, soils, sediments or sludges; cool to 4°C. For liquids; add 4 drops of concentrated hydrochloric acid and cool to 4°C. M.H.T. = 14 days.

QUALITY CONTROL: Analyze a reagent blank, matrix spike and matrix spike duplicate/duplicate for each analytical batch (up to 20 Samples). Demonstrate the purity of glassware and reagents by analyzing a reagent water method blank. Internal, surrogate and five concentration level calibration standards are used.

REFERENCE: Method 8010, SW-846, 3rd ed., Nov 1986.

PRIMARY NAME: Chlorobenzene Method 502.1

TITLE: Halogenated VOCs in Water

MATRIX: Drinking water (finished or any treatment stage) and raw source water.

CAS #: 108-90-7

APPLICATION: Method covers 40 halogenated VOCs. An inert gas is bubbled through a 5 mL water sample. Purged sample components are trapped in tube of sorbent materials. When purging is complete, sorbent tube is heated and backflushed with inert gas to desorb trapped sample onto a packed GC column where it is separated.

INTERFERENCES: During analysis, major contaminant sources are volatile materials in the lab and impurities in purging gas and sorbent trap. With high and low level samples, there can be carry-over contamination. Watch for methylene chloride, it permeates through PTFE tubing.

INSTRUMENTATION: Purge and Trap GC w/halide specific detector. (Two GC columns are recommended); Column #1: 1% SP-1000 on Carbopack B; Column #2: n-octane bonded on Poracil C.

RANGE: 8.0-505 mg/L (Drinking water) MDL: 0.005 mg/L in water

PRECISION: RSD = 9.3% @ 0.40 mg/L spike (S.L.) (18 Samples)

ACCURACY: Avg Recovery = 88% @ 0.40 mg/L spike (S.L.) (18 Samples)

SAMPLING METHOD: Use a 40-120 mL screw cap vial (pre-washed with detergent, rinsed with distilled water and oven dried at 105°C) with a PTFE-faced silicone septum. If residual chlorine is in the water add about 25 mg of ascorbic acid to each vial before sample collection. Collect bubble free samples.

STABILITY: Cool to 4°C; HCl to pH <2. M.H.T. = 14 Days.

QUALITY CONTROL: As an initial demonstration of lab accuracy and precision, analyze 4 to 7 replicates of a lab fortified blank containing analyte at 0.1-5 mg/L. Collect all samples in duplicate.

REFERENCE: Method 502.1, Volatile Halogenated Organic Compounds in Water by Purge and Trap GC, EPA 600/4-88/039.

PRIMARY NAME: Chlorobenzene Method 502.2

TITLE: VOCs in Water by Purge and Trap capillary column GC

MATRIX: Drinking water (finished or any treatment stage) and raw source water.

CAS #: 108-90-7

APPLICATION: Method covers 60 VOCs. An inert gas is bubbled through a 5 mL sample in a purging chamber. Purged sample components are trapped in tube of sorbent materials. When purging is done, sorbent tube is heated and backflushed w/helium to desorb trapped sample onto a capillary GC column.

INTERFERENCES: During analysis, major contaminant sources are volatile materials in the lab and impurities in purging gas and sorbent trap. With high and low level samples, there can be carry-over contamination. Beware of methylene chloride, it permeates through PTFE tubing.

INSTRUMENTATION: (Capillary Column) GC, w/PID and electrolytic conductivity detectors in series. Column #1: VOCOL wide bore column; Column #2: RTX-502.2 mega bore column; Column #3: DB-62 mega bore column.

RANGE: Approximately 0.02-200 μg/L

MDL: 0.01 μg/L (reagent C) (ELCD) 60 m × 0.75 mm VOCOL capillary column

PRECISION: RSD = 3.6% @ Spike of 10 μg/L (reagent C) (ELCD) 60 m × 0.75 mm VOCOL capillary column

ACCURACY: Recovery = 103% @ spike of 10 μg/L (reagent C) (ELCD) 60 m × 0.75 mm VOCOL capillary column

SAMPLING METHOD: Use a 40-120 mL screw cap vial (pre-washed with detergent, rinsed with distilled water and oven dried at 105°C) with a PTFE-faced silicone septum. If residual chlorine is in the water add about 25 mg of ascorbic acid to each vial before sample collection. Collect bubble free samples.

STABILITY: Cool to 4°C; HCl to pH <2. M.H.T. = 14 Days.

QUALITY CONTROL: As an initial demonstration of lab accuracy and

precision, analyze 4 to 7 replicates of a lab fortified blank containing analyte at 0.1-5 μg/L. Collect all samples in duplicate.

REFERENCE: Method 502.2, Volatile Organic Compounds in Water by Purge and Trap Capillary Column GC, EPA 600/4-88/039.

PRIMARY NAME: Chlorobenzene Method 503.1

TITLE: Aromatic and Unsaturated VOCs in Water

MATRIX: Drinking water (finished or any treatment stage) and raw source water.

CAS #: 108-90-7

APPLICATION: Method covers 28 aromatic and unsaturated VOCs. An inert gas is bubbled through a 5 mL water sample. Purged sample components are trapped in tube of sorbent materials. When purging is complete, sorbent tube is heated and backflushed with inert gas to desorb trapped sample onto a packed GC column.

INTERFERENCES: During analysis, major contaminant sources are volatile materials in the lab and impurities in purging gas and sorbent trap. With high and low level samples, there can be carry-over contamination. Excess water causes a negative baseline deflection.

INSTRUMENTATION: Purge and Trap GC w/photoionization detector. (Two GC columns are recommended); Column #1: 5% SP-1200 and 1.75% Bentone 34 on Supelcoport; 5% 1,2,3-tris(2-cyanoethoxy)propane on Chromosorb W.

RANGE: 2.2-600 μg/L (Drinking water) MDL: 0.004 μg/L in water

PRECISION: RSD = 12.1% @ 27.6 μg/L; 7 labs

ACCURACY: Avg Recovery = 98% @ 27.6 μg/L; 7 labs

SAMPLING METHOD: Use a 40-120 mL screw cap vial (pre-washed with detergent, rinsed with distilled water and oven dried at 105°C) with a PTFE-faced silicone septum. If residual chlorine is in the water add about 25 mg of ascorbic acid to each vial before sample collection. Collect bubble free samples.

STABILITY: Cool to 4°C; HCl to pH <2. M.H.T. = 14 Days.

QUALITY CONTROL: As an initial demonstration of lab accuracy and precision, analyze 4 to 7 replicates of a lab fortified blank containing analyte at 0.1-5 μg/L. Collect all samples in duplicate.

REFERENCE: Method 503.1, Volatile aromatic and unsaturated organic compounds in C by Purge and Trap GC, EPA 600/4-88/039.

PRIMARY NAME: Chlorobenzene Method 524.1

TITLE: VOCs in Water by Purge and Trap GC/MS

MATRIX: Drinking water (finished or any treatment stage) and raw source water.

CAS #: 108-90-7

APPLICATION: Method covers 51 VOCs. An inert gas is bubbled through a 25 mL water sample. Purged sample components are trapped in tube of sorbent materials. When purging is complete, sorbent tube is heated and backflushed with helium to desorb trapped sample onto a packed GC column.

INTERFERENCES: During analysis, major contaminant sources are volatile materials in the lab and impurities in purging gas and sorbent trap. With high and low level samples, there can be carry-over contamination. Watch for methylene chloride, it permeates through PTFE tubing.

INSTRUMENTATION: Gas Chromatography/Mass Spectrometry/Data System; 1% SP-1000 on Carbopack B column.

RANGE: Approximately 0.2-200 μg/L MDL: 0.1 μg/L

PRECISION: RSD = 4.6% @ 1.0 μg/L

ACCURACY: Mean Accuracy = 102% @ 1.0 μg/L

SAMPLING METHOD: Use a 60-120 mL screw cap vial (pre-washed with detergent, rinsed with distilled water and oven dried at 105°C) with a PTFE-faced silicone septum. If residual chlorine is in the water add about 25 mg of ascorbic acid to each vial before sample collection. Collect bubble free samples.

STABILITY: Cool to 4°C; HCl to pH <2. M.H.T. = 14 Days.

QUALITY CONTROL: As an initial demonstration of lab accuracy and precision, analyze 4 to 7 replicates of a lab fortified blank containing analyte at 0.2-5 μg/L. Collect all samples in duplicate.

REFERENCE: Method 524.1, VOCs in Water by Purge and Trap Gas Chromatography/Mass Spectrometry, EPA 600/4-88/039.

PRIMARY NAME: Chlorobenzene Method 524.2

TITLE: VOCs in Water Purge and Trap capillary column GC/MS

MATRIX: Drinking water (finished or any treatment stage) and raw source water.

CAS #: 108-90-7

APPLICATION: Method covers 60 VOCs. An inert gas is bubbled through a 25 mL water sample. Purged sample components are trapped in tube of sorbent materials. When purging is complete, sorbent tube is heated and backflushed with helium to desorb trapped sample onto a capillary GC column.

INTERFERENCES: During analysis, major contaminant sources are volatile materials in the lab and impurities in purging gas and sorbent trap. With high and low level samples, there can be carry-over contamination. Watch for methylene chloride, it permeates through PTFE tubing.

INSTRUMENTATION: (Capillary Column) Gas Chromatography/Mass Spectrometry/Data System. Column #1: VOCOL glass wide bore; Column #2: DB-624 fused silica; Column #3: DB-5 fused silica.

RANGE: Approximately 0.1-200 μg/L

MDL: 0.04 μg/L on wide bore capillary column.

PRECISION: RSD = 5.9% @ (0.1-10) μg/L (31 samples) wide bore capillary column.

ACCURACY: Recovery = 98% @ (0.1-10) μg/L (31 samples) wide bore capillary column.

SAMPLING METHOD: Use a 60-120 mL screw cap vial (pre-washed with detergent, rinsed with distilled water and oven dried at 105°C) with a PTFE-faced silicone septum. If residual chlorine is in the water add about 25 mg of ascorbic acid to each vial before sample collection. Collect bubble free samples.

STABILITY: Cool to 4°C; HCl to pH <2. M.H.T. = 14 Days.

QUALITY CONTROL: As an initial demonstration of lab accuracy and precision, analyze 4 to 7 replicates of a lab fortified blank containing analyte at 0.2-5 μg/L. Collect all samples in duplicate.

REFERENCE: Method 524.2, VOCs in C by Purge and Trap Capillary column GC/MS, EPA 600/4-88/039.

PRIMARY NAME: Chlorobenzene Method 601

TITLE: Purgeable Halocarbons MATRIX: Wastewater

CAS #: 108-90-7

APPLICATION: Method covers 29 purgeable halocarbons. (Method 624 provides GC/MS conditions appropriate for the qualitative and quantitative confirmation of results). Method describes conditions for a second GC column to confirm measurements made with primary column.

INTERFERENCES: Impurities in the purge gas and organic compounds outgassing from the plumbing ahead of the trap. With high and low level samples, there can be carry-over contamination. Diffusion of volatile organics through the septum seal into the sample.

INSTRUMENTATION: GC equipped with halide-specific detector. (With purge and trap unit).

RANGE: 8.0 To 500 μg/L. MDL: 0.25 μg/L.

PRECISION: 0.18X+1.21 μg/L (overall precision)

ACCURACY: 1.00C-1.23 μg/L (as recovery)

SAMPLING METHOD: 25 mL glass vial. Teflon lined septum.

STABILITY: Cool, 4°C, 0.008% sodium thiosulfate. M.H.T. = 14 Days.

QUALITY CONTROL: The laboratory must on an ongoing basis, spike at least 10% of the samples from each sample site being monitored to assess accuracy.

REFERENCE: Method 601, Federal Register Part VIII 40 CFR Part 136, Oct 26, 1984.

PRIMARY NAME: Chlorobenzene Method 602

TITLE: Purgeable Aromatics MATRIX: Wastewater

CAS #: 108-90-7

APPLICATION: Method covers 7 purgeable aromatics. (Method 624 provides GC/MS conditions appropriate for the qualitative and quantitative confirmation of results). Method describes conditions for a second GC column to confirm measurements made with primary column.

INTERFERENCES: Impurities in the purge gas and organic compounds outgassing from the plumbing ahead of the trap. With high and low level samples, there can be carry-over contamination. Diffusion of volatile organics through the septum seal into the sample.

INSTRUMENTATION: GC equipped with photoionization detector. (With purge and trap unit)

RANGE: 2.1 To 550 µg/L. MDL: 0.2 µg/L.

PRECISION: 0.17X+0.10 µg/L (overall precision)

ACCURACY: 0.95C+0.02 µg/L (as recovery)

SAMPLING METHOD: 25 mL glass vial. Teflon lined septum.

STABILITY: Cool, 4°C, 0.008% sodium thiosulfate. M.H.T. = 14 Days.

QUALITY CONTROL: The laboratory must on an ongoing basis, spike at

least 10% of the samples from each sample site being monitored to assess accuracy.

REFERENCE: Method 602, Federal Register Part VIII 40 CFR Part 136, Oct 26, 1984.

PRIMARY NAME: Chlorobenzene Method 624

TITLE: Purgeables MATRIX: Wastewater

CAS #: 108-90-7

APPLICATION: Method covers 31 purgeable organics. An inert gas is bubbled through a 5 mL water sample in a specially designed purging chamber. Here, purgeables are transferred from aqueous to gaseous phase, passed onto a sorbent column and trapped. Trap is heated and backflushed with inert gas to desorb purgeables onto a GC column, where purgeables are separated.

INTERFERENCES: Impurities in the purge gas, organic compounds out-gassing from the plumbing ahead of the trap, and solvent vapors in the laboratory. With high and low level samples, there can be carry-over contamination.

INSTRUMENTATION: GC/MS with purge and trap unit.

RANGE: 5-600 μg/L MDL: 6.0 μg/L

PRECISION: 0.26X-1.92 μg/L (overall precision)

ACCURACY: 0.98C+2.28 μg/L (as recovery)

SAMPLING METHOD: 25 mL glass vial. Teflon lined septum.

STABILITY: Cool, 4°C, 0.008% sodium thiosulfate. M.H.T. = 14 Days.

QUALITY CONTROL: The laboratory must on an ongoing basis, spike at least 5% of the samples from each sample site being monitored to assess accuracy.

REFERENCE: Method 624, Federal Register Part VIII 40 CFR Part 136, Oct 26, 1984.

PRIMARY NAME: Chlorobenzene

Method 8010

TITLE: Halogenated Volatile Organics

MATRIX: groundwater, soils, sludges water miscible liquid wastes, and non-water miscible wastes.

CAS #: 108-90-7

APPLICATION: This method is used for the analysis of 39 halogenated VOCs. Samples are analyzed using direct injection or purge and trap methods. Groundwater must be analyzed by the purge and trap method. The method provides an optional GC column which is used for analyte confirmation and that may help resolve analytes from interferences.

INTERFERENCES: There can be carry-over contamination with high and low level samples. Impurities may come from the purge and trap apparatus, organic compounds outgassing from the plumbing ahead of trap, diffusion of VOCs through the sample bottle septum during shipping or storage, or from solvent vapors in the lab.

INSTRUMENTATION: GC capable of on-column injections or purge and trap sample introduction and a halogen specific detector. Column 1: 8 foot by 0.1 inch 1% SP-1000 on Carbopack-B. Column 2: 6 foot by 0.1 inch bonded n-octane on Porasil-C.

RANGE: 8 to 500 µg/L (reagent water)

MDL: 0.25 µg/L (reagent water)

PRACTICAL QUANTITATION LIMIT FACTORS FOR MULTIPLYING TIMES MDL VALUE

Matrix	Multiplication Factor
Groundwater	10
Low-level soil	10
Water miscible liquid waste	500
High-level soil and sludge	1250
Non-water miscible waste	1250

PRECISION: 0.18X + 1.21 µg/L (overall precision)

ACCURACY: 1.00C - 1.23 µg/L (as recovery)

SAMPLING METHOD: For water and liquid samples; use glass 40 mL vials with Teflon lined septum caps and collect two vials per sample location with no headspace. For solids and concentrated waste samples; use widemouth glass bottles with Teflon liners.

STABILITY: For concentrated wastes, soils, sediments or sludges; cool to 4°C. For liquids; add 4 drops of concentrated hydrochloric acid and cool to 4°C. M.H.T. = 14 days.

QUALITY CONTROL: Analyze a reagent blank, matrix spike and matrix spike duplicate/duplicate for each analytical batch (up to 20 Samples). Demonstrate the purity of glassware and reagents by analyzing a reagent water method blank. Internal, surrogate and five concentration level calibration standards are used.

REFERENCE: Method 8010, SW-846, 3rd ed., Nov 1986.

PRIMARY NAME: Chlorobenzene

Method 8020

TITLE: Aromatic Volatile Organics

MATRIX: groundwater, soils, sludges water miscible liquid wastes, and non-water miscible wastes.

CAS #: 108-90-7

APPLICATION: This method is used to analyze for 8 aromatic VOCs. Samples are analyzed using direct injection or purge and trap methods. Groundwater must be analyzed by the purge and trap method. The method provides an optional GC column that is used for analyte confirmation and may also help resolve analytes from interferences.

INTERFERENCES: There can be carry-over contamination with high and low level samples. Impurities may come from the purge and trap apparatus, organic compounds outgassing from the plumbing ahead of trap, diffusion of VOCs through the sample bottle septum during shipping or storage, or from solvent vapors in the lab.

INSTRUMENTATION: GC capable of on-column injections or purge-and-trap sample introduction and a photoionization detector (PID). Column 1: 6 foot by 0.082 inch with 5% SP-1200 and 1.75% Bentone-34 on Supelcoport.

Column 2: 8 foot by 0.1 inch with 5% 1,2,3-tris(2-cyanoethoxy)propane on Chromosorb W-AW.

RANGE: 2.1 to 500 μg/L MDL: 0.2 μg/L (reagent water)

PRACTICAL QUANTITATION LIMIT FACTORS FOR MULTIPLYING TIMES MDL VALUE

Matrix	Multiplication Factor
Groundwater	10
Low-level soil	10
Water miscible liquid waste	500
High-level soil and sludge	1250
Non-water miscible waste	1250

PRECISION: 0.17X + 0.10 μg/L (overall precision)

ACCURACY: 0.95C + 0.02 μg/L (as recovery)

SAMPLING METHOD: For water and liquid samples use glass 40 mL vials with Teflon lined septum caps and collect two vials per sample location with no headspace. For solids and concentrated waste samples use widemouth glass bottles with Teflon liners. Cool all samples to 4°C

STABILITY: For concentrated wastes, soils, sediments or sludges cool to 4°C. For liquids, add 4 drops of concentrated hydrochloric acid and cool to 4°C. M.H.T. = 14 days.

QUALITY CONTROL: Analyze a reagent blank, matrix spike and matrix spike duplicate/duplicate for each analytical batch (up to 20 Samples). Demonstrate the purity of glassware and reagents by analyzing a reagent water method blank. Internal, surrogate and five concentration level calibration standards are used. The QC check sample concentrate should contain this compound at 10 μg/mL in methanol.

REFERENCE: Method 8020, SW-846, 3rd ed., Nov 1986.

PRIMARY NAME: 2-Chlorotoluene Method 502.1

TITLE: Halogenated VOCs in Water MATRIX: Drinking water (finished or any treatment stage) and raw source water.

CAS #: 95-49-8

APPLICATION: Method covers 40 halogenated VOCs. An inert gas is bubbled through a 5 mL water sample. Purged sample components are trapped in tube of sorbent materials. When purging is complete, sorbent tube is heated and backflushed with inert gas to desorb trapped sample onto a packed GC column where it is separated.

INTERFERENCES: During analysis, major contaminant sources are volatile materials in the lab and impurities in purging gas and sorbent trap. With high and low level samples, there can be carry-over contamination. Watch for methylene chloride, it permeates through PTFE tubing.

INSTRUMENTATION: Purge and Trap GC w/halide specific detector. (Two GC columns are recommended); Column #1: 1% SP-1000 on Carbopack B; Column #2: n-octane bonded on Poracil C.

RANGE: 8.0-505 μg/L (Drinking water) MDL: not determined.

PRECISION: RSD = 9.3% @ 0.40 μg/L spike (S.L.) (20 Samples)

ACCURACY: Avg Recovery = 85% @ 0.40 μg/L spike (S.L.) (20 Samples)

SAMPLING METHOD: Use a 40-120 mL screw cap vial (pre-washed with detergent, rinsed with distilled water and oven dried at 105°C) with a PTFE-faced silicone septum. If residual chlorine is in the water add about 25 mg of ascorbic acid to each vial before sample collection. Collect bubble free samples.

STABILITY: Cool to 4°C; HCl to pH <2. M.H.T. = 14 Days.

QUALITY CONTROL: As an initial demonstration of lab accuracy and precision, analyze 4 to 7 replicates of a lab fortified blank containing analyte at 0.1-5 μg/L. Collect all samples in duplicate.

REFERENCE: Method 502.1, Volatile Halogenated Organic Compounds in Water by Purge and Trap GC, EPA 600/4-88/039.

PRIMARY NAME: 2-Chlorotoluene Method 502.2

TITLE: VOCs in Water by Purge and Trap capillary column GC

MATRIX: Drinking water (finished or any treatment stage) and raw source water.

CAS #: 95-49-8

APPLICATION: Method covers 60 VOCs. An inert gas is bubbled through a 5 mL sample in a purging chamber. Purged sample components are trapped in tube of sorbent materials. When purging is done,sorbent tube is heated and backflushed w/helium to desorb trapped sample onto a capillary GC column.

INTERFERENCES: During analysis, major contaminant sources are volatile materials in the lab and impurities in purging gas and sorbent trap. With high and low level samples, there can be carry-over contamination. Beware of methylene chloride, it permeates through PTFE tubing.

INSTRUMENTATION: (capillary column) GC, w/PID and electrolytic conductivity detectors in series. Column #1: VOCOL wide bore column; Column #2: RTX-502.2 mega bore column; Column #3: DB-62 mega bore column.

RANGE: Approximately 0.02-200 µg/L

MDL: 0.01 µg/L (reagent C) (ELCD) 60 m × 0.75 mm VOCOL capillary column

PRECISION: RSD = 2.7% @ Spike of 10 µg/L (reagent C) (ELCD) 60 m × 0.75 mm VOCOL capillary column

ACCURACY: Recovery = 97% @ spike of 10 µg/L (reagent C) (ELCD) 60 m × 0.75 mm VOCOL capillary column

SAMPLING METHOD: Use a 40-120 mL screw cap vial (pre-washed with detergent, rinsed with distilled water and oven dried at 105°C) with a PTFE-faced silicone septum. If residual chlorine is in the water add about 25 mg of ascorbic acid to each vial before sample collection. Collect bubble free samples.

STABILITY: Cool to 4°C; HCl to pH <2.

M.H.T. = 14 Days.

QUALITY CONTROL: As an initial demonstration of lab accuracy and precision, analyze 4 to 7 replicates of a lab fortified blank containing analyte at 0.1-5 µg/L. Collect all samples in duplicate.

REFERENCE: Method 502.2, Volatile Organic Compounds in Water by Purge and Trap Capillary Column GC, EPA 600/4-88/039.

PRIMARY NAME: 2-Chlorotoluene Method 503.1

TITLE: Aromatic and Unsaturated VOCs in Water

MATRIX: Drinking water (finished or any treatment stage) and raw source water.

CAS #: 95-49-8

APPLICATION: Method covers 28 aromatic and unsaturated VOCs. An inert gas is bubbled through a 5 mL water sample. Purged sample components are trapped in tube of sorbent materials. When purging is complete, sorbent tube is heated and backflushed with inert gas to desorb trapped sample onto a packed GC column.

INTERFERENCES: During analysis, major contaminant sources are volatile materials in the lab and impurities in purging gas and sorbent trap. With high and low level samples, there can be carry-over contamination. Excess water causes a negative baseline deflection.

INSTRUMENTATION: Purge and Trap GC w/photoionization detector. (Two GC columns are recommended); Column #1: 5% SP-1200 and 1.75% Bentone 34 on Supelcoport; 5% 1,2,3-tris(2-cyanoethoxy)propane on Chromosorb W.

RANGE: 2.2-600 μg/L (Drinking water) MDL: 0.008 μg/L in water

PRECISION: not listed.

ACCURACY: not listed.

SAMPLING METHOD: Use a 40-120 mL screw cap vial (pre-washed with detergent, rinsed with distilled water and oven dried at 105°C) with a PTFE-faced silicone septum. If residual chlorine is in the water add about 25 mg of ascorbic acid to each vial before sample collection. Collect bubble free samples.

STABILITY: Cool to 4°C; HCl to pH <2. M.H.T. = 14 Days.

QUALITY CONTROL: As an initial demonstration of lab accuracy and precision, analyze 4 to 7 replicates of a lab fortified blank containing analyte at 0.1-5 μg/L. Collect all samples in duplicate.

REFERENCE: Method 503.1, Volatile Aromatic and Unsaturated Organic Compounds in C by Purge and Trap GC, EPA 600/4-88/039.

PRIMARY NAME: 2-Chlorotoluene

Method 524.1

TITLE: VOCs in Water by Purge and Trap GC/MS

MATRIX: Drinking water (finished or any treatment stage) and raw source water.

CAS #: 95-49-8

APPLICATION: Method covers 51 VOCs. An inert gas is bubbled through a 25 mL water sample. Purged sample components are trapped in tube of sorbent materials. When purging is complete, sorbent tube is heated and backflushed with helium to desorb trapped sample onto a packed GC column.

INTERFERENCES: During analysis, major contaminant sources are volatile materials in the lab and impurities in purging gas and sorbent trap. With high and low level samples, there can be carry-over contamination. Watch for methylene chloride, it permeates through PTFE tubing.

INSTRUMENTATION: Gas Chromatography/Mass Spectrometry/Data System; 1% SP-1000 on Carbopack B column.

RANGE: Approximately 0.2-200 µg/L

MDL: not determined.

PRECISION: not listed.

ACCURACY: not listed.

SAMPLING METHOD: Use a 60-120 mL screw cap vial (pre-washed with detergent, rinsed with distilled water and oven dried at 105°C) with a PTFE-faced silicone septum. If residual chlorine is in the water add about 25 mg of ascorbic acid to each vial before sample collection. Collect bubble free samples.

STABILITY: Cool to 4°C; HCl to pH <2.

M.H.T. = 14 Days.

QUALITY CONTROL: As an initial demonstration of lab accuracy and

precision, analyze 4 to 7 replicates of a lab fortified blank containing analyte at 0.2-5 µg/L. Collect all samples in duplicate.

REFERENCE: Method 524.1, VOCs in Water by Purge and Trap Gas Chromatography/Mass Spectrometry, EPA 600/4-88/039.

PRIMARY NAME: 2-Chlorotoluene

Method 524.2

TITLE: VOCs in Water Purge and Trap capillary column GC/MS

MATRIX: Drinking water (finished or any treatment stage) and raw source water.

CAS #: 95-49-8

APPLICATION: Method covers 60 VOCs. An inert gas is bubbled through a 25 mL water sample. Purged sample components are trapped in tube of sorbent materials. When purging is complete, sorbent tube is heated and backflushed with helium to desorb trapped sample onto a capillary GC column.

INTERFERENCES: During analysis, major contaminant sources are volatile materials in the lab and impurities in purging gas and sorbent trap. With high and low level samples, there can be carry-over contamination. Watch for methylene chloride, it permeates through PTFE tubing.

INSTRUMENTATION: (Capillary Column) Gas Chromatography/Mass Spectrometry/Data System. Column #1: VOCOL glass wide bore; Column #2: DB-624 fused silica; Column #3: DB-5 fused silica.

RANGE: Approximately 0.1-200 µg/L

MDL: 0.04 µg/L on wide bore capillary column.

PRECISION: RSD = 6.2% @ (0.1-10) µg/L (31 samples) wide bore capillary column.

ACCURACY: Recovery = 90% @ (0.1-10) µg/L (31 samples) wide bore capillary column.

SAMPLING METHOD: Use a 60-120 mL screw cap vial (pre-washed with detergent, rinsed with distilled water and oven dried at 105°C) with a PTFE-

faced silicone septum. If residual chlorine is in the water add about 25 mg of ascorbic acid to each vial before sample collection. Collect bubble free samples.

STABILITY: Cool to 4°C; HCl to pH <2. M.H.T. = 14 Days.

QUALITY CONTROL: As an initial demonstration of lab accuracy and precision, analyze 4 to 7 replicates of a lab fortified blank containing analyte at 0.2-5 µg/L. Collect all samples in duplicate.

REFERENCE: Method 524.2, VOCs in C by Purge and Trap Capillary column GC/MS, EPA 600/4-88/039.

PRIMARY NAME: 4-Chlorotoluene Method 502.1

TITLE: Halogenated VOCs in Water

MATRIX: Drinking water (finished or any treatment stage) and raw source water.

CAS #: 106-43-4

APPLICATION: Method covers 40 halogenated VOCs. An inert gas is bubbled through a 5 mL water sample. Purged sample components are trapped in tube of sorbent materials. When purging is complete, sorbent tube is heated and backflushed with inert gas to desorb trapped sample onto a packed GC column where it is separated.

INTERFERENCES: During analysis, major contaminant sources are volatile materials in the lab and impurities in purging gas and sorbent trap. With high and low level samples, there can be carry-over contamination. Watch for methylene chloride, it permeates through PTFE tubing.

INSTRUMENTATION: Purge and Trap GC w/halide specific detector. (Two GC columns are recommended); Column #1: 1% SP-1000 on Carbopack B; Column #2: n-octane bonded on Poracil C.

RANGE: 8.0-505 µg/L (Drinking water) MDL: not listed.

PRECISION: not listed (single laboratory or single analyst)

ACCURACY: not listed (single laboratory or single analyst)

SAMPLING METHOD: Use a 40-120 mL screw cap vial (pre-washed with detergent, rinsed with distilled water and oven dried at 105°C) with a PTFE-faced silicone septum. If residual chlorine is in the water add about 25 mg of ascorbic acid to each vial before sample collection. Collect bubble free samples.

STABILITY: Cool to 4°C; HCl to pH <2. M.H.T. = 14 Days.

QUALITY CONTROL: As an initial demonstration of lab accuracy and precision, analyze 4 to 7 replicates of a lab fortified blank containing analyte at 0.1-5 μg/L. Collect all samples in duplicate.

REFERENCE: Method 502.1, Volatile Halogenated Organic Compounds in Water by Purge and Trap GC, EPA 600/4-88/039.

PRIMARY NAME: 4-Chlorotoluene Method 502.2

TITLE: VOCs in Water by Purge and Trap capillary column GC

MATRIX: Drinking water (finished or any treatment stage) and raw source water.

CAS #: 106-43-4

APPLICATION: Method covers 60 VOCs. An inert gas is bubbled through a 5 mL sample in a purging chamber. Purged sample components are trapped in tube of sorbent materials. When purging is done, sorbent tube is heated and backflushed w/helium to desorb trapped sample onto a capillary GC column.

INTERFERENCES: During analysis, major contaminant sources are volatile materials in the lab and impurities in purging gas and sorbent trap. With high and low level samples, there can be carry-over contamination. Beware of methylene chloride, it permeates through PTFE tubing.

INSTRUMENTATION: (capillary column) GC, w/PID and electrolytic conductivity detectors in series. Column #1: VOCOL wide bore column; Column #2: RTX-502.2 mega bore column; Column #3: DB-62 mega bore column.

RANGE: Approximately 0.02-200 μg/L

MDL: 0.01 μg/L (reagent C) (ELCD) 60 m × 0.75 mm VOCOL capillary column

PRECISION: RSD = 3.2% @ spike of 10 μg/L (reagent C) (PID) 60 m × 0.75 mm VOCOL capillary column

ACCURACY: Recovery = 97% @ spike of 10 μg/L (reagent C) (PID) 60 m × 0.75 mm VOCOL capillary column

SAMPLING METHOD: Use a 40-120 mL screw cap vial (pre-washed with detergent, rinsed with distilled water and oven dried at 105°C) with a PTFE-faced silicone septum. If residual chlorine is in the water add about 25 mg of ascorbic acid to each vial before sample collection. Collect bubble free samples.

STABILITY: Cool to 4°C; HCl to pH <2. M.H.T. = 14 Days.

QUALITY CONTROL: As an initial demonstration of lab accuracy and precision, analyze 4 to 7 replicates of a lab fortified blank containing analyte at 0.1-5 μg/L. Collect all samples in duplicate.

REFERENCE: Method 502.2, Volatile Organic Compounds in Water by Purge and Trap Capillary Column GC, EPA 600/4-88/039.

PRIMARY NAME: 4-Chlorotoluene Method 503.1

TITLE: Aromatic and Unsaturated VOCs in Water

MATRIX: Drinking water (finished or any treatment stage) and raw source water.

CAS #: 106-43-4

APPLICATION: Method covers 28 aromatic and unsaturated VOCs. An inert gas is bubbled through a 5 mL water sample. Purged sample components are trapped in tube of sorbent materials. When purging is complete, sorbent tube is heated and backflushed with inert gas to desorb trapped sample onto a packed GC column.

INTERFERENCES: During analysis, major contaminant sources are volatile materials in the lab and impurities in purging gas and sorbent trap. With high and low level samples, there can be carry-over contamination. Excess water causes a negative baseline deflection.

INSTRUMENTATION: Purge and Trap GC w/photoionization detector.

(Two GC columns are recommended); Column #1: 5% SP-1200 and 1.75% Bentone 34 on Supelcoport; 5% 1,2,3-tris(2-cyanoethoxy)propane on Chromosorb W.

RANGE: 2.2-600 μg/L (Drinking water) MDL: not listed.

PRECISION: RSD = 5.0% @ 0.50 μg/L; 17 samples

ACCURACY: Avg Recovery = 91% @ 0.50 μg/L; 17 saples

SAMPLING METHOD: Use a 40-120 mL screw cap vial (pre-washed with detergent, rinsed with distilled water and oven dried at 105°C) with a PTFE-faced silicone septum. If residual chlorine is in the water add about 25 mg of ascorbic acid to each vial before sample collection. Collect bubble free samples.

STABILITY: Cool to 4°C; HCl to pH <2. M.H.T. = 14 Days.

QUALITY CONTROL: As an initial demonstration of lab accuracy and precision, analyze 4 to 7 replicates of a lab fortified blank containing analyte at 0.1-5 μg/L. Collect all samples in duplicate.

REFERENCE: Method 503.1, Volatile Aromatic and Unsaturated Organic Compounds in C by Purge and Trap GC, EPA 600/4-88/039.

PRIMARY NAME: 4-Chlorotoluene Method 524.1

TITLE: VOCs in Water by Purge and Trap GC/MS

MATRIX: Drinking water (finished or any treatment stage) and raw source water.

CAS #: 106-43-4

APPLICATION: Method covers 51 VOCs. An inert gas is bubbled through a 25 mL water sample. Purged sample components are trapped in tube of sorbent materials.When purging is complete, sorbent tube is heated and backflushed with helium to desorb trapped sample onto a packed GC column.

INTERFERENCES: During analysis, major contaminant sources are volatile materials in the lab and impurities in purging gas and sorbent trap. With

high and low level samples, there can be carry-over contamination. Watch for methylene chloride, it permeates through PTFE tubing.

INSTRUMENTATION: Gas Chromatography/Mass Spectrometry/Data System; 1% SP-1000 on Carbopack B column.

RANGE: Approximately 0.2-200 μg/L MDL: not determined.

PRECISION: not listed.

ACCURACY: not listed.

SAMPLING METHOD: Use a 60-120 mL screw cap vial (pre-washed with detergent, rinsed with distilled water and oven dried at 105°C) with a PTFE-faced silicone septum. If residual chlorine is in the water add about 25 mg of ascorbic acid to each vial before sample collection. Collect bubble free samples.

STABILITY: Cool to 4°C; HCl to pH <2. M.H.T. = 14 Days.

QUALITY CONTROL: As an initial demonstration of lab accuracy and precision, analyze 4 to 7 replicates of a lab fortified blank containing analyte at 0.2-5 μg/L. Collect all samples in duplicate.

REFERENCE: Method 524.1, VOCs in Water by Purge and Trap Gas Chromatography/Mass Spectrometry, EPA 600/4-88/039.

PRIMARY NAME: 4-Chlorotoluene Method 524.2

TITLE: VOCs in Water Purge and Trap capillary column GC/MS

MATRIX: Drinking water (finished or any treatment stage) and raw source water.

CAS #: 106-43-4

APPLICATION: Method covers 60 VOCs. An inert gas is bubbled through a 25 mL water sample. Purged sample components are trapped in tube of sorbent materials. When purging is complete, sorbent tube is heated and backflushed with helium to desorb trapped sample onto a capillary GC column.

INTERFERENCES: During analysis, major contaminant sources are volatile materials in the lab and impurities in purging gas and sorbent trap. With high and low level samples, there can be carry-over contamination. Watch for methylene chloride, it permeates through PTFE tubing.

INSTRUMENTATION: (capillary column) Gas Chromatography/Mass Spectrometry/Data System. Column #1: VOCOL glass wide bore; Column #2: DB-624 fused silica; Column #3: DB-5 fused silica.

RANGE: Approximately 0.1-200 µg/L

MDL: 0.06 µg/L on wide bore capillary column.

PRECISION: RSD = 8.3% @ (0.1-10) µg/L (31 samples) wide bore capillary column.

ACCURACY: Recovery = 99% @ (0.1-10) µg/L (31 samples) wide bore capillary column.

SAMPLING METHOD: Use a 60-120 mL screw cap vial (pre-washed with detergent, rinsed with distilled water and oven dried at 105°C) with a PTFE-faced silicone septum. If residual chlorine is in the water add about 25 mg of ascorbic acid to each vial before sample collection. Collect bubble free samples.

STABILITY: Cool to 4°C; HCl to pH <2.

M.H.T. = 14 Days.

QUALITY CONTROL: As an initial demonstration of lab accuracy and precision, analyze 4 to 7 replicates of a lab fortified blank containing analyte at 0.2-5 µg/L. Collect all samples in duplicate.

REFERENCE: Method 524.2, VOCs in C by Purge and Trap Capillary column GC/MS, EPA 600/4-88/039.

PRIMARY NAME: Dibromochloromethane

Method 502.1

TITLE: Halogenated VOCs in Water

MATRIX: Drinking water (finished or any treatment stage) and raw source water.

CAS #: 124-48-1

APPLICATION: Method covers 40 halogenated VOCs. An inert gas is bubbled through a 5 mL water sample. Purged sample components are trapped in tube of sorbent materials. When purging is complete, sorbent tube is heated and backflushed with inert gas to desorb trapped sample onto a packed GC column where it is separated.

INTERFERENCES: During analysis, major contaminant sources are volatile materials in the lab and impurities in purging gas and sorbent trap. With high and low level samples, there can be carry-over contamination. Watch for methylene chloride, it permeates through PTFE tubing.

INSTRUMENTATION: Purge and Trap GC w/halide specific detector. (Two GC columns are recommended); Column #1: 1% SP-1000 on Carbopack B; Column #2: n-octane bonded on Poracil C.

RANGE: 8.0-505 μg/L (Drinking water) MDL: 0.008 μg/L

PRECISION: RSD = 7.0% @ 0.20 μg/L spike (S.L.) (17 Samples)

ACCURACY: Avg Recovery = 95% @ 0.20 μg/L spike (S.L.) (17 Samples)

SAMPLING METHOD: Use a 40-120 mL screw cap vial (pre-washed with detergent, rinsed with distilled water and oven dried at 105°C) with a PTFE-faced silicone septum. If residual chlorine is in the water add about 25 mg of ascorbic acid to each vial before sample collection. Collect bubble free samples.

STABILITY: Cool to 4°C; HCl to pH <2. M.H.T. = 14 Days.

QUALITY CONTROL: As an initial demonstration of lab accuracy and precision, analyze 4 to 7 replicates of a lab fortified blank containing analyte at 0.1-5 μg/L. Collect all samples in duplicate.

REFERENCE: Method 502.1, Volatile Halogenated Organic Compounds in Water by Purge and Trap GC, EPA 600/4-88/039.

PRIMARY NAME: Dibromochloromethane Method 502.2

TITLE: VOCs in Water by Purge and Trap capillary column GC

MATRIX: Drinking water (finished or any treatment stage) and raw source water.

CAS #: 124-48-1

APPLICATION: Method covers 60 VOCs. An inert gas is bubbled through a 5 mL sample in a purging chamber. Purged sample components are trapped in tube of sorbent materials. When purging is done, sorbent tube is heated and backflushed w/helium to desorb trapped sample onto a capillary GC column.

INTERFERENCES: During analysis, major contaminant sources are volatile materials in the lab and impurities in purging gas and sorbent trap. With high and low level samples, there can be carry-over contamination. Beware of methylene chloride, it permeates through PTFE tubing.

INSTRUMENTATION: (capillary column) GC, w/PID and electrolytic conductivity detectors in series. Column #1: VOCOL wide bore column; Column #2: RTX-502.2 mega bore column; Column #3: DB-62 mega bore column.

RANGE: Approximately 0.02-200 μg/L

MDL: 0.03 μg/L (reagent C) (ELCD) 60 m × 0.75 mm VOCOL capillary column

PRECISION: RSD = 3.3% @ Spike of 10 μg/L (reagent C) (ELCD). 60 m × 0.75 mm VOCOL capillary column

ACCURACY: Recovery = 102% @ spike of 10 μg/L (reagent C) (ELCD) 60 m × 0.75 mm VOCOL capillary column

SAMPLING METHOD: Use a 40-120 mL screw cap vial (pre-washed with detergent, rinsed with distilled water and oven dried at 105°C) with a PTFE-faced silicone septum. If residual chlorine is in the water add about 25 mg of ascorbic acid to each vial before sample collection. Collect bubble free samples.

STABILITY: Cool to 4°C; HCl to pH <2.

M.H.T. = 14 Days.

QUALITY CONTROL: As an initial demonstration of lab accuracy and precision, analyze 4 to 7 replicates of a lab fortified blank containing analyte at 0.1-5 μg/L. Collect all samples in duplicate.

REFERENCE: Method 502.2, Volatile Organic Compounds in Water by Purge and Trap Capillary Column GC, EPA 600/4-88/039.

PRIMARY NAME: Dibromochloromethane

Method 524.1

TITLE: VOCs in Water by Purge and Trap GC/MS

MATRIX: Drinking water (finished or any treatment stage) and raw source water.

CAS #: 124-48-1

APPLICATION: Method covers 51 VOCs. An inert gas is bubbled through a 25 mL water sample. Purged sample components are trapped in tube of sorbent materials. When purging is complete, sorbent tube is heated and backflushed with helium to desorb trapped sample onto a packed GC column.

INTERFERENCES: During analysis, major contaminant sources are volatile materials in the lab and impurities in purging gas and sorbent trap. With high and low level samples, there can be carry-over contamination. Watch for methylene chloride, it permeates through PTFE tubing.

INSTRUMENTATION: Gas Chromatography/Mass Spectrometry/Data System; 1% SP-1000 on Carbopack B column.

RANGE: Approximately 0.2-200 µg/L

MDL: 0.4 µg/L

PRECISION: RSD = 15% @ 1.0 µg/L

ACCURACY: Mean Accuracy = 92% @ 1.0 µg/L

SAMPLING METHOD: Use a 60-120 mL screw cap vial (pre-washed with detergent, rinsed with distilled water and oven dried at 105°C) with a PTFE-faced silicone septum. If residual chlorine is in the water add about 25 mg of ascorbic acid to each vial before sample collection. Collect bubble free samples.

STABILITY: Cool to 4°C; HCl to pH <2.

M.H.T. = 14 Days.

QUALITY CONTROL: As an initial demonstration of lab accuracy and precision, analyze 4 to 7 replicates of a lab fortified blank containing analyte at 0.2-5 µg/L. Collect all samples in duplicate.

REFERENCE: Method 524.1, VOCs in Water by Purge and Trap Gas Chromatography/Mass Spectrometry, EPA 600/4-88/039.

PRIMARY NAME: Dibromochloromethane Method 524.2

TITLE: VOCs in Water Purge and Trap capillary column GC/MS

MATRIX: Drinking water (finished or any treatment stage) and raw source water.

CAS #: 124-48-1

APPLICATION: Method covers 60 VOCs. An inert gas is bubbled through a 25 mL water sample. Purged sample components are trapped in tube of sorbent materials. When purging is complete, sorbent tube is heated and backflushed with helium to desorb trapped sample onto a capillary GC column.

INTERFERENCES: During analysis, major contaminant sources are volatile materials in the lab and impurities in purging gas and sorbent trap. With high and low level samples, there can be carry-over contamination. Watch for methylene chloride, it permeates through PTFE tubing.

INSTRUMENTATION: (capillary column) Gas Chromatography/Mass Spectrometry/Data System. Column #1: VOCOL glass wide bore; Column #2: DB-624 fused silica; Column #3: DB-5 fused silica.

RANGE: Approximately 0.1-200 µg/L

MDL: 0.05 µg/L on wide bore capillary column.

PRECISION: RSD = 7.0% @ (0.1-10) µg/L (31 samples) wide bore capillary column.

ACCURACY: Recovery = 92% @ (0.1-10) µg/L (31 samples) wide bore capillary column.

SAMPLING METHOD: Use a 60-120 mL screw cap vial (pre-washed with detergent, rinsed with distilled water and oven dried at 105°C) with a PTFE-faced silicone septum. If residual chlorine is in the water add about 25 mg of ascorbic acid to each vial before sample collection. Collect bubble free samples.

STABILITY: Cool to 4°C; HCl to pH <2. M.H.T. = 14 Days.

QUALITY CONTROL: As an initial demonstration of lab accuracy and precision, analyze 4 to 7 replicates of a lab fortified blank containing analyte at 0.2-5 µg/L. Collect all samples in duplicate.

REFERENCE: Method 524.2, VOCs in Water by Purge and Trap Capillary column GC/MS, EPA 600/4-88/039.

PRIMARY NAME: Dibromochloromethane — Method 601

TITLE: Purgeable Halocarbons — MATRIX: Wastewater

CAS #: 124-48-1

APPLICATION: Method covers 29 purgeable halocarbons. (Method 624 provides GC/MS conditions appropriate for the qualitative and quantitative confirmation of results). Method describes conditions for a second GC column to confirm measurements made with primary column.

INTERFERENCES: Impurities in the purge gas and organic compounds outgassing from the plumbing ahead of the trap. With high and low level samples, there can be carry-over contamination. Diffusion of volatile organics through the septum seal into the sample.

INSTRUMENTATION: GC equipped with halide-specific detector. (With purge and trap unit).

RANGE: 8.0 To 500 µg/L. MDL: 0.09 µg/L.

PRECISION: 0.24X+1.68 µg/L (overall precision)

ACCURACY: 0.94C+2.72 µg/L (as recovery)

SAMPLING METHOD: 25 mL glass vial. Teflon lined septum.

STABILITY: Cool, 4°C, 0.008% sodium thiosulfate. M.H.T. = 14 Days.

QUALITY CONTROL: The laboratory must on an ongoing basis, spike at least 10% of the samples from each sample site being monitored to assess accuracy.

REFERENCE: Method 601, Federal Register Part VIII 40 CFR Part 136, Oct 26, 1984.

PRIMARY NAME: Dibromochloromethane Method 624

TITLE: Purgeables MATRIX: Wastewater

CAS #: 124-48-1

APPLICATION: Method covers 31 purgeable organics. An inert gas is bubbled through a 5 mL water sample in a specially designed purging chamber. Here, purgeables are transferred from aqueous to gaseous phase, passed onto a sorbent column and trapped. Trap is heated and backflushed with inert gas to desorb purgeables onto a GC column, where purgeables are separated.

INTERFERENCES: Impurities in the purge gas, organic compounds outgassing from the plumbing ahead of the trap, and solvent vapors in the laboratory. With high and low level samples, there can be carry-over contamination.

INSTRUMENTATION: GC/MS with purge and trap unit.

RANGE: 5-600 μg/L MDL: 3.1 μg/L

PRECISION: 0.17X+0.49 μg/L (overall precision)

ACCURACY: 1.01C-0.03 μg/L (as recovery)

SAMPLING METHOD: 25 mL glass vial. Teflon lined septum.

STABILITY: Cool, 4°C, 0.008% sodium thiosulfate. M.H.T. = 14 Days.

QUALITY CONTROL: The laboratory must on an ongoing basis, spike at least 5% of the samples from each sample site being monitored to assess accuracy.

REFERENCE: Method 624, Federal Register Part VIII 40 CFR Part 136, Oct 26, 1984.

PRIMARY NAME: Dibromochloromethane

Method 8010

TITLE: Halogenated Volatile Organics

MATRIX: groundwater, soils, sludges water miscible liquid wastes, and non-water miscible wastes.

CAS #: 124-48-1

APPLICATION: This method is used for the analysis of 39 halogenated VOCs. Samples are analyzed using direct injection or purge and trap methods. Groundwater must be analyzed by the purge and trap method. The method provides an optional GC column which is used for analyte confirmation and that may help resolve analytes from interferences.

INTERFERENCES: There can be carry-over contamination with high and low level samples. Impurities may come from the purge and trap apparatus, organic compounds outgassing from the plumbing ahead of trap, diffusion of VOCs through the sample bottle septum during shipping or storage, or from solvent vapors in the lab.

INSTRUMENTATION: GC capable of on-column injections or purge and trap sample introduction and a halogen specific detector. Column 1: 8 foot by 0.1 inch 1% SP-1000 on Carbopack-B. Column 2: 6 foot by 0.1 inch bonded n-octane on Porasil-C.

RANGE: 8 to 500 μg/L (reagent water)

MDL: 0.09 μg/L (reagent water)

PRACTICAL QUANTITATION LIMIT FACTORS FOR MULTIPLYING TIMES MDL VALUE

Matrix	Multiplication Factor
Groundwater	10
Low-level soil	10
Water miscible liquid waste	500
High-level soil and sludge	1250
Non-water miscible waste	1250

PRECISION: 0.24X + 1.68 μg/L (overall precision)

ACCURACY: 0.94C + 2.72 μg/L (as recovery)

SAMPLING METHOD: For water and liquid samples; use glass 40 mL vials with Teflon lined septum caps and collect two vials per sample location with no headspace. For solids and concentrated waste samples; use widemouth glass bottles with Teflon liners.

STABILITY: For concentrated wastes, soils, sediments or sludges; cool to 4°C. For liquids; add 4 drops of concentrated hydrochloric acid and cool to 4°C. M.H.T. = 14 days.

QUALITY CONTROL: Analyze a reagent blank, matrix spike and matrix spike duplicate/duplicate for each analytical batch (up to 20 Samples). Demonstrate the purity of glassware and reagents by analyzing a reagent water method blank. Internal, surrogate and five concentration level calibration standards are used.

REFERENCE: Method 8010, SW-846, 3rd ed., Nov 1986.

PRIMARY NAME: 1,2-Dibromo-3-chloropropane Method 502.2

TITLE: VOCs in Water by Purge and Trap capillary column GC

MATRIX: Drinking water (finished or any treatment stage) and raw source water.

CAS #: 96-12-8

APPLICATION: Method covers 60 VOCs. An inert gas is bubbled through a 5 mL sample in a purging chamber. Purged sample components are trapped in tube of sorbent materials. When purging is done, sorbent tube is heated and backflushed w/helium to desorb trapped sample onto a capillary GC column.

INTERFERENCES: During analysis, major contaminant sources are volatile materials in the lab and impurities in purging gas and sorbent trap. With high and low level samples, there can be carry-over contamination. Beware of methylene chloride, it permeates through PTFE tubing.

INSTRUMENTATION: (Capillary Column) GC, w/PID and electrolytic conductivity detectors in series. Column #1: VOCOL wide bore column; Column #2: RTX-502.2 mega bore column; Column #3: DB-62 mega bore column.

RANGE: Approximately 0.02-200 μg/L

MDL: 3.0 μg/L (reagent C) (ELCD) 60 m × 0.75 mm VOCOL capillary column

PRECISION: RSD = 11.3% @ Spike of 10 μg/L (reagent C) (ELCD) 60 m × 0.75 mm VOCOL capillary column

ACCURACY: Recovery = 86% @ spike of 10 μg/L (reagent C) (ELCD) 60 m × 0.75 mm VOCOL capillary column

SAMPLING METHOD: Use a 40-120 mL screw cap vial (pre-washed with detergent, rinsed with distilled water and oven dried at 105°C) with a PTFE-faced silicone septum. If residual chlorine is in the water add about 25 mg of ascorbic acid to each vial before sample collection. Collect bubble free samples.

STABILITY: Cool to 4°C; HCl to pH <2.

M.H.T. = 14 Days.

QUALITY CONTROL: As an initial demonstration of lab accuracy and precision, analyze 4 to 7 replicates of a lab fortified blank containing analyte at 0.1-5 μg/L. Collect all samples in duplicate.

REFERENCE: Method 502.2, Volatile Organic Compounds in Water by Purge and Trap Capillary Column GC, EPA 600/4-88/039.

PRIMARY NAME: 1,2-Dibromo-3-chloropropane

Method 504

TITLE: (EDB)and(DBCP) in Water by Microextraction

MATRIX: Finished drinking water and unfinished ground-water.

CAS #: 96-12-8

APPLICATION: 35 mL Of sample are extracted with 2ml of hexane. 2ul Of the extract are injected into a GC with a linearized electron capture detector for separation and analysis. Aqueous calibration standards are run in same manner to compensate for possible extraction losses.

INTERFERENCES: Impurities contained in the extracting solvent usually account for the majority of analytical problems. (Run solvent blanks as checks). Store interference-free solvents in area free of chlorinated solvents. Extraction technique extracts polars with non-polars.

INSTRUMENTATION: GC with EC detector and capillary column splitless injector. Use two GC columns. Column #1: Durawax-DX3 fused silica; Column #2: DB-1 fused silica; Column #3: RTX-Volatiles wide bore.

RANGE: 0.03-200 μg/L MDL: 0.01 μg/L

PRECISION: (overall) S = 0.143X - 0.00 (RW); S = 0.160X + 0.006 (groundwater)

ACCURACY: (recovery) X = 0.987C - 0.00 (RW); X = 0.972C + 0.007 (groundwater)

SAMPLING METHOD: Use a 40-120 mL screw cap vial (pre-washed with detergent, rinsed with distilled water and oven dried at 105°C) with a PTFE-faced silicone septum. If residual chlorine is in the water add about 25 mg of ascorbic acid to each vial before sample collection. Collect bubble free samples.

STABILITY: Cool to 4°C; add 4 drops of 10% sodium thiosulfate and 4 drops of hydrochloric acid. M.H.T. = 28 Days.

QUALITY CONTROL: As an initial demonstration of lab accuracy and precision, analyze 4 to 7 replicates of a lab fortified blank containing analyte at 0.1-5 mg/L. control. The frequency of the qc check standard analyses is equivalent to 5% of all samples analyzed. Collect all samps in duplicate.

REFERENCE: Method 504, (EDB) and (DBCP) in Water by Microextraction and GC, EPA 600/4-88/039.

PRIMARY NAME: 1,2-Dibromo-3-chloropropane Method 524.1

TITLE: VOCs in Water by Purge and Trap GC/MS

MATRIX: Drinking water (finished or any treatment stage) and raw source water.

CAS # : 96-12-8

APPLICATION: Method covers 51 VOCs. An inert gas is bubbled through a 25 mL water sample. Purged sample components are trapped in tube of sorbent materials. When purging is complete, sorbent tube is heated and backflushed with helium to desorb trapped sample onto a packed GC column.

INTERFERENCES: During analysis, major contaminant sources are volatile materials in the lab and impurities in purging gas and sorbent trap. With high and low level samples, there can be carry-over contamination. Watch for methylene chloride, it permeates through PTFE tubing.

INSTRUMENTATION: Gas Chromatography/Mass Spectrometry/Data System; 1% SP-1000 on Carbopack B column.

RANGE: Approximately 0.2-200 μg/L MDL: 2 μg/L

PRECISION: RSD = 18% @ 3.5 μg/L

ACCURACY: Mean Accuracy = 100% @ 3.5 μg/L

SAMPLING METHOD: Use a 60-120 mL screw cap vial (pre-washed with detergent, rinsed with distilled water and oven dried at 105°C) with a PTFE-faced silicone septum. If residual chlorine is in the water add about 25 mg of ascorbic acid to each vial before sample collection. Collect bubble free samples.

STABILITY: Cool to 4°C; HCl to pH <2. M.H.T. = 14 Days.

QUALITY CONTROL: As an initial demonstration of lab accuracy and precision, analyze 4 to 7 replicates of a lab fortified blank containing analyte at 0.2-5 μg/L. Collect all samples in duplicate.

REFERENCE: Method 524.1, VOCs in Water by Purge and Trap Gas chromatography/mass spectrometry, EPA 600/4-88/039.

PRIMARY NAME: 1,2-Dibromo-3-chloropropane Method 524.2

TITLE: VOCs in Water Purge and Trap capillary column GC/MS

MATRIX: Drinking water (finished or any treatment stage) and raw source water.

CAS # : 96-12-8

APPLICATION: Method covers 60 VOCs. An inert gas is bubbled through a 25 mL water sample. Purged sample components are trapped in tube of sorbent materials. When purging is complete, sorbent tube is heated and backflushed with helium to desorb trapped sample onto a capillary GC column.

INTERFERENCES: During analysis, major contaminant sources are volatile materials in the lab and impurities in purging gas and sorbent trap. With high and low level samples, there can be carry-over contamination. Watch for methylene chloride, it permeates through PTFE tubing.

INSTRUMENTATION: (Capillary Column) Gas Chromatography/Mass Spectrometry/Data System. Column #1: VOCOL glass wide bore; Column #2: DB-624 fused silica; Column #3: DB-5 fused silica.

RANGE: Approximately 0.1-200 μg/L

MDL: 0.26 μg/L on wide bore capillary column.

PRECISION: RSD = 19.9% @ (0.5-10) μg/L (24 samples) wide bore capillary column.

ACCURACY: Recovery = 83% @ (0.5-10) μg/L (24 samples) wide bore capillary column.

SAMPLING METHOD: Use a 60-120 mL screw cap vial (pre-washed with detergent, rinsed with distilled water and oven dried at 105°C) with a PTFE-faced silicone septum. If residual chlorine is in the water add about 25 mg of ascorbic acid to each vial before sample collection. Collect bubble free samples.

STABILITY: Cool to 4°C; HCl to pH <2.

M.H.T. = 14 Days.

QUALITY CONTROL: As an initial demonstration of lab accuracy and precision, analyze 4 to 7 replicates of a lab fortified blank containing analyte at 0.2-5 μg/L. Collect all samples in duplicate.

REFERENCE: Method 524.2, VOCs in Water by Purge and Trap Capillary column GC/MS, EPA 600/4-88/039.

PRIMARY NAME: 1,2-Dibromoethane (EDB) Method 502.1

TITLE: Halogenated VOCs in Water

MATRIX: Drinking water (finished or any treatment stage) and raw source water.

CAS # : 106-93-4

APPLICATION: Method covers 40 halogenated VOCs. An inert gas is bubbled through a 5 mL water sample. Purged sample components are trapped in tube of sorbent materials. When purging is complete, sorbent tube is heated and backflushed with inert gas to desorb trapped sample onto a packed GC column where it is separated.

INTERFERENCES: During analysis, major contaminant sources are volatile materials in the lab and impurities in purging gas and sorbent trap. With high and low level samples, there can be carry-over contamination.Watch for methylene chloride, it permeates through PTFE tubing.

INSTRUMENTATION: Purge and Trap GC w/halide specific detector. (Two GC columns are recommended); Column #1: 1% SP-1000 on Carbopack B; Column #2: n-octane bonded on Poracil C.

RANGE: 8.0-505 µg/L (Drinking water) MDL: 0.04 µg/L in water

PRECISION: RSD = 12.5% @ 0.40 µg/L spike (S.L.) (18 Samples)

ACCURACY: Avg Recovery = 93% @ 0.40 µg/L spike (S.L.) (18 Samples)

SAMPLING METHOD: Use a 40-120 mL screw cap vial (pre-washed with detergent, rinsed with distilled water and oven dried at 105°C) with a PTFE-faced silicone septum. If residual chlorine is in the water add about 25 mg of ascorbic acid to each vial before sample collection. Collect bubble free samples.

STABILITY: Cool to 4°C; HCl to pH <2. M.H.T. = 14 Days.

QUALITY CONTROL: As an initial demonstration of lab accuracy and precision, analyze 4 to 7 replicates of a lab fortified blank containing analyte at 0.1-5 µg/L. Collect all samples in duplicate.

REFERENCE: Method 502.1, Volatile Halogenated Organic Compounds in Water by Purge and Trap GC, EPA 600/4-88/039.

PRIMARY NAME: 1,2-Dibromoethane (EDB) Method 502.2

TITLE: VOCs in Water by Purge and Trap capillary column GC

MATRIX: Drinking water (finished or any treatment stage) and raw source water.

CAS # : 106-93-4

APPLICATION: Method covers 60 VOCs. An inert gas is bubbled through a 5 mL sample in a purging chamber. Purged sample components are trapped in tube of sorbent materials. When purging is done, sorbent tube is heated and backflushed w/helium to desorb trapped sample onto a capillary GC column.

INTERFERENCES: During analysis, major contaminant sources are volatile materials in the lab and impurities in purging gas and sorbent trap. With high and low level samples, there can be carry-over contamination. Beware of methylene chloride, it permeates through PTFE tubing.

INSTRUMENTATION: (Capillary Column) GC, w/PID and electrolytic conductivity detectors in series. Column #1: VOCOL wide bore column; Column #2: RTX-502.2 mega bore column; Column #3: DB-62 mega bore column.

RANGE: Approximately 0.02-200 µg/L

MDL: 0.8 µg/L (reagent C) (ELCD) 60 m x 0.75 mm VOCOL capillary column

PRECISION: RSD = 2.8% @ Spike of 10 µg/L (reagent C) (ELCD) 60 m x 0.75 mm VOCOL capillary column

ACCURACY: Recovery = 97% @ spike of 10 µg/L (reagent C) (ELCD) 60 m x 0.75 mm VOCOL capillary column

SAMPLING METHOD: Use a 40-120 mL screw cap vial (pre-washed with detergent, rinsed with distilled water and oven dried at 105°C) with a PTFE-faced silicone septum. If residual chlorine is in the water add about 25 mg of ascorbic acid to each vial before sample collection. Collect bubble free samples.

STABILITY: Cool to 4°C; HCl to pH <2. M.H.T. = 14 Days.

QUALITY CONTROL: As an initial demonstration of lab accuracy and

precision, analyze 4 to 7 replicates of a lab fortified blank containing analyte at 0.1-5 μg/L. Collect all samples in duplicate.

REFERENCE: Method 502.2, Volatile Organic Compounds in Water by Purge and Trap Capillary Column GC, EPA 600/4-88/039.

PRIMARY NAME: 1,2-Dibromoethane (EDB) Method 504

TITLE: (EDB)and(DBCP) in Water by Microextraction

MATRIX: Finished drinking water and groundwater.

CAS # : 106-93-4

APPLICATION: 35 mL Of sample are extracted with 2ml of hexane. 2μl Of the extract are injected into a GC with a linearized electron capture detector for separation and analysis. Aqueous calibration standards are run in same manner to compensate for possible extraction losses.

INTERFERENCES: Impurities contained in the extracting solvent usually account for the majority of analytical problems. (Run solvent blanks as checks). EDB at low concentrations may be masked by high levels of dibromochloromethane. Extraction technique extracts polars with non-polars.

INSTRUMENTATION: GC with EC detector and capillary column splitless injector. Use two GC columns. Column #1: Durawax-DX3 fused silica; Column #2: DB-1 fused silica; Column #3: RTX-Volatiles wide bore.

RANGE: 0.03-200 μg/L

MDL: 0.01 μg/L

PRECISION: (overall) S = 0.075X + 0.008 (RW); S = 0.102X + 0.006 (groundwater)

ACCURACY: (recovery) X = 1.072C - 0.006 (RW); X = 1.077C - 0.001 (groundwater)

SAMPLING Method: 40 mL screw cap vials, PTFE-faced silicon septum.

STABILITY: Cool to 4°C; add 4 drops of 10% sodium thiosulfate and 4 drops of hydrochloric acid. M.H.T. = 28 Days.

QUALITY CONTROL: Laboratory must, on an ongoing basis, demonstrate through analyses of lab fortified blanks that the operation of the measurement system is in control. The frequency of the lab fortified blank analyses must be equivalent to 10% of all samples analyzed.

REFERENCE: Method 504, (EDB) and (DBCP) in Water by Microextraction and GC, EPA 600/4-88/039.

PRIMARY NAME: 1,2-Dibromoethane (EDB) Method 524.1

TITLE: VOCs in Water by Purge and Trap GC/MS

MATRIX: Drinking water (finished or any treatment stage) and raw source water.

CAS # : 106-93-4

APPLICATION: Method covers 51 VOCs. An inert gas is bubbled through a 25 mL water sample. Purged sample components are trapped in tube of sorbent materials. When purging is complete, sorbent tube is heated and backflushed with helium to desorb trapped sample onto a packed GC column.

INTERFERENCES: During analysis, major contaminant sources are volatile materials in the lab and impurities in purging gas and sorbent trap. With high and low level samples, there can be carry-over contamination. Watch for methylene chloride, it permeates through PTFE tubing.

INSTRUMENTATION: Gas Chromatography/Mass Spectrometry/Data System; 1% SP-1000 on Carbopack B column.

RANGE: Approximately 0.2-200 μg/L MDL: 0.4 μg/L

PRECISION: RSD = 14% @ 1.0 μg/L

ACCURACY: Mean Accuracy = 93% @ 1.0 μg/L

SAMPLING METHOD: Use a 60-120 mL screw cap vial (pre-washed with detergent, rinsed with distilled water and oven dried at 105°C) with a PTFE-faced silicone septum. If residual chlorine is in the water add about 25 mg of ascorbic acid to each vial before sample collection. Collect bubble free samples.

STABILITY: Cool to 4°C; HCl to pH <2. M.H.T. = 14 Days.

QUALITY CONTROL: As an initial demonstration of lab accuracy and precision, analyze 4 to 7 replicates of a lab fortified blank containing analyte at 0.2-5 μg/L. Collect all samples in duplicate.

REFERENCE: Method 524.1, VOCs in Water by Purge and Trap Gas chromatography/mass spectrometry, EPA 600/4-88/039.

PRIMARY NAME: 1,2-Dibromoethane (EDB) Method 524.2

TITLE: VOCs in Water (Purge and Trap capillary column GC/MS

MATRIX: Drinking water (finished or any treatment stage) and raw source water.

CAS # : 106-93-4

APPLICATION: Method covers 60 VOCs. An inert gas is bubbled through a 25 mL water sample. Purged sample components are trapped in tube of sorbent materials. When purging is complete, sorbent tube is heated and backflushed with helium to desorb trapped sample onto a capillary GC column.

INTERFERENCES: During analysis, major contaminant sources are volatile materials in the lab and impurities in purging gas and sorbent trap. With high and low level samples,there can be carry-over contamination. Watch for methylene chloride, it permeates through PTFE tubing.

INSTRUMENTATION: (Capillary Column) Gas Chromatography/Mass Spectrometry/Data System. Column #1: VOCOL glass wide bore; Column #2: DB-624 fused silica; Column #3: DB-5 fused silica.

RANGE: Approximately 0.1-200 μg/L

MDL: 0.06 μg/L on wide bore capillary column.

PRECISION: RSD = 3.9% @ (0.5-10) μg/L (24 samples) wide bore capillary column.

ACCURACY: Recovery = 102% @ (0.5-10) μg/L (24 samples) wide bore capillary column.

SAMPLING METHOD: Use a 60-120 mL screw cap vial (pre-washed with detergent, rinsed with distilled water and oven dried at 105°C) with a PTFE-faced silicone septum. If residual chlorine is in the water add about 25 mg of ascorbic acid to each vial before sample collection. Collect bubble free samples.

STABILITY: Cool to 4°C; HCl to pH <2. M.H.T. = 14 Days.

QUALITY CONTROL: As an initial demonstration of lab accuracy and precision, analyze 4 to 7 replicates of a lab fortified blank containing analyte at 0.2-5 µg/L. Collect all samples in duplicate.

REFERENCE: Method 524.2, VOCs in Water by Purge and Trap Capillary column GC/MS, EPA 600/4-88/039.

PRIMARY NAME: Dibromomethane Method 502.1

TITLE: Halogenated VOCs in Water

MATRIX: Drinking water (finished or any treatment stage) and raw source water.

CAS # : 74-95-3

APPLICATION: Method covers 40 halogenated VOCs. An inert gas is bubbled through a 5 mL water sample. Purged sample components are trapped in tube of sorbent materials. When purging is complete, sorbent tube is heated and backflushed with inert gas to desorb trapped sample onto a packed GC column where it is separated.

INTERFERENCES: During analysis, major contaminant sources are volatile materials in the lab and impurities in purging gas and sorbent trap. With high and low level samples, there can be carry-over contamination. Watch for methylene chloride, it permeates through PTFE tubing.

INSTRUMENTATION: Purge and Trap GC w/halide specific detector. (Two GC columns are recommended); Column #1: 1% SP-1000 on Carbopack B; Column #2: n-octane bonded on Poracil C.

RANGE: 8.0-505 µg/L (Drinking water) MDL: not determined.

PRECISION: RSD = 8.0% @ 0.40 μg/L spike (S.L.) (5 Samples)

ACCURACY: Avg Recovery = 100% @ 0.40 μg/L spike (S.L.) (5 Samples)

SAMPLING METHOD: Use a 40-120 mL screw cap vial (pre-washed with detergent, rinsed with distilled water and oven dried at 105°C) with a PTFE-faced silicone septum. If residual chlorine is in the water add about 25 mg of ascorbic acid to each vial before sample collection. Collect bubble free samples.

STABILITY: Cool to 4°C; HCl to pH <2. M.H.T. = 14 Days.

QUALITY CONTROL: As an initial demonstration of lab accuracy and precision, analyze 4 to 7 replicates of a lab fortified blank containing analyte at 0.1-5 μg/L. Collect all samples in duplicate.

REFERENCE: Method 502.1, Volatile Halogenated Organic Compounds in Water by Purge and Trap GC, EPA 600/4-88/039.

PRIMARY NAME: Dibromomethane Method 502.2

TITLE: VOCs in Water by Purge and Trap capillary column GC

MATRIX: Drinking water (finished or any treatment stage) and raw source water.

CAS # : 74-95-3

APPLICATION: Method covers 60 VOCs. An inert gas is bubbled through a 5 mL sample in a purging chamber. Purged sample components are trapped in tube of sorbent materials. When purging is done, sorbent tube is heated and backflushed w/helium to desorb trapped sample onto a capillary GC column.

INTERFERENCES: During analysis, major contaminant sources are volatile materials in the lab and impurities in purging gas and sorbent trap. With high and low level samples, there can be carry-over contamination. Beware of methylene chloride, it permeates through PTFE tubing.

INSTRUMENTATION: (capillary column) GC, w/PID and electrolytic conductivity detectors in series. Column #1: VOCOL wide bore column;

Column #2: RTX-502.2 mega bore column; Column #3: DB-62 mega bore column.

RANGE: Approximately 0.02-200 μg/L

MDL: 2.2 μg/L (reagent C) (ELCD) 60 m x 0.75 mm VOCOL capillary column

PRECISION: RSD = 6.7% @ Spike of 10 μg/L (reagent C) (ELCD). 60 m x 0.75 mm VOCOL capillary column

ACCURACY: Recovery = 109% @ spike of 10 μg/L (reagent C) (ELCD) 60 m x 0.75 mm VOCOL capillary column

SAMPLING METHOD: Use a 40-120 mL screw cap vial (pre-washed with detergent, rinsed with distilled water and oven dried at 105°C) with a PTFE-faced silicone septum. If residual chlorine is in the water add about 25 mg of ascorbic acid to each vial before sample collection. Collect bubble free samples.

STABILITY: Cool to 4°C; HCl to pH <2. M.H.T. = 14 Days.

QUALITY CONTROL: As an initial demonstration of lab accuracy and precision, analyze 4 to 7 replicates of a lab fortified blank containing analyte at 0.1-5 μg/L. Collect all samples in duplicate.

REFERENCE: Method 502.2, Volatile Organic Compounds in Water by Purge and Trap Capillary Column GC, EPA 600/4-88/039.

PRIMARY NAME: Dibromomethane Method 524.1

TITLE: VOCs in Water by Purge and Trap GC/MS

MATRIX: Drinking water (finished or any treatment stage) and raw source water.

CAS # : 74-95-3

APPLICATION: Method covers 51 VOCs. An inert gas is bubbled through a 25 mL water sample. Purged sample components are trapped in tube of sorbent materials. When purging is complete, sorbent tube is heated and backflushed with helium to desorb trapped sample onto a packed GC column.

INTERFERENCES: During analysis, major contaminant sources are volatile materials in the lab and impurities in purging gas and sorbent trap. With high and low level samples, there can be carry-over contamination. Watch for methylene chloride, it permeates through PTFE tubing.

INSTRUMENTATION: Gas Chromatography/Mass Spectrometry/Data System; 1% SP-1000 on Carbopack B column.

RANGE: Approximately 0.2-200 μg/L MDL: 0.3 μg/L

PRECISION: RSD = 12% @ 1.0 μg/L

ACCURACY: Mean Accuracy = 94% @ 1.0 μg/L

SAMPLING METHOD: Use a 60-120 mL screw cap vial (pre-washed with detergent, rinsed with distilled water and oven dried at 105°C) with a PTFE-faced silicone septum. If residual chlorine is in the water add about 25 mg of ascorbic acid to each vial before sample collection. Collect bubble free samples.

STABILITY: Cool to 4°C; HCl to pH <2. M.H.T. = 14 Days.

QUALITY CONTROL: As an initial demonstration of lab accuracy and precision, analyze 4 to 7 replicates of a lab fortified blank containing analyte at 0.2-5 μg/L. Collect all samples in duplicate.

REFERENCE: Method 524.1, VOCs in Water by Purge and Trap Gas Chromatography/Mass Spectrometry, EPA 600/4-88/039.

PRIMARY NAME: Dibromomethane Method 524.2

TITLE: VOCs in Water Purge and Trap capillary column GC/MS

MATRIX: Drinking water (finished or any treatment stage) and raw source water.

CAS # : 74-95-3

APPLICATION: Method covers 60 VOCs. An inert gas is bubbled through a 25 mL water sample. Purged sample components are trapped in tube of sorbent materials. When purging is complete, sorbent tube is heated and backflushed with helium to desorb trapped sample onto a capillary GC column.

INTERFERENCES: During analysis, major contaminant sources are volatile materials in the lab and impurities in purging gas and sorbent trap. With high and low level samples, there can be carry-over contamination. Watch for methylene chloride, it permeates through PTFE tubing.

INSTRUMENTATION: (capillary column) Gas Chromatography/Mass Spectrometry/Data System. Column #1: VOCOL glass wide bore; Column #2: DB-624 fused silica; Column #3: DB-5 fused silica.

RANGE: Approximately 0.1-200 µg/L

MDL: 0.24 µg/L on wide bore capillary column.

PRECISION: RSD = 5.6% @ (0.5-10) µg/L (24 samples) wide bore capillary column.

ACCURACY: Recovery = 100% @ (0.5-10) µg/L (24 samples) wide bore capillary column.

SAMPLING METHOD: Use a 60-120 mL screw cap vial (pre-washed with detergent, rinsed with distilled water and oven dried at 105°C) with a PTFE-faced silicone septum. If residual chlorine is in the water add about 25 mg of ascorbic acid to each vial before sample collection. Collect bubble free samples.

STABILITY: Cool to 4°C; HCl to pH <2.

M.H.T. = 14 Days.

QUALITY CONTROL: As an initial demonstration of lab accuracy and precision, analyze 4 to 7 replicates of a lab fortified blank containing analyte at 0.2-5 µg/L. Collect all samples in duplicate.

REFERENCE: Method 524.2, VOCs in Water by Purge and Trap Capillary column GC/MS, EPA 600/4-88/039.

PRIMARY NAME: 1,2-Dichlorobenzene

Method 502.1

TITLE: Halogenated VOCs in Water

MATRIX: Drinking water (finished or any treatment stage) and raw source water.

CAS # : 95-50-1

APPLICATION: Method covers 40 halogenated VOCs. An inert gas is

bubbled through a 5 mL water sample. Purged sample components are trapped in tube of sorbent materials. When purging is complete, sorbent tube is heated and backflushed with inert gas to desorb trapped sample onto a packed GC column where it is separated.

INTERFERENCES: During analysis, major contaminant sources are volatile materials in the lab and impurities in purging gas and sorbent trap. With high and low level samples,there can be carry-over contamination. Watch for methylene chloride, it permeates through PTFE tubing.

INSTRUMENTATION: Purge and Trap GC w/halide specific detector. (Two GC columns are recommended); Column #1: 1% SP-1000 on Carbopack B; Column #2: n-octane bonded on Poracil C.

RANGE: 8.0-505 µg/L (Drinking water) MDL: not determined.

PRECISION: RSD = 13% @ 0.40 µg/L spike (S.L.) (21 Samples)

ACCURACY: Avg Recovery = 95% @ 0.40 µg/L spike (S.L.) (21 Samples)

SAMPLING METHOD: Use a 40-120 mL screw cap vial (pre-washed with detergent, rinsed with distilled water and oven dried at 105°C) with a PTFE-faced silicone septum. If residual chlorine is in the water add about 25 mg of ascorbic acid to each vial before sample collection. Collect bubble free samples.

STABILITY: Cool to 4°C; HCl to pH <2. M.H.T. = 14 Days.

QUALITY CONTROL: As an initial demonstration of lab accuracy and precision, analyze 4 to 7 replicates of a lab fortified blank containing analyte at 0.1-5 µg/L. Collect all samples in duplicate.

REFERENCE: Method 502.1, Volatile Halogenated Organic Compounds in Water by Purge and Trap GC, EPA 600/4-88/039.

PRIMARY NAME: 1,2-Dichlorobenzene Method 502.2

TITLE: VOCs in Water by Purge and Trap capillary column GC

MATRIX: Drinking water (finished or any treatment stage) and raw source water.

CAS # : 95-50-1

APPLICATION: Method covers 60 VOCs. An inert gas is bubbled through a 5 mL sample in a purging chamber. Purged sample components are trapped in tube of sorbent materials. When purging is done, sorbent tube is heated and backflushed w/helium to desorb trapped sample onto a capillary GC column.

INTERFERENCES: During analysis, major contaminant sources are volatile materials in the lab and impurities in purging gas and sorbent trap. With high and low level samples, there can be carry-over contamination. Beware of methylene chloride, it permeates through PTFE tubing.

INSTRUMENTATION: (capillary column) GC, w/PID and electrolytic conductivity detectors in series. Column #1: VOCOL wide bore column; Column #2: RTX-502.2 mega bore column; Column #3: DB-62 mega bore column.

RANGE: Approximately 0.02-200 µg/L

MDL: 0.02 µg/L (reagent C) (ELCD) 60 m x 0.75 mm VOCOL capillary column

PRECISION: RSD = 1.5% @ Spike of 10 µg/L (reagent C) (ELCD) 60 m x 0.75 mm VOCOL capillary column

ACCURACY: Recovery = 100% @ spike of 10 µg/L (reagent C) (ELCD) 60 m x 0.75 mm VOCOL capillary column

SAMPLING METHOD: Use a 40-120 mL screw cap vial (pre-washed with detergent, rinsed with distilled water and oven dried at 105°C) with a PTFE-faced silicone septum. If residual chlorine is in the water add about 25 mg of ascorbic acid to each vial before sample collection. Collect bubble free samples.

STABILITY: Cool to 4°C; HCl to pH <2.

M.H.T. = 14 Days.

QUALITY CONTROL: As an initial demonstration of lab accuracy and precision, analyze 4 to 7 replicates of a lab fortified blank containing analyte at 0.1-5 µg/L. Collect all samples in duplicate.

REFERENCE: Method 502.2, Volatile Organic Compounds in Water by Purge and Trap Capillary Column GC, EPA 600/4-88/039.

PRIMARY NAME: 1,2-Dichlorobenzene Method 503.1

TITLE: Aromatic and Unsaturated VOCs in Water

MATRIX: Drinking water (finished or any treatment stage) and raw source water.

CAS # : 95-50-1

APPLICATION: Method covers 28 aromatic and unsaturated VOCs. An inert gas is bubbled through a 5 mL water sample. Purged sample components are trapped in tube of sorbent materials. When purging is complete, sorbent tube is heated and backflushed with inert gas to desorb trapped sample onto a packed GC column.

INTERFERENCES: During analysis, major contaminant sources are volatile materials in the lab and impurities in purging gas and sorbent trap. With high and low level samples, there can be carry-over contamination. Excess water causes a negative baseline deflection.

INSTRUMENTATION: Purge and Trap GC w/photoionization detector. (Two GC columns are recommended); Column #1: 5% SP-1200 and 1.75% Bentone 34 on Supelcoport; 5% 1,2,3-tris(2-cyanoethoxy)propane on Chromosorb W.

RANGE: 2.2-600 µg/L (Drinking water) MDL: 0.02 µg/L in water

PRECISION: RSD = 18.8% @ 19.4 µg/L; 4 labs

ACCURACY: Avg Recovery = 85% @ 19.4 µg/L; 4 labs

SAMPLING METHOD: Use a 40-120 mL screw cap vial (pre-washed with detergent, rinsed with distilled water and oven dried at 105°C) with a PTFE-faced silicone septum. If residual chlorine is in the water add about 25 mg of ascorbic acid to each vial before sample collection. Collect bubble free samples.

STABILITY: Cool to 4°C; HCl to pH <2. M.H.T. = 14 Days.

QUALITY CONTROL: As an initial demonstration of lab accuracy and precision, analyze 4 to 7 replicates of a lab fortified blank containing analyte at 0.1-5 µg/L. Collect all samples in duplicate.

REFERENCE: Method 503.1, Volatile Aromatic and Unsaturated Organic Compounds in C by Purge and Trap GC, EPA 600/4-88/039.

PRIMARY NAME: 1,2-Dichlorobenzene Method 524.1

TITLE: VOCs in Water by Purge and Trap GC/MS

MATRIX: Drinking water (finished or any treatment stage) and raw source water.

CAS # : 95-50-1

APPLICATION: Method covers 51 VOCs. An inert gas is bubbled through a 25 mL water sample. Purged sample components are trapped in tube of sorbent materials. When purging is complete, sorbent tube is heated and backflushed with helium to desorb trapped sample onto a packed GC column.

INTERFERENCES: During analysis, major contaminant sources are volatile materials in the lab and impurities in purging gas and sorbent trap. With high and low level samples, there can be carry-over contamination. Watch for methylene chloride, it permeates through PTFE tubing.

INSTRUMENTATION: Gas Chromatography/Mass Spectrometry/Data System; 1% SP-1000 on Carbopack B column.

RANGE: Approximately 0.2-200 μg/L MDL: 1.0 μg/L

PRECISION: RSD = 7.0% @ 5.0 μg/L

ACCURACY: Mean Accuracy = 100% @ 5.0 μg/L

SAMPLING METHOD: Use a 60-120 mL screw cap vial (pre-washed with detergent, rinsed with distilled water and oven dried at 105°C) with a PTFE-faced silicone septum. If residual chlorine is in the water add about 25 mg of ascorbic acid to each vial before sample collection. Collect bubble free samples.

STABILITY: Cool to 4°C; HCl to pH <2. M.H.T. = 14 Days.

QUALITY CONTROL: As an initial demonstration of lab accuracy and

precision, analyze 4 to 7 replicates of a lab fortified blank containing analyte at 0.2-5 μg/L. Collect all samples in duplicate.

REFERENCE: Method 524.1, VOCs in Water by Purge and Trap Gas Chromatography/Mass Spectrometry, EPA 600/4-88/039.

PRIMARY NAME: 1,2-Dichlorobenzene

Method 524.2

TITLE: VOCs in Water Purge and Trap capillary column GC/MS

MATRIX: Drinking water (finished or any treatment stage) and raw source water.

CAS # : 95-50-1

APPLICATION: Method covers 60 VOCs. An inert gas is bubbled through a 25 mL water sample. Purged sample components are trapped in tube of sorbent materials. When purging is complete, sorbent tube is heated and backflushed with helium to desorb trapped sample onto a capillary GC column.

INTERFERENCES: During analysis, major contaminant sources are volatile materials in the lab and impurities in purging gas and sorbent trap. With high and low level samples, there can be carry-over contamination. Watch for methylene chloride, it permeates through PTFE tubing.

INSTRUMENTATION: (Capillary Column) Gas Chromatography/Mass Spectrometry/Data System. Column #1: VOCOL glass wide bore; Column #2: DB-624 fused silica; Column #3: DB-5 fused silica.

RANGE: Approximately 0.1-200 μg/L

MDL: 0.03 μg/L on wide bore capillary column.

PRECISION: RSD = 6.2% @ (0.1-10) μg/L (31 samples) wide bore capillary column.

ACCURACY: Recovery = 93% @ (0.1-10) μg/L (31 samples) wide bore capillary column.

SAMPLING METHOD: Use a 60-120 mL screw cap vial (pre-washed with detergent, rinsed with distilled water and oven dried at 105°C) with a PTFE-

faced silicone septum. If residual chlorine is in the water add about 25 mg of ascorbic acid to each vial before sample collection. Collect bubble free samples.

STABILITY: Cool to 4°C; HCl to pH <2. M.H.T. = 14 Days.

QUALITY CONTROL: As an initial demonstration of lab accuracy and precision, analyze 4 to 7 replicates of a lab fortified blank containing analyte at 0.2-5 μg/L. Collect all samples in duplicate.

REFERENCE: Method 524.2, VOCs in Water by Purge and Trap Capillary column GC/MS, EPA 600/4-88/039.

PRIMARY NAME: 1,2-Dichlorobenzene Method 601

TITLE: Purgeable Halocarbons MATRIX: Wastewater

CAS # : 95-50-1

APPLICATION: Method covers 29 purgeable halocarbons. (Method 624 provides GC/MS conditions appropriate for the qualitative and quantitative confirmation of results). Method describes conditions for a second GC column to confirm measurements made with primary column.

INTERFERENCES: Impurities in the purge gas and organic compounds outgassing from the plumbing ahead of the trap. With high and low level samples, there can be carry-over contamination. Diffusion of volatile organics through the septum seal into the sample.

INSTRUMENTATION: GC equipped with halide-specific detector. (With purge and trap unit).

RANGE: 8.0 To 500 μg/L. MDL: 0.15 μg/L.

PRECISION: 0.13X+6.13 μg/L (overall precision)

ACCURACY: 0.93C+1.70 μg/L (as recovery)

SAMPLING METHOD: 25 mL glass vial. Teflon lined septum.

STABILITY: Cool, 4°C, 0.008% sodium thiosulfate. M.H.T. = 14 Days.

QUALITY CONTROL: The laboratory must on an ongoing basis, spike at least 10% of the samples from each sample site being monitored to assess accuracy.

REFERENCE: Method 601, Federal Register Part VIII 40 CFR Part 136, Oct 26, 1984.

PRIMARY NAME: 1,2-Dichlorobenzene Method 602

TITLE: Purgeable Aromatics MATRIX: Wastewater

CAS # : 95-50-1

APPLICATION: Method covers 7 purgeable aromatics. (Method 624 provides GC/MS conditions appropriate for the qualitative and quantitative confirmation of results). Method describes conditions for a second GC column to confirm measurements made with primary column.

INTERFERENCES: Impurities in the purge gas and organic compounds outgassing from the plumbing ahead of the trap. With high and low level samples, there can be carry-over contamination. Diffusion of volatile organics through the septum seal into the sample.

INSTRUMENTATION: GC equipped with photoionization detector. (With purge and trap unit)

RANGE: 2.1 To 550 µg/L. MDL: 0.4 µg/L.

PRECISION: 0.22X+0.53 µg/L (overall precision)

ACCURACY: 0.93C+0.52 µg/L (as recovery)

SAMPLING METHOD: 25 mL glass vial. Teflon lined septum.

STABILITY: Cool, 4°C, 0.008% sodium thiosulfate. M.H.T. = 14 Days.

QUALITY CONTROL: The laboratory must on an ongoing basis, spike at least 10% of the samples from each sample site being monitored to assess accuracy.

REFERENCE: Method 602, Federal Register Part VIII 40 CFR Part 136, Oct 26, 1984.

PRIMARY NAME: 1,2-Dichlorobenzene Method 624

TITLE: Purgeables MATRIX: Wastewater

CAS # : 95-50-1

APPLICATION: Method covers 31 purgeable organics. An inert gas is bubbled through a 5 mL water sample in a specially designed purging chamber. Here, purgeables are transferred from aqueous to gaseous phase, passed onto a sorbent column and trapped. Trap is heated and backflushed with inert gas to desorb purgeables onto a GC column, where purgeables are separated.

INTERFERENCES: Impurities in the purge gas, organic compounds outgassing from the plumbing ahead of the trap, and solvent vapors in the laboratory. With high and low level samples, there can be carry-over contamination.

INSTRUMENTATION: GC/MS with purge and trap unit.

RANGE: 5-600 μg/L MDL: not determined

PRECISION: 0.30X-1.20 μg/L (overall precision)

ACCURACY: 0.94C+4.47 μg/L (as recovery)

SAMPLING METHOD: 25 mL glass vial. Teflon lined septum.

STABILITY: Cool, 4°C, 0.008% sodium thiosulfate. M.H.T. = 14 Days.

QUALITY CONTROL: The laboratory must on an ongoing basis, spike at least 5% of the samples from each sample site being monitored to assess accuracy.

REFERENCE: Method 624, Federal Register Part VIII 40 CFR Part 136, Oct 26, 1984.

PRIMARY NAME: 1,2-Dichlorobenzene

Method 8010

TITLE: Halogenated Volatile Organics

MATRIX: groundwater, soils, sludges water miscible liquid wastes,

CAS # : 95-50-1 and non-water miscible wastes

APPLICATION: This method is used for the analysis of 39 halogenated VOCs. Samples are analyzed using direct injection or purge and trap methods. Groundwater must be analyzed by the purge and trap method. The method provides an optional GC column which is used for analyte confirmation and that may help resolve analytes from interferences.

INTERFERENCES: There can be carry-over contamination with high and low level samples. Impurities may come from the purge and trap apparatus, organic compounds outgassing from the plumbing ahead of trap, diffusion of VOCs through the sample bottle septum during shipping or storage, or from solvent vapors in the lab.

INSTRUMENTATION: GC capable of on-column injections or purge and trap sample introduction and a halogen specific detector. Column 1: 8 foot by 0.1 inch 1% SP-1000 on Carbopack-B. Column 2: 6 foot by 0.1 inch bonded n-octane on Porasil-C.

RANGE: 8 to 500 μg/L (reagent water)

MDL: 0.15 μg/L (reagent water)

PRACTICAL QUANTITATION LIMIT FACTORS FOR MULTIPLYING TIMES MDL VALUE

Matrix	Multiplication Factor
Groundwater	10
Low-level soil	10
Water miscible liquid waste	500
High-level soil and sludge	1250
Non-water miscible waste	1250

PRECISION: 0.13X + 6.13 μg/L (overall precision)

ACCURACY: 0.93C + 1.70 μg/L (as recovery)

SAMPLING METHOD: For water and liquid samples; use glass 40 mL vials with Teflon lined septum caps and collect two vials per sample location with no headspace. For solids and concentrated waste samples; use widemouth glass bottles with Teflon liners.

STABILITY: For concentrated wastes, soils, sediments or sludges; cool to 4°C. For liquids; add 4 drops of concentrated hydrochloric acid and cool to 4°C. M.H.T. = 14 days.

QUALITY CONTROL: Analyze a reagent blank, matrix spike and matrix spike duplicate/duplicate for each analytical batch (up to 20 Samples). Demonstrate the purity of glassware and reagents by analyzing a reagent water method blank. Internal, surrogate and five concentration level calibration standards are used.

REFERENCE: Method 8010, SW-846, 3rd ed., Nov 1986.

PRIMARY NAME: 1,2-Dichlorobenzene

Method 8020

TITLE: Aromatic Volatile Organics

MATRIX: groundwater, soils, sludges water miscible liquid wastes, and non-water miscible wastes.

CAS # : 95-50-1

APPLICATION: This method is used to analyze for 8 aromatic VOCs. Samples are analyzed using direct injection or purge and trap methods. Groundwater must be analyzed by the purge and trap method. The method provides an optional GC column that is used for analyte confirmation and may also help resolve analytes from interferences.

INTERFERENCES: There can be carry-over contamination with high and low level samples. Impurities may come from the purge and trap apparatus, organic compounds outgassing from the plumbing ahead of trap, diffusion of VOCs through the sample bottle septum during shipping or storage, or from solvent vapors in the lab.

INSTRUMENTATION: GC capable of on-column injections or purge-and-trap sample introduction and a photoionization detector (PID). Column 1: 6 foot by 0.082 inch with 5% SP-1200 and 1.75% Bentone-34 on Supelcoport. Column 2: 8 foot by 0.1 inch with 5% 1,2,3-tris(2-cyanoethoxy)propane on Chromosorb W-AW.

RANGE: 2.1 to 500µg/L MDL: 0.4 µg/L (reagent water)

PRACTICAL QUANTITATION LIMIT FACTORS FOR MULTIPLYING TIMES MDL VALUE

Matrix	Multiplication Factor
Groundwater	10
Low-level soil	10
Water miscible liquid waste	500
High-level soil and sludge	1250
Non-water miscible waste	1250

PRECISION: 0.22X + 0.53 µg/L (overall precision)

ACCURACY: 0.93C + 0.52 µg/L (as recovery)

SAMPLING METHOD: For water and liquid samples use glass 40 mL vials with Teflon lined septum caps and collect two vials per sample location with no headspace. For solids and concentrated waste samples use widemouth glass bottles with Teflon liners. Cool all samples to 4°C

STABILITY: For concentrated wastes, soils, sediments or sludges cool to 4°C. For liquids, add 4 drops of concentrated hydrochloric acid and cool to 4°C. M.H.T. = 14 days.

QUALITY CONTROL: Analyze a reagent blank, matrix spike and matrix spike duplicate/duplicate for each analytical batch (up to 20 Samples). Demonstrate the purity of glassware and reagents by analyzing a reagent water method blank. Internal, surrogate and five concentration level calibration standards are used. The QC check sample concentrate should contain this compound at 10 mg/mL in methanol.

REFERENCE: Method 8020, SW-846, 3rd ed., Nov 1986.

PRIMARY NAME: 1,3-Dichlorobenzene Method 502.1

TITLE: Halogenated VOCs in Water

MATRIX: Drinking water (finished or any treatment stage) and raw source water.

CAS # : 541-73-1

APPLICATION: Method covers 40 halogenated VOCs. An inert gas is bubbled through a 5 mL water sample. Purged sample components are trapped in tube of sorbent materials. When purging is complete, sorbent tube is heated and backflushed with inert gas to desorb trapped sample onto a packed GC column where it is separated.

INTERFERENCES: During analysis, major contaminant sources are volatile materials in the lab and impurities in purging gas and sorbent trap. With high and low level samples, there can be carry-over contamination. Watch for methylene chloride, it permeates through PTFE tubing.

INSTRUMENTATION: Purge and Trap GC w/halide specific detector. (Two GC columns are recommended); Column #1: 1% SP-1000 on Carbopack B; Column #2: n-octane bonded on Poracil C.

RANGE: 8.0-505 μg/L (Drinking water) MDL: not determined.

PRECISION: RSD = 8.3% @ 0.40 μg/L spike (S.L.) (21 Samples)

ACCURACY: Avg Recovery = 95% @ 0.40 μg/L spike (S.L.) (21 Samples)

SAMPLING METHOD: Use a 40-120 mL screw cap vial (pre-washed with detergent, rinsed with distilled water and oven dried at 105°C) with a PTFE-faced silicone septum. If residual chlorine is in the water add about 25 mg of ascorbic acid to each vial before sample collection. Collect bubble free samples.

STABILITY: Cool to 4°C; HCl to pH <2. M.H.T. = 14 Days.

QUALITY CONTROL: As an initial demonstration of lab accuracy and precision, analyze 4 to 7 replicates of a lab fortified blank containing analyte at 0.1-5 μg/L. Collect all samples in duplicate.

REFERENCE: Method 502.1, Volatile Halogenated Organic Compounds in Water by Purge and Trap GC, EPA 600/4-88/039.

PRIMARY NAME: 1,3-Dichlorobenzene

Method 502.2

TITLE: VOCs in Water by Purge and Trap capillary column GC

MATRIX: Drinking water (finished or any treatment stage) and raw source water.

CAS # : 541-73-1

APPLICATION: Method covers 60 VOCs. An inert gas is bubbled through a 5 mL sample in a purging chamber. Purged sample components are trapped in tube of sorbent materials. When purging is done, sorbent tube is heated and backflushed w/helium to desorb trapped sample onto a capillary GC column.

INTERFERENCES: During analysis, major contaminant sources are volatile materials in the lab and impurities in purging gas and sorbent trap. With high and low level samples, there can be carry-over contamination. Beware of methylene chloride, it permeates through PTFE tubing.

INSTRUMENTATION: (capillary column) GC, w/PID and electrolytic conductivity detectors in series. Column #1: VOCOL wide bore column; Column #2: RTX-502.2 mega bore column; Column #3: DB-62 mega bore column.

RANGE: Approximately 0.02-200 μg/L

MDL: 0.02 μg/L (reagent C) (PID) 60 m x 0.75 mm VOCOL capillary column

PRECISION: RSD = 4.0% @ Spike of 10 μg/L (reagent C) (PID) 60 m x 0.75 mm VOCOL capillary column

ACCURACY: Recovery = 106% @ spike of 10 μg/L (reagent C) (PID) 60 m x 0.75 mm VOCOL capillary column

SAMPLING METHOD: Use a 40-120 mL screw cap vial (pre-washed with detergent, rinsed with distilled water and oven dried at 105°C) with a PTFE-faced silicone septum. If residual chlorine is in the water add about 25 mg of ascorbic acid to each vial before sample collection. Collect bubble free samples.

STABILITY: Cool to 4°C; HCl to pH <2. M.H.T. = 14 Days.

QUALITY CONTROL: As an initial demonstration of lab accuracy and precision, analyze 4 to 7 replicates of a lab fortified blank containing analyte at 0.1-5 μg/L. Collect all samples in duplicate.

REFERENCE: Method 502.2, Volatile Organic Compounds in Water by Purge and Trap Capillary Column GC, EPA 600/4-88/039.

PRIMARY NAME: 1,3-Dichlorobenzene Method 503.1

TITLE: Aromatic and Unsaturated VOCs in Water

MATRIX: Drinking water (finished or any treatment stage) and raw source water.

CAS # : 541-73-1

APPLICATION: Method covers 28 aromatic and unsaturated VOCs. An inert gas is bubbled through a 5 mL water sample. Purged sample components are trapped in tube of sorbent materials. When purging is complete, sorbent tube is heated and backflushed with inert gas to desorb trapped sample onto a packed GC column.

INTERFERENCES: During analysis, major contaminant sources are volatile materials in the lab and impurities in purging gas and sorbent trap. With high and low level samples,there can be carry-over contamination. Excess water causes a negative baseline deflection.

INSTRUMENTATION: Purge and Trap GC w/photoionization detector. (Two GC columns are recommended); Column #1: 5% SP-1200 and 1.75% Bentone 34 on Supelcoport; 5% 1,2,3-tris(2-cyanoethoxy)propane on Chromosorb W.

RANGE: 2.2-600 μg/L (Drinking water) MDL: 0.006 μg/L in water

PRECISION: RSD = 8.5% @ 0.50 μg/L; 19 samples

ACCURACY: Avg Recovery = 91% @ 0.50 μg/L; 19 samples

SAMPLING METHOD: Use a 40-120 mL screw cap vial (pre-washed with detergent, rinsed with distilled water and oven dried at 105°C) with a PTFE-

faced silicone septum. If residual chlorine is in the water add about 25 mg of ascorbic acid to each vial before sample collection. Collect bubble free samples.

STABILITY: Cool to 4°C; HCl to pH <2. M.H.T. = 14 Days.

QUALITY CONTROL: As an initial demonstration of lab accuracy and precision, analyze 4 to 7 replicates of a lab fortified blank containing analyte at 0.1-5 µg/L. Collect all samples in duplicate.

REFERENCE: Method 503.1, Volatile Aromatic and Unsaturated Organic Compounds in C by Purge and Trap GC, EPA 600/4-88/039.

PRIMARY NAME: 1,3-Dichlorobenzene Method 524.1

TITLE: VOCs in Water by Purge and Trap GC/MS

MATRIX: Drinking water (finished or any treatment stage) and raw source water.

CAS # : 541-73-1

APPLICATION: Method covers 51 VOCs. An inert gas is bubbled through a 25 mL water sample. Purged sample components are trapped in tube of sorbent materials. When purging is complete, sorbent tube is heated and backflushed with helium to desorb trapped sample onto a packed GC column.

INTERFERENCES: During analysis, major contaminant sources are volatile materials in the lab and impurities in purging gas and sorbent trap. With high and low level samples, there can be carry-over contamination. Watch for methylene chloride, it permeates through PTFE tubing.

INSTRUMENTATION: Gas Chromatography/Mass Spectrometry/Data System; 1% SP-1000 on Carbopack B column.

RANGE: Approximately 0.2-200 µg/L MDL: not listed.

PRECISION: not listed.

ACCURACY: not listed.

SAMPLING METHOD: Use a 60-120 mL screw cap vial (pre-washed with detergent, rinsed with distilled water and oven dried at 105°C) with a PTFE-faced silicone septum. If residual chlorine is in the water add about 25 mg of ascorbic acid to each vial before sample collection. Collect bubble free samples.

STABILITY: Cool to 4°C; HCl to pH <2. M.H.T. = 14 Days.

QUALITY CONTROL: As an initial demonstration of lab accuracy and precision, analyze 4 to 7 replicates of a lab fortified blank containing analyte at 0.2-5 μg/L. Collect all samples in duplicate.

REFERENCE: Method 524.1, VOCs in Water by Purge and Trap Gas Chromatography/Mass Spectrometry, EPA 600/4-88/039.

PRIMARY NAME: 1,3-Dichlorobenzene Method 524.2

TITLE: VOCs in Water Purge and Trap capillary column GC/MS

MATRIX: Drinking water (finished or any treatment stage) and raw source water.

CAS # : 541-73-1

APPLICATION: Method covers 60 VOCs. An inert gas is bubbled through a 25 mL water sample. Purged sample components are trapped in tube of sorbent materials. When purging is complete, sorbent tube is heated and backflushed with helium to desorb trapped sample onto a capillary GC column.

INTERFERENCES: During analysis, major contaminant sources are volatile materials in the lab and impurities in purging gas and sorbent trap. With high and low level samples, there can be carry-over contamination. Watch for methylene chloride, it permeates through PTFE tubing.

INSTRUMENTATION: (Capillary Column) Gas Chromatography/Mass Spectrometry/Data System. Column #1: VOCOL glass wide bore; Column #2: DB-624 fused silica; Column #3: DB-5 fused silica.

RANGE: Approximately 0.1-200 μg/L

MDL: 0.12 μg/L on wide bore capillary column.

PRECISION: RSD = 6.9% @ (0.5-10) μg/L (24 samples) wide bore capillary column.

ACCURACY: Recovery = 99% @ (0.5-10) μg/L (24 samples) wide bore capillary column.

SAMPLING METHOD: Use a 60-120 mL screw cap vial (pre-washed with detergent, rinsed with distilled water and oven dried at 105°C) with a PTFE-faced silicone septum. If residual chlorine is in the water add about 25 mg of ascorbic acid to each vial before sample collection. Collect bubble free samples.

STABILITY: Cool to 4°C; HCl to pH <2. M.H.T. = 14 Days.

QUALITY CONTROL: As an initial demonstration of lab accuracy and precision, analyze 4 to 7 replicates of a lab fortified blank containing analyte at 0.2-5 μg/L. Collect all samples in duplicate.

REFERENCE: Method 524.2, VOCs in Water by Purge and Trap Capillary column GC/MS, EPA 600/4-88/039.

PRIMARY NAME: 1,3-Dichlorobenzene Method 601

TITLE: Purgeable Halocarbons MATRIX: Wastewater

CAS # : 541-73-1

APPLICATION: Method covers 29 purgeable halocarbons. (Method 624 provides GC/MS conditions appropriate for the qualitative and quantitative confirmation of results). Method describes conditions for a second GC column to confirm measurements made with primary column.

INTERFERENCES: Impurities in the purge gas and organic compounds outgassing from the plumbing ahead of the trap. With high and low level samples, there can be carry-over contamination. Diffusion of volatile organics through the septum seal into the sample.

INSTRUMENTATION: GC equipped with halide-specific detector. (With purge and trap unit).

RANGE: 8.0 To 500 μg/L. MDL: 0.32 μg/L.

PRECISION: 0.26X+2.34 μg/L (overall precision)

ACCURACY: 0.95C+0.43 μg/L (as recovery)

SAMPLING METHOD: 25 mL glass vial. Teflon lined septum.

STABILITY: Cool, 4°C, 0.008% sodium thiosulfate. M.H.T. = 14 Days.

QUALITY CONTROL: The laboratory must on an ongoing basis, spike at least 10% of the samples from each sample site being monitored to assess accuracy.

REFERENCE: Method 601, Federal Register Part VIII 40 CFR Part 136, Oct 26, 1984.

PRIMARY NAME: 1,3-Dichlorobenzene Method 602

TITLE: Purgeable Aromatics MATRIX: Wastewater

CAS # : 541-73-1

APPLICATION: Method covers 7 purgeable aromatics. (Method 624 provides GC/MS conditions appropriate for the qualitative and quantitative confirmation of results).Method describes conditions for a second GC column to confirm measurements made with primary column.

INTERFERENCES: Impurities in the purge gas and organic compounds outgassing from the plumbing ahead of the trap. With high and low level samples, there can be carry-over contamination. Diffusion of volatile organics through the septum seal into the sample.

INSTRUMENTATION: GC equipped with photoionization detector. (With purge and trap unit)

RANGE: 2.1 To 550 μg/L. MDL: 0.4 μg/L.

PRECISION: 0.19X+0.09 μg/L (overall precision)

ACCURACY: 0.96C-0.04 μg/L (as recovery)

SAMPLING METHOD: 25 mL glass vial. Teflon lined septum.

STABILITY: Cool, 4°C, 0.008% sodium thiosulfate. M.H.T. = 14 Days.

QUALITY CONTROL: The laboratory must on an ongoing basis, spike at least 10% of the samples from each sample site being monitored to assess accuracy.

REFERENCE: Method 602, Federal Register Part VIII 40 CFR Part 136, Oct 26, 1984.

PRIMARY NAME: 1,3-Dichlorobenzene Method 624

TITLE: Purgeables MATRIX: Wastewater

CAS # : 541-73-1

APPLICATION: Method covers 31 purgeable organics. An inert gas is bubbled through a 5 mL water sample in a specially designed purging chamber. Here, purgeables are transferred from aqueous to gaseous phase, passed onto a sorbent column and trapped. Trap is heated and backflushed with inert gas to desorb purgeables onto a GC column, where purgeables are separated.

INTERFERENCES: Impurities in the purge gas, organic compounds outgassing from the plumbing ahead of the trap, and solvent vapors in the laboratory. With high and low level samples, there can be carry-over contamination.

INSTRUMENTATION: GC/MS with purge and trap unit.

RANGE: 5-600 μg/L MDL: not determined

PRECISION: 0.18X-0.82 μg/L (overall precision)

ACCURACY: 1.06C+1.68 μg/L (as recovery)

SAMPLING METHOD: 25 mL glass vial. Teflon lined septum.

STABILITY: Cool, 4°C, 0.008% sodium thiosulfate. M.H.T. = 14 Days.

QUALITY CONTROL: The laboratory must on an ongoing basis, spike at least 5% of the samples from each sample site being monitored to assess accuracy.

REFERENCE: Method 624, Federal Register Part VIII 40 CFR Part 136, Oct 26, 1984.

PRIMARY NAME: 1,3-Dichlorobenzene

Method 8010

TITLE: Halogenated Volatile Organics

MATRIX: groundwater, soils, sludges water miscible liquid wastes, and non-water miscible wastes.

CAS # : 541-73-1

APPLICATION: This method is used for the analysis of 39 halogenated VOCs. Samples are analyzed using direct injection or purge and trap methods. Groundwater must be analyzed by the purge and trap method. The method provides an optional GC column which is used for analyte confirmation and that may help resolve analytes from interferences.

INTERFERENCES: There can be carry-over contamination with high and low level samples. Impurities may come from the purge and trap apparatus, organic compounds outgassing from the plumbing ahead of trap, diffusion of VOCs through the sample bottle septum during shipping or storage, or from solvent vapors in the lab.

INSTRUMENTATION: GC capable of on-column injections or purge and trap sample introduction and a halogen specific detector. Column 1: 8 foot by 0.1 inch 1% SP-1000 on Carbopack-B. Column 2: 6 foot by 0.1 inch bonded n-octane on Porasil-C.

RANGE: 8 to 500 μg/L (reagent water)

MDL: 0.32 μg/L (reagent water)

PRACTICAL QUANTITATION LIMIT FACTORS FOR MULTIPLYING TIMES MDL VALUE

Matrix	Multiplication Factor
Groundwater	10
Low-level soil	10
Water miscible liquid waste	500
High-level soil and sludge	1250
Non-water miscible waste	1250

PRECISION: 0.26X + 2.34 μg/L (overall precision)

ACCURACY: 0.95C + 0.43 μg/L (as recovery)

SAMPLING METHOD: For water and liquid samples; use glass 40 mL vials with Teflon lined septum caps and collect two vials per sample location with no headspace. For solids and concentrated waste samples; use widemouth glass bottles with Teflon liners.

STABILITY: For concentrated wastes, soils, sediments or sludges; cool to 4°C. For liquids; add 4 drops of concentrated hydrochloric acid and cool to 4°C. M.H.T. = 14 days.

QUALITY CONTROL: Analyze a reagent blank, matrix spike and matrix spike duplicate/duplicate for each analytical batch (up to 20 Samples). Demonstrate the purity of glassware and reagents by analyzing a reagent water method blank. Internal, surrogate and five concentration level calibration standards are used.

REFERENCE: Method 8010, SW-846, 3rd ed., Nov 1986.

PRIMARY NAME: 1,3-Dichlorobenzene — Method 8020

TITLE: Aromatic Volatile Organics

MATRIX: groundwater, soils, sludges water miscible liquid wastes, and non-water miscible wastes.

CAS # : 541-73-1

APPLICATION: This method is used to analyze for 8 aromatic VOCs. Samples are analyzed using direct injection or purge and trap methods. Groundwater must be analyzed by the purge and trap method. The method provides an optional GC column that is used for analyte confirmation and may also help resolve analytes from interferences.

INTERFERENCES: There can be carry-over contamination with high and low level samples. Impurities may come from the purge and trap apparatus, organic compounds outgassing from the plumbing ahead of trap, diffusion of VOCs through the sample bottle septum during shipping or storage, or from solvent vapors in the lab.

INSTRUMENTATION: GC capable of on-column injections or purge-and-trap sample introduction and a photoionization detector (PID). Column 1: 6 foot by 0.082 inch with 5% SP-1200 and 1.75% Bentone-34 on Supelcoport. Column 2: 8 foot by 0.1 inch with 5% 1,2,3-tris(2-cyanoethoxy)propane on Chromosorb W-AW.

RANGE: 2.1 to 500 μg/L MDL: 0.4 μg/L (reagent water)

PRACTICAL QUANTITATION LIMIT FACTORS FOR MULTIPLYING TIMES MDL VALUE

Matrix	Multiplication Factor
Groundwater	10
Low-level soil	10
Water miscible liquid waste	500
High-level soil and sludge	1250
Non-water miscible waste	1250

PRECISION: 0.19X + 0.09 μg/L (overall precision)

ACCURACY: 0.96C + 0.04 μg/L (as recovery)

SAMPLING METHOD: For water and liquid samples use glass 40 mL vials with Teflon lined septum caps and collect two vials per sample location with no headspace. For solids and concentrated waste samples use widemouth glass bottles with Teflon liners. Cool all samples to 4°C

STABILITY: For concentrated wastes, soils, sediments or sludges cool to 4°C. For liquids, add 4 drops of concentrated hydrochloric acid and cool to 4°C. M.H.T. = 14 days.

QUALITY CONTROL: Analyze a reagent blank, matrix spike and matrix spike duplicate/duplicate for each analytical batch (up to 20 Samples). Demonstrate the purity of glassware and reagents by analyzing a reagent water method blank. Internal, surrogate and five concentration level calibration standards are used. The QC check sample concentrate should contain this compound at 10 mg/mL in methanol.

REFERENCE: Method 8020, SW-846, 3rd ed., Nov 1986.

PRIMARY NAME: 1,4-Dichlorobenzene Method 502.1

TITLE: Halogenated VOCs in Water

MATRIX: Drinking water (finished or any treatment stage) and raw source water.

CAS # : 106-46-7

APPLICATION: Method covers 40 halogenated VOCs. An inert gas is bubbled through a 5 mL water sample. Purged sample components are trapped in tube of sorbent materials. When purging is complete, sorbent tube is heated and backflushed with inert gas to desorb trapped sample onto a packed GC column where it is separated.

INTERFERENCES: During analysis, major contaminant sources are volatile materials in the lab and impurities in purging gas and sorbent trap. With high and low level samples, there can be carry-over contamination. Watch for methylene chloride, it permeates through PTFE tubing.

INSTRUMENTATION: Purge and Trap GC w/halide specific detector. (Two GC columns are recommended); Column #1: 1% SP-1000 on Carbopack B; Column #2: n-octane bonded on Poracil C.

RANGE: 8.0-505 µg/L (Drinking water) MDL: not determined.

PRECISION: RSD = 13% @ 0.40 µg/L spike (S.L.) (20 Samples)

ACCURACY: Avg Recovery = 90% @ 0.40 µg/L spike (S.L.) (20 Samples)

SAMPLING METHOD: Use a 40-120 mL screw cap vial (pre-washed with detergent, rinsed with distilled water and oven dried at 105°C) with a PTFE-faced silicone septum. If residual chlorine is in the water add about 25 mg of ascorbic acid to each vial before sample collection. Collect bubble free samples.

STABILITY: Cool to 4°C; HCl to pH <2. M.H.T. = 14 Days.

QUALITY CONTROL: As an initial demonstration of lab accuracy and precision, analyze 4 to 7 replicates of a lab fortified blank containing analyte at 0.1-5 µg/L. Collect all samples in duplicate.

REFERENCE: Method 502.1, Volatile Halogenated Organic Compounds in Water by Purge and Trap GC, EPA 600/4-88/039.

PRIMARY NAME: 1,4-Dichlorobenzene

Method 502.2

TITLE: VOCs in Water by Purge and Trap capillary column GC

MATRIX: Drinking water (finished or any treatment stage) and raw source water.

CAS # : 106-46-7

APPLICATION: Method covers 60 VOCs. An inert gas is bubbled through a 5 mL sample in a purging chamber. Purged sample components are trapped in tube of sorbent materials. When purging is done, sorbent tube is heated and backflushed w/helium to desorb trapped sample onto a capillary GC column.

INTERFERENCES: During analysis, major contaminant sources are volatile materials in the lab and impurities in purging gas and sorbent trap. With high and low level samples, there can be carry-over contamination. Beware of methylene chloride, it permeates through PTFE tubing.

INSTRUMENTATION: (Capillary Column) GC, w/PID and electrolytic conductivity detectors in series. Column #1: VOCOL wide bore column; Column #2: RTX-502.2 mega bore column; Column #3: DB-62 mega bore column.

RANGE: Approximately 0.02-200 μg/L

MDL: 0.01 μg/L (reagent C) (PID) 60 m x 0.75 mm VOCOL capillary column

PRECISION: RSD = 2.3% @ Spike of 10 μg/L (reagent C) (ELCD) 60 m x 0.75 mm VOCOL capillary column

ACCURACY: Recovery = 98% @ spike of 10 μg/L (reagent C) (ELCD) 60 m x 0.75 mm VOCOL capillary column

SAMPLING METHOD: Use a 40-120 mL screw cap vial (pre-washed with detergent, rinsed with distilled water and oven dried at 105°C) with a PTFE-faced silicone septum. If residual chlorine is in the water add about 25 mg of ascorbic acid to each vial before sample collection. Collect bubble free samples.

STABILITY: Cool to 4°C; HCl to pH <2.

M.H.T. = 14 Days.

QUALITY CONTROL: As an initial demonstration of lab accuracy and

precision, analyze 4 to 7 replicates of a lab fortified blank containing analyte at 0.1-5 μg/L. Collect all samples in duplicate.

REFERENCE: Method 502.2, Volatile Organic Compounds in Water by Purge and Trap Capillary Column GC, EPA 600/4-88/039.

PRIMARY NAME: 1,4-Dichlorobenzene Method 503.1

TITLE: Aromatic and Unsaturated VOCs in Water

MATRIX: Drinking water (finished or any treatment stage) and raw source water.

CAS # : 106-46-7

APPLICATION: Method covers 28 aromatic and unsaturated VOCs. An inert gas is bubbled through a 5 mL water sample. Purged sample components are trapped in tube of sorbent materials. When purging is complete, sorbent tube is heated and backflushed with inert gas to desorb trapped sample onto a packed GC column.

INTERFERENCES: During analysis, major contaminant sources are volatile materials in the lab and impurities in purging gas and sorbent trap. With high and low level samples, there can be carry-over contamination. Excess water causes a negative baseline deflection.

INSTRUMENTATION: Purge and Trap GC w/photoionization detector. (Two GC columns are recommended); Column #1: 5% SP-1200 and 1.75% Bentone 34 on Supelcoport; 5% 1,2,3-tris(2-cyanoethoxy)propane on Chromosorb W.

RANGE: 2.2-600 μg/L (Drinking water) MDL: 0.006 μg/L in water

PRECISION: RSD = 22.8% @ 68.6 μg/L; 5 labs

ACCURACY: Avg Recovery = 91% @ 68.6 μg/L; 5 labs

SAMPLING METHOD: Use a 40-120 mL screw cap vial (pre-washed with detergent, rinsed with distilled water and oven dried at 105°C) with a PTFE-faced silicone septum. If residual chlorine is in the water add about 25 mg of ascorbic acid to each vial before sample collection. Collect bubble free samples.

STABILITY: Cool to 4°C; HCl to pH <2. M.H.T. = 14 Days.

QUALITY CONTROL: As an initial demonstration of lab accuracy and precision, analyze 4 to 7 replicates of a lab fortified blank containing analyte at 0.1-5 µg/L. Collect all samples in duplicate.

REFERENCE: Method 503.1, Volatile aromatic and unsaturated organic compounds in Water by Purge and Trap GC, EPA 600/4-88/039.

PRIMARY NAME: 1,4-Dichlorobenzene Method 524.1

TITLE: VOCs in Water by Purge and Trap GC/MS

MATRIX: Drinking water (finished or any treatment stage) and raw source water.

CAS # : 106-46-7

APPLICATION: Method covers 51 VOCs. An inert gas is bubbled through a 25 mL water sample. Purged sample components are trapped in tube of sorbent materials. When purging is complete, sorbent tube is heated and backflushed with helium to desorb trapped sample onto a packed GC column.

INTERFERENCES: During analysis, major contaminant sources are volatile materials in the lab and impurities in purging gas and sorbent trap. With high and low level samples, there can be carry-over contamination. Watch for methylene chloride, it permeates through PTFE tubing.

INSTRUMENTATION: Gas Chromatography/Mass Spectrometry/Data System; 1% SP-1000 on Carbopack B column.

RANGE: Approximately 0.2-200 µg/L MDL: 2 µg/L

PRECISION: RSD = 13% @ 5.0 µg/L

ACCURACY: Mean Accuracy = 112% @ 5.0 µg/L

SAMPLING METHOD: Use a 60-120 mL screw cap vial (pre-washed with detergent, rinsed with distilled water and oven dried at 105°C) with a PTFE-faced silicone septum. If residual chlorine is in the water add about 25 mg of ascorbic acid to each vial before sample collection. Collect bubble free samples.

STABILITY: Cool to 4°C; HCl to pH <2. M.H.T. = 14 Days.

QUALITY CONTROL: As an initial demonstration of lab accuracy and precision, analyze 4 to 7 replicates of a lab fortified blank containing analyte at 0.2-5 μg/L. Collect all samples in duplicate.

REFERENCE: Method 524.1, VOCs in Water by Purge and Trap Gas Chromatography/Mass Spectrometry, EPA 600/4-88/039.

PRIMARY NAME: 1,4-Dichlorobenzene Method 524.2

TITLE: VOCs in Water (Purge and Trap) capillary column GC/MS

MATRIX: Drinking water (finished or any treatment stage) and raw source water.

CAS # : 106-46-7

APPLICATION: Method covers 60 VOCs. An inert gas is bubbled through a 25 mL water sample. Purged sample components are trapped in tube of sorbent materials. When purging is complete, sorbent tube is heated and backflushed with helium to desorb trapped sample onto a capillary GC column.

INTERFERENCES: During analysis, major contaminant sources are volatile materials in the lab and impurities in purging gas and sorbent trap. With high and low level samples, there can be carry-over contamination. Watch for methylene chloride, it permeates through PTFE tubing.

INSTRUMENTATION: (Capillary Column) Gas Chromatography/Mass Spectrometry/Data System. Column #1: VOCOL glass wide bore; Column #2: DB-624 fused silica; Column #3: DB-5 fused silica.

RANGE: Approximately 0.1-200 μg/L

MDL: 0.03 μg/L on wide bore capillary column.

PRECISION: RSD = 6.4% @ (0.2-20) μg/L (31 samples) wide bore capillary column.

ACCURACY: Recovery = 103% @ (0.2-20) μg/L (31 samples) wide bore capillary column.

SAMPLING METHOD: Use a 60-120 mL screw cap vial (pre-washed with detergent, rinsed with distilled water and oven dried at 105°C) with a PTFE-faced silicone septum. If residual chlorine is in the water add about 25 mg of ascorbic acid to each vial before sample collection. Collect bubble free samples.

STABILITY: Cool to 4°C; HCl to pH <2. M.H.T. = 14 Days.

QUALITY CONTROL: As an initial demonstration of lab accuracy and precision, analyze 4 to 7 replicates of a lab fortified blank containing analyte at 0.2-5 μg/L. Collect all samples in duplicate.

REFERENCE: Method 524.2, VOCs in Water by Purge and Trap Capillary column GC/MS, EPA 600/4-88/039.

PRIMARY NAME: 1,4-Dichlorobenzene Method 601

TITLE: Purgeable Halocarbons MATRIX: Wastewater

CAS # : 106-46-7

APPLICATION: Method covers 29 purgeable halocarbons. (Method 624 provides GC/MS conditions appropriate for the qualitative and quantitative confirmation of results). Method describes conditions for a second GC column to confirm measurements made with primary column.

INTERFERENCES: Impurities in the purge gas and organic compounds outgassing from the plumbing ahead of the trap. With high and low level samples, there can be carry-over contamination. Diffusion of volatile organics through the septum seal into the sample.

INSTRUMENTATION: GC equipped with halide-specific detector. (With purge and trap unit).

RANGE: 8.0 To 500 μg/L. MDL: 0.24 μg/L.

PRECISION: 0.20X+0.41 μg/L (overall precision)

ACCURACY: 0.93C-0.09 μg/L (as recovery)

SAMPLING METHOD: 25 mL glass vial. Teflon lined septum.

STABILITY: Cool, 4°C, 0.008% sodium thiosulfate. M.H.T. = 14 Days.

QUALITY CONTROL: The laboratory must on an ongoing basis, spike at least 10% of the samples from each sample site being monitored to assess accuracy.

REFERENCE: Method 601, Federal Register Part VIII 40 CFR Part 136, Oct 26, 1984.

PRIMARY NAME: 1,4-Dichlorobenzene Method 602

TITLE: Purgeable Aromatics MATRIX: Wastewater

CAS # : 106-46-7

APPLICATION: Method covers 7 purgeable aromatics. (Method 624 provides GC/MS conditions appropriate for the qualitative and quantitative confirmation of results).Method describes conditions for a second GC column to confirm measurements made with primary column.

INTERFERENCES: Impurities in the purge gas and organic compounds outgassing from the plumbing ahead of the trap. With high and low level samples, there can be carry-over contamination. Diffusion of volatile organics through the septum seal into the sample.

INSTRUMENTATION: GC equipped with photoionization detector. (With purge and trap unit)

RANGE: 2.1 To 550 μg/L. MDL: 0.3 μg/L.

PRECISION: 0.20X+0.41 μg/L (overall precision)

ACCURACY: 0.93C+0.09 μg/L (as recovery)

SAMPLING METHOD: 25 mL glass vial. Teflon lined septum.

STABILITY: Cool, 4°C, 0.008% sodium thiosulfate. M.H.T. = 14 Days.

QUALITY CONTROL: The laboratory must on an ongoing basis, spike at least 10% of the samples from each sample site being monitored to assess accuracy.

REFERENCE: Method 602, Federal Register Part VIII 40 CFR Part 136, Oct 26, 1984.

PRIMARY NAME: 1,4-Dichlorobenzene Method 624

TITLE: Purgeables MATRIX: Wastewater

CAS # : 106-46-7

APPLICATION: Method covers 31 purgeable organics. An inert gas is bubbled through a 5 mL water sample in a specially designed purging chamber. Here, purgeables are transferred from aqueous to gaseous phase, passed onto a sorbent column and trapped. Trap is heated and backflushed with inert gas to desorb purgeables onto a GC column, where purgeables are separated.

INTERFERENCES: Impurities in the purge gas, organic compounds outgassing from the plumbing ahead of the trap, and solvent vapors in the laboratory. With high and low level samples, there can be carry-over contamination.

INSTRUMENTATION: GC/MS with purge and trap unit.

RANGE: 5-600 µg/L MDL: not determined

PRECISION: 0.30X-1.20 µg/L (overall precision)

ACCURACY: 0.94C+4.47 µg/L (as recovery)

SAMPLING METHOD: 25 mL glass vial. Teflon lined septum.

STABILITY: Cool, 4°C, 0.008% sodium thiosulfate. M.H.T. = 14 Days.

QUALITY CONTROL: The laboratory must on an ongoing basis, spike at least 5% of the samples from each sample site being monitored to assess accuracy.

REFERENCE: Method 624, Federal Register Part VIII 40 CFR Part 136, Oct 26, 1984.

PRIMARY NAME: 1,4-Dichlorobenzene

Method 8010

TITLE: Halogenated Volatile Organics

MATRIX: groundwater, soils, sludges water miscible liquid wastes, and non-water miscible wastes.

CAS # : 106-46-7

APPLICATION: This method is used for the analysis of 39 halogenated VOCs. Samples are analyzed using direct injection or purge and trap methods. Groundwater must be analyzed by the purge and trap method. The method provides an optional GC column which is used for analyte confirmation and that may help resolve analytes from interferences.

INTERFERENCES: There can be carry-over contamination with high and low level samples. Impurities may come from the purge and trap apparatus, organic compounds outgassing from the plumbing ahead of trap, diffusion of VOCs through the sample bottle septum during shipping or storage, or from solvent vapors in the lab.

INSTRUMENTATION: GC capable of on-column injections or purge and trap sample introduction and a halogen specific detector. Column 1: 8 foot by 0.1 inch 1% SP-1000 on Carbopack-B. Column 2: 6 foot by 0.1 inch bonded n-octane on Porasil-C.

RANGE: 8 to 500 μg/L (reagent water)

MDL: 0.24 μg/L (reagent water)

PRACTICAL QUANTITATION LIMIT FACTORS FOR MULTIPLYING TIMES MDL VALUE

Matrix	Multiplication Factor
Groundwater	10
Low-level soil	10
Water miscible liquid waste	500
High-level soil and sludge	1250
Non-water miscible waste	1250

PRECISION: 0.20X + 0.41 μg/L (overall precision)

ACCURACY: 0.93C - 0.09 μg/L (as recovery)

SAMPLING METHOD: For water and liquid samples; use glass 40 mL vials with Teflon lined septum caps and collect two vials per sample location with no headspace. For solids and concentrated waste samples; use widemouth glass bottles with Teflon liners.

STABILITY: For concentrated wastes, soils, sediments or sludges; cool to 4°C. For liquids; add 4 drops of concentrated hydrochloric acid and cool to 4°C. M.H.T. = 14 days.

QUALITY CONTROL: Analyze a reagent blank, matrix spike and matrix spike duplicate/duplicate for each analytical batch (up to 20 Samples). Demonstrate the purity of glassware and reagents by analyzing a reagent water method blank. Internal, surrogate and five concentration level calibration standards are used.

REFERENCE: Method 8010, SW-846, 3rd ed., Nov 1986.

PRIMARY NAME: 1,4-Dichlorobenzene Method 8020

TITLE: Aromatic Volatile Organics

MATRIX: groundwater, soils, sludges water miscible liquid wastes, and non-water miscible wastes.

CAS # : 106-46-7

APPLICATION: This method is used to analyze for 8 aromatic VOCs. Samples are analyzed using direct injection or purge and trap methods. Groundwater must be analyzed by the purge and trap method. The method provides an optional GC column that is used for analyte confirmation and may also help resolve analytes from interferences.

INTERFERENCES: There can be carry-over contamination with high and low level samples. Impurities may come from the purge and trap apparatus, organic compounds outgassing from the plumbing ahead of trap, diffusion of VOCs through the sample bottle septum during shipping or storage, or from solvent vapors in the lab.

INSTRUMENTATION: GC capable of on-column injections or purge-and-trap sample introduction and a photoionization detector (PID). Column 1: 6 foot by 0.082 inch with 5% SP-1200 and 1.75% Bentone-34 on Supelcoport.

Column 2: 8 foot by 0.1 inch with 5% 1,2,3-tris(2-cyanoethoxy)propane on Chromosorb W-AW.

RANGE: 2.1 to 500 μg/L MDL: 0.3 μg/L (reagent water)

PRACTICAL QUANTITATION LIMIT FACTORS FOR MULTIPLYING TIMES MDL VALUE

Matrix	Multiplication Factor
Groundwater	10
Low-level soil	10
Water miscible liquid waste	500
High-level soil and sludge	1250
Non-water miscible waste	1250

PRECISION: 0.20X + 0.41 μg/L (overall precision)

ACCURACY: 0.93C + 0.09 μg/L (as recovery)

SAMPLING METHOD: For water and liquid samples use glass 40 mL vials with Teflon lined septum caps and collect two vials per sample location with no headspace. For solids and concentrated waste samples use widemouth glass bottles with Teflon liners. Cool all samples to 4°C

STABILITY: For concentrated wastes, soils, sediments or sludges cool to 4°C. For liquids, add 4 drops of concentrated hydrochloric acid and cool to 4°C. M.H.T. = 14 days.

QUALITY CONTROL: Analyze a reagent blank, matrix spike and matrix spike duplicate/duplicate for each analytical batch (up to 20 Samples). Demonstrate the purity of glassware and reagents by analyzing a reagent water method blank. Internal, surrogate and five concentration level calibration standards are used. The QC check sample concentrate should contain this compound at 10 μg/mL in methanol.

REFERENCE: Method 8020, SW-846, 3rd ed., Nov 1986.

PRIMARY NAME: Dichlorodifluoromethane Method 502.1

TITLE: Halogenated VOCs in Water MATRIX: Drinking water (finished or any treatment stage) and raw source water.

CAS # : 75-71-8

APPLICATION: Method covers 40 halogenated VOCs. An inert gas is bubbled through a 5 mL water sample. Purged sample components are trapped in tube of sorbent materials. When purging is complete, sorbent tube is heated and backflushed with inert gas to desorb trapped sample onto a packed GC column where it is separated.

INTERFERENCES: During analysis, major contaminant sources are volatile materials in the lab and impurities in purging gas and sorbent trap. With high and low level samples, there can be carry-over contamination. Watch for methylene chloride, it permeates through PTFE tubing.

INSTRUMENTATION: Purge and Trap GC w/halide specific detector. (Two GC columns are recommended); Column #1: 1% SP-1000 on Carbopack B; Column #2: n-octane bonded on Poracil C.

RANGE: 8.0-505 µg/L (Drinking water) MDL: not determined.

PRECISION: RSD = 20% @ 0.40 µg/L spike (S.L.) (12 Samples)

ACCURACY: Avg Recovery = 103% @ 0.40 µg/L spike (S.L.) (12 Samples)

SAMPLING METHOD: Use a 40-120 mL screw cap vial (pre-washed with detergent, rinsed with distilled water and oven dried at 105°C) with a PTFE-faced silicone septum. If residual chlorine is in the water add about 25 mg of ascorbic acid to each vial before sample collection. Collect bubble free samples.

STABILITY: Cool to 4°C; HCl to pH <2. M.H.T. = 14 Days.

QUALITY CONTROL: As an initial demonstration of lab accuracy and precision, analyze 4 to 7 replicates of a lab fortified blank containing analyte at 0.1-5 µg/L. Collect all samples in duplicate.

REFERENCE: Method 502.1, Volatile Halogenated Organic Compounds in Water by Purge and Trap GC, EPA 600/4-88/039.

PRIMARY NAME: Dichlorodifluoromethane Method 502.2

TITLE: VOCs in Water by Purge and Trap capillary column GC

MATRIX: Drinking water (finished or any treatment stage) and raw source water.

CAS # : 75-71-8

APPLICATION: Method covers 60 VOCs. An inert gas is bubbled through a 5 mL sample in a purging chamber. Purged sample components are trapped in tube of sorbent materials. When purging is done, sorbent tube is heated and backflushed w/helium to desorb trapped sample onto a capillary GC column.

INTERFERENCES: During analysis, major contaminant sources are volatile materials in the lab and impurities in purging gas and sorbent trap. With high and low level samples, there can be carry-over contamination. Beware of methylene chloride, it permeates through PTFE tubing.

INSTRUMENTATION: (capillary column) GC, w/PID and electrolytic conductivity detectors in series. Column #1: VOCOL wide bore column; Column #2: RTX-502.2 mega bore column; Column #3: DB-62 mega bore column.

RANGE: Approximately 0.02-200 µg/L

MDL: 0.05 µg/L (reagent C) (ELCD) 60 m x 0.75 mm VOCOL capillary column

PRECISION: RSD = 6.6% @ Spike of 10 µg/L (reagent C) (ELCD). 60 m x 0.75 mm VOCOL capillary column

ACCURACY: Recovery = 89% @ spike of 10 µg/L (reagent C) (ELCD) 60 m x 0.75 mm VOCOL capillary column

SAMPLING METHOD: Use a 40-120 mL screw cap vial (pre-washed with detergent, rinsed with distilled water and oven dried at 105°C) with a PTFE-faced silicone septum. If residual chlorine is in the water add about 25 mg of ascorbic acid to each vial before sample collection. Collect bubble free samples.

STABILITY: Cool to 4°C; HCl to pH <2.

M.H.T. = 14 Days.

QUALITY CONTROL: As an initial demonstration of lab accuracy and precision, analyze 4 to 7 replicates of a lab fortified blank containing analyte at 0.1-5 µg/L. Collect all samples in duplicate.

REFERENCE: Method 502.2, Volatile Organic Compounds in Water by Purge and Trap Capillary Column GC, EPA 600/4-88/039.

PRIMARY NAME: Dichlorodifluoromethane Method 524.1

TITLE: VOCs in μ by Purge and Trap GC/MS

MATRIX: Drinking water (finished or any treatment stage) and raw source water.

CAS # : 75-71-8

APPLICATION: Method covers 51 VOCs. An inert gas is bubbled through a 25 mL water sample. Purged sample components are trapped in tube of sorbent materials. When purging is complete, sorbent tube is heated and backflushed with helium to desorb trapped sample onto a packed GC column.

INTERFERENCES: During analysis, major contaminant sources are volatile materials in the lab and impurities in purging gas and sorbent trap. With high and low level samples, there can be carry-over contamination. Watch for methylene chloride, it permeates through PTFE tubing.

INSTRUMENTATION: Gas Chromatography/Mass Spectrometry/Data System; 1% SP-1000 on Carbopack B column.

RANGE: Approximately 0.2-200 μg/L MDL: 0.3 μg/L

PRECISION: RSD = 12% @ 1.0 μg/L

ACCURACY: Mean Accuracy = 96% @ 1.0 μg/L

SAMPLING METHOD: Use a 60-120 mL screw cap vial (pre-washed with detergent, rinsed with distilled water and oven dried at 105°C) with a PTFE-faced silicone septum. If residual chlorine is in the water add about 25 mg of ascorbic acid to each vial before sample collection. Collect bubble free samples.

STABILITY: Cool to 4°C; HCl to pH <2. M.H.T. = 14 Days.

QUALITY CONTROL: As an initial demonstration of lab accuracy and precision, analyze 4 to 7 replicates of a lab fortified blank containing analyte at 0.2-5 μg/L. Collect all samples in duplicate.

REFERENCE: Method 524.1, VOCs in Water by Purge and Trap Gas Chromatography/Mass Spectrometry, EPA 600/4-88/039.

PRIMARY NAME: Dichlorodifluoromethane Method 524.2

TITLE: VOCs inWater Purge and Trap capillary column GC/MS

MATRIX: Drinking water (finished or any treatment stage) and raw source water.

CAS # : 75-71-8

APPLICATION: Method covers 60 VOCs. An inert gas is bubbled through a 25 mL water sample. Purged sample components are trapped in tube of sorbent materials. When purging is complete, sorbent tube is heated and backflushed with helium to desorb trapped sample onto a capillary GC column.

INTERFERENCES: During analysis, major contaminant sources are volatile materials in the lab and impurities in purging gas and sorbent trap. With high and low level samples, there can be carry-over contamination. Watch for methylene chloride, it permeates through PTFE tubing.

INSTRUMENTATION: (capillary column) Gas Chromatography/Mass Spectrometry/Data System. Column #1: VOCOL glass wide bore; Column #2: DB-624 fused silica; Column #3: DB-5 fused silica.

RANGE: Approximately 0.1-200 µg/L

MDL: 0.10 µg/L on wide bore capillary column.

PRECISION: RSD = 7.7% @ (0.5-10) µg/L (18 samples) wide bore capillary column.

ACCURACY: Recovery = 90% @ (0.5-10) µg/L (18 samples) wide bore capillary column.

SAMPLING METHOD: Use a 60-120 mL screw cap vial (pre-washed with detergent, rinsed with distilled water and oven dried at 105°C) with a PTFE-faced silicone septum. If residual chlorine is in the water add about 25 mg of ascorbic acid to each vial before sample collection. Collect bubble free samples.

STABILITY: Cool to 4°C; HCl to pH <2. M.H.T. = 14 Days.

QUALITY CONTROL: As an initial demonstration of lab accuracy and pre-

cision, analyze 4 to 7 replicates of a lab fortified blank containing analyte at 0.2-5 μg/L. Collect all samples in duplicate.

REFERENCE: Method 524.2, VOCs in Water by Purge and Trap Capillary column GC/MS, EPA 600/4-88/039.

PRIMARY NAME: Dichlorodifluoromethane Method 601

TITLE: Purgeable Halocarbons MATRIX: Wastewater

CAS # : 75-71-8

APPLICATION: Method covers 29 purgeable halocarbons. Method describes conditions for a second GC column to confirm measurements made with primary column.

INTERFERENCES: Impurities in the purge gas and organic compounds outgassing from the plumbing ahead of the trap. With high and low level samples, there can be carry-over contamination. Diffusion of volatile organics through the septum seal into the sample.

INSTRUMENTATION: GC equipped with halide-specific detector. (With purge and trap unit).

RANGE: 8.0 To 500 μg/L. MDL: 1.81 μg/L

PRECISION: not listed

ACCURACY: not listed

SAMPLING METHOD: 25 mL glass vial. Teflon lined septum.

STABILITY: Cool, 4°C, 0.008% sodium thiosulfate. M.H.T. = 14 Days.

QUALITY CONTROL: The laboratory must on an ongoing basis, spike at least 10% of the samples from each sample site being monitored to assess accuracy.

REFERENCE: Method 601, Federal Register Part VIII 40 CFR Part 136, Oct 26, 1984.

PRIMARY NAME: 1,2,3-Trichlorobenzene

Method 502.2

TITLE: VOCs in Water by Purge and Trap capillary column GC

MATRIX: Drinking water (finished or any treatment stage) and raw source water.

CAS # : 87-61-6

APPLICATION: Method covers 60 VOCs. An inert gas is bubbled through a 5 mL sample in a purging chamber. Purged sample components are trapped in tube of sorbent materials. When purging is done, sorbent tube is heated and backflushed w/helium to desorb trapped sample onto a capillary GC column.

INTERFERENCES: During analysis, major contaminant sources are volatile materials in the lab and impurities in purging gas and sorbent trap. With high and low level samples, there can be carry-over contamination. Beware of methylene chloride, it permeates through PTFE tubing.

INSTRUMENTATION: (Capillary Column) GC, w/PID and electrolytic conductivity detectors in series. Column #1: VOCOL wide bore column; Column #2: RTX-502.2 mega bore column; Column #3: DB-62 mega bore column.

RANGE: Approximately 0.02-200 μg/L

MDL: 0.03 μg/L (reagent C) (ELCD) 60 m x 0.75 mm VOCOL capillary column

PRECISION: RSD = 3.1% @ Spike of 10 μg/L (reagent C) (ELCD) 60 m x 0.75 mm VOCOL capillary column

ACCURACY: Recovery = 98% @ spike of 10 μg/L (reagent C) (ELCD) 60 m x 0.75 mm VOCOL capillary column

SAMPLING METHOD: Use a 40-120 mL screw cap vial (pre-washed with detergent, rinsed with distilled water and oven dried at 105°C) with a PTFE-faced silicone septum. If residual chlorine is in the water add about 25 mg of ascorbic acid to each vial before sample collection. Collect bubble free samples.

STABILITY: Cool to 4°C; HCl to pH <2.

M.H.T. = 14 Days.

QUALITY CONTROL: As an initial demonstration of lab accuracy and

precision, analyze 4 to 7 replicates of a lab fortified blank containing analyte at 0.1-5 μg/L. Collect all samples in duplicate.

REFERENCE: Method 502.2, Volatile Organic Compounds in Water by Purge and Trap Capillary Column GC, EPA 600/4-88/039.

PRIMARY NAME: 1,2,3-Trichlorobenzene Method 503.1

TITLE: Aromatic and Unsaturated VOCs in Water

MATRIX: Drinking water (finished or any treatment stage) and raw source water.

CAS # : 87-61-6

APPLICATION: Method covers 28 aromatic and unsaturated VOCs. An inert gas is bubbled through a 5 mL water sample. Purged sample components are trapped in tube of sorbent materials. When purging is complete, sorbent tube is heated and backflushed with inert gas to desorb trapped sample onto a packed GC column.

INTERFERENCES: During analysis, major contaminant sources are volatile materials in the lab and impurities in purging gas and sorbent trap. With high and low level samples, there can be carry-over contamination. Excess water causes a negative baseline deflection.

INSTRUMENTATION: Purge and Trap GC w/photoionization detector. (Two GC columns are recommended); Column #1: 5% SP-1200 and 1.75% Bentone 34 on Supelcoport; 5% 1,2,3-tris(2-cyanoethoxy)propane on Chromosorb W.

RANGE: 2.2-600 μg/L (Drinking water) MDL: 0.03 μg/L in water

PRECISION: RSD = 10.4% @ 0.50 μg/L

ACCURACY: Avg Recovery = 85% @ 0.50 μg/L

SAMPLING METHOD: Use a 40-120 mL screw cap vial (pre-washed with detergent, rinsed with distilled water and oven dried at 105°C) with a PTFE-faced silicone septum. If residual chlorine is in the water add about 25 mg of ascorbic acid to each vial before sample collection. Collect bubble free samples.

STABILITY: Cool to 4°C; HCl to pH <2. M.H.T. = 14 Days.

QUALITY CONTROL: As an initial demonstration of lab accuracy and precision, analyze 4 to 7 replicates of a lab fortified blank containing analyte at 0.1-5 μg/L. Collect all samples in duplicate.

REFERENCE: Method 503.1, Volatile aromatic and unsaturated organic compounds in Water by Purge and Trap GC, EPA 600/4-88/039.

PRIMARY NAME: 1,2,3-Trichlorobenzene Method 524.2

TITLE: VOCs in Water Purge and Trap capillary column GC/MS

MATRIX: Drinking water (finished or any treatment stage) and raw source water.

CAS # : 87-61-6

APPLICATION: Method covers 60 VOCs. An inert gas is bubbled through a 25 mL water sample. Purged sample components are trapped in tube of sorbent materials. When purging is complete, sorbent tube is heated and backflushed with helium to desorb trapped sample onto a capillary GC column.

INTERFERENCES: During analysis, major contaminant sources are volatile materials in the lab and impurities in purging gas and sorbent trap. With high and low level samples, there can be carry-over contamination. Watch for methylene chloride, it permeates through PTFE tubing.

INSTRUMENTATION: (Capillary Column) Gas Chromatography/Mass Spectrometry/Data System. Column #1: VOCOL glass wide bore; Column #2: DB-624 fused silica; Column #3: DB-5 fused silica.

RANGE: Approximately 0.1-200 μg/L

MDL: 0.03 μg/L on wide bore capillary column.

PRECISION: RSD = 8.6% @ (0.5-10) μg/L (18 samples) wide bore capillary column.

ACCURACY: Recovery = 109% @ (0.5-10) μg/L (18 samples) wide bore capillary column.

SAMPLING METHOD: Use a 60-120 mL screw cap vial (pre-washed with detergent, rinsed with distilled water and oven dried at 105°C) with a PTFE-faced silicone septum. If residual chlorine is in the water add about 25 mg of ascorbic acid to each vial before sample collection. Collect bubble free samples.

STABILITY: Cool to 4°C; HCl to pH <2. M.H.T. = 14 Days.

QUALITY CONTROL: As an initial demonstration of lab accuracy and precision, analyze 4 to 7 replicates of a lab fortified blank containing analyte at 0.2-5 μg/L. Collect all samples in duplicate.

REFERENCE: Method 524.2, VOCs in C by Purge and Trap Capillary column GC/MS, EPA 600/4-88/039.

PRIMARY NAME: 1,2,4-Trichlorobenzene Method 502.2

TITLE: VOCs in Water by Purge and Trap capillary column GC

MATRIX: Drinking water (finished or any treatment stage) and raw source water.

CAS # : 120-82-1

APPLICATION: Method covers 60 VOCs. An inert gas is bubbled through a 5 mL sample in a purging chamber. Purged sample components are trapped in tube of sorbent materials. When purging is done, sorbent tube is heated and backflushed w/helium to desorb trapped sample onto a capillary GC column.

INTERFERENCES: During analysis, major contaminant sources are volatile materials in the lab and impurities in purging gas and sorbent trap. With high and low level samples, there can be carry-over contamination. Beware of methylene chloride, it permeates through PTFE tubing.

INSTRUMENTATION: (Capillary Column) GC, w/PID and electrolytic conductivity detectors in series. Column #1: VOCOL wide bore column; Column #2: RTX-502.2 mega bore column; Column #3: DB-62 mega bore column.

RANGE: Approximately 0.02-200 μg/L

MDL: 0.03 μg/L (reagent C) (PID) 60 m x 0.75 mm VOCOL capillary column

PRECISION: RSD = 2.1% @ Spike of 10 µg/L (reagent C) (ELCD) 60 m x 0.75 mm VOCOL capillary column

ACCURACY: Recovery = 102% @ spike of 10 µg/L (reagent C) (ELCD) 60 m x 0.75 mm VOCOL capillary column

SAMPLING METHOD: Use a 40-120 mL screw cap vial (pre-washed with detergent, rinsed with distilled water and oven dried at 105°C) with a PTFE-faced silicone septum. If residual chlorine is in the water add about 25 mg of ascorbic acid to each vial before sample collection. Collect bubble free samples.

STABILITY: Cool to 4°C; HCl to pH <2. M.H.T. = 14 Days.

QUALITY CONTROL: As an initial demonstration of lab accuracy and precision, analyze 4 to 7 replicates of a lab fortified blank containing analyte at 0.1-5 µg/L. Collect all samples in duplicate.

REFERENCE: Method 502.2, Volatile Organic Compounds in Water by Purge and Trap Capillary Column GC, EPA 600/4-88/039.

PRIMARY NAME: 1,2,4-Trichlorobenzene Method 503.1

TITLE: Aromatic and Unsaturated VOCs in Water

MATRIX: Drinking water (finished or any treatment stage) and raw source water.

CAS # : 120-82-1

APPLICATION: Method covers 28 aromatic and unsaturated VOCs. An inert gas is bubbled through a 5 mL water sample. Purged sample components are trapped in tube of sorbent materials. When purging is complete, sorbent tube is heated and backflushed with inert gas to desorb trapped sample onto a packed GC column.

INTERFERENCES: During analysis, major contaminant sources are volatile materials in the lab and impurities in purging gas and sorbent trap. With high and low level samples, there can be carry-over contamination. Excess water causes a negative baseline deflection.

INSTRUMENTATION: Purge and Trap GC w/photoionization detector. (Two GC columns are recommended); Column #1: 5% SP-1200 and 1.75% Bentone 34 on Supelcoport; 5% 1,2,3-tris(2-cyanoethoxy)propane on Chromosorb W.

RANGE: 2.2-600 µg/L (Drinking water) MDL: 0.03 µg/L in water

PRECISION: RSD = 14.3% @ 80.8 µg/L

ACCURACY: Avg Recovery = 96% @ 80.8 µg/L

SAMPLING METHOD: Use a 40-120 mL screw cap vial (pre-washed with detergent, rinsed with distilled water and oven dried at 105°C) with a PTFE-faced silicone septum. If residual chlorine is in the water add about 25 mg of ascorbic acid to each vial before sample collection. Collect bubble free samples.

STABILITY: Cool to 4°C; HCl to pH <2. M.H.T. = 14 Days.

QUALITY CONTROL: As an initial demonstration of lab accuracy and precision, analyze 4 to 7 replicates of a lab fortified blank containing analyte at 0.1-5 µg/L. Collect all samples in duplicate.

REFERENCE: Method 503.1, Volatile Aromatic and Unsaturated Organic Compounds in C by Purge and Trap GC, EPA 600/4-88/039.

PRIMARY NAME: 1,2,4-Trichlorobenzene Method 524.2

TITLE: VOCs in Water Purge and Trap capillary column GC/MS

MATRIX: Drinking water (finished or any treatment stage) and raw source water.

CAS # : 120-82-1

APPLICATION: Method covers 60 VOCs. An inert gas is bubbled through a 25 mL water sample. Purged sample components are trapped in tube of sorbent materials. When purging is complete, sorbent tube is heated and backflushed with helium to desorb trapped sample onto a capillary GC column.

INTERFERENCES: During analysis, major contaminant sources are volatile materials in the lab and impurities in purging gas and sorbent trap. With high and low level samples, there can be carry-over contamination. Watch for methylene chloride, it permeates through PTFE tubing.

INSTRUMENTATION: (Capillary Column) Gas Chromatography/Mass Spectrometry/Data System. Column #1: VOCOL glass wide bore; Column #2: DB-624 fused silica; Column #3: DB-5 fused silica.

RANGE: Approximately 0.1-200 μg/L

MDL: 0.04 μg/L on wide bore capillary column.

PRECISION: RSD = 8.3% @ (0.5-10) μg/L (18 samples) wide bore capillary column.

ACCURACY: Recovery = 108% @ (0.5-10) μg/L (18 samples) wide bore capillary column.

SAMPLING METHOD: Use a 60-120 mL screw cap vial (pre-washed with detergent, rinsed with distilled water and oven dried at 105°C) with a PTFE-faced silicone septum. If residual chlorine is in the water add about 25 mg of ascorbic acid to each vial before sample collection. Collect bubble free samples.

STABILITY: Cool to 4°C; HCl to pH <2.

M.H.T. = 14 Days.

QUALITY CONTROL: As an initial demonstration of lab accuracy and precision, analyze 4 to 7 replicates of a lab fortified blank containing analyte at 0.2-5 μg/L. Collect all samples in duplicate.

REFERENCE: Method 524.2, VOCs in Water by Purge and Trap Capillary column GC/MS, EPA 600/4-88/039.

PRIMARY NAME: Trichlorofluoromethane

Method 502.1

TITLE: Halogenated VOCs in Water

MATRIX: Drinking water (finished or any treatment stage) and raw source water.

CAS # : 75-69-4

APPLICATION: Method covers 40 halogenated VOCs. An inert gas is

bubbled through a 5 mL water sample. Purged sample components are trapped in tube of sorbent materials. When purging is complete, sorbent tube is heated and backflushed with inert gas to desorb trapped sample onto a packed GC column where it is separated.

INTERFERENCES: During analysis, major contaminant sources are volatile materials in the lab and impurities in purging gas and sorbent trap. With high and low level samples, there can be carry-over contamination. Watch for methylene chloride, it permeates through PTFE tubing.

INSTRUMENTATION: Purge and Trap GC w/halide specific detector. (Two GC columns are recommended); Column #1: 1% SP-1000 on Carbopack B; Column #2: n-octane bonded on Poracil C.

RANGE: 8.0-505 µg/L (Drinking water) MDL: not determined.

PRECISION: RSD = 9.3% @ 0.40 µg/L spike (S.L.) (21 Samples)

ACCURACY: Avg Recovery = 90% @ 0.40 µg/L spike (S.L.) (21 Samples)

SAMPLING METHOD: Use a 40-120 mL screw cap vial (pre-washed with detergent, rinsed with distilled water and oven dried at 105°C) with a PTFE-faced silicone septum. If residual chlorine is in the water add about 25 mg of ascorbic acid to each vial before sample collection. Collect bubble free samples.

STABILITY: Cool to 4°C; HCl to pH <2. M.H.T. = 14 Days.

QUALITY CONTROL: As an initial demonstration of lab accuracy and precision, analyze 4 to 7 replicates of a lab fortified blank containing analyte at 0.1-5 µg/L. Collect all samples in duplicate.

REFERENCE: Method 502.1, Volatile Halogenated Organic Compounds in Water by Purge and Trap GC, EPA 600/4-88/039.

PRIMARY NAME: Trichlorofluoromethane Method 502.2

TITLE: VOCs in Water by Purge and Trap capillary column GC

MATRIX: Drinking water (finished or any treatment stage) and raw source water.

CAS # : 75-69-4

APPLICATION: Method covers 60 VOCs. An inert gas is bubbled through a 5 mL sample in a purging chamber. Purged sample components are trapped in tube of sorbent materials. When purging is done, sorbent tube is heated and backflushed w/helium to desorb trapped sample onto a capillary GC column.

INTERFERENCES: During analysis, major contaminant sources are volatile materials in the lab and impurities in purging gas and sorbent trap. With high and low level samples, there can be carry-over contamination. Beware of methylene chloride, it permeates through PTFE tubing.

INSTRUMENTATION: (capillary column) GC, w/PID and electrolytic conductivity detectors in series. Column #1: VOCOL wide bore column; Column #2: RTX-502.2 mega bore column; Column #3: DB-62 mega bore column.

RANGE: Approximately 0.02-200 µg/L

MDL: 0.03 µg/L (reagent C) (ELCD) 60 m x 0.75 mm VOCOL capillary column

PRECISION: RSD = 3.5% @ Spike of 10 µg/L (reagent C) (ELCD). 60 m x 0.75 mm VOCOL capillary column

ACCURACY: Recovery = 96% @ spike of 10 µg/L (reagent C) (ELCD) 60 m x 0.75 mm VOCOL capillary column

SAMPLING METHOD: Use a 40-120 mL screw cap vial (pre-washed with detergent, rinsed with distilled water and oven dried at 105°C) with a PTFE-faced silicone septum. If residual chlorine is in the water add about 25 mg of ascorbic acid to each vial before sample collection. Collect bubble free samples.

STABILITY: Cool to 4°C; HCl to pH <2.

M.H.T. = 14 Days.

QUALITY CONTROL: As an initial demonstration of lab accuracy and precision, analyze 4 to 7 replicates of a lab fortified blank containing analyte at 0.1-5 µg/L. Collect all samples in duplicate.

REFERENCE: Method 502.2, Volatile Organic Compounds in Water by Purge and Trap Capillary Column GC, EPA 600/4-88/039.

PRIMARY NAME: Trichlorofluoromethane Method 524.1

TITLE: VOCs in Water by Purge and Trap GC/MS

MATRIX: Drinking water (finished or any treatment stage) and raw source water.

CAS # : 75-69-4

APPLICATION: Method covers 51 VOCs. An inert gas is bubbled through a 25 mL water sample. Purged sample components are trapped in tube of sorbent materials. When purging is complete, sorbent tube is heated and backflushed with helium to desorb trapped sample onto a packed GC column.

INTERFERENCES: During analysis, major contaminant sources are volatile materials in the lab and impurities in purging gas and sorbent trap. With high and low level samples, there can be carry-over contamination. Watch for methylene chloride, it permeates through PTFE tubing.

INSTRUMENTATION: Gas Chromatography/mass spectrometry/Data System; 1% SP-1000 on Carbopack B column.

RANGE: Approximately 0.2-200 μg/L MDL: 0.2 μg/L

PRECISION: RSD = 6.6% @ 1.0 μg/L

ACCURACY: Mean Accuracy = 109% @ 1.0 μg/L

SAMPLING METHOD: Use a 60-120 mL screw cap vial (pre-washed with detergent, rinsed with distilled water and oven dried at 105°C) with a PTFE-faced silicone septum. If residual chlorine is in the water add about 25 mg of ascorbic acid to each vial before sample collection. Collect bubble free samples.

STABILITY: Cool to 4°C; HCl to pH <2. M.H.T. = 14 Days.

QUALITY CONTROL: As an initial demonstration of lab accuracy and precision, analyze 4 to 7 replicates of a lab fortified blank containing analyte at 0.2-5 μg/L. Collect all samples in duplicate.

REFERENCE: Method 524.1, VOCs in Water by Purge and Trap Gas Chromatography/Mass Spectometry, EPA 600/4-88/039.

PRIMARY NAME: Trichlorofluoromethane

Method 524.2

TITLE: VOCs in Water Purge and Trap capillary column GC/MS

MATRIX: Drinking water (finished or any treatment stage) and raw source water.

CAS # : 75-69-4

APPLICATION: Method covers 60 VOCs. An inert gas is bubbled through a 25 mL water sample. Purged sample components are trapped in tube of sorbent materials. When purging is complete, sorbent tube is heated and backflushed with helium to desorb trapped sample onto a capillary GC column.

INTERFERENCES: During analysis, major contaminant sources are volatile materials in the lab and impurities in purging gas and sorbent trap. With high and low level samples, there can be carry-over contamination. Watch for methylene chloride, it permeates through PTFE tubing.

INSTRUMENTATION: (capillary column) Gas Chromatography/Mass Spectometry/Data System. Column #1: VOCOL glass wide bore; Column #2: DB-624 fused silica; Column #3: DB-5 fused silica.

RANGE: Approximately 0.1-200 µg/L

MDL: 0.08 µg/L on wide bore capillary column.

PRECISION: RSD = 8.1% @ (0.5-10) µg/L (24 samples) wide bore capillary column.

ACCURACY: Recovery = 89% @ (0.5-10) µg/L (24 samples) wide bore capillary column.

SAMPLING METHOD: Use a 60-120 mL screw cap vial (pre-washed with detergent, rinsed with distilled water and oven dried at 105°C) with a PTFE-faced silicone septum. If residual chlorine is in the water add about 25 mg of ascorbic acid to each vial before sample collection. Collect bubble free samples.

STABILITY: Cool to 4°C; HCl to pH <2.

M.H.T. = 14 Days.

QUALITY CONTROL: As an initial demonstration of lab accuracy and precision, analyze 4 to 7 replicates of a lab fortified blank containing analyte at 0.2-5 μg/L. Collect all samples in duplicate.

REFERENCE: Method 524.2, VOCs in Water by Purge and Trap Capillary column GC/MS, EPA 600/4-88/039.

PRIMARY NAME: Trichlorofluoromethane Method 601

TITLE: Purgeable Halocarbons MATRIX: Wastewater

CAS # : 75-69-4

APPLICATION: Method covers 29 purgeable halocarbons. (Method 624 provides GC/MS conditions appropriate for the qualitative and quantitative confirmation of results). Method describes conditions for a second GC column to confirm measurements made with primary column.

INTERFERENCES: Impurities in the purge gas and organic compounds outgassing from the plumbing ahead of the trap. With high and low level samples, there can be carry-over contamination. Diffusion of volatile organics through the septum seal into the sample.

INSTRUMENTATION: GC equipped with halide-specific detector. (With purge and trap unit).

RANGE: 8.0 To 500 μg/L. MDL: not determined

PRECISION: 0.26X+0.91 μg/L (overall precision)

ACCURACY: 0.89C-0.07 μg/L (as recovery)

SAMPLING METHOD: 25 mL glass vial. Teflon lined septum.

STABILITY: Cool, 4°C, 0.008% sodium thiosulfate. M.H.T. = 14 Days.

QUALITY CONTROL: The laboratory must on an ongoing basis, spike at least 10% of the samples from each sample site being monitored to assess accuracy.

REFERENCE: Method 601, Federal Register Part VIII 40 CFR Part 136, Oct 26, 1984.

PRIMARY NAME: Trichlorofluoromethane Method 624

TITLE: Purgeables MATRIX: Wastewater

CAS # : 75-69-4

APPLICATION: Method covers 31 purgeable organics. An inert gas is bubbled through a 5 mL water sample in a specially designed purging chamber. Here, purgeables are transferred from aqueous to gaseous phase, passed onto a sorbent column and trapped. Trap is heated and backflushed with inert gas to desorb purgeables onto a GC column, where purgeables are separated.

INTERFERENCES: Impurities in the purge gas, organic compounds outgassing from the plumbing ahead of the trap, and solvent vapors in the laboratory. With high and low level samples, there can be carry-over contamination.

INSTRUMENTATION: GC/MS with purge and trap unit.

RANGE: 5-600 μg/L MDL: not determined

PRECISION: 0.34X-0.39 μg/L (overall precision)

ACCURACY: 0.99C+0.39 μg/L (as recovery)

SAMPLING METHOD: 25 mL glass vial. Teflon lined septum.

STABILITY: Cool, 4°C, 0.008% sodium thiosulfate. M.H.T. = 14 Days.

QUALITY CONTROL: The laboratory must on an ongoing basis, spike at least 5% of the samples from each sample site being monitored to assess accuracy.

REFERENCE: Method 624, Federal Register Part VIII 40 CFR Part 136, Oct 26, 1984.

PRIMARY NAME: Trichlorofluoromethane Method 8010

TITLE: Halogenated Volatile Organics MATRIX: groundwater, soils, sludges water miscible liquid wastes, and non-water miscible wastes.

CAS # : 75-69-4

APPLICATION: This method is used for the analysis of 39 halogenated VOCs. Samples are analyzed using direct injection or purge and trap methods. Groundwater must be analyzed by the purge and trap method. The method provides an optional GC column which is used for analyte confirmation and that may help resolve analytes from interferences.

INTERFERENCES: There can be carry-over contamination with high and low level samples. Impurities may come from the purge and trap apparatus, organic compounds outgassing from the plumbing ahead of trap, diffusion of VOCs through the sample bottle septum during shipping or storage, or from solvent vapors in the lab.

INSTRUMENTATION: GC capable of on-column injections or purge and trap sample introduction and a halogen specific detector. Column 1: 8 foot by 0.1 inch 1% SP-1000 on Carbopack-B. Column 2: 6 foot by 0.1 inch bonded n-octane on Porasil-C.

RANGE: 8 to 500 μg/L (reagent water)   MDL: Not determined

PRACTICAL QUANTITATION LIMIT FACTORS FOR MULTIPLYING TIMES MDL VALUE

Matrix	Multiplication Factor
Groundwater	10
Low-level soil	10
Water miscible liquid waste	500
High-level soil and sludge	1250
Non-water miscible waste	1250

PRECISION: 0.26X + 0.91 μg/L (overall precision)

ACCURACY: 0.89C - 0.07 μg/L (as recovery)

SAMPLING METHOD: For water and liquid samples; use glass 40 mL vials with Teflon lined septum caps and collect two vials per sample location with no headspace. For solids and concentrated waste samples; use widemouth glass bottles with Teflon liners.

STABILITY: For concentrated wastes, soils, sediments or sludges; cool to 4°C. For liquids; add 4 drops of concentrated hydrochloric acid and cool to 4°C. M.H.T. = 14 days.

QUALITY CONTROL: Analyze a reagent blank, matrix spike and matrix spike duplicate/duplicate for each analytical batch (up to 20 Samples). Demonstrate the purity of glassware and reagents by analyzing a reagent water method blank. Internal, surrogate and five concentration level calibration standards are used.

REFERENCE: Method 8010, SW-846, 3rd ed., Nov 1986.

Chapter 3
Nonhalogenated Volatile Organic Compunds

PRIMARY NAME: Acetonitrile

Method 8030

TITLE: Other Nonhalogenated VOCs

MATRIX: groundwater, soils, sludges water miscible liquid wastes, and non-water miscible wastes.

CAS # : 75-05-8

APPLICATION: This method is used for the analysis of 3 nonhalogenated VOCs. Samples are analyzed using direct injection or purge and trap methods. Groundwater must be analyzed by the purge and trap method. The method provides an optional GC column which is used for analyte confirmation and that may help resolve analytes from interferences.

INTERFERENCES: There can be carry-over contamination with high and low level samples. Impurities may come from the purge and trap apparatus, organic compounds outgassing from the plumbing ahead of trap, diffusion of VOCs through the sample bottle septum during shipping or storage, or from solvent vapors in the lab.

INSTRUMENTATION: GC capable of on-column injections or purge and trap sample introduction and a flame ionization detector (FID). Column 1: 10 foot by 2 mm with Porapak-QS. Column 2: 6 foot by 0.1 inch with Chromosorb 101.

RANGE: 5 to 100 μg/L

MDL: 0.7 μg/L (reagent water)

PRACTICAL QUANTITATION LIMIT FACTORS FOR MULTIPLYING TIMES MDL VALUE

Matrix	Multiplication Factor
Groundwater	10
Low-level soil	10
Water miscible liquid waste	500
High-level soil and sludge	1250
Non-water miscible waste	1250

PRECISION: Not determined

ACCURACY: Not determined

SAMPLING METHOD: Use two glass 40 mL vials with Teflon lined septum caps per sample location and collect samples with no headspace.

STABILITY: Adjust to pH 4 - 5 and cool to 4°C. M.H.T. = 14 days.

QUALITY CONTROL: Analyze a reagent blank, matrix spike and matrix spike duplicate/duplicate for each analytical batch (up to 20 Samples). Demonstrate the purity of glassware and reagents by analyzing a reagent water method blank. Internal, surrogate and five concentration level calibration standards are used. The QC check sample concentrate should contain this compound at 25 μg/mL in reagent water.

REFERENCE: Method 8030, SW-846, 3rd ed., Nov 1986.

PRIMARY NAME: Acrolein Method 8030

TITLE: Other Nonhalogenated VOCs

MATRIX: groundwater, soils, sludges water miscible liquid wastes, 107-02-8 and non-water miscible wastes

CAS # :

APPLICATION: This method is used for the analysis of 3 nonhalogenated VOCs. Samples are analyzed using direct injection or purge and trap methods. Groundwater must be analyzed by the purge and trap method. The method provides an optional GC column which is used for analyte confirmation and that may help resolve analytes from interferences.

INTERFERENCES: There can be carry-over contamination with high and low level samples. Impurities may come from the purge and trap apparatus, organic compounds outgassing from the plumbing ahead of trap, diffusion of VOCs through the sample bottle septum during shipping or storage, or from solvent vapors in the lab.

INSTRUMENTATION: GC capable of on-column injections or purge and trap sample introduction and a flame ionization detector (FID). Column 1: 10 foot by 2 mm with Porapak-QS. Column 2: 6 foot by 0.1 inch with Chromosorb 101.

RANGE: 5 to 100 µg/L MDL: 0.7 µg/L (reagent water)

PRACTICAL QUANTITATION LIMIT FACTORS FOR MULTIPLYING TIMES MDL VALUE

Matrix	Multiplication Factor
Groundwater	10
Low-level soil	10
Water miscible liquid waste	500
High-level soil and sludge	1250
Non-water miscible waste	1250

PRECISION (as standard deviation): 0.7 µg/L, reagent water and 0.8 µg/L, POTW at 50 µg/L spike. 1.1 µg/L, industrial water at 100 µg/L spike.

ACCURACY (as % recovery): 103%, reagent water and 89%,POTW at 50 µg/L spike. 9%, industrial wastewater at 100 µg/L spike.

SAMPLING METHOD: Use two glass 40 mL vials with Teflon lined septum caps per sample location and collect samples with no headspace.

STABILITY: Adjust to pH 4 - 5 and cool to 4°C. M.H.T. = 14 days.

QUALITY CONTROL: Analyze a reagent blank, matrix spike and matrix spike duplicate/duplicate for each analytical batch (up to 20 Samples). Demonstrate the purity of glassware and reagents by analyzing a reagent water method blank. Internal, surrogate and five concentration level calibration standards are used. The QC check sample concentrate should contain this compound at 25 µg/mL in reagent water.

REFERENCE: Method 8030, SW-846, 3rd ed., Nov 1986.

PRIMARY NAME: Acrylamide

Method 8015

TITLE: Nonhalogenated Volatile Organics

MATRIX: groundwater, soils, sludges water miscible liquid wastes, and non-water miscible wastes.

CAS # : 79-06-1

APPLICATION: This method is used for the analysis of 6 nonhalogenated VOCs. Samples are analyzed using direct injection or purge and trap methods. Groundwater must be analyzed by the purge and trap method. The method provides an optional GC column that may help resolve analytes from interferences and which is also used for analyte confirmation.

INTERFERENCES: There can be carry-over contamination with high and low level samples. Impurities may come from the purge and trap apparatus, organic compounds outgassing from the plumbing ahead of trap, diffusion of VOCs through the sample bottle septum during shipping or storage, or from solvent vapors in the lab.

INSTRUMENTATION: GC capable of on-column injections or purge-and-trap sample introduction and a flame ionization detector (FID). Column 1: an 8 foot by 0.1 inch 1% SP-1000 on Carbopack-B. Column 2: a 6 foot by 0.1 inch bonded n-octane on Porasil-C.

RANGE: Not available

MDL: Not available

PRECISION: Not available

ACCURACY: Not available

SAMPLING METHOD: For water and liquid samples; use glass 40 mL vials with Teflon lined septum caps and collect two vials per sample location with no headspace. For solids and concentrated waste samples; use widemouth glass bottles with Teflon liners. Cool all samples to 4°C.

STABILITY: For concentrated wastes, soils, sediments or sludges; cool to 4°C. For liquids; add 4 drops of concentrated hydrochloric acid, cool to 4°C. M.H.T. = 14 days.

QUALITY CONTROL: Analyze a reagent blank, matrix spike and matrix spike duplicate/duplicate for each analytical batch (up to 20 Samples). Demonstrate the purity of glassware and reagents by analyzing a reagent water method blank. Internal, surrogate and five concentration level calibration standards are used.

REFERENCE: Method 8015, SW-846, 3rd ed., Nov 1986.

PRIMARY NAME: Acrylonitrile

Method 8030

TITLE: Other Nonhalogenated VOCs

MATRIX: groundwater, soils, sludges water miscible liquid wastes, 107-13-1 and non-water miscible wastes.

CAS # :

APPLICATION: This method is used for the analysis of 3 nonhalogenated VOCs. Samples are analyzed using direct injection or purge and trap methods. Groundwater must be analyzed by the purge and trap method. The method provides an optional GC column which is used for analyte confirmation and that may help resolve analytes from interferences.

INTERFERENCES: There can be carry-over contamination with high and low level samples. Impurities may come from the purge and trap apparatus, organic compounds outgassing from the plumbing ahead of trap, diffusion of VOCs through the sample bottle septum during shipping or storage, or from solvent vapors in the lab.

INSTRUMENTATION: GC capable of on-column injections or purge and trap sample introduction and a flame ionization detector (FID). Column 1: 10 foot by 2 mm with Porapak-QS. Column 2: 6 foot by 0.1 inch with Chromosorb 101.

RANGE: 5 to 100 μg/L

MDL: 0.5 μg/L (reagent water)

PRACTICAL QUANTITATION LIMIT FACTORS FOR MULTIPLYING TIMES MDL VALUE

Matrix	Multiplication Factor
Groundwater	10
Low-level soil	10
Water miscible liquid waste	500
High-level soil and sludge	1250
Non-water miscible waste	1250

PRECISION (as standard deviation): 1.5 μg/L, reagent water at 50 μg/L spike; 1.5 μg/L, POTW and 3.2 μg/L industrial water at 100 μg/L spike.

ACCURACY (as % recovery): 103%, reagent water at 50 μg/L spike; 101% POTW and 104% industrial water at 100 μg/L spike.

SAMPLING METHOD: Use two glass 40 mL vials with Teflon lined septum caps per sample location and collect samples with no headspace.

STABILITY: Adjust to pH 4 - 5 and cool to 4°C. M.H.T. = 14 days.

QUALITY CONTROL: Analyze a reagent blank, matrix spike and matrix spike duplicate/duplicate for each analytical batch (up to 20 Samples). Demonstrate the purity of glassware and reagents by analyzing a reagent water method blank. Internal, surrogate and five concentration level calibration standards are used. The QC check sample concentrate should contain this compound at 25 μg/mL in reagent water.

REFERENCE: Method 8030, SW-846, 3rd ed., Nov 1986.

PRIMARY NAME: Benzene Method 502.2

TITLE: VOCs in H_2O by Purge and Trap capillary column GC

MATRIX: Drinking water (finished or any treatment stage) and raw source water.

CAS # : 71-43-2

APPLICATION: Method covers 60 VOCs. An inert gas is bubbled through a 5 mL sample in a purging chamber. Purged sample components are trapped in tube of sorbent materials. When purging is done, sorbent tube is heated and backflushed w/helium to desorb trapped sample onto a capillary GC column.

INTERFERENCES: During analysis, major contaminant sources are volatile materials in the lab and impurities in purging gas and sorbent trap. With high and low level samples, there can be carry-over contamination. Beware of methylene chloride, it permeates through PTFE tubing.

INSTRUMENTATION: (capillary column) GC, w/PID and electrolytic conductivity detectors in series. Column #1: VOCOL wide bore; Column #2: RTX-502.2 mega bore; Column #3: DB-62 mega bore.

RANGE: Approximately 0.02-200 µg/L.

MDL: 0.01 µg/L (reagent H_2O) (PID).

PRECISION: RSD = 1.2% @ spike of 10 µg/L (reagent H_2O) (PID).

ACCURACY: Recovery = 99% @ spike of 10 µg/L (reagent H_2O) (PID)

SAMPLING METHOD: Use a 40-120 mL screw cap vial (pre-washed with detergent, rinsed with distilled water and oven dried at 105°C) with a PTFE-faced silicone septum. If residual chlorine is in the water add about 25 mg of ascorbic acid to each vial before sample collection. Collect bubble free samples.

STABILITY: Cool to 4°C; HCl to pH <2.

M.H.T. = 14 Days.

QUALITY CONTROL: As initial demonstration of lab accuracy and precision, analyze 4 to 7 replicates of a lab fortified blank containing the analyte at 0.1-5 µg/L. Collect all samples in duplicate.

REFERENCE: Method 502.2, Volatile Organic Compounds in Water by Purge and Trap Capillary Column GC, EPA 600/4-88/039.

PRIMARY NAME: Benzene

Method 503.1

TITLE: Aromatic and Unsaturated VOCs in Water

MATRIX: Drinking water (finished or any treatment stage) and raw source water.

CAS # : 71-43-2

APPLICATION: Method covers 28 aromatic and unsaturated VOCs. An inert gas is bubbled through a 5 mL water sample. Purged sample components are trapped in tube of sorbent materials. When purging is complete, sorbent tube is heated and backflushed with inert gas to desorb trapped sample onto a packed GC column.

INTERFERENCES: During analysis, major contaminant sources are volatile materials in the lab and impurities in purging gas and sorbent trap. With high and low level samples, there can be carry-over contamination. Excess water causes a negative baseline deflection.

INSTRUMENTATION: Purge and Trap GC w/photoionization detector. (Two GC columns are recommended); Column #1: 5% SP-1200 and 1.75% Bentone 34 on Supelcoport; Column #2: 1,2,3-tris(2-cyanoethoxy)propane on Chromosorb W.

RANGE: 2.2-600 µg/L. (Drinking water) MDL: 0.02 µg/L in water

PRECISION: RSD = 11.8% @ 47.0 µg/L conc., 10 labs

ACCURACY: Avg Recovery = 100% @ 47.0 µg/L conc.; 10 labs

SAMPLING METHOD: Use a 40-120 mL screw cap vial (pre-washed with detergent, rinsed with distilled water and oven dried at 105°C) with a PTFE-faced silicone septum. If residual chlorine is in the water add about 25 mg of ascorbic acid to each vial before sample collection. Collect bubble free samples.

STABILITY: Cool to 4°C; HCl to pH <2. M.H.T. = 14 Days.

QUALITY CONTROL: As initial demonstration of lab accuracy and precision, analyze 4 to 7 replicates of a lab fortified blank containing the analyte at 0.1-5 µg/L. Collect all samples in duplicate.

REFERENCE: Method 503.1, Volatile Aromatic and Unsaturated Organic Compounds in H_2O by Purge and Trap GC, EPA 600/4-88/039.

PRIMARY NAME: Benzene Method 524.1

TITLE: VOCs in H_2O by Purge and Trap GC/MS

MATRIX: Drinking water (finished or any treatment stage) and raw source water.

CAS # : 71-43-2

APPLICATION: Method covers 48 VOCs. An inert gas is bubbled through a 25 mL water sample. Purged sample components are trapped in tube of sorbent materials. When purging is complete, sorbent tube is heated and backflushed with helium to desorb trapped sample onto a packed GC column.

INTERFERENCES: During analysis, major contaminant sources are volatile materials in the lab and impurities in purging gas and sorbent trap. With high and low level samples, there can be carry-over contamination. Watch for methylene chloride, it permeates through PTFE tubing.

INSTRUMENTATION: Gas Chromatography/Mass Spectrometry/Data System; 1% SP-1000 on Carbopack B.

RANGE: Approximately 0.1-200 µg/L. MDL: 0.1 µg/L

PRECISION: RSD = 3.7% @ 1.0 µg/L

ACCURACY: Mean Accuracy = 97% @ 1.0 µg/L

SAMPLING METHOD: Use a 60-120 mL screw cap vial (pre-washed with detergent, rinsed with distilled water and oven dried at 105°C) with a PTFE-faced silicone septum. If residual chlorine is in the water add about 25 mg of ascorbic acid to each vial before sample collection. Collect bubble free samples.

STABILITY: Cool to 4°C; HCl to pH <2. M.H.T. = 14 Days.

QUALITY CONTROL: As initial demonstration of lab accuracy and precision, analyze 4 to 7 replicates of a lab fortified blank containing the analyte at 0.2-5 µg/L. Collect all samples in duplicate.

REFERENCE: Method 524.1, VOCs in Water by Purge and Trap Gas Chromatography/Mass Spectrometry, EPA 600/4-88/039.

PRIMARY NAME: Benzene Method 524.2

TITLE: VOCs in H_2O Purge and Trap capillary column GC/MS

MATRIX: Drinking water (finished or any treatment stage) and raw source water.

CAS # : 71-43-2

APPLICATION: Method covers 60 VOCs. An inert gas is bubbled through a 25 mL water sample. Purged sample components are trapped in tube of sorbent materials. When purging is complete, sorbent tube is heated and backflushed with helium to desorb trapped sample onto a capillary GC column.

INTERFERENCES: During analysis, major contaminant sources are volatile materials in the lab and impurities in purging gas and sorbent trap. With high and low level samples, there can be carry-over contamination. Watch for methylene chloride, it permeates through PTFE tubing.

INSTRUMENTATION: (Capillary Column) Gas Chromatography/Mass Spectrometry/Data System; Column #1: VOCOL glass wide bore capillary column; Column #2: DB-624 fused silica column; Column #3: DB-5 fused silica column.

RANGE: Approximately 0.02-200 µg/L.

MDL: 0.04 µg/L on wide bore capillary column.

PRECISION: RSD = 5.7% @ (0.1-10) µg/L (31 samples) wide bore capillary column.

ACCURACY: Recovery = 97% @ (0.1-10) µg/L (31 samples) wide bore capillary column.

SAMPLING METHOD: Use a 60-120 mL screw cap vial (pre-washed with detergent, rinsed with distilled water and oven dried at 105°C) with a PTFE-faced silicone septum. If residual chlorine is in the water add about 25 mg of ascorbic acid to each vial before sample collection. Collect bubble free samples.

STABILITY: Cool to 4°C; HCl to pH <2.

M.H.T. = 14 Days.

QUALITY CONTROL: As initial demonstration of lab accuracy and precision, analyze 4 to 7 replicates of a lab fortified blank containing the analyte at 0.2-5 µg/L. Collect all samples in duplicate.

REFERENCE: Method 524.2, VOCs in H_2O by Purge and Trap Capillary column GC/MS, EPA 600/4-88/039.

PRIMARY NAME: Benzene Method 602

TITLE: Purgeable Aromatics MATRIX: Wastewater

CAS # : 71-43-2

APPLICATION: Method covers 7 purgeable aromatics. (Method 624 provides GC/MS conditions appropriate for the qualitative and quantitative confirmation of results). Method describes conditions for a second GC column to confirm measurements made with primary column.

INTERFERENCES: Impurities in the purge gas and organic compounds outgassing from the plumbing ahead of the trap. With high and low level samples, there can be carry-over contamination. Diffusion of volatile organics through the septum seal into the sample.

INSTRUMENTATION: GC equipped with photoionization detector. (With purge and trap unit)

RANGE: 2.1 To 550 μg/L. MDL: 0.2 μg/L.

PRECISION: 0.21X+0.56 μg/L (overall precision)

ACCURACY: 0.92C+0.57 μg/L (as recovery)

SAMPLING METHOD: 25 mL glass vial. Teflon lined septum.

STABILITY: Cool, 4°C, 0.008% sodium thiosulfate, HCl to pH 2. M.H.T. = 14 Days.

QUALITY CONTROL: The laboratory must on an ongoing basis, spike at least 10% of the samples from each sample site being monitored to assess accuracy.

REFERENCE: Method 602, Federal Register Part VIII 40 CFR Part 136, Oct 26, 1984.

PRIMARY NAME: Benzene Method 624

TITLE: Purgeables MATRIX: Wastewater

CAS # : 71-43-2

APPLICATION: Method covers 31 purgeable organics. An inert gas is bubbled through a 5 mL water sample in a specially designed purging chamber. Here, purgeables are transferred from aqueous to gaseous phase, passed onto a sorbent column and trapped. Trap is heated and backflushed with inert gas to desorb purgeables onto a GC column, where purgeables are separated.

INTERFERENCES: Impurities in the purge gas, organic compounds outgassing from the plumbing ahead of the trap, and solvent vapors in the laboratory. With high and low level samples, there can be carry-over contamination.

INSTRUMENTATION: GC/MS with purge and trap unit.

RANGE: 5-600 μg/L MDL: 4.4 μg/L

PRECISION: 0.25X-1.33 μg/L (overall precision)

ACCURACY: 0.93C+2.00 μg/L (as recovery)

SAMPLING METHOD: 25 mL glass vial. Teflon lined septum.

STABILITY: Cool, 4°C, 0.008% sodium thiosulfate. HCl to pH 2. M.H.T. = 14 Days.

QUALITY CONTROL: The laboratory must on an ongoing basis, spike at least 5% of the samples from each sample site being monitored to assess accuracy.

REFERENCE: Method 624, Federal Register Part VIII 40 CFR Part 136, Oct 26, 1984.

PRIMARY NAME: Benzene Method 8020

TITLE: Aromatic Volatile Organics

MATRIX: groundwater, soils, sludges water miscible liquid wastes, and non-water miscible wastes

CAS # : 71-43-2

APPLICATION: This method is used to analyze for 8 aromatic VOCs. Samples are analyzed using direct injection or purge and trap methods. Groundwater must be analyzed by the purge and trap method. The method provides an optional GC column that is used for analyte confirmation and may also help resolve analytes from interferences.

INTERFERENCES: There can be carry-over contamination with high and low level samples. Impurities may come from the purge and trap apparatus, organic compounds outgassing from the plumbing ahead of trap, diffusion of VOCs through the sample bottle septum during shipping or storage, or from solvent vapors in the lab.

INSTRUMENTATION: GC capable of on-column injections or purge-and-trap sample introduction and a photoionization detector (PID). Column 1: 6 foot by 0.082 inch with 5% SP-1200 and 1.75% Bentone-34 on Supelcoport. Column 2: 8 foot by 0.1 inch with 5% 1,2,3-tris(2-cyanoethoxy)propane on Chromosorb W-AW.

RANGE: 2.1 to 500 μg/L

MDL: 0.2 μg/L (reagent water)

PRACTICAL QUANTITATION LIMIT FACTORS FOR MULTIPLYING TIMES MDL VALUE

Matrix	Multiplication Factor
Groundwater	10
Low-level soil	10
Water miscible liquid waste	500
High-level soil and sludge	1250
Non-water miscible waste	1250

PRECISION: 0.21X + 0.56 μg/L (overall precision)

ACCURACY: 0.92C + 0.57 μg/L (as recovery)

SAMPLING METHOD: For water and liquid samples use glass 40 mL vials with Teflon lined septum caps and collect two vials per sample location with no headspace. For solids and concentrated waste samples use widemouth glass bottles with Teflon liners. Cool all samples to 4°C.

STABILITY: For concentrated wastes, soils, sediments or sludges cool to 4°C.

For liquids, add 4 drops of concentrated hydrochloric acid and cool to 4°C. M.H.T. = 14 days.

QUALITY CONTROL: Analyze a reagent blank, matrix spike and matrix spike duplicate/duplicate for each analytical batch (up to 20 Samples). Demonstrate the purity of glassware and reagents by analyzing a reagent water method blank. Internal, surrogate and five concentration level calibration standards are used. The QC check sample concentrate should contain this compound at 10 μg/mL in methanol.

REFERENCE: Method 8020, SW-846, 3rd ed., Nov 1986.

PRIMARY NAME: n-Butylbenzene

Method 502.2

TITLE: VOCs in H_2O by Purge and Trap capillary column GC

MATRIX: Drinking water (finished or any treatment stage) and raw source water.

CAS # : 104-51-8

APPLICATION: Method covers 60 VOCs. An inert gas is bubbled through a 5 mL sample in a purging chamber. Purged sample components are trapped in tube of sorbent materials. When purging is done, sorbent tube is heated and backflushed w/helium to desorb trapped sample onto a capillary GC column.

INTERFERENCES: During analysis, major contaminant sources are volatile materials in the lab and impurities in purging gas and sorbent trap. With high and low level samples, there can be carry-over contamination. Beware of methylene chloride, it permeates through PTFE tubing.

INSTRUMENTATION: (Capillary Column) GC, w/PID and electrolytic conductivity detectors in series. Column #1: VOCOL wide bore; Column #2: RTX-502.2 mega bore; Column #3: DB-62 mega bore.

RANGE: Approximately 0.02-200 μg/L.

MDL: 0.02 μg/L (reagent H_2O) (PID)

PRECISION: RSD = 4.4% @ Spike of 10 μg/L (reagent H_2O) (PID).

ACCURACY: Recovery = 100% @ spike of 10 μg/L (reagent H_2O) (PID)

SAMPLING METHOD: Use a 40-120 mL screw cap vial (pre-washed with detergent, rinsed with distilled water and oven dried at 105°C) with a PTFE-faced silicone septum. If residual chlorine is in the water add about 25 mg of ascorbic acid to each vial before sample collection. Collect bubble free samples.

STABILITY: Cool to 4°C; HCl to pH <2. M.H.T. = 14 Days.

QUALITY CONTROL: As initial demonstration of lab accuracy and precision, analyze 4 to 7 replicates of a lab fortified blank containing the analyte at 0.1-5 μg/L. Collect all samples in duplicate.

REFERENCE: Method 502.2, Volatile Organic Compounds in Water by Purge and Trap Capillary Column GC, EPA 600/4-88/039.

PRIMARY NAME: n-Butylbenzene Method 503.1

TITLE: Aromatic and Unsaturated VOCs in Water

MATRIX: Drinking water (finished or any treatment stage) and raw source water.

CAS # : 104-51-8

APPLICATION: Method covers 28 aromatic and unsaturated VOCs. An inert gas is bubbled through a 5 mL water sample. Purged sample components are trapped in tube of sorbent materials. When purging is complete, sorbent tube is heated and backflushed with inert gas to desorb trapped sample onto a packed GC column.

INTERFERENCES: During analysis, major contaminant sources are volatile materials in the lab and impurities in purging gas and sorbent trap. With high and low level samples, there can be carry-over contamination. Excess water causes a negative baseline deflection.

INSTRUMENTATION: Purge and Trap GC w/photoionization detector. (Two GC columns are recommended); Column #1: 5% SP-1200 and 1.75% Bentone 34 on Supelcoport; Column #2: 1,2,3-tris(2-cyanoethoxy)propane on Chromosorb W.

RANGE: 2.2-600 μg/L. (Drinking water) MDL: 0.02 μg/L in water

PRECISION: RSD = 15.7% @ 0.40 μg/L conc.; 7 samples

ACCURACY: Avg Recovery = 78% @ 0.40 μg/L conc.; 7 samples

SAMPLING METHOD: Use a 40-120 mL screw cap vial (pre-washed with detergent, rinsed with distilled water and oven dried at 105°C) with a PTFE-faced silicone septum. If residual chlorine is in the water add about 25 mg of ascorbic acid to each vial before sample collection. Collect bubble free samples.

STABILITY: Cool to 4°C; HCl to pH <2. M.H.T. = 14 Days.

QUALITY CONTROL: As initial demonstration of lab accuracy and precision, analyze 4 to 7 replicates of a lab fortified blank containing the analyte at 0.1-5 μg/L. Collect all samples in duplicate.

REFERENCE: Method 503.1, Volatile Aromatic and Unsaturated Organic Compounds in H_2O by Purge and Trap GC, EPA 600/4-88/039.

PRIMARY NAME: n-Butylbenzene Method 524.2

TITLE: VOCs in H_2O Purge and Trap capillary column GC/MS

MATRIX: Drinking water (finished or any treatment stage) and raw source water.

CAS # : 104-51-8

APPLICATION: Method covers 60 VOCs. An inert gas is bubbled through a 25 mL water sample. Purged sample components are trapped in tube of sorbent materials. When purging is complete, sorbent tube is heated and backflushed with helium to desorb trapped sample onto a capillary GC column.

INTERFERENCES: During analysis, major contaminant sources are volatile materials in the lab and impurities in purging gas and sorbent trap. With high and low level samples, there can be carry-over contamination. Watch for methylene chloride, it permeates through PTFE tubing.

INSTRUMENTATION: (capillary column) Gas Chromatography/Mass

Spectometry/Data System; Column #1: VOCOL glass wide bore capillary column; Column #2: DB-624 fused silica column; Column #3: DB-5 fused silica column.

RANGE: Approximately 0.02-200 μg/L.

MDL: 0.11 μg/L on wide bore capillary column.

PRECISION: RSD = 7.6% @ (0.5-10) μg/L (18 samples) wide bore capillary column.

ACCURACY: Recovery = 100% @ (0.5-10) μg/L (18 samples) wide bore capillary column.

SAMPLING METHOD: Use a 60-120 mL screw cap vial (pre-washed with detergent, rinsed with distilled water and oven dried at 105°C) with a PTFE-faced silicone septum. If residual chlorine is in the water add about 25 mg of ascorbic acid to each vial before sample collection. Collect bubble free samples.

STABILITY: Cool to 4°C; HCl to pH <2.

M.H.T. = 14 Days.

QUALITY CONTROL: As initial demonstration of lab accuracy and precision, analyze 4 to 7 replicates of a lab fortified blank containing the analyte at 0.2-5 μg/L. Collect all samples in duplicate.

REFERENCE: Method 524.2, VOCs in H_2O by Purge and Trap Capillary column GC/MS, EPA 600/4-88/039.

PRIMARY NAME: sec-Butylbenzene

Method 502.2

TITLE: VOCs in H_2O by Purge and Trap capillary column GC

MATRIX: Drinking water (finished or any treatment stage) and raw source water.

CAS # : 135-98-8

APPLICATION: Method covers 60 VOCs. An inert gas is bubbled through a 5 mL sample in a purging chamber. Purged sample components are trapped in tube of sorbent materials. When purging is done, sorbent tube is heated and backflushed w/helium to desorb trapped sample onto a capillary GC column.

INTERFERENCES: During analysis, major contaminant sources are volatile materials in the lab and impurities in purging gas and sorbent trap. With high and low level samples, there can be carry-over contamination. Beware of methylene chloride, it permeates through PTFE tubing.

INSTRUMENTATION: (capillary column) GC, w/PID and electrolytic conductivity detectors in series. Column #1: VOCOL wide bore; Column #2: RTX-502.2 mega bore; Column #3: DB-62 mega bore.

RANGE: Approximately 0.02-200 μg/L.

MDL: 0.02 μg/L (reagent H_2O) (PID)

PRECISION: RSD = 2.7% @ Spike of 10 μg/L (reagent H_2O) (PID).

ACCURACY: Recovery = 97% @ spike of 10 μg/L (reagent H_2O) (PID)

SAMPLING METHOD: Use a 40-120 mL screw cap vial (pre-washed with detergent, rinsed with distilled water and oven dried at 105°C) with a PTFE-faced silicone septum. If residual chlorine is in the water add about 25 mg of ascorbic acid to each vial before sample collection. Collect bubble free samples.

STABILITY: Cool to 4°C; HCl to pH <2.

M.H.T. = 14 Days.

QUALITY CONTROL: As initial demonstration of lab accuracy and precision, analyze 4 to 7 replicates of a lab fortified blank containing the analyte at 0.1-5 μg/L. Collect all samples in duplicate.

REFERENCE: Method 502.2, Volatile Organic Compounds in Water by Purge and Trap Capillary Column GC, EPA 600/4-88/039.

PRIMARY NAME: sec-Butylbenzene

Method 503.1

TITLE: Aromatic and Unsaturated VOCs in Water

MATRIX: Drinking water (finished or any treatment stage) and raw source water.

CAS # : 135-98-8

APPLICATION: Method covers 28 aromatic and unsaturated VOCs. An inert gas is bubbled through a 5 mL water sample. Purged sample components are trapped in tube of sorbent materials. When purging is complete, sorbent tube is heated and backflushed with inert gas to desorb trapped sample onto a packed GC column.

INTERFERENCES: During analysis, major contaminant sources are volatile materials in the lab and impurities in purging gas and sorbent trap. With high and low level samples, there can be carry-over contamination. Excess water causes a negative baseline deflection.

INSTRUMENTATION: Purge and Trap GC w/photoionization detector. (Two GC columns are recommended); Column #1: 5% SP-1200 and 1.75% Bentone 34 on Supelcoport; Column #2: 1,2,3-tris(2-cyanoethoxy)propane on Chromosorb W.

RANGE: 2.2-600 μg/L. (Drinking water) MDL: 0.02 μg/L in water

PRECISION: RSD = 11.0% @ 0.40 μg/L conc.; 7 samples

ACCURACY: Avg Recovery = 80% @ 0.40 μg/L conc.; 7 samples

SAMPLING METHOD: Use a 40-120 mL screw cap vial (pre-washed with detergent, rinsed with distilled water and oven dried at 105°C) with a PTFE-faced silicone septum. If residual chlorine is in the water add about 25 mg of ascorbic acid to each vial before sample collection. Collect bubble free samples.

STABILITY: Cool to 4°C; HCl to pH <2. M.H.T. = 14 Days.

QUALITY CONTROL: As initial demonstration of lab accuracy and precision, analyze 4 to 7 replicates of a lab fortified blank containing the analyte at 0.1-5 μg/L. Collect all samps in duplicate.

REFERENCE: Method 503.1, Volatile Aromatic and Unsaturated Organic Compounds in H_2O by Purge and Trap GC, EPA 600/4-88/039.

PRIMARY NAME: sec-Butylbenzene Method 524.2

TITLE: VOCs in H_2O Purge and Trap capillary column GC/MS

MATRIX: Drinking water (finished or any treatment stage) and raw source water.

CAS # : 135-98-8

APPLICATION: Method covers 60 VOCs. An inert gas is bubbled through a 25 mL water sample. Purged sample components are trapped in tube of sorbent materials. When purging is complete, sorbent tube is heated and backflushed with helium to desorb trapped sample onto a capillary GC column.

INTERFERENCES: During analysis, major contaminant sources are volatile materials in the lab and impurities in purging gas and sorbent trap. With high and low level samples, there can be carry-over contamination. Watch for methylene chloride, it permeates through PTFE tubing.

INSTRUMENTATION: (capillary column) Gas Chromatography/Mass Spectometry/Data System; Column #1: VOCOL glass wide bore capillary column; Column #2: DB-624 fused silica column; Column #3: DB-5 fused silica column.

RANGE: Approximately 0.02-200 μg/L.

MDL: 0.13 μg/L on wide bore capillary column.

PRECISION: RSD = 7.6% @ (0.5-10) μg/L (16 samples) wide bore capillary column.

ACCURACY: Recovery = 100% @ (0.5-10) μg/L (16 samples) wide bore capillary column.

SAMPLING METHOD: Use a 60-120 mL screw cap vial (pre-washed with detergent, rinsed with distilled water and oven dried at 105°C) with a PTFE-faced silicone septum. If residual chlorine is in the water add about 25 mg of ascorbic acid to each vial before sample collection. Collect bubble free samples.

STABILITY: Cool to 4°C; HCl to pH <2.

M.H.T. = 14 Days.

QUALITY CONTROL: As initial demonstration of lab accuracy and precision, analyze 4 to 7 replicates of a lab fortified blank containing the analyte at 0.2-5 μg/L. Collect all samples in duplicate.

REFERENCE: Method 524.2, VOCs in H_2O by Purge and Trap Capillary column GC/MS, EPA 600/4-88/039.

PRIMARY NAME: tert-Butylbenzene

Method 502.2

TITLE: VOCs in H_2O by Purge and Trap capillary column GC

MATRIX: Drinking water (finished or any treatment stage) and raw source water.

CAS # : 98-06-6

APPLICATION: Method covers 60 VOCs. An inert gas is bubbled through a 5 mL sample in a purging chamber. Purged sample components are trapped in tube of sorbent materials. When purging is done, sorbent tube is heated and backflushed w/helium to desorb trapped sample onto a capillary GC column.

INTERFERENCES: During analysis, major contaminant sources are volatile materials in the lab and impurities in purging gas and sorbent trap. With high and low level samples, there can be carry-over contamination. Beware of methylene chloride, it permeates through PTFE tubing.

INSTRUMENTATION: (capillary column) GC, w/PID and electrolytic conductivity detectors in series. Column #1: VOCOL wide bore; Column #2: RTX-502.2 mega bore; Column #3: DB-62 mega bore.

RANGE: Approximately 0.02-200 µg/L.

MDL: 0.06 µg/L (reagent H_2O) (PID)

PRECISION: RSD = 2.3% @ Spike of 10 µg/L (reagent H_2O) (PID).

ACCURACY: Recovery = 98% @ spike of 10 µg/L (reagent H_2O) (PID)

SAMPLING METHOD: Use a 40-120 mL screw cap vial (pre-washed with detergent, rinsed with distilled water and oven dried at 105°C) with a PTFE-faced silicone septum. If residual chlorine is in the water add about 25 mg of ascorbic acid to each vial before sample collection. Collect bubble free samples.

STABILITY: Cool to 4°C; HCl to pH <2.

M.H.T. = 14 Days.

QUALITY CONTROL: As initial demonstration of lab accuracy and precision, analyze 4 to 7 replicates of a lab fortified blank containing the analyte at 0.1-5 µg/L. Collect all samples in duplicate.

REFERENCE: Method 502.2, Volatile Organic Compounds in Water by Purge and Trap Capillary Column GC, EPA 600/4-88/039.

PRIMARY NAME: tert-Butylbenzene

Method 503.1

TITLE: Aromatic and Unsaturated VOCs in Water

MATRIX: Drinking water (finished or any treatment stage) and raw source water.

CAS # : 98-06-6

APPLICATION: Method covers 28 aromatic and unsaturated VOCs. An inert gas is bubbled through a 5 mL water sample. Purged sample components are trapped in tube of sorbent materials. When purging is complete, sorbent tube is heated and backflushed with inert gas to desorb trapped sample onto a packed GC column.

INTERFERENCES: During analysis, major contaminant sources are volatile materials in the lab and impurities in purging gas and sorbent trap. With high and low level samples, there can be carry-over contamination. Excess water causes a negative baseline deflection.

INSTRUMENTATION: Purge and Trap GC w/photoionization detector. (Two GC columns are recommended); Column #1: 5% SP-1200 and 1.75% Bentone 34 on Supelcoport; Column #2: 1,2,3-tris(2-cyanoethoxy)propane on Chromosorb W.

RANGE: 2.2-600 μg/L. (Drinking water)

MDL: 0.006 μg/L in water

PRECISION: RSD = 8.7% @ 0.40 μg/L conc.; 7 samples

ACCURACY: Avg Recovery = 88% @ 0.40 μg/L conc.; 7 samples

SAMPLING METHOD: Use a 40-120 mL screw cap vial (pre-washed with detergent, rinsed with distilled water and oven dried at 105°C) with a PTFE-faced silicone septum. If residual chlorine is in the water add about 25 mg of ascorbic acid to each vial before sample collection. Collect bubble free samples.

STABILITY: Cool to 4°C; HCl to pH <2.

M.H.T. = 14 Days.

QUALITY CONTROL: As initial demonstration of lab accuracy and precision, analyze 4 to 7 replicates of a lab fortified blank containing the analyte at 0.1-5 µg/L. Collect all samples in duplicate.

REFERENCE: Method 503.1, Volatile Aromatic and Unsaturated Organic Compounds in H_2O by Purge and Trap GC, EPA 600/4-88/039.

PRIMARY NAME: tert-Butylbenzene

Method 524.2

TITLE: VOCs in H_2O Purge and Trap capillary column GC/MS

MATRIX: Drinking water (finished or any treatment stage) and raw source water.

CAS # : 98-06-6

APPLICATION: Method covers 60 VOCs. An inert gas is bubbled through a 25 mL water sample. Purged sample components are trapped in tube of sorbent materials. When purging is complete, sorbent tube is heated and backflushed with helium to desorb trapped sample onto a capillary GC column.

INTERFERENCES: During analysis, major contaminant sources are volatile materials in the lab and impurities in purging gas and sorbent trap. With high and low level samples, there can be carry-over contamination. Watch for methylene chloride, it permeates through PTFE tubing.

INSTRUMENTATION: (capillary column) Gas Chromatography/Mass Spectometry/Data System; Column #1: VOCOL glass wide bore capillary column; Column #2: DB-624 fused silica column; Column #3: DB-5 fused silica column.

RANGE: Approximately 0.02-200 µg/L.

MDL: 0.14 µg/L on wide bore capillary column.

PRECISION: RSD = 7.3% @ (0.5-10) µg/L (18 samples) wide bore capillary column.

ACCURACY: Recovery = 102% @ (0.5-10) µg/L (18 samples) wide bore capillary column.

SAMPLING METHOD: Use a 60-120 mL screw cap vial (pre-washed with detergent, rinsed with distilled water and oven dried at 105°C) with a PTFE-faced silicone septum. If residual chlorine is in the water add about 25 mg of ascorbic acid to each vial before sample collection. Collect bubble free samples.

STABILITY: Cool to 4°C; HCl to pH <2. M.H.T. = 14 Days.

QUALITY CONTROL: As initial demonstration of lab accuracy and precision, analyze 4 to 7 replicates of a lab fortified blank containing the analyte at 0.2-5 µg/L. Collect all samples in duplicate.

REFERENCE: Method 524.2, VOCs in H_2O by Purge and Trap Capillary Column GC/MS, EPA 600/4-88/039.

PRIMARY NAME: Diethyl ether Method 8015

TITLE: Nonhalogenated Volatile Organics

MATRIX: groundwater, soils, sludges water miscible liquid wastes, and non-water miscible wastes.

CAS # : 60-29-7

APPLICATION: This method is used for the analysis of 6 nonhalogenated VOCs. Samples are analyzed using direct injection or purge and trap methods. Groundwater must be analyzed by the purge and trap method. The method provides an optional GC column that may help resolve analytes from interferences and which is also used for analyte confirmation.

INTERFERENCES: There can be carry-over contamination with high and low level samples. Impurities may come from the purge and trap apparatus, organic compounds outgassing from the plumbing ahead of trap, diffusion of VOCs through the sample bottle septum during shipping or storage, or from solvent vapors in the lab.

INSTRUMENTATION: GC capable of on-column injections or purge-and-trap sample introduction and a flame ionization detector (FID). Column 1: an 8 foot by 0.1 inch 1% SP-1000 on Carbopack-B column. Column 2: a 6 foot by 0.1 inch bonded n-octane on Porasil-C.

RANGE: Not available MDL: Not available

PRECISION: Not available

ACCURACY: Not available

SAMPLING METHOD: For water and liquid samples; use glass 40 mL vials with Teflon-lined septum caps and collect two vials per sample location with no headspace. For solids and concentrated waste samples; use widemouth glass bottles with Teflon liners. Cool all samples to 4°C.

STABILITY: For concentrated wastes, soils, sediments or sludges; cool to 4°C. For liquids; add 4 drops of concentrated hydrochloric acid, cool to 4°C. M.H.T. = 14 days.

QUALITY CONTROL: Analyze a reagent blank, matrix spike and matrix spike duplicate/duplicate for each analytical batch (up to 20 Samples). Demonstrate the purity of glassware and reagents by analyzing a reagent water method blank. Internal, surrogate and five concentration level calibration standards are used.

REFERENCE: Method 8015, SW-846, 3rd ed., Nov 1986.

PRIMARY NAME: Ethanol Method 8015

TITLE: Nonhalogenated Volatile Organics

MATRIX: groundwater, soils, sludges water miscible liquid wastes, and non-water miscible wastes.

CAS # : 64-17-5

APPLICATION: This method is used for the analysis of 6 nonhalogenated VOCs. Samples are analyzed using direct injection or purge and trap methods. Groundwater must be analyzed by the purge and trap method. The method provides an optional GC column that may help resolve analytes from interferences and which is also used for analyte confirmation.

INTERFERENCES: There can be carry-over contamination with high and low level samples. Impurities may come from the purge and trap apparatus, organic compounds outgassing from the plumbing ahead of trap, diffusion of VOCs through the sample bottle septum during shipping or storage, or from solvent vapors in the lab.

INSTRUMENTATION: GC capable of on-column injections or purge-and-trap sample introduction and a flame ionization detector (FID). Column 1: an 8 foot by 0.1 inch 1% SP-1000 on Carbopack-B column. Column 2: a 6 foot by 0.1 inch bonded n-octane on Porasil-C.

RANGE: Not available

MDL: Not available

PRECISION: Not available

ACCURACY: Not available

SAMPLING METHOD: For water and liquid samples; use glass 40 mL vials with Teflon-lined septum caps and collect two vials per sample location with no headspace. For solids and concentrated waste samples; use widemouth glass bottles with Teflon liners. Cool all samples to 4°C.

STABILITY: For concentrated wastes, soils, sediments or sludges; cool to 4°C. For liquids; add 4 drops of concentrated hydrochloric acid, cool to 4°C. M.H.T. = 14 days.

QUALITY CONTROL: Analyze a reagent blank, matrix spike and matrix spike duplicate/duplicate for each analytical batch (up to 20 Samples). Demonstrate the purity of glassware and reagents by analyzing a reagent water method blank. Internal, surrogate and five concentration level calibration standards are used.

REFERENCE: Method 8015, SW-846, 3rd ed., Nov 1986.

PRIMARY NAME: Ethylbenzene

Method 502.2

TITLE: VOCs in H_2O by Purge and Trap capillary column GC

MATRIX: Drinking water (finished or any treatment stage) and raw source water.

CAS # : 100-41-4

APPLICATION: Method covers 60 VOCs. An inert gas is bubbled through a 5 mL sample in a purging chamber. Purged sample components are trapped in tube of sorbent materials. When purging is done, sorbent tube is heated and backflushed w/helium to desorb trapped sample onto a capillary GC column.

INTERFERENCES: During analysis, major contaminant sources are volatile materials in the lab and impurities in purging gas and sorbent trap. With high and low level samples, there can be carry-over contamination. Beware of methylene chloride, it permeates through PTFE tubing.

INSTRUMENTATION: (capillary column) GC, w/PID and electrolytic conductivity detectors in series. Column #1: VOCOL wide bore; Column #2: RTX-502.2 mega bore; Column #3: DB-62 mega bore.

RANGE: Approximately 0.02-200 µg/L. MDL: 0.01 µg/L (reagent H_2O) (PID)

PRECISION: RSD = 1.4% @ Spike of 10 µg/L (reagent H_2O) (PID).

ACCURACY: Recovery = 101% @ spike of 10 µg/L (reagent H_2O) (PID)

SAMPLING METHOD: Use a 40-120 mL screw cap vial (pre-washed with detergent, rinsed with distilled water and oven dried at 105°C) with a PTFE-faced silicone septum. If residual chlorine is in the water add about 25 mg of ascorbic acid to each vial before sample collection. Collect bubble free samples.

STABILITY: Cool to 4°C; HCl to pH <2. M.H.T. = 14 Days.

QUALITY CONTROL: As initial demonstration of lab accuracy and precision, analyze 4 to 7 replicates of a lab fortified blank containing the analyte at 0.1-5 µg/L. Collect all samples in duplicate.

REFERENCE: Method 502.2, Volatile Organic Compounds in Water by Purge and Trap Capillary Column GC, EPA 600/4-88/039.

PRIMARY NAME: Ethylbenzene Method 503.1

TITLE: Aromatic and Unsaturated VOCs in Water

MATRIX: Drinking water (finished or any treatment stage) and raw source water.

CAS # : 100-41-4

APPLICATION: Method covers 28 aromatic and unsaturated VOCs. An inert gas is bubbled through a 5 mL water sample. Purged sample components are trapped in tube of sorbent materials.When purging is complete, sorbent tube is heated and backflushed with inert gas to desorb trapped sample onto a packed GC column.

INTERFERENCES: During analysis, major contaminant sources are volatile materials in the lab and impurities in purging gas and sorbent trap. With high and low level samples, there can be carry-over contamination. Excess water causes a negative baseline deflection.

INSTRUMENTATION: Purge and Trap GC w/photoionization detector. (Two GC columns are recommended); Column #1: 5% SP-1200 and 1.75% Bentone 34 on Supelcoport; Column #2: 1,2,3-tris(2-cyanoethoxy)propane on Chromosorb W.

RANGE: 2.2-600 μg/L. (Drinking water) MDL: 0.002 μg/L in water

PRECISION: RSD = 8.5% @ 0.40 μg/L conc.; 7 samples

ACCURACY: Avg Recovery = 93% @ 0.40 μg/L conc.; 7 samples

SAMPLING METHOD: Use a 40-120 mL screw cap vial (pre-washed with detergent, rinsed with distilled water and oven dried at 105°C) with a PTFE-faced silicone septum. If residual chlorine is in the water add about 25 mg of ascorbic acid to each vial before sample collection. Collect bubble free samples.

STABILITY: Cool to 4°C; HCl to pH <2. M.H.T. = 14 Days.

QUALITY CONTROL: As initial demonstration of lab accuracy and precision, analyze 4 to 7 replicates of a lab fortified blank containing the analyte at 0.1-5 μg/L. Collect all samples in duplicate.

REFERENCE: Method 503.1, Volatile Aromatic and Unsaturated Organic Compounds in H_2O by Purge and Trap GC, EPA 600/4-88/039.

PRIMARY NAME: Ethylbenzene Method 524.1

TITLE: VOCs in H_2O by Purge and Trap GC/MS

MATRIX: Drinking water (finished or any treatment stage) and raw source water.

CAS # : 100-41-4

APPLICATION: Method covers 48 VOCs. An inert gas is bubbled through a 25 mL water sample. Purged sample components are trapped in tube of sorbent materials. When purging is complete, sorbent tube is heated and backflushed with helium to desorb trapped sample onto a packed GC column.

INTERFERENCES: During analysis, major contaminant sources are volatile materials in the lab and impurities in purging gas and sorbent trap. With high and low level samples, there can be carry-over contamination. Watch for methylene chloride, it permeates through PTFE tubing.

INSTRUMENTATION: Gas Chromatography/Mass Spectrometry/Data System; 1% SP-1000 on Carbopack B.

RANGE: Approximately 0.1-200 µg/L. MDL: not listed.

PRECISION: not listed.

ACCURACY: not listed.

SAMPLING METHOD: Use a 60-120 mL screw cap vial (pre-washed with detergent, rinsed with distilled water and oven dried at 105°C) with a PTFE-faced silicone septum. If residual chlorine is in the water add about 25 mg of ascorbic acid to each vial before sample collection. Collect bubble free samples.

STABILITY: Cool to 4°C; HCl to pH <2. M.H.T. = 14 Days.

QUALITY CONTROL: As initial demonstration of lab accuracy and precision, analyze 4 to 7 replicates of a lab fortified blank containing the analyte at 0.2-5 µg/L. Collect all samples in duplicate.

REFERENCE: Method 524.1, VOCs in Water by Purge and Trap Gas Chromatography/Mass Spectrometry, EPA 600/4-88/039.

PRIMARY NAME: Ethylbenzene Method 524.2

TITLE: VOCs in H_2O Purge and Trap capillary column GC/MS

MATRIX: Drinking water (finished or any treatment stage) and raw source water.

CAS # : 100-41-4

APPLICATION: Method covers 60 VOCs. An inert gas is bubbled through a 25 mL water sample. Purged sample components are trapped in tube of sorbent materials. When purging is complete, sorbent tube is heated and backflushed with helium to desorb trapped sample onto a capillary GC column.

INTERFERENCES: During analysis, major contaminant sources are volatile materials in the lab and impurities in purging gas and sorbent trap. With high and low level samples, there can be carry-over contamination. Watch for methylene chloride, it permeates through PTFE tubing.

INSTRUMENTATION: (Capillary Column) Gas Chromatography/Mass Spectrometry/Data System; Column #1: VOCOL glass wide bore capillary column; Column #2: DB-624 fused silica column; Column #3: DB-5 fused silica column.

RANGE: Approximately 0.02-200 µg/L.

MDL: 0.06 µg/L on wide bore capillary column.

PRECISION: RSD = 8.6% @ (0.1-10) µg/L (31 samples) wide bore capillary column.

ACCURACY: Recovery = 99% @ (0.1-10) µg/L (31 samples) wide bore capillary column.

SAMPLING METHOD: Use a 60-120 mL screw cap vial (pre-washed with detergent, rinsed with distilled water and oven dried at 105°C) with a PTFE-faced silicone septum. If residual chlorine is in the water add about 25 mg of ascorbic acid to each vial before sample collection. Collect bubble free samples.

STABILITY: Cool to 4°C; HCl to pH <2.

M.H.T. = 14 Days.

QUALITY CONTROL: As initial demonstration of lab accuracy and precision, analyze 4 to 7 replicates of a lab fortified blank containing the analyte at 0.2-5 µg/L. Collect all samples in duplicate.

REFERENCE: Method 524.2, VOCs in H_2O by Purge and Trap Capillary column GC/MS, EPA 600/4-88/039.

PRIMARY NAME: Ethylbenzene

Method 602

TITLE: Purgeable Aromatics

MATRIX: Wastewater

CAS # : 100-41-4

APPLICATION: Method covers 7 purgeable aromatics. (Method 624 provides GC/MS conditions appropriate for the qualitative and quantitative confirmation of results).Method describes conditions for a second GC column to confirm measurements made with primary column.

INTERFERENCES: Impurities in the purge gas and organic compounds outgassing from the plumbing ahead of the trap. With high and low level samples, there can be carry-over contamination. Diffusion of volatile organics through the septum seal into the sample.

INSTRUMENTATION: GC equipped with photoionization detector. (With purge and trap unit)

RANGE: 2.1 To 550 μg/L. MDL: 0.2 μg/L.

PRECISION: 0.26X+0.23 μg/L (overall precision)

ACCURACY: 0.94C+0.31 μg/L (as recovery)

SAMPLING METHOD: 25 mL glass vial. Teflon lined septum.

STABILITY: Cool, 4°C, 0.008% sodium thiosulfate. HCl to pH 2. M.H.T. = 14 Days.

QUALITY CONTROL: The laboratory must on an ongoing basis, spike at least 10% of the samples from each sample site being monitored to assess accuracy.

REFERENCE: Method 602, Federal Register Part VIII 40 CFR Part 136, Oct 26, 1984.

PRIMARY NAME: Ethylbenzene Method 624

TITLE: Purgeables MATRIX: Wastewater

CAS # : 100-41-4

APPLICATION: Method covers 31 purgeable organics. An inert gas is bubbled through a 5 mL water sample in a specially designed purging chamber. Here, purgeables are transferred from aqueous to gaseous phase,

passed onto a sorbent column and trapped. Trap is heated and backflushed with inert gas to desorb purgeables onto a GC column, where purgeables are separated.

INTERFERENCES: Impurities in the purge gas, organic compounds outgassing from the plumbing ahead of the trap, and solvent vapors in the laboratory. With high and low level samples, there can be carry-over contamination.

INSTRUMENTATION: GC/MS with purge and trap unit.

RANGE: 5-600 µg/L MDL: 7.2 µg/L

PRECISION: 0.26X-1.72 µg/L (overall precision)

ACCURACY: 0.98C+2.48 µg/L (as recovery)

SAMPLING METHOD: 25 mL glass vial. Teflon lined septum.

STABILITY: Cool, 4°C, 0.008% sodium thiosulfate. HCl to pH 2. M.H.T. = 14 Days.

QUALITY CONTROL: The laboratory must on an ongoing basis, spike at least 5% of the samples from each sample site being monitored to assess accuracy.

REFERENCE: Method 624, Federal Register Part VIII 40 CFR Part 136, Oct 26, 1984.

PRIMARY NAME: Ethylbenzene Method 8020

TITLE: Aromatic Volatile Organics

MATRIX: groundwater, soils, sludges water miscible liquid wastes, and non-water miscible wastes.

CAS # : 100-41-4

APPLICATION: This method is used to analyze for 8 aromatic VOCs. Samples are analyzed using direct injection or purge and trap methods. Groundwater must be analyzed by the purge and trap method. The method provides an optional GC column that is used for analyte confirmation and may also help resolve analytes from interferences.

INTERFERENCES: There can be carry-over contamination with high and low level samples. Impurities may come from the purge and trap apparatus, organic compounds outgassing from the plumbing ahead of trap, diffusion of VOCs through the sample bottle septum during shipping or storage, or from solvent vapors in the lab.

INSTRUMENTATION: GC capable of on-column injections or purge-and-trap sample introduction and a photoionization detector (PID). Column 1: 6 foot by 0.082 inch with 5% SP-1200 and 1.75% Bentone-34 on Supelcoport. Column 2: 8 foot by 0.1 inch with 5% 1,2,3-tris(2-cyanoethoxy)propane on Chromosorb W-AW.

RANGE: 2.1 to 500 μg/L MDL: 0.2 μg/L (reagent water)

PRACTICAL QUANTITATION LIMIT FACTORS FOR MULTIPLYING TIMES MDL VALUE

Matrix	Multiplication Factor
Groundwater	10
Low-level soil	10
Water miscible liquid waste	500
High-level soil and sludge	1250
Non-water miscible waste	1250

PRECISION: 0.26X + 0.23 μg/L (overall precision)

ACCURACY: 0.94C + 0.31 μg/L (as recovery)

SAMPLING METHOD: For water and liquid samples use glass 40 mL vials with Teflon lined septum caps and collect two vials per sample location with no headspace. For solids and concentrated waste samples use widemouth glass bottles with Teflon liners. Cool all samples to 4°C.

STABILITY: For concentrated wastes, soils, sediments or sludges cool to 4°C. For liquids, add 4 drops of concentrated hydrochloric acid and cool to 4°C. M.H.T. = 14 days.

QUALITY CONTROL: Analyze a reagent blank, matrix spike and matrix spike duplicate/duplicate for each analytical batch (up to 20 Samples). Demonstrate the purity of glassware and reagents by analyzing a reagent water method blank. Internal, surrogate and five concentration level calibration standards are used. The QC check sample concentrate should contain this compound at 10 µg/mL in methanol.

REFERENCE: Method 8020, SW-846, 3rd ed., Nov 1986.

PRIMARY NAME: Isopropylbenzene

Method 502.2

TITLE: VOCs in H_2O by Purge and Trap capillary column GC

MATRIX: Drinking water (finished or any treatment stage) and raw source water.

CAS # : 98-82-8

APPLICATION: Method covers 60 VOCs. An inert gas is bubbled through a 5 mL sample in a purging chamber. Purged sample components are trapped in tube of sorbent materials. When purging is done, sorbent tube is heated and backflushed w/helium to desorb trapped sample onto a capillary GC column.

INTERFERENCES: During analysis, major contaminant sources are volatile materials in the lab and impurities in purging gas and sorbent trap. With high and low level samples, there can be carry-over contamination. Beware of methylene chloride, it permeates through PTFE tubing.

INSTRUMENTATION: (capillary column) GC, w/PID and electrolytic conductivity detectors in series. Column #1: VOCOL wide bore; Column #2: RTX-502.2 mega bore; Column #3: DB-62 mega bore.

RANGE: Approximately 0.02-200 µg/L.

MDL: 0.05 µg/L (reagent H_2O) (PID)

PRECISION: RSD = 0.9% @ Spike of 10 µg/L (reagent H_2O) (PID).

ACCURACY: Recovery = 98% @ spike of 10 µg/L (reagent H_2O) (PID)

SAMPLING METHOD: Use a 40-120 mL screw cap vial (pre-washed with detergent, rinsed with distilled water and oven dried at 105°C) with a PTFE-faced silicone septum. If residual chlorine is in the water add about 25 mg of ascorbic acid to each vial before sample collection. Collect bubble free samples.

STABILITY: Cool to 4°C; HCl to pH <2. M.H.T. = 14 Days.

QUALITY CONTROL: As initial demonstration of lab accuracy and precision, analyze 4 to 7 replicates of a lab fortified blank containing the analyte at 0.1-5 µg/L. Collect all samples in duplicate.

REFERENCE: Method 502.2, Volatile Organic Compounds in Water by Purge and Trap Capillary Column GC, EPA 600/4-88/039.

PRIMARY NAME: Isopropylbenzene Method 503.1

TITLE: Aromatic and Unsaturated VOCs in Water

MATRIX: Drinking water (finished or any treatment stage) and raw source water.

CAS # : 98-82-8

APPLICATION: Method covers 28 aromatic and unsaturated VOCs. An inert gas is bubbled through a 5 mL water sample. Purged sample components are trapped in tube of sorbent materials. When purging is complete, sorbent tube is heated and backflushed with inert gas to desorb trapped sample onto a packed GC column.

INTERFERENCES: During analysis, major contaminant sources are volatile materials in the lab and impurities in purging gas and sorbent trap. With high and low level samples, there can be carry-over contamination. Excess water causes a negative baseline deflection.

INSTRUMENTATION: Purge and Trap GC w/photoionization detector. (Two GC columns are recommended); Column #1: 5% SP-1200 and 1.75% Bentone 34 on Supelcoport; Column #2: 1,2,3-tris(2-cyanoethoxy)propane on Chromosorb W.

RANGE: 2.2-600 µg/L. (Drinking water) MDL: 0.005 µg/L in water

PRECISION: RSD = 8.7% @ 0.40 µg/L conc.; 7 samples

ACCURACY: Avg Recovery = 88% @ 0.40 µg/L conc.; 7 samples

SAMPLING METHOD: Use a 40-120 mL screw cap vial (pre-washed with detergent, rinsed with distilled water and oven dried at 105°C) with a PTFE-faced silicone septum. If residual chlorine is in the water add about 25 mg of

ascorbic acid to each vial before sample collection. Collect bubble free samples.

STABILITY: Cool to 4°C; HCl to pH <2. M.H.T. = 14 Days.

QUALITY CONTROL: As initial demonstration of lab accuracy and precision, analyze 4 to 7 replicates of a lab fortified blank containing the analyte at 0.1-5 µg/L. Collect all samples in duplicate.

REFERENCE: Method 503.1, Volatile Aromatic and Unsaturated Organic Compounds in H_2O by Purge and Trap GC, EPA 600/4-88/039.

PRIMARY NAME: Isopropylbenzene Method 524.2

TITLE: VOCs in H_2O Purge and Trap capillary column GC/MS

MATRIX: Drinking water (finished or any treatment stage) and raw source water.

CAS # : 98-82-8

APPLICATION: Method covers 60 VOCs. An inert gas is bubbled through a 25 mL water sample. Purged sample components are trapped in tube of sorbent materials. When purging is complete, sorbent tube is heated and backflushed with helium to desorb trapped sample onto a capillary GC column.

INTERFERENCES: During analysis, major contaminant sources are volatile materials in the lab and impurities in purging gas and sorbent trap. With high and low level samples, there can be carry-over contamination. Watch for methylene chloride, it permeates through PTFE tubing.

INSTRUMENTATION: (Capillary Column) Gas Chromatography/Mass Spectrometry/Data System; Column #1: VOCOL glass wide bore capillary column; Column #2: DB-624 fused silica column; Column #3: DB-5 fused silica column.

RANGE: Approximately 0.02-200 µg/L.

MDL: 0.15 µg/L on wide bore capillary column.

PRECISION: RSD = 7.6% @ (0.5-10) µg/L (16 samples) wide bore capillary column.

ACCURACY: Recovery = 101% @ (0.5-10) µg/L (16 samples) wide bore capillary column.

SAMPLING METHOD: Use a 60-120 mL screw cap vial (pre-washed with detergent, rinsed with distilled water and oven dried at 105°C) with a PTFE-faced silicone septum. If residual chlorine is in the water add about 25 mg of ascorbic acid to each vial before sample collection. Collect bubble free samples.

STABILITY: Cool to 4°C; HCl to pH <2. M.H.T. = 14 Days.

QUALITY CONTROL: As initial demonstration of lab accuracy and precision, analyze 4 to 7 replicates of a lab fortified blank containing the analyte at 0.2-5 µg/L. Collect all samples in duplicate.

REFERENCE: Method 524.2, VOCs in H_2O by Purge and Trap Capillary column GC/MS, EPA 600/4-88/039.

PRIMARY NAME: 4-Isopropyltoluene Method 502.2

TITLE: VOCs in H_2O by Purge and Trap capillary column GC

MATRIX: Drinking water (finished or any treatment stage) and raw source water.

CAS # : 99-87-6

APPLICATION: Method covers 60 VOCs. An inert gas is bubbled through a 5 mL sample in a purging chamber. Purged sample components are trapped in tube of sorbent materials. When purging is done, sorbent tube is heated and backflushed w/helium to desorb trapped sample onto a capillary GC column.

INTERFERENCES: During analysis, major contaminant sources are volatile materials in the lab and impurities in purging gas and sorbent trap. With high and low level samples, there can be carry-over contamination. Beware of methylene chloride, it permeates through PTFE tubing.

INSTRUMENTATION: (capillary column) GC, w/PID and electrolytic conductivity detectors in series. Column #1: VOCOL wide bore; Column #2: RTX-502.2 mega bore; Column #3: DB-62 mega bore.

RANGE: Approximately 0.02-200 µg/L.

MDL: 0.01 µg/L (reagent H_2O) (PID)

PRECISION: RSD = 2.4% @ Spike of 10 µg/L (reagent H_2O) (PID).

ACCURACY: Recovery = 98% @ spike of 10 µg/L (reagent H_2O) (PID)

SAMPLING METHOD: Use a 40-120 mL screw cap vial (pre-washed with detergent, rinsed with distilled water and oven dried at 105°C) with a PTFE-faced silicone septum. If residual chlorine is in the water add about 25 mg of ascorbic acid to each vial before sample collection. Collect bubble free samples.

STABILITY: Cool to 4°C; HCl to pH <2.

M.H.T. = 14 Days.

QUALITY CONTROL: As initial demonstration of lab accuracy and precision, analyze 4 to 7 replicates of a lab fortified blank containing the analyte at 0.1-5 µg/L. Collect all samples in duplicate.

REFERENCE: Method 502.2, Volatile Organic Compounds in Water by Purge and Trap Capillary Column GC, EPA 600/4-88/039.

PRIMARY NAME: 4-Isopropyltoluene

Method 503.1

TITLE: Aromatic and Unsaturated VOCs in Water

MATRIX: Drinking water (finished or any treatment stage) and raw source water.

CAS # : 99-87-6

APPLICATION: Method covers 28 aromatic and unsaturated VOCs. An inert gas is bubbled through a 5 mL water sample. Purged sample components are trapped in tube of sorbent materials. When purging is complete, sorbent tube is heated and backflushed with inert gas to desorb trapped sample onto a packed GC column.

INTERFERENCES: During analysis, major contaminant sources are volatile materials in the lab and impurities in purging gas and sorbent trap. With high and low level samples, there can be carry-over contamination. Excess water causes a negative baseline deflection.

INSTRUMENTATION: Purge and Trap GC w/photoionization detector. (Two GC columns are recommended); Column #1: 5% SP-1200 and 1.75% Bentone 34 on Supelcoport; Column #2: 1,2,3-tris(2-cyanoethoxy)propane on Chromosorb W.

RANGE: 2.2-600 μg/L. (Drinking water) MDL: 0.009 μg/L in water

PRECISION: not listed.

ACCURACY: not listed.

SAMPLING METHOD: Use a 40-120 mL screw cap vial (pre-washed with detergent, rinsed with distilled water and oven dried at 105°C) with a PTFE-faced silicone septum. If residual chlorine is in the water add about 25 mg of ascorbic acid to each vial before sample collection. Collect bubble free samples.

STABILITY: Cool to 4°C; HCl to pH <2. M.H.T. = 14 Days.

QUALITY CONTROL: As initial demonstration of lab accuracy and precision, analyze 4 to 7 replicates of a lab fortified blank containing the analyte at 0.1-5 μg/L. Collect all samples in duplicate.

REFERENCE: Method 503.1, Volatile Aromatic and Unsaturated Organic Compounds in H_2O by Purge and Trap GC, EPA 600/4-88/039.

PRIMARY NAME: 4-Isopropyltoluene Method 524.2

TITLE: VOCs in H_2O Purge and Trap capillary column GC/MS

MATRIX: Drinking water (finished or any treatment stage) and raw source water.

CAS # : 99-87-6

APPLICATION: Method covers 60 VOCs. An inert gas is bubbled through a 25 mL water sample. Purged sample components are trapped in tube of sorbent materials. When purging is complete, sorbent tube is heated and backflushed with helium to desorb trapped sample onto a capillary GC column.

INTERFERENCES: During analysis, major contaminant sources are volatile materials in the lab and impurities in purging gas and sorbent trap. With high and low level samples, there can be carry-over contamination. Watch for methylene chloride, it permeates through PTFE tubing.

INSTRUMENTATION: (capillary column) Gas Chromatography/Mass Spectometry/Data System; Column #1: VOCOL glass wide bore capillary column; Column #2: DB-624 fused silica column; Column #3: DB-5 fused silica column.

RANGE: Approximately 0.02-200 μg/L.

MDL: 0.12 μg/L on wide bore capillary column.

PRECISION: RSD = 6.7% @ (0.1-10) μg/L (23 samples) wide bore capillary column.

ACCURACY: Recovery = 99% @ (0.1-10) μg/L (23 samples) wide bore capillary column.

SAMPLING METHOD: Use a 60-120 mL screw cap vial (pre-washed with detergent, rinsed with distilled water and oven dried at 105°C) with a PTFE-faced silicone septum. If residual chlorine is in the water add about 25 mg of ascorbic acid to each vial before sample collection. Collect bubble free samples.

STABILITY: Cool to 4°C; HCl to pH <2.

M.H.T. = 14 Days.

QUALITY CONTROL: As initial demonstration of lab accuracy and precision, analyze 4 to 7 replicates of a lab fortified blank containing the analyte at 0.2-5 μg/L. Collect all samples in duplicate.

REFERENCE: Method 524.2, VOCs in H_2O by Purge and Trap Capillary column GC/MS, EPA 600/4-88/039.

PRIMARY NAME: Methyl ethyl ketone (MEK)

Method 8015

TITLE: Nonhalogenated Volatile Organics

MATRIX: groundwater, soils, sludges water miscible liquid wastes, and non-water miscible wastes.

CAS # : 78-93-3

APPLICATION: This method is used for the analysis of 6 nonhalogenated VOCs. Samples are analyzed using direct injection or purge and trap methods. Groundwater must be analyzed by the purge and trap method. The method provides an optional GC column that may help resolve analytes from interferences and which is also used for analyte confirmation.

INTERFERENCES: There can be carry-over contamination with high and low level samples. Impurities may come from the purge and trap apparatus, organic compounds outgassing from the plumbing ahead of trap, diffusion of VOCs through the sample bottle septum during shipping or storage, or from solvent vapors in the lab.

INSTRUMENTATION: GC capable of on-column injections or purge-and-trap sample introduction and a flame ionization detector (FID). Column 1: an 8 foot by 0.1 inch 1% SP-1000 on Carbopack-B column. Column 2: a 6 foot by 0.1 inch bonded n-octane on Porasil-C.

RANGE: Not available

MDL: Not available

PRECISION: Not available

ACCURACY: Not available

SAMPLING METHOD: For water and liquid samples; use glass 40 mL vials with Teflon-lined septum caps and collect two vials per sample location with no headspace. For solids and concentrated waste samples; use widemouth glass bottles with Teflon liners. Cool all samples to 4°C.

STABILITY: For concentrated wastes, soils, sediments or sludges; cool to 4°C. For liquids; add 4 drops of concentrated hydrochloric acid, cool to 4°C. M.H.T. = 14 days.

QUALITY CONTROL: Analyze a reagent blank, matrix spike and matrix spike duplicate/duplicate for each analytical batch (up to 20 Samples). Demonstrate the purity of glassware and reagents by analyzing a reagent water method blank. Internal, surrogate and five concentration level calibration standards are used.

REFERENCE: Method 8015, SW-846, 3rd ed., Nov 1986.

PRIMARY NAME: Methyl isobutyl ketone (MIBK) Method 8015

TITLE: Nonhalogenated Volatile Organics

MATRIX: groundwater, soils, sludges water miscible liquid wastes, and non-water miscible wastes.

CAS # : 108-10-1

APPLICATION: This method is used for the analysis of 6 nonhalogenated VOCs. Samples are analyzed using direct injection or purge and trap methods. Groundwater must be analyzed by the purge and trap method. The method provides an optional GC column that may help resolve analytes from interferences and which is also used for analyte confirmation.

INTERFERENCES: There can be carry-over contamination with high and low level samples. Impurities may come from the purge and trap apparatus, organic compounds outgassing from the plumbing ahead of trap, diffusion of VOCs through the sample bottle septum during shipping or storage, or from solvent vapors in the lab.

INSTRUMENTATION: GC capable of on-column injections or purge-and-trap sample introduction and a flame ionization detector (FID). Column 1: an 8 foot by 0.1 inch 1% SP-1000 on Carbopack-B column. Column 2: a 6 foot by 0.1 inch bonded n-octane on Porasil-C.

RANGE: Not available

MDL: Not available

PRECISION: Not available

ACCURACY: Not available

SAMPLING METHOD: For water and liquid samples; use glass 40 mL vials with Teflon-lined septum caps and collect two vials per sample location with no headspace. For solids and concentrated waste samples; use widemouth glass bottles with Teflon liners. Cool all samples to 4°C.

STABILITY: For concentrated wastes, soils, sediments or sludges; cool to 4°C. For liquids; add 4 drops of concentrated hydrochloric acid, cool to 4°C. M.H.T. = 14 days.

QUALITY CONTROL: Analyze a reagent blank, matrix spike and matrix spike duplicate/duplicate for each analytical batch (up to 20 Samples). Demonstrate the purity of glassware and reagents by analyzing a reagent water method blank. Internal, surrogate and five concentration level calibration standards are used.

REFERENCE: Method 8015, SW-846, 3rd ed., Nov 1986.

PRIMARY NAME: Naphthalene

Method 502.2

TITLE: VOCs in H2O by Purge and Trap capillary column GC

MATRIX: Drinking water (finished or any treatment stage) and raw source water.

CAS # : 91-20-3

APPLICATION: Method covers 60 VOCs. An inert gas is bubbled through a 5 mL sample in a purging chamber. Purged sample components are trapped in tube of sorbent materials. When purging is done, sorbent tube is heated and backflushed w/helium to desorb trapped sample onto a capillary GC column.

INTERFERENCES: During analysis, major contaminant sources are volatile materials in the lab and impurities in purging gas and sorbent trap. With high and low level samples, there can be carry-over contamination. Beware of methylene chloride, it permeates through PTFE tubing.

INSTRUMENTATION: (capillary column) GC, w/PID and electrolytic conductivity detectors in series. Column #1: VOCOL wide bore; Column #2: RTX-502.2 mega bore; Column #3: DB-62 mega bore.

RANGE: Approximately 0.02-200 µg/L.

MDL: 0.06 µg/L (reagent H_2O) (PID)

PRECISION: RSD = 6.2% @ Spike of 10 µg/L (reagent H_2O) (PID).

ACCURACY: Recovery = 102% @ spike of 10 µg/L (reagent H_2O) (PID)

SAMPLING METHOD: Use a 40-120 mL screw cap vial (pre-washed with detergent, rinsed with distilled water and oven dried at 105°C) with a PTFE-faced silicone septum. If residual chlorine is in the water add about 25 mg of ascorbic acid to each vial before sample collection. Collect bubble free samples.

STABILITY: Cool to 4°C; HCl to pH <2. M.H.T. = 14 Days.

QUALITY CONTROL: As initial demonstration of lab accuracy and precision, analyze 4 to 7 replicates of a lab fortified blank containing the analyte at 0.1-5 μg/L. Collect all samples in duplicate.

REFERENCE: Method 502.2, Volatile Organic Compounds in Water by Purge and Trap Capillary Column GC, EPA 600/4-88/039.

PRIMARY NAME: Naphthalene Method 503.1

TITLE: Aromatic and Unsaturated VOCs in Water

MATRIX: Drinking water (finished or any treatment stage) and raw source water.

CAS # : 91-20-3

APPLICATION: Method covers 28 aromatic and unsaturated VOCs. An inert gas is bubbled through a 5 mL water sample. Purged sample components are trapped in tube of sorbent materials. When purging is complete, sorbent tube is heated and backflushed with inert gas to desorb trapped sample onto a packed GC column.

INTERFERENCES: During analysis, major contaminant sources are volatile materials in the lab and impurities in purging gas and sorbent trap. With high and low level samples, there can be carry-over contamination. Excess water causes a negative baseline deflection.

INSTRUMENTATION: Purge and Trap GC w/photoionization detector. (Two GC columns are recommended); Column #1: 5% SP-1200 and 1.75% Bentone 34 on Supelcoport; Column #2: 1,2,3-tris(2-cyanoethoxy)propane on Chromosorb W.

RANGE: 2.2-600 μg/L. (Drinking water) MDL: 0.04 μg/L in water

PRECISION: RSD = 14.8% @ 0.50 μg/L conc.; 16 samples

ACCURACY: Avg Recovery = 92% @ 0.50 μg/L conc.; 16 samples

SAMPLING METHOD: Use a 40-120 mL screw cap vial (pre-washed with detergent, rinsed with distilled water and oven dried at 105°C) with a PTFE-

faced silicone septum. If residual chlorine is in the water add about 25 mg of ascorbic acid to each vial before sample collection. Collect bubble free samples.

STABILITY: Cool to 4°C; HCl to pH <2. M.H.T. = 14 Days.

QUALITY CONTROL: As initial demonstration of lab accuracy and precision, analyze 4 to 7 replicates of a lab fortified blank containing the analyte at 0.1-5 μg/L. Collect all samples in duplicate.

REFERENCE: Method 503.1, Volatile Aromatic and Unsaturated Organic Compounds in H_2O by Purge and Trap GC, EPA 600/4-88/039.

PRIMARY NAME: Naphthalene Method 524.2

TITLE: VOCs in H_2O Purge and Trap capillary column GC/MS

MATRIX: Drinking water (finished or any treatment stage) and raw source water.

CAS # : 91-20-3

APPLICATION: Method covers 60 VOCs. An inert gas is bubbled through a 25 mL water sample. Purged sample components are trapped in tube of sorbent materials. When purging is complete, sorbent tube is heated and backflushed with helium to desorb trapped sample onto a capillary GC column.

INTERFERENCES: During analysis, major contaminant sources are volatile materials in the lab and impurities in purging gas and sorbent trap. With high and low level samples, there can be carry-over contamination. Watch for methylene chloride, it permeates through PTFE tubing.

INSTRUMENTATION: (Capillary Column) Gas Chromatography/Mass Spectrometry/Data System; Column #1: VOCOL glass wide bore capillary column; Column #2: DB-624 fused silica column; Column #3: DB-5 fused silica column.

RANGE: Approximately 0.02-200 μg/L.

MDL: 0.04 μg/L on wide bore capillary column.

PRECISION: RSD = 8.2% @ (0.1-100) μg/L (31 samples) wide bore capillary column.

ACCURACY: Recovery = 104% @ (0.1-100) μg/L (31 samples) wide bore capillary column.

SAMPLING METHOD: Use a 60-120 mL screw cap vial (pre-washed with detergent, rinsed with distilled water and oven dried at 105°C) with a PTFE-faced silicone septum. If residual chlorine is in the water add about 25 mg of ascorbic acid to each vial before sample collection. Collect bubble free samples.

STABILITY: Cool to 4°C; HCl to pH <2. M.H.T. = 14 Days.

QUALITY CONTROL: As initial demonstration of lab accuracy and precision, analyze 4 to 7 replicates of a lab fortified blank containing the analyte at 0.2-5 μg/L. Collect all samples in duplicate.

REFERENCE: Method 524.2, VOCs in H_2O by Purge and Trap Capillary column GC/MS, EPA 600/4-88/039.

PRIMARY NAME: Paraldehyde Method 8015

TITLE: Nonhalogenated Volatile Organics

MATRIX: groundwater, soils, sludges water miscible liquid wastes, and non-water miscible wastes.

CAS # : 123-63-7

APPLICATION: This Method is used for the analysis of 6 nonhalogenated VOCs. Samples are analyzed using direct injection or purge and trap Methods. Groundwater must be analyzed by the purge and trap Method. The Method provides an optional GC column that may help resolve analytes from interferences and which is also used for analyte confirmation.

INTERFERENCES: There can be carry-over contamination with high and low level samples. Impurities may come from the purge and trap apparatus, organic compounds outgassing from the plumbing ahead of trap, diffusion of VOCs through the sample bottle septum during shipping or storage, or from solvent vapors in the lab.

INSTRUMENTATION: GC capable of on-column injections or purge-and-trap sample introduction and a flame ionization detector (FID). Column 1: an

8 foot by 0.1 inch 1% SP-1000 on Carbopack-B column. Column 2: a 6 foot by 0.1 inch bonded n-octane on Porasil-C.

RANGE: Not available

MDL: Not available

PRECISION: Not available

ACCURACY: Not available

SAMPLING Method: For water and liquid samples; use glass 40 mL vials with Teflon-lined septum caps and collect two vials per sample location with no headspace. For solids and concentrated waste samples; use widemouth glass bottles with Teflon liners. Cool all samples to 4°C.

STABILITY: For concentrated wastes, soils, sediments or sludges; cool to 4°C. For liquids; add 4 drops of concentrated hydrochloric acid, cool to 4°C. M.H.T. = 14 days.

QUALITY CONTROL: Analyze a reagent blank, matrix spike and matrix spike duplicate/duplicate for each analytical batch (up to 20 Samples). Demonstrate the purity of glassware and reagents by analyzing a reagent water Method blank. Internal, surrogate and five concentration level calibration standards are used.

REFERENCE: Method 8015, SW-846, 3rd ed., Nov 1986.

PRIMARY NAME: n-Propylbenzene

Method 502.2

TITLE: VOCs in H_2O by Purge and Trap capillary column GC

MATRIX: Drinking water (finished or any treatment stage) and raw source water.

CAS # : 103-65-1

APPLICATION: Method covers 60 VOCs. An inert gas is bubbled through a 5 mL sample in a purging chamber. Purged sample components are trapped in tube of sorbent materials. When purging is done, sorbent tube is heated and backflushed w/helium to desorb trapped sample onto a capillary GC column.

INTERFERENCES: During analysis, major contaminant sources are volatile materials in the lab and impurities in purging gas and sorbent trap. With

high and low level samples, there can be carry-over contamination. Beware of methylene chloride, it permeates through PTFE tubing.

INSTRUMENTATION: (Capillary Column) GC, w/PID and electrolytic conductivity detectors in series. Column #1: VOCOL wide bore; Column #2: RTX-502.2 mega bore; Column #3: DB-62 mega bore.

RANGE: Approximately 0.02-200 µg/L.

MDL: 0.01 µg/L (reagent H_2O) (PID)

PRECISION: RSD = 2.0% @ Spike of 10 µg/L (reagent H_2O) (PID).

ACCURACY: Recovery = 103% @ spike of 10 µg/L (reagent H_2O) (PID)

SAMPLING Method: Use a 40-120 mL screw cap vial (pre-washed with detergent, rinsed with distilled water and oven dried at 105°C) with a PTFE-faced silicone septum. If residual chlorine is in the water add about 25 mg of ascorbic acid to each vial before sample collection. Collect bubble free samples.

STABILITY: Cool to 4°C; HCl to pH <2.

M.H.T. = 14 Days.

QUALITY CONTROL: As initial demonstration of lab accuracy and precision, analyze 4 to 7 replicates of a lab fortified blank containing the analyte at 0.1-5 µg/L. Collect all samples in duplicate.

REFERENCE: Method 502.2, Volatile Organic Compounds in Water by Purge and Trap Capillary Column GC, EPA 600/4-88/039.

PRIMARY NAME: n-Propylbenzene

Method 503.1

TITLE: Aromatic and Unsaturated VOCs in Water

MATRIX: Drinking water (finished or any treatment stage) and raw source water.

CAS # : 103-65-1

APPLICATION: Method covers 28 aromatic and unsaturated VOCs. An inert gas is bubbled through a 5 mL water sample. Purged sample components are trapped in tube of sorbent materials. When purging is complete, sorbent tube is heated and backflushed with inert gas to desorb trapped sample onto a packed GC column.

INTERFERENCES: During analysis, major contaminant sources are volatile materials in the lab and impurities in purging gas and sorbent trap. With high and low level samples, there can be carry-over contamination. Excess water causes a negative baseline deflection.

INSTRUMENTATION: Purge and Trap GC w/photoionization detector. (Two GC columns are recommended); Column #1: 5% SP-1200 and 1.75% Bentone 34 on Supelcoport; Column #2: 1,2,3-tris(2-cyanoethoxy)propane on Chromosorb W.

RANGE: 2.2-600 µg/L. (Drinking water) MDL: 0.009 µg/L in water

PRECISION: RSD = 9.3% @ 0.40 µg/L conc.; 7 samples

ACCURACY: Avg Recovery = 83% @ 0.40 µg/L conc.; 7 samples

SAMPLING Method: Use a 40-120 mL screw cap vial (pre-washed with detergent, rinsed with distilled water and oven dried at 105°C) with a PTFE-faced silicone septum. If residual chlorine is in the water add about 25 mg of ascorbic acid to each vial before sample collection. Collect bubble free samples.

STABILITY: Cool to 4°C; HCl to pH <2. M.H.T. = 14 Days.

QUALITY CONTROL: As initial demonstration of lab accuracy and precision, analyze 4 to 7 replicates of a lab fortified blank containing the analyte at 0.1-5 µg/L. Collect all samples in duplicate.

REFERENCE: Method 503.1, Volatile Aromatic and Unsaturated Organic Compounds in H_2O by Purge and Trap GC, EPA 600/4-88/039.

PRIMARY NAME: n-Propylbenzene Method 524.2

TITLE: VOCs in H_2O Purge and Trap capillary column GC/MS

MATRIX: Drinking water (finished or any treatment stage) and raw source water.

CAS # : 103-65-1

APPLICATION: Method covers 60 VOCs. An inert gas is bubbled through a 25 mL water sample. Purged sample components are trapped in tube of sorbent materials. When purging is complete, sorbent tube is heated and backflushed with helium to desorb trapped sample onto a capillary GC column.

INTERFERENCES: During analysis, major contaminant sources are volatile materials in the lab and impurities in purging gas and sorbent trap. With high and low level samples, there can be carry-over contamination. Watch for methylene chloride, it permeates through PTFE tubing.

INSTRUMENTATION: (capillary column) Gas Chromatography/Mass Spectometry/Data System; Column #1: VOCOL glass wide bore capillary column; Column #2: DB-624 fused silica column; Column #3: DB-5 fused silica column.

RANGE: Approximately 0.02-200 µg/L.

MDL: 0.04 µg/L on wide bore capillary column.

PRECISION: RSD = 5.8% @ (0.1-10) µg/L (31 samples) wide bore capillary column.

ACCURACY: Recovery = 100% @ (0.1-10) µg/L (31 samples) wide bore capillary column.

SAMPLING Method: Use a 60-120 mL screw cap vial (pre-washed with detergent, rinsed with distilled water and oven dried at 105°C) with a PTFE-faced silicone septum. If residual chlorine is in the water add about 25 mg of ascorbic acid to each vial before sample collection. Collect bubble free samples.

STABILITY: Cool to 4°C; HCl to pH <2.

M.H.T. = 14 Days.

QUALITY CONTROL: As initial demonstration of lab accuracy and precision, analyze 4 to 7 replicates of a lab fortified blank containing the analyte at 0.2-5 µg/L. Collect all samples in duplicate.

REFERENCE: Method 524.2, VOCs in H_2O by Purge and Trap Capillary column GC/MS, EPA 600/4-88/039.

PRIMARY NAME: Styrene

Method 502.2

TITLE: VOCs in H_2O by Purge and Trap capillary column GC

MATRIX: Drinking water (finished or any treatment stage) and raw source water.

CAS # : 100-42-5

APPLICATION: Method covers 60 VOCs. An inert gas is bubbled through a 5 mL sample in a purging chamber. Purged sample components are trapped in tube of sorbent materials. When purging is done,sorbent tube is heated and backflushed w/helium to desorb trapped sample onto a capillary GC column.

INTERFERENCES: During analysis, major contaminant sources are volatile materials in the lab and impurities in purging gas and sorbent trap. With high and low level samples, there can be carry-over contamination. Beware of methylene chloride, it permeates through PTFE tubing.

INSTRUMENTATION: (Capillary Column) GC, w/PID and electrolytic conductivity detectors in series. Column #1: VOCOL wide bore; Column #2: RTX-502.2 mega bore; Column #3: DB-62 mega bore.

RANGE: Approximately 0.02-200 μg/L.

MDL: 0.01 μg/L (reagent H_2O) (PID)

PRECISION: RSD = 1.3% @ Spike of 10 μg/L (reagent H_2O) (PID).

ACCURACY: Recovery = 104% @ spike of 10 μg/L (reagent H_2O) (PID)

SAMPLING Method: Use a 40-120 mL screw cap vial (pre-washed with detergent, rinsed with distilled water and oven dried at 105°C) with a PTFE-faced silicone septum. If residual chlorine is in the water add about 25 mg of ascorbic acid to each vial before sample collection. Collect bubble free samples.

STABILITY: Cool to 4°C; HCl to pH <2.

M.H.T. = 14 Days.

QUALITY CONTROL: As initial demonstration of lab accuracy and precision, analyze 4 to 7 replicates of a lab fortified blank containing the analyte at 0.1-5 μg/L. Collect all samples in duplicate.

REFERENCE: Method 502.2, Volatile Organic Compounds in Water by Purge and Trap Capillary Column GC, EPA 600/4-88/039.

PRIMARY NAME: Styrene

Method 503.1

TITLE: Aromatic and Unsaturated VOCs in Water

MATRIX: Drinking water (finished or any treatment stage) and raw source water.

CAS # : 100-42-5

APPLICATION: Method covers 28 aromatic and unsaturated VOCs. An inert gas is bubbled through a 5 mL water sample. Purged sample components are trapped in tube of sorbent materials. When purging is complete, sorbent tube is heated and backflushed with inert gas to desorb trapped sample onto a packed GC column.

INTERFERENCES: During analysis, major contaminant sources are volatile materials in the lab and impurities in purging gas and sorbent trap. With high and low level samples, there can be carry-over contamination. Excess water causes a negative baseline deflection.

INSTRUMENTATION: Purge and Trap GC w/photoionization detector. (Two GC columns are recommended); Column #1: 5% SP-1200 and 1.75% Bentone 34 on Supelcoport; Column #2: 1,2,3-tris(2-cyanoethoxy)propane on Chromosorb W.

RANGE: 2.2-600 µg/L. (Drinking water) MDL: 0.008 µg/L in water

PRECISION: not listed.

ACCURACY: not listed.

SAMPLING Method: Use a 40-120 mL screw cap vial (pre-washed with detergent, rinsed with distilled water and oven dried at 105°C) with a PTFE-faced silicone septum. If residual chlorine is in the water add about 25 mg of ascorbic acid to each vial before sample collection. Collect bubble free samples.

STABILITY: Cool to 4°C; HCl to pH <2. M.H.T. = 14 Days.

QUALITY CONTROL: As initial demonstration of lab accuracy and precision, analyze 4 to 7 replicates of a lab fortified blank containing the analyte at 0.1-5 µg/L. Collect all samples in duplicate.

REFERENCE: Method 503.1, Volatile Aromatic and Unsaturated Organic Compounds in H_2O by Purge and Trap GC, EPA 600/4-88/039.

PRIMARY NAME: Styrene Method 524.1

TITLE: VOCs in H_2O by Purge and Trap GC/MS

MATRIX: Drinking water (finished or any treatment stage) and raw source water.

CAS # : 100-42-5

APPLICATION: Method covers 48 VOCs. An inert gas is bubbled through a 25 mL water sample. Purged sample components are trapped in tube of sorbent materials. When purging is complete, sorbent tube is heated and backflushed with helium to desorb trapped sample onto a packed GC column.

INTERFERENCES: During analysis, major contaminant sources are volatile materials in the lab and impurities in purging gas and sorbent trap. With high and low level samples, there can be carry-over contamination. Watch for methylene chloride, it permeates through PTFE tubing.

INSTRUMENTATION: Gas Chromatography/Mass Spectrometry/Data System; 1% SP-1000 on Carbopack B.

RANGE: Approximately 0.1-200 μg/L. MDL: 0.2 μg/L

PRECISION: RSD = 6% @ 1.0 μg/L

ACCURACY: Mean Accuracy = 120% @ 1.0 μg/L

SAMPLING Method: Use a 60-120 mL screw cap vial (pre-washed with detergent, rinsed with distilled water and oven dried at 105°C) with a PTFE-faced silicone septum. If residual chlorine is in the water add about 25 mg of ascorbic acid to each vial before sample collection. Collect bubble free samples. about 25 mg ascorbic acid to each vial before collecting sample.

STABILITY: Cool to 4°C; HCl to pH <2. M.H.T. = 14 Days.

QUALITY CONTROL: As initial demonstration of lab accuracy and precision, analyze 4 to 7 replicates of a lab fortified blank containing the analyte at 0.2-5 μg/L. Collect all samples in duplicate.

REFERENCE: Method 524.1, VOCs in Water by Purge and Trap Gas Chromatography/Mass Spectrometry, EPA 600/4-88/039.

PRIMARY NAME: Styrene Method 524.2

TITLE: VOCs in H_2O Purge and Trap capillary column GC/MS

MATRIX: Drinking water (finished or any treatment stage) and raw source water.

CAS # : 100-42-5

APPLICATION: Method covers 60 VOCs. An inert gas is bubbled through a 25 mL water sample. Purged sample components are trapped in tube of sorbent materials. When purging is complete, sorbent tube is heated and backflushed with helium to desorb trapped sample onto a capillary GC column.

INTERFERENCES: During analysis, major contaminant sources are volatile materials in the lab and impurities in purging gas and sorbent trap. With high and low level samples, there can be carry-over contamination. Watch for methylene chloride, it permeates through PTFE tubing.

INSTRUMENTATION: (Capillary Column) Gas Chromatography/Mass Spectrometry/Data System; Column #1: VOCOL glass wide bore capillary column; Column #2: DB-624 fused silica column; Column #3: DB-5 fused silica column.

RANGE: Approximately 0.02-200 µg/L.

MDL: 0.04 µg/L on wide bore capillary column.

PRECISION: RSD = 7.2% @ (0.1-100) µg/L (39 samples) wide bore capillary column.

ACCURACY: Recovery = 102% @ (0.1-100) µg/L (39 samples) wide bore capillary column.

SAMPLING Method: Use a 60-120 mL screw cap vial (pre-washed with detergent, rinsed with distilled water and oven dried at 105°C) with a PTFE-faced silicone septum. If residual chlorine is in the water add about 25 mg of ascorbic acid to each vial before sample collection. Collect bubble free samples.

STABILITY: Cool to 4°C; HCl to pH <2.

M.H.T. = 14 Days.

QUALITY CONTROL: As initial demonstration of lab accuracy and precision, analyze 4 to 7 replicates of a lab fortified blank containing the analyte at 0.2-5 µg/L. Collect all samples in duplicate.

REFERENCE: Method 524.2, VOCs in H_2O by Purge and Trap Capillary column GC/MS, EPA 600/4-88/039.

PRIMARY NAME: Toluene

Method 502.2

TITLE: VOCs in H_2O by Purge and Trap capillary column GC

MATRIX: Drinking water (finished or any treatment stage) and raw source water.

CAS # : 108-88-3

APPLICATION: Method covers 60 VOCs. An inert gas is bubbled through a 5 mL sample in a purging chamber. Purged sample components are trapped in tube of sorbent materials. When purging is done, sorbent tube is heated and backflushed w/helium to desorb trapped sample onto a capillary GC column.

INTERFERENCES: During analysis, major contaminant sources are volatile materials in the lab and impurities in purging gas and sorbent trap. With high and low level samples, there can be carry-over contamination. Beware of methylene chloride, it permeates through PTFE tubing.

INSTRUMENTATION: (capillary column) GC, w/PID and electrolytic conductivity detectors in series. Column #1: VOCOL wide bore; Column #2: RTX-502.2 mega bore; Column #3: DB-62 mega bore.

RANGE: Approximately 0.02-200 µg/L.

MDL: 0.01 µg/L (reagent H_2O) (PID)

PRECISION: RSD = 0.8% @ Spike of 10 µg/L (reagent H_2O) (PID).

ACCURACY: Recovery = 99% @ spike of 10 µg/L (reagent H_2O) (PID)

SAMPLING Method: Use a 40-120 mL screw cap vial (pre-washed with detergent, rinsed with distilled water and oven dried at 105°C) with a PTFE-faced silicone septum. If residual chlorine is in the water add about 25 mg of ascorbic acid to each vial before sample collection. Collect bubble free samples.

STABILITY: Cool to 4°C; HCl to pH <2.

M.H.T. = 14 Days.

QUALITY CONTROL: As initial demonstration of lab accuracy and precision, analyze 4 to 7 replicates of a lab fortified blank containing the analyte at 0.1-5 µg/L. Collect all samples in duplicate.

REFERENCE: Method 502.2, Volatile Organic Compounds in Water by Purge and Trap Capillary Column GC, EPA 600/4-88/039.

PRIMARY NAME: Toluene

Method 503.1

TITLE: Aromatic and Unsaturated VOCs in Water

MATRIX: Drinking water (finished or any treatment stage) and raw source water.

CAS # : 108-88-3

APPLICATION: Method covers 28 aromatic and unsaturated VOCs. An inert gas is bubbled through a 5 mL water sample. Purged sample components are trapped in tube of sorbent materials. When purging is complete, sorbent tube is heated and backflushed with inert gas to desorb trapped sample onto a packed GC column.

INTERFERENCES: During analysis, major contaminant sources are volatile materials in the lab and impurities in purging gas and sorbent trap. With high and low level samples, there can be carry-over contamination. Excess water causes a negative baseline deflection.

INSTRUMENTATION: Purge and Trap GC w/photoionization detector. (Two GC columns are recommended); Column #1: 5% SP-1200 and 1.75% Bentone 34 on Supelcoport; Column #2: 1,2,3-tris(2-cyanoethoxy)propane on Chromosorb W.

RANGE: 2.2-600 μg/L. (Drinking water)

MDL: 0.02 μg/L in water

PRECISION: RSD = 6.6% @ 0.40 μg/L conc.; 13 samples

ACCURACY: Avg Recovery = 94% @ 0.40 μg/L conc.; 13 samples

SAMPLING Method: Use a 40-120 mL screw cap vial (pre-washed with detergent, rinsed with distilled water and oven dried at 105°C) with a PTFE-faced silicone septum. If residual chlorine is in the water add about 25 mg of ascorbic acid to each vial before sample collection. Collect bubble free samples.

STABILITY: Cool to 4°C; HCl to pH <2.

M.H.T. = 14 Days.

QUALITY CONTROL: As initial demonstration of lab accuracy and precision, analyze 4 to 7 replicates of a lab fortified blank containing the analyte at 0.1-5 μg/L. Collect all samples in duplicate.

REFERENCE: Method 503.1, Volatile Aromatic and Unsaturated Organic Compounds in H_2O by Purge and Trap GC, EPA 600/4-88/039.

PRIMARY NAME: Toluene

Method 524.1

TITLE: VOCs in H_2O by Purge and Trap GC/MS

MATRIX: Drinking water (finished or any treatment stage) and raw source water.

CAS # : 108-88-3

APPLICATION: Method covers 48 VOCs. An inert gas is bubbled through a 25 mL water sample. Purged sample components are trapped in tube of sorbent materials. When purging is complete, sorbent tube is heated and backflushed with helium to desorb trapped sample onto a packed GC column.

INTERFERENCES: During analysis, major contaminant sources are volatile materials in the lab and impurities in purging gas and sorbent trap. With high and low level samples, there can be carry-over contamination. Watch for methylene chloride, it permeates through PTFE tubing.

INSTRUMENTATION: Gas Chromatography/Mass Spectrometry/Data System; 1% SP-1000 on Carbopack B.

RANGE: Approximately 0.1-200 μg/L.

MDL: 0.1 μg/L

PRECISION: RSD = 4.1% @ 1.0 μg/L conc (8 samples)

ACCURACY: Mean Accuracy = 105% @ 1.0 μg/L

SAMPLING Method: Use a 60-120 mL screw cap vial (pre-washed with detergent, rinsed with distilled water and oven dried at 105°C) with a PTFE-faced silicone septum. If residual chlorine is in the water add about 25 mg of ascorbic acid to each vial before sample collection. Collect bubble free samples.

STABILITY: Cool to 4°C; HCl to pH <2. M.H.T. = 14 Days.

QUALITY CONTROL: As initial demonstration of lab accuracy and precision, analyze 4 to 7 replicates of a lab fortified blank containing the analyte at 0.2-5 μg/L. Collect all samples in duplicate.

REFERENCE: Method 524.1, VOCs in Water by Purge and Trap Gas Chromatography/Mass Spectrometry, EPA 600/4-88/039.

PRIMARY NAME: Toluene Method 524.2

TITLE: VOCs in H_2O Purge and Trap capillary column GC/MS

MATRIX: Drinking water (finished or any treatment stage) and raw source water.

CAS # : 108-88-3

APPLICATION: Method covers 60 VOCs. An inert gas is bubbled through a 25 mL water sample. Purged sample components are trapped in tube of sorbent materials. When purging is complete, sorbent tube is heated and backflushed with helium to desorb trapped sample onto a capillary GC column.

INTERFERENCES: During analysis, major contaminant sources are volatile materials in the lab and impurities in purging gas and sorbent trap. With high and low level samples, there can be carry-over contamination. Watch for methylene chloride, it permeates through PTFE tubing.

INSTRUMENTATION: (capillary column) Gas Chromatography/Mass Spectrometry/Data System; Column #1: VOCOL glass wide bore capillary column; Column #2: DB-624 fused silica column; Column #3: DB-5 fused silica column.

RANGE: Approximately 0.02-200 μg/L.

MDL: 0.11 μg/L on wide bore capillary column.

PRECISION: RSD = 8.0% @ (0.5-10) μg/L (18 samples) wide bore capillary column.

ACCURACY: Recovery = 102% @ (0.5-10) μg/L (18 samples) wide bore capillary column.

SAMPLING Method: Use a 60-120 mL screw cap vial (pre-washed with detergent, rinsed with distilled water and oven dried at 105°C) with a PTFE-faced silicone septum. If residual chlorine is in the water add about 25 mg of ascorbic acid to each vial before sample collection. Collect bubble free samples.

STABILITY: Cool to 4°C; HCl to pH <2. M.H.T. = 14 Days.

QUALITY CONTROL: As initial demonstration of lab accuracy and precision, analyze 4 to 7 replicates of a lab fortified blank containing the analyte at 0.2-5 μg/L. Collect all samples in duplicate.

REFERENCE: Method 524.2, VOCs in H_2O by Purge and Trap Capillary column GC/MS, EPA 600/4-88/039.

PRIMARY NAME: Toluene Method 602

TITLE: Purgeable Aromatics MATRIX: Wastewater

CAS # : 108-88-3

APPLICATION: Method covers 7 purgeable aromatics. (Method 624 provides GC/MS conditions appropriate for the qualitative and quantitative confirmation of results). Method describes conditions for a second GC column to confirm measurements made with primary column.

INTERFERENCES: Impurities in the purge gas and organic compounds outgassing from the plumbing ahead of the trap. With high and low level samples, there can be carry-over contamination. Diffusion of volatile organics through the septum seal into the sample.

INSTRUMENTATION: GC equipped with photoionization detector. (With purge and trap unit)

RANGE: 2.1 To 550 μg/L. MDL: 0.2 μg/L.

PRECISION: 0.18X+0.71 μg/L (overall precision)

ACCURACY: 0.94C+0.65 μg/L (as recovery)

SAMPLING Method: 25 mL glass vial. Teflon lined septum.

STABILITY: Cool, 4°C, 0.008% sodium thiosulfate. HCl to pH 2. M.H.T. = 14 Days.

QUALITY CONTROL: The laboratory must on an ongoing basis, spike at least 10% of the samples from each sample site being monitored to assess accuracy.

REFERENCE: Method 602, Federal Register Part VIII 40 CFR Part 136, Oct 26, 1984.

PRIMARY NAME: Toluene — Method 624

TITLE: Purgeables — MATRIX: Wastewater

CAS # : 108-88-3

APPLICATION: Method covers 31 purgeable organics. An inert gas is bubbled through a 5 mL water sample in a specially designed purging chamber. Here, purgeables are transferred from aqueous to gaseous phase, passed onto a sorbent column and trapped. Trap is heated and backflushed with inert gas to desorb purgeables onto a GC column, where purgeables are separated.

INTERFERENCES: Impurities in the purge gas, organic compounds outgassing from the plumbing ahead of the trap, and solvent vapors in the laboratory. With high and low level samples, there can be carry-over contamination.

INSTRUMENTATION: GC/MS with purge and trap unit.

RANGE: 5-600 μg/L — MDL: 6.0 μg/L

PRECISION: 0.22X-1.71 μg/L (overall precision)

ACCURACY: 0.98C+2.03 μg/L (as recovery)

SAMPLING Method: 25 mL glass vial. Teflon lined septum.

STABILITY: Cool, 4°C, 0.008% sodium thiosulfate. HCl to pH 2. M.H.T. = 14 Days.

QUALITY CONTROL: The laboratory must on an ongoing basis, spike at least 5% of the samples from each sample site being monitored to assess accuracy.

REFERENCE: Method 624, Federal Register Part VIII 40 CFR Part 136, Oct 26, 1984.

PRIMARY NAME: Toluene

Method 8020

TITLE: Aromatic Volatile Organics

MATRIX: groundwater, soils, sludges water miscible liquid wastes, and non-water miscible wastes.

CAS # : 108-88-3

APPLICATION: This Method is used to analyze for 8 aromatic VOCs. Samples are analyzed using direct injection or purge and trap methods. Groundwater must be analyzed by the purge and trap method. The method provides an optional GC column that is used for analyte confirmation and may also help resolve analytes from interferences.

INTERFERENCES: There can be carry-over contamination with high and low level samples. Impurities may come from the purge and trap apparatus, organic compounds outgassing from the plumbing ahead of trap, diffusion of VOCs through the sample bottle septum during shipping or storage, or from solvent vapors in the lab.

INSTRUMENTATION: GC capable of on-column injections or purge-and-trap sample introduction and a photoionization detector (PID). Column 1: 6 foot by 0.082 inch with 5% SP-1200 and 1.75% Bentone-34 on Supelcoport. Column 2: 8 foot by 0.1 inch with 5% 1,2,3-tris(2-cyanoethoxy)propane on Chromosorb W-AW.

RANGE: 2.1 to 500 μg/L

MDL: 0.2 μg/L (reagent water)

PRACTICAL QUANTITATION LIMIT FACTORS FOR MULTIPLYING TIMES MDL VALUE

Matrix	Multiplication Factor
Groundwater	10
Low-level soil	10
Water miscible liquid waste	500
High-level soil and sludge	1250
Non-water miscible waste	1250

PRECISION: 0.18X - 0.71 µg/L (overall precision)

ACCURACY: 0.94C + 0.65 µg/L (as recovery)

SAMPLING Method: For water and liquid samples use glass 40 mL vials with Teflon lined septum caps and collect two vials per sample location with no headspace. For solids and concentrated waste samples use widemouth glass bottles with Teflon liners. Cool all samples to 4°C.

STABILITY: For concentrated wastes, soils, sediments or sludges cool to 4°C. For liquids, add 4 drops of concentrated hydrochloric acid and cool to 4°C. M.H.T. = 14 days.

QUALITY CONTROL: Analyze a reagent blank, matrix spike and matrix spike duplicate/duplicate for each analytical batch (up to 20 Samples). Demonstrate the purity of glassware and reagents by analyzing a reagent water method blank. Internal, surrogate and five concentration level calibration standards are used. The QC check sample concentrate should contain this compound at 10 µg/mL in methanol.

REFERENCE: Method 8020, SW-846, 3rd ed., Nov 1986.

PRIMARY NAME: 1,2,4-Trimethylbenzene

Method 502.2

TITLE: VOCs in H_2O by Purge and Trap capillary column GC

MATRIX: Drinking water (finished or any treatment stage) and raw source water.

CAS # : 95-63-6

APPLICATION: Method covers 60 VOCs. An inert gas is bubbled through a 5 mL sample in a purging chamber. Purged sample components are trapped in tube of sorbent materials. When purging is done, sorbent tube is heated and backflushed w/helium to desorb trapped sample onto a capillary GC column.

INTERFERENCES: During analysis, major contaminant sources are volatile materials in the lab and impurities in purging gas and sorbent trap. With high and low level samples, there can be carry-over contamination. Beware of methylene chloride, it permeates through PTFE tubing.

INSTRUMENTATION: (Capillary Column) GC, w/PID and electrolytic conductivity detectors in series. Column #1: VOCOL wide bore; Column #2: RTX-502.2 mega bore; Column #3: DB-62 mega bore.

RANGE: Approximately 0.02-200 µg/L. MDL: 0.05 µg/L (reagent H_2O) (PID)

PRECISION: RSD = 1.2% @ Spike of 10 µg/L (reagent H_2O) (PID).

ACCURACY: Recovery = 99% @ spike of 10 µg/L (reagent H_2O) (PID)

SAMPLING Method: Use a 40-120 mL screw cap vial (pre-washed with detergent, rinsed with distilled water and oven dried at 105°C) with a PTFE-faced silicone septum. If residual chlorine is in the water add about 25 mg of ascorbic acid to each vial before sample collection. Collect bubble free samples.

STABILITY: Cool to 4°C; HCl to pH <2. M.H.T. = 14 Days.

QUALITY CONTROL: As initial demonstration of lab accuracy and precision, analyze 4 to 7 replicates of a lab fortified blank containing the analyte at 0.1-5 µg/L. Collect all samples in duplicate.

REFERENCE: Method 502.2, Volatile Organic Compounds in Water by Purge and Trap Capillary Column GC, EPA 600/4-88/039.

PRIMARY NAME: 1,2,4-Trimethylbenzene Method 503.1

TITLE: Aromatic and Unsaturated VOCs in Water

MATRIX: Drinking water (finished or any treatment stage) and raw source water.

CAS # : 95-63-6

APPLICATION: Method covers 28 aromatic and unsaturated VOCs. An inert gas is bubbled through a 5 mL water sample. Purged sample components are trapped in tube of sorbent materials. When purging is complete, sorbent tube is heated and backflushed with inert gas to desorb trapped sample onto a packed GC column.

INTERFERENCES: During analysis, major contaminant sources are volatile materials in the lab and impurities in purging gas and sorbent trap. With high and low level samples, there can be carry-over contamination. Excess water causes a negative baseline deflection.

INSTRUMENTATION: Purge and Trap GC w/photoionization detector. (Two GC columns are recommended); Column #1: 5% SP-1200 and 1.75% Bentone 34 on Supelcoport; Column #2: 1,2,3-tris(2-cyanoethoxy)propane on Chromosorb W.

RANGE: 2.2-600 μg/L. (Drinking water) MDL: 0.006 μg/L. in water

PRECISION: RSD = 8.7% @ 0.40 μg/L conc. 7 samples

ACCURACY: Avg Recovery = 75% @ 0.40 μg/L conc.; 7 samples

SAMPLING Method: Use a 40-120 mL screw cap vial (pre-washed with detergent, rinsed with distilled water and oven dried at 105°C) with a PTFE-faced silicone septum. If residual chlorine is in the water add about 25 mg of ascorbic acid to each vial before sample collection. Collect bubble free samples.

STABILITY: Cool to 4°C; HCl to pH <2. M.H.T. = 14 Days.

QUALITY CONTROL: As initial demonstration of lab accuracy and precision, analyze 4 to 7 replicates of a lab fortified blank containing the analyte at 0.1-5 μg/L. QC criteria. Collect all samps in duplicate.

REFERENCE: Method 503.1, Volatile Aromatic and Unsaturated Organic Compounds in H_2O by Purge and Trap GC, EPA 600/4-88/039.

PRIMARY NAME: 1,2,4-Trimethylbenzene — Method 524.2

TITLE: VOCs in H_2O Purge and Trap capillary column GC/MS

MATRIX: Drinking water (finished or any treatment stage) and raw source water.

CAS # : 95-63-6

APPLICATION: Method covers 60 VOCs. An inert gas is bubbled through a 25 mL water sample. Purged sample components are trapped in tube of sorbent materials. When purging is complete, sorbent tube is heated and backflushed with helium to desorb trapped sample onto a capillary GC column.

INTERFERENCES: During analysis, major contaminant sources are volatile materials in the lab and impurities in purging gas and sorbent trap. With high and low level samples, there can be carry-over contamination. Watch for methylene chloride, it permeates through PTFE tubing.

INSTRUMENTATION: (Capillary Column) Gas Chromatography/Mass Spectrometry/Data System; Column #1: VOCOL glass wide bore capillary column; Column #2: DB-624 fused silica column; Column #3: DB-5 fused silica column.

RANGE: Approximately 0.02-200 µg/L.

MDL: 0.13 µg/L on wide bore capillary column.

PRECISION: RSD = 8.1% @ (0.5-10) µg/L (18 samples) wide bore capillary column.

ACCURACY: Recovery = 99% @ (0.5-10) µg/L (18 samples) wide bore capillary column.

SAMPLING Method: Use a 60-120 mL screw cap vial (pre-washed with detergent, rinsed with distilled water and oven dried at 105°C) with a PTFE-faced silicone septum. If residual chlorine is in the water add about 25 mg of ascorbic acid to each vial before sample collection. Collect bubble free samples.

STABILITY: Cool to 4°C; HCl to pH <2. M.H.T. = 14 Days.

QUALITY CONTROL: As initial demonstration of lab accuracy and precision, analyze 4 to 7 replicates of a lab fortified blank containing the analyte at 0.2-5 μg/L. Collect all samples in duplicate.

REFERENCE: Method 524.2, VOCs in H_2O by Purge and Trap Capillary column GC/MS, EPA 600/4-88/039.

PRIMARY NAME: 1,3,5-Trimethylbenzene Method 502.2

TITLE: VOCs in H_2O by Purge and Trap capillary column GC

MATRIX: Drinking water (finished or any treatment stage) and raw source water.

CAS # : 108-67-8

APPLICATION: Method covers 60 VOCs. An inert gas is bubbled through a 5 mL sample in a purging chamber. Purged sample components are trapped in tube of sorbent materials. When purging is done, sorbent tube is heated and backflushed w/helium to desorb trapped sample onto a capillary GC column.

INTERFERENCES: During analysis, major contaminant sources are volatile materials in the lab and impurities in purging gas and sorbent trap. With high and low level samples, there can be carry-over contamination. Beware of methylene chloride, it permeates through PTFE tubing.

INSTRUMENTATION: (Capillary Column) GC, w/PID and electrolytic conductivity detectors in series. Column #1: VOCOL wide bore; Column #2: RTX-502.2 mega bore; Column #3: DB-62 mega bore.

RANGE: Approximately 0.02-200 μg/L.

MDL: 0.01 μg/L (reagent H_2O) (PID)

PRECISION: RSD = 1.4% @ Spike of 10 μg/L (reagent H_2O) (PID).

ACCURACY: Recovery = 101% @ spike of 10 μg/L (reagent H_2O) (PID)

SAMPLING Method: Use a 40-120 mL screw cap vial (pre-washed with detergent, rinsed with distilled water and oven dried at 105°C) with a PTFE-

faced silicone septum. If residual chlorine is in the water add about 25 mg of ascorbic acid to each vial before sample collection. Collect bubble free samples.

STABILITY: Cool to 4°C; HCl to pH <2. M.H.T. = 14 Days.

QUALITY CONTROL: As initial demonstration of lab accuracy and precision, analyze 4 to 7 replicates of a lab fortified blank containing the analyte at 0.1-5 µg/L. Collect all samples in duplicate.

REFERENCE: Method 502.2, Volatile Organic Compounds in Water by Purge and Trap Capillary Column GC, EPA 600/4-88/039.

PRIMARY NAME: 1,3,5-Trimethylbenzene Method 503.1

TITLE: Aromatic and Unsaturated VOCs in Water

MATRIX: Drinking water (finished or any treatment stage) and raw source water.

CAS # : 108-67-8

APPLICATION: Method covers 28 aromatic and unsaturated VOCs. An inert gas is bubbled through a 5 mL water sample. Purged sample components are trapped in tube of sorbent materials. When purging is complete, sorbent tube is heated and backflushed with inert gas to desorb trapped sample onto a packed GC column.

INTERFERENCES: During analysis, major contaminant sources are volatile materials in the lab and impurities in purging gas and sorbent trap. With high and low level samples, there can be carry-over contamination. Excess water causes a negative baseline deflection.

INSTRUMENTATION: Purge and Trap GC w/photoionization detector. (Two GC columns are recommended); Column #1: 5% SP-1200 and 1.75% Bentone 34 on Supelcoport; Column #2: 1,2,3-tris(2-cyanoethoxy)propane on Chromosorb W.

RANGE: 2.2-600 µg/L. (Drinking water) MDL: 0.003 µg/L in water

PRECISION: RSD = 8.7% @ 0.50 µg/L conc.; 10 samples

ACCURACY: Avg Recovery = 92% @ 0.50 µg/L conc.; 10 samples

SAMPLING Method: Use a 40-120 mL screw cap vial (pre-washed with detergent, rinsed with distilled water and oven dried at 105°C) with a PTFE-faced silicone septum. If residual chlorine is in the water add about 25 mg of ascorbic acid to each vial before sample collection. Collect bubble free samples.

STABILITY: Cool to 4°C; HCl to pH <2. M.H.T. = 14 Days.

QUALITY CONTROL: As initial demonstration of lab accuracy and precision, analyze 4 to 7 replicates of a lab fortified blank containing the analyte at 0.1-5 µg/L. QC criteria. Collect all samples in duplicate.

REFERENCE: Method 503.1, Volatile Aromatic and Unsaturated Organic Compounds in H_2O by Purge and Trap GC, EPA 600/4-88/039.

PRIMARY NAME: 1,3,5-Trimethylbenzene Method 524.2

TITLE: VOCs in H_2O Purge and Trap capillary column GC/MS

MATRIX: Drinking water (finished or any treatment stage) and raw source water.

CAS # : 108-67-8

APPLICATION: Method covers 60 VOCs. An inert gas is bubbled through a 25 mL water sample. Purged sample components are trapped in tube of sorbent materials. When purging is complete, sorbent tube is heated and backflushed with helium to desorb trapped sample onto a capillary GC column.

INTERFERENCES: During analysis, major contaminant sources are volatile materials in the lab and impurities in purging gas and sorbent trap. With high and low level samples, there can be carry-over contamination. Watch for methylene chloride, it permeates through PTFE tubing.

INSTRUMENTATION: (Capillary Column) Gas Chromatography/Mass Spectrometry/Data System; Column #1: VOCOL glass wide bore capillary column; Column #2: DB-624 fused silica column; Column #3: DB-5 fused silica column.

RANGE: Approximately 0.02-200 µg/L.

MDL: 0.05 µg/L on wide bore capillary column.

PRECISION: RSD = 7.4% @ (0.5-10) µg/L (23 samples) wide bore capillary column.

ACCURACY: Recovery = 92% @ (0.5-10) µg/L (23 samples) wide bore capillary column.

SAMPLING Method: Use a 60-120 mL screw cap vial (pre-washed with detergent, rinsed with distilled water and oven dried at 105°C) with a PTFE-faced silicone septum. If residual chlorine is in the water add about 25 mg of ascorbic acid to each vial before sample collection. Collect bubble free samples.

STABILITY: Cool to 4°C; HCl to pH <2.

M.H.T. = 14 Days.

QUALITY CONTROL: As initial demonstration of lab accuracy and precision, analyze 4 to 7 replicates of a lab fortified blank containing the analyte at 0.2-5 µg/L. Collect all samples in duplicate.

REFERENCE: Method 524.2, VOCs in H_2O by Purge and Trap Capillary column GC/MS, EPA 600/4-88/039.

PRIMARY NAME: m-Xylene

Method 502.2

TITLE: VOCs in H_2O by Purge and Trap capillary column GC

MATRIX: Drinking water (finished or any treatment stage) and raw source water.

CAS # : 108-38-3

APPLICATION: Method covers 60 VOCs. An inert gas is bubbled through a 5 mL sample in a purging chamber. Purged sample components are trapped in tube of sorbent materials. When purging is done, sorbent tube is heated and backflushed w/helium to desorb trapped sample onto a capillary GC column.

INTERFERENCES: During analysis, major contaminant sources are volatile materials in the lab and impurities in purging gas and sorbent trap. With

high and low level samples, there can be carry-over contamination. Beware of methylene chloride, it permeates through PTFE tubing.

INSTRUMENTATION: (capillary column) GC, w/PID and electrolytic conductivity detectors in series. Column #1: VOCOL wide bore; Column #2: RTX-502.2 mega bore; Column #3: DB-62 mega bore.

RANGE: Approximately 0.02-200 μg/L. MDL: 0.01 μg/L (reagent H_2O) (PID)

PRECISION: RSD = 1.4% @ Spike of 10 μg/L (reagent H_2O) (PID).

ACCURACY: Recovery = 100% @ spike of 10 μg/L (reagent H_2O) (PID)

SAMPLING Method: Use a 40-120 mL screw cap vial (pre-washed with detergent, rinsed with distilled water and oven dried at 105°C) with a PTFE-faced silicone septum. If residual chlorine is in the water add about 25 mg of ascorbic acid to each vial before sample collection. Collect bubble free samples.

STABILITY: Cool to 4°C; HCl to pH <2. M.H.T. = 14 Days.

QUALITY CONTROL: As initial demonstration of lab accuracy and precision, analyze 4 to 7 replicates of a lab fortified blank containing the analyte at 0.1-5 μg/L. Collect all samples in duplicate.

REFERENCE: Method 502.2, Volatile Organic Compounds in Water by Purge and Trap Capillary Column GC, EPA 600/4-88/039.

PRIMARY NAME: m-Xylene Method 503.1

TITLE: Aromatic and Unsaturated VOCs in Water

MATRIX: Drinking water (finished or any treatment stage) and raw source water.

CAS # : 108-38-3

APPLICATION: Method covers 28 aromatic and unsaturated VOCs. An inert gas is bubbled through a 5 mL water sample. Purged sample components are trapped in tube of sorbent materials. When purging is complete, sorbent tube is heated and backflushed with inert gas to desorb trapped sample onto a packed GC column.

INTERFERENCES: During analysis, major contaminant sources are volatile materials in the lab and impurities in purging gas and sorbent trap. With high and low level samples, there can be carry-over contamination. Excess water causes a negative baseline deflection.

INSTRUMENTATION: Purge and Trap GC w/photoionization detector. (Two GC columns are recommended); Column #1: 5% SP-1200 and 1.75% Bentone 34 on Supelcoport; Column #2: 1,2,3-tris(2-cyanoethoxy)propane on Chromosorb W.

RANGE: 2.2-600 µg/L. (Drinking water) MDL: 0.004 µg/L in water

PRECISION: RSD = 7.7% @ 0.40 µg/L conc.; 7 samples

ACCURACY: Avg Recovery = 90% @ 0.40 µg/L conc.; 7 samples

SAMPLING Method: Use a 40-120 mL screw cap vial (pre-washed with detergent, rinsed with distilled water and oven dried at 105°C) with a PTFE-faced silicone septum. If residual chlorine is in the water add about 25 mg of ascorbic acid to each vial before sample collection. Collect bubble free samples.

STABILITY: Cool to 4°C; HCl to pH <2. M.H.T. = 14 Days.

QUALITY CONTROL: As initial demonstration of lab accuracy and precision, analyze 4 to 7 replicates of a lab fortified blank containing the analyte at 0.1-5 µg/L. Collect all samples in duplicate.

REFERENCE: Method 503.1, Volatile Aromatic and Unsaturated Organic Compounds in H_2O by Purge and Trap GC, EPA 600/4-88/039.

PRIMARY NAME: m-Xylene Method 524.1

TITLE: VOCs in H_2O by Purge and Trap GC/MS

MATRIX: Drinking water (finished or any treatment stage) and raw source water.

CAS # : 108-38-3

APPLICATION: Method covers 48 VOCs. An inert gas is bubbled through a 25 mL water sample. Purged sample components are trapped in tube of sorbent materials. When purging is complete, sorbent tube is heated and backflushed with helium to desorb trapped sample onto a packed GC column.

INTERFERENCES: During analysis, major contaminant sources are volatile materials in the lab and impurities in purging gas and sorbent trap. With high and low level samples, there can be carry-over contamination. Watch for methylene chloride, it permeates through PTFE tubing.

INSTRUMENTATION: Gas Chromatography/Mass Spectometry/Data System; 1% SP-1000 on Carbopack B.

RANGE: Approximately 0.1-200 µg/L. MDL: not determined.

PRECISION: not listed.

ACCURACY: not listed.

SAMPLING Method: Use a 60-120 mL screw cap vial (pre-washed with detergent, rinsed with distilled water and oven dried at 105°C) with a PTFE-faced silicone septum. If residual chlorine is in the water add about 25 mg of ascorbic acid to each vial before sample collection. Collect bubble free samples.

STABILITY: Cool to 4°C; HCl to pH <2. M.H.T. = 14 Days.

QUALITY CONTROL: As initial demonstration of lab accuracy and precision, analyze 4 to 7 replicates of a lab fortified blank containing the analyte at 0.2-5 µg/L. Collect all samples in duplicate.

REFERENCE: Method 524.1, VOCs in Water by Purge and Trap Gas Chromatography/Mass Spectometry, EPA 600/4-88/039.

PRIMARY NAME: m-Xylene Method 524.2

TITLE: VOCs in H_2O Purge and Trap capillary column GC/MS

MATRIX: Drinking water (finished or any treatment stage) and raw source water.

CAS # : 108-38-3

APPLICATION: Method covers 60 VOCs. An inert gas is bubbled through a 25 mL water sample. Purged sample components are trapped in tube of sorbent materials. When purging is complete, sorbent tube is heated and backflushed with helium to desorb trapped sample onto a capillary GC column.

INTERFERENCES: During analysis, major contaminant sources are volatile materials in the lab and impurities in purging gas and sorbent trap. With high and low level samples, there can be carry-over contamination. Watch for methylene chloride, it permeates through PTFE tubing.

INSTRUMENTATION: (capillary column) Gas Chromatography/Mass Spectometry/Data System; Column #1: VOCOL glass wide bore capillary column; Column #2: DB-624 fused silica column; Column #3: DB-5 fused silica column.

RANGE: Approximately 0.02-200 μg/L.

MDL: 0.05 μg/L on wide bore capillary column.

PRECISION: RSD = 6.5% @ (0.1-10) μg/L (31 samples) wide bore capillary column.

ACCURACY: Recovery = 97% @ (0.1-10) μg/L (31 samples) wide bore capillary column.

SAMPLING Method: Use a 60-120 mL screw cap vial (pre-washed with detergent, rinsed with distilled water and oven dried at 105°C) with a PTFE-faced silicone septum. If residual chlorine is in the water add about 25 mg of ascorbic acid to each vial before sample collection. Collect bubble free samples.

STABILITY: Cool to 4°C; HCl to pH <2.

M.H.T. = 14 Days.

QUALITY CONTROL: As initial demonstration of lab accuracy and precision, analyze 4 to 7 replicates of a lab fortified blank containing the analyte at 0.2-5 μg/L. Collect all samples in duplicate.

REFERENCE: Method 524.2, VOCs in H_2O by Purge and Trap Capillary column GC/MS, EPA 600/4-88/039.

PRIMARY NAME: o-Xylene

Method 502.2

TITLE: VOCs in H_2O by Purge and Trap capillary column GC

MATRIX: Drinking water (finished or any treatment stage) and raw source water.

CAS # : 95-47-6

APPLICATION: Method covers 60 VOCs. An inert gas is bubbled through a 5 mL sample in a purging chamber. Purged sample components are trapped in tube of sorbent materials. When purging is done, sorbent tube is heated and backflushed w/helium to desorb trapped sample onto a capillary GC column.

INTERFERENCES: During analysis, major contaminant sources are volatile materials in the lab and impurities in purging gas and sorbent trap. With high and low level samples, there can be carry-over contamination. Beware of methylene chloride, it permeates through PTFE tubing.

INSTRUMENTATION: (capillary column) GC, w/PID and electrolytic conductivity detectors in series. Column #1: VOCOL wide bore; Column #2: RTX-502.2 mega bore; Column #3: DB-62 mega bore.

RANGE: Approximately 0.02-200 μg/L.

MDL: 0.02 μg/L (reagent H_2O) (PID)

PRECISION: RSD = 0.8% @ Spike of 10 μg/L (reagent H_2O) (PID).

ACCURACY: Recovery = 99% @ spike of 10 μg/L (reagent H_2O) (PID)

SAMPLING Method: Use a 40-120 mL screw cap vial (pre-washed with detergent, rinsed with distilled water and oven dried at 105°C) with a PTFE-faced silicone septum. If residual chlorine is in the water add about 25 mg of ascorbic acid to each vial before sample collection. Collect bubble free samples.

STABILITY: Cool to 4°C; HCl to pH <2.

M.H.T. = 14 Days.

QUALITY CONTROL: As initial demonstration of lab accuracy and precision, analyze 4 to 7 replicates of a lab fortified blank containing the analyte at 0.1-5 μg/L. Collect all samples in duplicate.

REFERENCE: Method 502.2, Volatile Organic Compounds in Water by Purge and Trap Capillary Column GC, EPA 600/4-88/039.

PRIMARY NAME: o-Xylene

Method 503.1

TITLE: Aromatic and Unsaturated VOCs in Water

MATRIX: Drinking water (finished or any treatment stage) and raw source water.

CAS # : 95-47-6

APPLICATION: Method covers 28 aromatic and unsaturated VOCs. An inert gas is bubbled through a 5 mL water sample. Purged sample components are trapped in tube of sorbent materials. When purging is complete, sorbent tube is heated and backflushed with inert gas to desorb trapped sample onto a packed GC column.

INTERFERENCES: During analysis, major contaminant sources are volatile materials in the lab and impurities in purging gas and sorbent trap. With high and low level samples, there can be carry-over contamination. Excess water causes a negative baseline deflection.

INSTRUMENTATION: Purge and Trap GC w/photoionization detector. (Two GC columns are recommended); Column #1: 5% SP-1200 and 1.75% Bentone 34 on Supelcoport; Column #2: 1,2,3-tris(2-cyanoethoxy)propane on Chromosorb W.

RANGE: 2.2-600 µg/L. (Drinking water)

MDL: 0.004 µg/L in water

PRECISION: RSD = 7.2% @ 0.40 µg/L conc.; 7 samples

ACCURACY: Avg Recovery = 90% @ 0.40 µg/L conc.; 7 samples

SAMPLING Method: Use a 40-120 mL screw cap vial (pre-washed with detergent, rinsed with distilled water and oven dried at 105°C) with a PTFE-faced silicone septum. If residual chlorine is in the water add about 25 mg of ascorbic acid to each vial before sample collection. Collect bubble free samples.

STABILITY: Cool to 4°C; HCl to pH <2.

M.H.T. = 14 Days.

QUALITY CONTROL: As initial demonstration of lab accuracy and precision, analyze 4 to 7 replicates of a lab fortified blank containing the analyte at 0.1-5 µg/L. Collect all samples in duplicate.

REFERENCE: Method 503.1, Volatile Aromatic and Unsaturated Organic Compounds in H_2O by Purge and Trap GC, EPA 600/4-88/039.

PRIMARY NAME: o-Xylene

Method 524.1

TITLE: VOCs in H_2O by Purge and Trap GC/MS

MATRIX: Drinking water (finished or any treatment stage) and raw source water.

CAS # : 95-47-6

APPLICATION: Method covers 48 VOCs. An inert gas is bubbled through a 25 mL water sample. Purged sample components are trapped in tube of sorbent materials. When purging is complete, sorbent tube is heated and backflushed with helium to desorb trapped sample onto a packed GC column.

INTERFERENCES: During analysis, major contaminant sources are volatile materials in the lab and impurities in purging gas and sorbent trap. With high and low level samples, there can be carry-over contamination. Watch for methylene chloride, it permeates through PTFE tubing.

INSTRUMENTATION: Gas Chromatography/Mass Spectometry/Data System; 1% SP-1000 on Carbopack B.

RANGE: Approximately 0.1-200 µg/L.

MDL: 0.2 µg/L

PRECISION: RSD = 6.7% @ 1.0 µg/L conc (8 samples)

ACCURACY: Mean Accuracy = 102% @ 1.0 µg/L

SAMPLING Method: Use a 60-120 mL screw cap vial (pre-washed with detergent, rinsed with distilled water and oven dried at 105°C) with a PTFE-faced silicone septum. If residual chlorine is in the water add about 25 mg of ascorbic acid to each vial before sample collection. Collect bubble free samples.

STABILITY: Cool to 4°C; HCl to pH <2. M.H.T. = 14 Days.

QUALITY CONTROL: As initial demonstration of lab accuracy and precision, analyze 4 to 7 replicates of a lab fortified blank containing the analyte at 0.2-5 μg/L. Collect all samples in duplicate.

REFERENCE: Method 524.1, VOCs in Water by Purge and Trap Gas Chromatography/Mass Spectometry, EPA 600/4-88/039.

PRIMARY NAME: o-Xylene Method 524.2

TITLE: VOCs in H_2O Purge and Trap capillary column GC/MS

MATRIX: Drinking water (finished or any treatment stage) and raw source water.

CAS # : 95-47-6

APPLICATION: Method covers 60 VOCs. An inert gas is bubbled through a 25 mL water sample. Purged sample components are trapped in tube of sorbent materials. When purging is complete, sorbent tube is heated and backflushed with helium to desorb trapped sample onto a capillary GC column.

INTERFERENCES: During analysis, major contaminant sources are volatile materials in the lab and impurities in purging gas and sorbent trap. With high and low level samples, there can be carry-over contamination. Watch for methylene chloride, it permeates through PTFE tubing.

INSTRUMENTATION: (capillary column) Gas Chromatography/Mass Spectometry/Data System; Column #1: VOCOL glass wide bore capillary column; Column #2: DB-624 fused silica column; Column #3: DB-5 fused silica column.

RANGE: Approximately 0.02-200 μg/L.

MDL: 0.11 μg/L on wide bore capillary column.

PRECISION: RSD = 7.2% @ (0.1-31) μg/L (18 samples) wide bore capillary column.

ACCURACY: Recovery = 103% @ (0.1-31) μg/L (18 samples) wide bore capillary column.

SAMPLING Method: Use a 60-120 mL screw cap vial (pre-washed with detergent, rinsed with distilled water and oven dried at 105°C) with a PTFE-faced silicone septum. If residual chlorine is in the water add about 25 mg of ascorbic acid to each vial before sample collection. Collect bubble free samples.

STABILITY: Cool to 4°C; HCl to pH <2. M.H.T. = 14 Days.

QUALITY CONTROL: As initial demonstration of lab accuracy and precision, analyze 4 to 7 replicates of a lab fortified blank containing the analyte at 0.2-5 μg/L. Collect all samples in duplicate.

REFERENCE: Method 524.2, VOCs in H_2O by Purge and Trap Capillary column GC/MS, EPA 600/4-88/039.

PRIMARY NAME: p-Xylene Method 502.2

TITLE: VOCs in H_2O by Purge and Trap capillary column GC

MATRIX: Drinking water (finished or any treatment stage) and raw source water.

CAS # : 106-42-3

APPLICATION: Method covers 60 VOCs. An inert gas is bubbled through a 5 mL sample in a purging chamber. Purged sample components are trapped in tube of sorbent materials. When purging is done, sorbent tube is heated and backflushed w/helium to desorb trapped sample onto a capillary GC column.

INTERFERENCES: During analysis, major contaminant sources are volatile materials in the lab and impurities in purging gas and sorbent trap. With high and low level samples, there can be carry-over contamination. Beware of methylene chloride, it permeates through PTFE tubing.

INSTRUMENTATION: (capillary column) GC, w/PID and electrolytic conductivity detectors in series. Column #1: VOCOL wide bore; Column #2: RTX-502.2 mega bore; Column #3: DB-62 mega bore.

RANGE: Approximately 0.02-200 μg/L.

MDL: 0.01 μg/L (reagent H_2O) (PID)

PRECISION: RSD = 0.9% @ Spike of 10 μg/L (reagent H_2O) (PID).

ACCURACY: Recovery = 99% @ spike of 10 µg/L (reagent H_2O) (PID)

SAMPLING Method: Use a 40-120 mL screw cap vial (pre-washed with detergent, rinsed with distilled water and oven dried at 105°C) with a PTFE-faced silicone septum. If residual chlorine is in the water add about 25 mg of ascorbic acid to each vial before sample collection. Collect bubble free samples.

STABILITY: Cool to 4°C; HCl to pH <2. M.H.T. = 14 Days.

QUALITY CONTROL: As initial demonstration of lab accuracy and precision, analyze 4 to 7 replicates of a lab fortified blank containing the analyte at 0.1-5 µg/L. Collect all samples in duplicate.

REFERENCE: Method 502.2, Volatile Organic Compounds in Water by Purge and Trap Capillary Column GC, EPA 600/4-88/039.

PRIMARY NAME: p-Xylene Method 503.1

TITLE: Aromatic and Unsaturated VOCs in Water

MATRIX: Drinking water (finished or any treatment stage) and raw source water.

CAS # : 106-42-3

APPLICATION: Method covers 28 aromatic and unsaturated VOCs. An inert gas is bubbled through a 5 mL water sample. Purged sample components are trapped in tube of sorbent materials. When purging is complete, sorbent tube is heated and backflushed with inert gas to desorb trapped sample onto a packed GC column.

INTERFERENCES: During analysis, major contaminant sources are volatile materials in the lab and impurities in purging gas and sorbent trap. With high and low level samples, there can be carry-over contamination. Excess water causes a negative baseline deflection.

INSTRUMENTATION: Purge and Trap GC w/photoionization detector. (Two GC columns are recommended); Column #1: 5% SP-1200 and 1.75% Bentone 34 on Supelcoport; Column #2: 1,2,3-tris(2-cyanoethoxy)propane on Chromosorb W.

RANGE: 2.2-600 µg/L. (Drinking water) MDL: 0.002 µg/L in water

PRECISION: RSD = 8.7% @ 0.40 µg/L conc.; 7 samples

ACCURACY: Avg Recovery = 85% @ 0.40 µg/L conc.; 7 samples

SAMPLING Method: Use a 40-120 mL screw cap vial (pre-washed with detergent, rinsed with distilled water and oven dried at 105°C) with a PTFE-faced silicone septum. If residual chlorine is in the water add about 25 mg of ascorbic acid to each vial before sample collection. Collect bubble free samples.

STABILITY: Cool to 4°C; HCl to pH <2. M.H.T. = 14 Days.

QUALITY CONTROL: As initial demonstration of lab accuracy and precision, analyze 4 to 7 replicates of a lab fortified blank containing the analyte at 0.1-5 µg/L. Collect all samples in duplicate.

REFERENCE: Method 503.1, Volatile Aromatic and Unsaturated Organic Compounds in H_2O by Purge and Trap GC, EPA 600/4-88/039.

PRIMARY NAME: p-Xylene Method 524.1

TITLE: VOCs in H_2O by Purge and Trap GC/MS

MATRIX: Drinking water (finished or any treatment stage) and raw source water.

CAS # : 106-42-3

APPLICATION: Method covers 48 VOCs. An inert gas is bubbled through a 25 mL water sample. Purged sample components are trapped in tube of sorbent materials.When purging is complete, sorbent tube is heated and backflushed with helium to desorb trapped sample onto a packed GC column.

INTERFERENCES: During analysis, major contaminant sources are volatile materials in the lab and impurities in purging gas and sorbent trap. With high and low level samples, there can be carry-over contamination. Watch for methylene chloride, it permeates through PTFE tubing.

INSTRUMENTATION: Gas Chromatography/Mass Spectometry/Data System; 1% SP-1000 on Carbopack B.

RANGE: Approximately 0.1-200 µg/L. MDL: 0.3 µg/L

PRECISION: RSD = 4.2% @ 1.0 μg/L

ACCURACY: Mean Accuracy = 111% @1.0 μg/L

SAMPLING Method: Use a 60-120 mL screw cap vial (pre-washed with detergent, rinsed with distilled water and oven dried at 105°C) with a PTFE-faced silicone septum. If residual chlorine is in the water add about 25 mg of ascorbic acid to each vial before sample collection. Collect bubble free samples.

STABILITY: Cool to 4°C; HCl to pH <2. M.H.T. = 14 Days.

QUALITY CONTROL: As initial demonstration of lab accuracy and precision, analyze 4 to 7 replicates of a lab fortified blank containing the analyte at 0.2-5 μg/L. Collect all samples in duplicate.

REFERENCE: Method 524.1, VOCs in Water by Purge and Trap Gas Chromatography/Mass Spectometry, EPA 600/4-88/039.

PRIMARY NAME: p-Xylene Method 524.2

TITLE: VOCs in H_2O Purge and Trap capillary column GC/MS

MATRIX: Drinking water (finished or any treatment stage) and raw source water.

CAS # : 106-42-3

APPLICATION: Method covers 60 VOCs. An inert gas is bubbled through a 25 mL water sample. Purged sample components are trapped in tube of sorbent materials. When purging is complete, sorbent tube is heated and backflushed with helium to desorb trapped sample onto a capillary GC column.

INTERFERENCES: During analysis, major contaminant sources are volatile materials in the lab and impurities in purging gas and sorbent trap. With high and low level samples, there can be carry-over contamination. Watch for methylene chloride, it permeates through PTFE tubing.

INSTRUMENTATION: (capillary column) Gas Chromatography/Mass Spectometry/Data System; Column #1: VOCOL glass wide bore capillary column; Column #2: DB-624 fused silica column; Column #3: DB-5 fused silica column.

RANGE: Approximately 0.02-200 μg/L.

MDL: 0.13 μg/L on wide bore capillary column.

PRECISION: RSD = 7.7% @ (0.5-10) μg/L (18 samples) wide bore capillary column.

ACCURACY: Recovery = 104% @ (0.5-10) μg/L (18 samples) wide bore capillary column.

SAMPLING Method: Use a 60-120 mL screw cap vial (pre-washed with detergent, rinsed with distilled water and oven dried at 105°C) with a PTFE-faced silicone septum. If residual chlorine is in the water add about 25 mg of ascorbic acid to each vial before sample collection. Collect bubble free samples.

STABILITY: Cool to 4°C; HCl to pH <2.

M.H.T. = 14 Days.

QUALITY CONTROL: As initial demonstration of lab accuracy and precision, analyze 4 to 7 replicates of a lab fortified blank containing the analyte at 0.2-5 μg/L. Collect all samples in duplicate.

REFERENCE: Method 524.2, VOCs in H_2O by Purge and Trap Capillary column GC/MS, EPA 600/4-88/039.

PRIMARY NAME: Xylenes

Method 8020

TITLE: Aromatic Volatile Organics

MATRIX: groundwater, soils, sludges water miscible liquid wastes, and non-water miscible wastes.

CAS # :

APPLICATION: This Method is used to analyze for 8 aromatic VOCs. Samples are analyzed using direct injection or purge and trap Methods. Groundwater must be analyzed by the purge and trap Method. The Method provides an optional GC column that is used for analyte confirmation and may also help resolve analytes from interferences.

INTERFERENCES: There can be carry-over contamination with high and low level samples. Impurities may come from the purge and trap apparatus,

organic compounds outgassing from the plumbing ahead of trap, diffusion of VOCs through the sample bottle septum during shipping or storage, or from solvent vapors in the lab.

INSTRUMENTATION: GC capable of on-column injections or purge-and-trap sample introduction and a photoionization detector (PID). Column 1: 6 foot by 0.082 inch with 5% SP-1200 and 1.75% Bentone-34 on Supelcoport. Column 2: 8 foot by 0.1 inch with 5% 1,2,3-tris(2-cyanoethoxy)propane on Chromosorb W-AW.

RANGE: 2.1 to 500 μg/L MDL: Not available

PRACTICAL QUANTITATION LIMIT FACTORS FOR MULTIPLYING TIMES MDL VALUE:

Matrix	Multiplication Factor
Groundwater	10
Low-level soil	10
Water miscible liquid waste	500
High-level soil and sludge	1250
Non-water miscible waste	1250

PRECISION: Not available

ACCURACY: Not available

SAMPLING Method: For water and liquid samples use glass 40 mL vials with Teflon lined septum caps and collect two vials per sample location with no headspace. For solids and concentrated waste samples use widemouth glass bottles with Teflon liners. Cool all samples to 4°C.

STABILITY: For concentrated wastes, soils, sediments or sludges cool to 4°C. For liquids, add 4 drops of concentrated hydrochloric acid and cool to 4°C. M.H.T. = 14 days.

QUALITY CONTROL: Analyze a reagent blank, matrix spike and matrix spike duplicate/duplicate for each analytical batch (up to 20 Samples). Demonstrate the purity of glassware and reagents by analyzing a reagent water Method blank. Internal, surrogate and five concentration level calibration standards are used. The QC check sample concentrate should contain this compound at 10 μg/mL in methanol.

REFERENCE: Method 8020, SW-846, 3rd ed., Nov 1986.

Chapter 4
Semivolatile Organic Compounds

PRIMARY NAME: Acenaphthene — Method 1625,B

TITLE: Semivolatile Organic Compounds — MATRIX: Wastewater

CAS # : 83-32-9

APPLICATION: Method covers determination of 82 semivolatile compounds by isotope dilution GC/MS. Stable isotopically labeled analogs are added to a 1L wastewater sample. Sample is extracted @ pH 12-13 and then pH <2 with methylene chloride. The extract is dried and concentrated to 1ml volume.

INTERFERENCES: Solvents, reagents, glassware and other processing hardware may yield artifacts and/or elevated baselines causing misinterpretation of chromatograms and spectra. Interferences coextracted from samples will vary some from source to source of wastewater.

INSTRUMENTATION: GC/MS with data system

RANGE: Not listed — MDL: 10 μg (with no interferences)

PRECISION: 21 μg/L (initial)

ACCURACY: 79-134 μg/L (initial)

SAMPLING METHOD: Amber glass bottle. Teflon-lined screw cap.

STABILITY: Cool to 4°C, add 0.008% sodium thiosulfate if samples have any residual chlorine. M.H.T. = 7 days to extract; 40 days to analyze

QUALITY CONTROL: The laboratory shall, on an ongoing basis, demonstrate, through calibration verification and analysis of precision standard, that the analysis is in control. An internal standard is added to the extract and the extract is injected into a GC/MS

REFERENCE: Method 1625,B, Federal Register Part VIII 40 CFR Part 136, Oct 26, 1984.

PRIMARY NAME: Acenaphthene Method 625

TITLE: Base/Neutrals and Acids MATRIX: Wastewater

CAS # : 83-32-9

APPLICATION: Method covers the determination of 72 organic compounds that are partitioned into an organic solvent and are amenable to gas chromatography. Samples are extracted twice with methylene chloride, at pH 11 and pH <2. Next the extract is dried, concentrated to 1 mL and analyzed. Acenaphthene is a base/neutral extractable compound in the pH 11 fraction.

INTERFERENCES: Contaminants in solvents, reagents, glassware, and other processing hardware may interfere. Matrix interferences may be caused by coextracted contaminants. Packed GC columns recommended for the basic fraction may not exhibit sufficient resolution of all analytes.

INSTRUMENTATION: GC/MS. Automatic sampler (optional).

RANGE: 5-1300 μg/L. MDL: 1.9 μg/L.

PRECISION: 0.21X - 0.67 μg/L (overall precision).

ACCURACY: 0.96C + 0.19 μg/L (as recovery).

SAMPLING METHOD: Amber glass bottle with a Teflon lined screw cap.

STABILITY: Cool to 4°C, add 0.008% sodium thiosulfate if samples have any residual chlorine. Store in the dark.
M.H.T. = 7 days to extract; 40 days to analyze

QUALITY CONTROL: The laboratory must, on an ongoing basis, spike at least 5% of the samples from each sample site being monitored to assess accuracy.

REFERENCE: Method 625, Federal Register Part VIII 40 CFR Part 136, Oct 26, 1984.

PRIMARY NAME: Acenaphthene Method 8100

TITLE: Polynuclear Aromatic Hydrocarbons MATRIX: groundwater, soils, sludges water miscible liquid wastes, and non-water miscible wastes.

CAS # : 83-32-9

APPLICATION: This method is used for the analysis of various PAHs. Samples are extracted, concentrated and analyzed using direct injection of both neat and diluted organic liquids. The method provides two optional GC columns that are better than Column 1 and that may help resolve analytes from interferences.

INTERFERENCES: Solvents, reagents and glassware may introduce artifacts. Other interferences may come from coextracted compounds from samples.

INSTRUMENTATION: GC capable of on-column injections and a flame ionization detector (FID). Column 1: a 1.8 meter by 2 mm 3% OV-17 on Chromosorb W-AW-DCMS column. Column 2: a 30 meter by 0.25 mm SE-54 fused silica capillary colunm. Column 3: a 30 meter by 0.32 mm SE-54 fused silica capillary column.

RANGE: 0.1 to 425 μg/L MDL: Not reported

PRACTICAL QUANTITATION LIMIT FACTORS FOR MULTIPLYING TIMES FID MDL VALUE

Not available

PRECISION: 0.53X + 1.32 μg/L (overall precision)

ACCURACY: 0.52C + 0.54 μg/L (as recovery)

SAMPLING METHOD: Use 8 oz. widemouth glass bottles with Teflon lined caps for concentrated waste samples, soils, sediments and sludges. Use 1 or 2 1/2 gallon amber glass bottles with Teflon lined caps for liquid (water) samples.

STABILITY: Cool soil, sediment, sludge and liquid samples to 4°C. If residual chlorine is present in liquid samples add 3 mL of 10% sodium thiosulfate per gallon of sample and cool to 4°C.
M.H.T. = 14 days for concentrated waste, soil, sediment or sludge.
M.H.T. = 7 days for liquid samples.
All extracts must be analyzed within 40 days.

QUALITY CONTROL: A quality control check sample concentrate containing each analyte of interest is required. The QC check sample concentrate may be prepared from pure standard materials or purchased as certified solutions. Use appropriate trip, matrix, control site, method, reagent and solvent blanks. Internal, surrogate and five concentration level calibration standards are used. The quality control check sample concentrate should contain acenaphthene at 100 μg/mL in acetonitrile.

REFERENCE: Method 8100, SW-846, 3rd ed., Nov 1986.

PRIMARY NAME: Acenaphthene Method 8310

TITLE: Polynuclear Aromatic Hydrocarbons

MATRIX: groundwater, soils, sludges water miscible liquid wastes, and non-water miscible wastes.

CAS # : 83-32-9

APPLICATION: This method is used for the analysis of 16 polynuclear aromatic hydrocarbons (PAHs). Samples are extracted, concentrated and analyzed using HPLC with detection by UV and fluorescence detectors.

INTERFERENCES: Solvents, reagents and glassware may introduce artifacts. Other interferences may come from coextracted compounds from samples.

INSTRUMENTATION: HPLC with a gradient pumping system and a 250 mm by 2.6 mm reverse phase HC-ODS Sil-X 5-micron particle size column. The UV detector uses an excitation wavelength of 254 nm coupled to the fluorescence detector. The fluorescence detector uses an excitation wavelength of 280 nm and emission greater than 389 nm cutoff with dispersive optics.

RANGE: 0.1 to 425 μg/L

MDL: 1.8 μg/L (UV detector; reagent water)

PRACTICAL QUANTITATION LIMIT FACTORS FOR MULTIPLYING TIMES FID MDL VALUE

Matrix	Multiplication Factor
Groundwater	10
Low-level soil by sonication with GPC cleanup	670
High-level soil and sludge by sonication	10,000
Non-water miscible waste	100,000

PRECISION: 0.53X + 1.32 μg/L (overall precision)

ACCURACY: 0.52C + 0.54 μg/L (as recovery)

SAMPLING METHOD: Use 8 oz. widemouth glass bottles with Teflon lined caps for concentrated waste samples, soils, sediments and sludges. Use 1 or 2 1/2 gallon amber glass bottles with Teflon lined caps for liquid (water) samples.

STABILITY: Cool soil, sediment, sludge and liquid samples to 4°C. If residual chlorine is present in liquid samples add 3 mL of 10% sodium thiosulfate per gallon of sample and cool to 4°C.

M.H.T. = 14 days for concentrated waste, soil, sediment or sludge.

M.H.T. = 7 days for liquid samples.

All extracts must be analyzed within 40 days.

QUALITY CONTROL: Internal, surrogate and five concentration level calibration standards are used. The calibration standards must be used with the analytical method blank. A quality control check sample concentrate containing acenaphthene at 100 μg/mL is required. The QC check sample concentrate may be prepared from pure standard materials or purchased as certified solutions. Use appropriate trip, matrix, control site, method, reagent and solvent blanks.

REFERENCE: Method 8310, SW-846, 3rd ed., Nov 1986.

PRIMARY NAME: Acenaphthylene — Method 8100

TITLE: Polynuclear Aromatic Hydrocarbons

MATRIX: groundwater, soils, sludges water miscible liquid wastes, and non-water miscible wastes.

CAS # : 208-96-8

APPLICATION: This method is used for the analysis of various PAHs. Samples are extracted, concentrated and analyzed using direct injection of both neat and diluted organic liquids. The method provides two optional GC columns that are better than Column 1 and that may help resolve analytes from interferences.

INTERFERENCES: Solvents, reagents and glassware may introduce artifacts. Other interferences may come from coextracted compounds from samples.

INSTRUMENTATION: GC capable of on-column injections and a flame ionization detector (FID). Column 1: a 1.8 meter by 2 mm 3% OV-17 on Chromosorb W-AW-DCMS column. Column 2: a 30 meter by 0.25 mm SE-54 fused silica capillary colunm. Column 3: a 30 meter by 0.32 mm SE-54 fused silica capillary column.

RANGE: 0.1 to 425 μg/L

MDL: Not reported

PRACTICAL QUANTITATION LIMIT FACTORS FOR MULTIPLYING TIMES FID MDL VALUE

Not available

PRECISION: 0.42X + 0.52 μg/L (overall precision)

ACCURACY: 0.69C - 1.89 μg/L (as recovery)

SAMPLING METHOD: Use 8 oz. widemouth glass bottles with Teflon lined caps for concentrated waste samples, soils, sediments and sludges. Use 1 or 2 1/2 gallon amber glass bottles with Teflon lined caps for liquid (water) samples.

STABILITY: Cool soil, sediment, sludge and liquid samples to 4°C. If residual chlorine is present in liquid samples add 3 mL of 10% sodium thiosulfate per gallon of sample and cool to 4°C.

M.H.T. = 14 days for concentrated waste, soil, sediment or sludge.

M.H.T. = 7 days for liquid samples.

All extracts must be analyzed within 40 days.

QUALITY CONTROL: A quality control check sample concentrate containing each analyte of interest is required. The QC check sample concentrate may be prepared from pure standard materials or purchased as certified solutions. Use appropriate trip, matrix, control site, method, reagent and solvent blanks. Internal, surrogate and five concentration level calibration standards are used. The quality control check sample concentrate should contain acenaphthylene at 100 μg/mL in acetonitrile.

REFERENCE: Method 8100, SW-846, 3rd ed., Nov 1986.

PRIMARY NAME: Acenaphthylene Method 8310

TITLE: Polynuclear Aromatic Hydrocarbons

MATRIX: groundwater, soils, sludges water miscible liquid wastes, and non-water miscible wastes.

CAS # : 208-96-8

APPLICATION: This method is used for the analysis of 16 polynuclear aromatic hydrocarbons (PAHs). Samples are extracted, concentrated and analyzed using HPLC with detection by UV and fluorescence detectors.

INTERFERENCES: Solvents, reagents and glassware may introduce artifacts. Other interferences may come from coextracted compounds from samples.

INSTRUMENTATION: HPLC with a gradient pumping system and a 250 mm by 2.6 mm reverse phase HC-ODS Sil-X 5-micron particle size column. The UV detector uses an excitation wavelength of 254 nm coupled to the fluorescence detector. The fluorescence detector uses an excitation wavelength of 280 nm and emission greater than 389 nm cutoff with dispersive optics.

RANGE: 0.1 to 425 µg/L

MDL: 2.3 µg/L (UV detector; reagent water)

PRACTICAL QUANTITATION LIMIT FACTORS FOR MULTIPLYING TIMES FID MDL VALUE

Matrix	Multiplication Factor
Groundwater	10
Low-level soil by sonication with GPC cleanup	670
High-level soil and sludge by sonication	10,000
Non-water miscible waste	100,000

PRECISION: 0.42X + 0.52 µg/L (overall precision)

ACCURACY: 0.69C - 1.89 µg/L (as recovery)

SAMPLING METHOD: Use 8 oz. widemouth glass bottles with Teflon lined caps for concentrated waste samples, soils, sediments and sludges. Use 1 or 2 1/2 gallon amber glass bottles with Teflon lined caps for liquid (water) samples.

STABILITY: Cool soil, sediment, sludge and liquid samples to 4°C. If residual chlorine is present in liquid samples add 3 mL of 10% sodium thiosulfate per gallon of sample and cool to 4°C.

M.H.T. = 14 days for concentrated waste, soil, sediment or sludge.
M.H.T. = 7 days for liquid samples.
All extracts must be analyzed within 40 days.

QUALITY CONTROL: Internal, surrogate and five concentration level calibration standards are used. The calibration standards must be used with the analytical method blank. A quality control check sample concentrate containing acenaphthylene at 100 μg/mL is required. The QC check sample concentrate may be prepared from pure standard materials or purchased as certified solutions. Use appropriate trip, matrix, control site, method, reagent and solvent blanks.

REFERENCE: Method 8310, SW-846, 3rd ed., Nov 1986.

PRIMARY NAME: Anthracene Method 8100

TITLE: Polynuclear Aromatic Hydrocarbons MATRIX: groundwater, soils, sludges water miscible liquid wastes, and non-water miscible wastes.

CAS # : 120-12-7

APPLICATION: This method is used for the analysis of various PAHs. Samples are extracted, concentrated and analyzed using direct injection of both neat and diluted organic liquids. The method provides two optional GC columns that are better than Column 1 and that may help resolve analytes from interferences.

INTERFERENCES: Solvents, reagents and glassware may introduce artifacts. Other interferences may come from coextracted compounds from samples.

INSTRUMENTATION: GC capable of on-column injections and a flame ionization detector (FID). Column 1: a 1.8 meter by 2 mm 3% OV-17 on Chromosorb W-AW-DCMS column. Column 2: a 30 meter by 0.25 mm SE-54 fused silica capillary colunm. Column 3: a 30 meter by 0.32 mm SE-54 fused silica capillary column.

RANGE: 0.1 to 425 μg/L MDL: Not reported

PRACTICAL QUANTITATION LIMIT FACTORS FOR MULTIPLYING TIMES FID MDL VALUE

Not available

PRECISION: 0.41X + 0.45 μg/L (overall precision)

ACCURACY: 0.63C - 1.26 μg/L (as recovery)

SAMPLING METHOD: Use 8 oz. widemouth glass bottles with Teflon lined caps for concentrated waste samples, soils, sediments and sludges. Use 1 or 2 1/2 gallon amber glass bottles with Teflon lined caps for liquid (water) samples.

STABILITY: Cool soil, sediment, sludge and liquid samples to 4°C. If residual chlorine is present in liquid samples add 3 mL of 10% sodium thiosulfate per gallon of sample and cool to 4°C.

M.H.T. = 14 days for concentrated waste, soil, sediment or sludge.

M.H.T. = 7 days for liquid samples.

All extracts must be analyzed within 40 days.

QUALITY CONTROL: A quality control check sample concentrate containing each analyte of interest is required. The QC check sample concentrate may be prepared from pure standard materials or purchased as certified solutions. Use appropriate trip, matrix, control site, method, reagent and solvent blanks. Internal, surrogate and five concentration level calibration standards are used. The quality control check sample concentrate should contain anthracene at 100 μg/mL in acetonitrile.

REFERENCE: Method 8100, SW-846, 3rd ed., Nov 1986.

PRIMARY NAME: Anthracene — Method 8310

TITLE: Polynuclear Aromatic Hydrocarbons

MATRIX: groundwater, soils, sludges water miscible liquid wastes, and non-water miscible wastes.

CAS # : 120-12-7

APPLICATION: This method is used for the analysis of 16 polynuclear aromatic hydrocarbons (PAHs). Samples are extracted, concentrated and analyzed using HPLC with detection by UV and fluorescence detectors.

INTERFERENCES: Solvents, reagents and glassware may introduce artifacts. Other interferences may come from coextracted compounds from samples.

INSTRUMENTATION: HPLC with a gradient pumping system and a 250 mm by 2.6 mm reverse phase HC-ODS Sil-X 5-micron particle size column. The fluorescence detector uses an excitation wavelength of 280 nm and emission greater than 389 nm cutoff with dispersive optics.

RANGE: 0.1 to 425 µg/L

MDL: 0.66 µg/L (Fluorescence; reagent water)

PRACTICAL QUANTITATION LIMIT FACTORS FOR MULTIPLYING TIMES FID MDL VALUE

Matrix	Multiplication Factor
Groundwater	10
Low-level soil by sonication with GPC cleanup	670
High-level soil and sludge by sonication	10,000
Non-water miscible waste	100,000

PRECISION: 0.41X + 0.45 µg/L (overall precision)

ACCURACY: 0.63C - 1.26 µg/L (as recovery)

SAMPLING METHOD: Use 8 oz. widemouth glass bottles with Teflon lined caps for concentrated waste samples, soils, sediments and sludges. Use 1 or 2 1/2 gallon amber glass bottles with Teflon lined caps for liquid (water) samples.

STABILITY: Cool soil, sediment, sludge and liquid samples to 4°C. If residual chlorine is present in liquid samples add 3 mL of 10% sodium thiosulfate per gallon of sample and cool to 4°C.

M.H.T. = 14 days for concentrated waste, soil, sediment or sludge.

M.H.T. = 7 days for liquid samples.

All extracts must be analyzed within 40 days.

QUALITY CONTROL: Internal, surrogate and five concentration level calibration standards are used. The calibration standards must be used with the analytical method blank. A quality control check sample concentrate containing anthracene at 100 μg/mL is required. The QC check sample concentrate may be prepared from pure standard materials or purchased as certified solutions. Use appropriate trip, matrix, control site, method, reagent and solvent blanks.

REFERENCE: Method 8310, SW-846, 3rd ed., Nov 1986.

PRIMARY NAME: Benzo(a)anthracene Method 8100

TITLE: Polynuclear Aromatic Hydrocarbons

MATRIX: groundwater, soils, sludges water miscible liquid wastes, and non-water miscible wastes.

CAS # : 56-55-3

APPLICATION: This method is used for the analysis of various PAHs. Samples are extracted, concentrated and analyzed using direct injection of both neat and diluted organic liquids. The method provides two optional GC columns that are better than Column 1 and that may help resolve analytes from interferences.

INTERFERENCES: Solvents, reagents and glassware may introduce artifacts. Other interferences may come from coextracted compounds from samples.

INSTRUMENTATION: GC capable of on-column injections and a flame ionization detector (FID). Column 1: a 1.8 meter by 2 mm 3% OV-17 on Chromosorb W-AW-DCMS column. Column 2: a 30 meter by 0.25 mm SE-54 fused silica capillary colunm. Column 3: a 30 meter by 0.32 mm SE-54 fused silica capillary column.

RANGE: 0.1 to 425 μg/L MDL: Not reported

PRACTICAL QUANTITATION LIMIT FACTORS FOR MULTIPLYING TIMES FID MDL VALUE

Not available

PRECISION: 0.34X + 0.02 μg/L (overall precision)

ACCURACY: 0.73C + 0.05 μg/L (as recovery)

SAMPLING METHOD: Use 8 oz. widemouth glass bottles with Teflon lined caps for concentrated waste samples, soils, sediments and sludges. Use 1 or 2 1/2 gallon amber glass bottles with Teflon lined caps for liquid (water) samples.

STABILITY: Cool soil, sediment, sludge and liquid samples to 4°C. If residual chlorine is present in liquid samples add 3 mL of 10% sodium thiosulfate per gallon of sample and cool to 4°C.

M.H.T. = 14 days for concentrated waste, soil, sediment or sludge.

M.H.T. = 7 days for liquid samples.

All extracts must be analyzed within 40 days.

QUALITY CONTROL: A quality control check sample concentrate containing each analyte of interest is required. The QC check sample concentrate may be prepared from pure standard materials or purchased as certified solutions. Use appropriate trip, matrix, control site, method, reagent and solvent blanks. Internal, surrogate and five concentration level calibration standards are used. The quality control check sample concentrate should contain benzo(a)anthracene at 10 μg/mL in acetonitrile.

REFERENCE: Method 8100, SW-846, 3rd ed., Nov 1986.

PRIMARY NAME: Benzo(a)anthracene Method 8310

TITLE: Polynuclear Aromatic Hydrocarbons

MATRIX: groundwater, soils, sludges water miscible liquid wastes, and non-water miscible wastes.

CAS # : 56-55-3

APPLICATION: This method is used for the analysis of 16 polynuclear aromatic hydrocarbons (PAHs). Samples are extracted, concentrated and analyzed using HPLC with detection by UV and fluorescence detectors.

INTERFERENCES: Solvents, reagents and glassware may introduce artifacts. Other interferences may come from coextracted compounds from samples.

INSTRUMENTATION: HPLC with a gradient pumping system and a 250 mm by 2.6 mm reverse phase HC-ODS Sil-X 5-micron particle size column. The fluorescence detector uses an excitation wavelength of 280 nm and emission greater than 389 nm cutoff with dispersive optics.

RANGE: 0.1 to 425 µg/L

MDL: 0.013 µg/L (Fluorescence; reagent water)

PRACTICAL QUANTITATION LIMIT FACTORS FOR MULTIPLYING TIMES FID MDL VALUE

Matrix	Multiplication Factor
Groundwater	10
Low-level soil by sonication with GPC cleanup	670
High-level soil and sludge by sonication	10,000
Non-water miscible waste	100,000

PRECISION: 0.34X + 0.02 µg/L (overall precision)

ACCURACY: 0.73C + 0.05 µg/L (as recovery)

SAMPLING METHOD: Use 8 oz. widemouth glass bottles with Teflon lined caps for concentrated waste samples, soils, sediments and sludges. Use 1 or 2 1/2 gallon amber glass bottles with Teflon lined caps for liquid (water) samples.

STABILITY: Cool soil, sediment, sludge and liquid samples to 4°C. If residual chlorine is present in liquid samples add 3 mL of 10% sodium thiosulfate per gallon of sample and cool to 4°C.

M.H.T. = 14 days for concentrated waste, soil, sediment or sludge.

M.H.T. = 7 days for liquid samples.

All extracts must be analyzed within 40 days.

QUALITY CONTROL: Internal, surrogate and five concentration level calibration standards are used. The calibration standards must be used with the analytical method blank. A quality control check sample concentrate containing benzo(a)anthracene at 10 µg/mL is required. The QC check sample concentrate may be prepared from pure standard materials or purchased as certified solutions. Use appropriate trip, matrix, control site, method, reagent and solvent blanks.

REFERENCE: Method 8310, SW-846, 3rd ed., Nov 1986.

PRIMARY NAME: Benzo(a)pyrene Method 8100

TITLE: Polynuclear Aromatic Hydrocarbons MATRIX: groundwater, soils, sludges water miscible liquid wastes, and non-water miscible wastes.

CAS # : 50-32-8

APPLICATION: This method is used for the analysis of various PAHs. Samples are extracted, concentrated and analyzed using direct injection of both neat and diluted organic liquids. The method provides two optional GC columns that are better than Column 1 and that may help resolve analytes from interferences.

INTERFERENCES: Solvents, reagents and glassware may introduce artifacts. Other interferences may come from coextracted compounds from samples.

INSTRUMENTATION: GC capable of on-column injections and a flame ionization detector (FID). Column 1: a 1.8 meter by 2 mm 3% OV-17 on Chromosorb W-AW-DCMS column. Column 2: a 30 meter by 0.25 mm SE-54 fused silica capillary colunm. Column 3: a 30 meter by 0.32 mm SE-54 fused silica capillary column.

RANGE: 0.1 to 425 μg/L MDL: Not reported

PRACTICAL QUANTITATION LIMIT FACTORS FOR MULTIPLYING TIMES FID MDL VALUE

Not available

PRECISION: 0.53X + 0.01 μg/L (overall precision)

ACCURACY: 0.56C + 0.01 μg/L (as recovery)

SAMPLING METHOD: Use 8 oz. widemouth glass bottles with Teflon lined caps for concentrated waste samples, soils, sediments and sludges. Use 1 or 2 1/2 gallon amber glass bottles with Teflon lined caps for liquid (water) samples.

STABILITY: Cool soil, sediment, sludge and liquid samples to 4°C. If residual chlorine is present in liquid samples add 3 mL of 10% sodium thiosulfate per gallon of sample and cool to 4°C.

M.H.T. = 14 days for concentrated waste, soil, sediment or sludge

M.H.T. = 7 days for liquid samples.

All extracts must be analyzed within 40 days.

QUALITY CONTROL: A quality control check sample concentrate containing each analyte of interest is required. The QC check sample concentrate may be prepared from pure standard materials or purchased as certified solutions. Use appropriate trip, matrix, control site, method, reagent and solvent blanks. Internal, surrogate and five concentration level calibration standards are used. The quality control check sample concentrate should contain benzo(a)pyrene at 10 μg/mL in acetonitrile.

REFERENCE: Method 8100, SW-846, 3rd ed., Nov 1986.

PRIMARY NAME: Benzo(a)pyrene Method 8310

TITLE: Polynuclear Aromatic Hydrocarbons

MATRIX: groundwater, soils, sludges water miscible liquid wastes, and non-water miscible wastes.

CAS # : 50-32-8

APPLICATION: This method is used for the analysis of 16 polynuclear aromatic hydrocarbons (PAHs). Samples are extracted, concentrated and analyzed using HPLC with detection by UV and fluorescence detectors.

INTERFERENCES: Solvents, reagents and glassware may introduce artifacts. Other interferences may come from coextracted compounds from samples.

INSTRUMENTATION: HPLC with a gradient pumping system and a 250 mm by 2.6 mm reverse phase HC-ODS Sil-X 5-micron particle size column. The fluorescence detector uses an excitation wavelength of 280 nm and emission greater than 389 nm cutoff with dispersive optics.

RANGE: 0.1 to 425 μg/L

MDL: 0.023 μg/L (Fluorescence; reagent water)

PRACTICAL QUANTITATION LIMIT FACTORS FOR MULTIPLYING TIMES FID MDL VALUE

Matrix	Multiplication Factor
Groundwater	10
Low-level soil by sonication with GPC cleanup	670
High-level soil and sludge by sonication	10,000
Non-water miscible waste	100,000

PRECISION: 0.53X - 0.01 μg/L (overall precision)

ACCURACY: 0.56C + 0.01 μg/L (as recovery)

SAMPLING METHOD: Use 8 oz. widemouth glass bottles with Teflon lined caps for concentrated waste samples, soils, sediments and sludges. Use 1 or 2 1/2 gallon amber glass bottles with Teflon lined caps for liquid (water) samples.

STABILITY: Cool soil, sediment, sludge and liquid samples to 4°C. If residual chlorine is present in liquid samples add 3 mL of 10% sodium thiosulfate per gallon of sample and cool to 4°C.

M.H.T. = 14 days for concentrated waste, soil, sediment or sludge.

M.H.T. = 7 days for liquid samples.

All extracts must be analyzed within 40 days.

QUALITY CONTROL: Internal, surrogate and five concentration level calibration standards are used. The calibration standards must be used with the analytical method blank. A quality control check sample concentrate containing benzo(a)pyrene at 10 μg/mL is required. The QC check sample concentrate may be prepared from pure standard materials or purchased as certified solutions. Use appropriate trip, matrix, control site, method, reagent and solvent blanks.

REFERENCE: Method 8310, SW-846, 3rd ed., Nov 1986.

PRIMARY NAME: Benzo(b)fluoranthene Method 8100

TITLE: Polynuclear Aromatic Hydrocarbons MATRIX: groundwater, soils, sludges water miscible liquid wastes, and non-water miscible wastes.

CAS # : 205-99-2

APPLICATION: This method is used for the analysis of various PAHs. Samples are extracted, concentrated and analyzed using direct injection of both neat and diluted organic liquids. The method provides two optional GC columns that are better than Column 1 and that may help resolve analytes from interferences.

INTERFERENCES: Solvents, reagents and glassware may introduce artifacts. Other interferences may come from coextracted compounds from samples.

INSTRUMENTATION: GC capable of on-column injections and a flame ionization detector (FID). Column 1: a 1.8 meter by 2 mm 3% OV-17 on Chromosorb W-AW-DCMS column. Column 2: a 30 meter by 0.25 mm SE-54 fused silica capillary colunm. Column 3: a 30 meter by 0.32 mm SE-54 fused silica capillary column.

RANGE: 0.1 to 425 μg/L MDL: Not reported

PRACTICAL QUANTITATION LIMIT FACTORS FOR MULTIPLYING TIMES FID MDL VALUE

Not available

PRECISION: 0.38X - 0.00 μg/L (overall precision)

ACCURACY: 0.78C + 0.01 μg/L (as recovery)

SAMPLING METHOD: Use 8 oz. widemouth glass bottles with Teflon lined

caps for concentrated waste samples, soils, sediments and sludges. Use 1 or 2 1/2 gallon amber glass bottles with Teflon lined caps for liquid (water) samples.

STABILITY: Cool soil, sediment, sludge and liquid samples to 4°C. If residual chlorine is present in liquid samples add 3 mL of 10% sodium thiosulfate per gallon of sample and cool to 4°C.

M.H.T. = 14 days for concentrated waste, soil, sediment or sludge.

M.H.T. = 7 days for liquid samples.

All extracts must be analyzed within 40 days.

QUALITY CONTROL: A quality control check sample concentrate containing each analyte of interest is required. The QC check sample concentrate may be prepared from pure standard materials or purchased as certified solutions. Use appropriate trip, matrix, control site, method, reagent and solvent blanks. Internal, surrogate and five concentration level calibration standards are used. The quality control check sample concentrate should contain benzo(b)fluoranthene at 10 μg/mL in acetonitrile.

REFERENCE: Method 8100, SW-846, 3rd ed., Nov 1986.

PRIMARY NAME: Benzo(b)fluoranthene Method 8310

TITLE: Polynuclear Aromatic Hydrocarbons

MATRIX: groundwater, soils, sludges water miscible liquid wastes, and non-water miscible wastes.

CAS # : 205-99-2

APPLICATION: This method is used for the analysis of 16 polynuclear aromatic hydrocarbons (PAHs). Samples are extracted, concentrated and analyzed using HPLC with detection by UV and fluorescence detectors.

INTERFERENCES: Solvents, reagents and glassware may introduce artifacts. Other interferences may come from coextracted compounds from samples.

INSTRUMENTATION: HPLC with a gradient pumping system and a 250 mm by 2.6 mm reverse phase HC-ODS Sil-X 5-micron particle size column. The fluorescence detector uses an excitation wavelength of 280 nm and emission greater than 389 nm cutoff with dispersive optics.

RANGE: 0.1 to 425 μg/L

MDL: 0.018 μg/L (Fluorescence; reagent water)

PRACTICAL QUANTITATION LIMIT FACTORS FOR MULTIPLYING TIMES FID MDL VALUE

Matrix	Multiplication Factor
Groundwater	10
Low-level soil by sonication with GPC cleanup	670
High-level soil and sludge by sonication	10,000
Non-water miscible waste	100,000

PRECISION: 0.38X - 0.00 μg/L (overall precision)

ACCURACY: 0.78C + 0.01 μg/L (as recovery)

SAMPLING METHOD: Use 8 oz. widemouth glass bottles with Teflon lined caps for concentrated waste samples, soils, sediments and sludges. Use 1 or 2 1/2 gallon amber glass bottles with Teflon lined caps for liquid (water) samples.

STABILITY: Cool soil, sediment, sludge and liquid samples to 4°C. If residual chlorine is present in liquid samples add 3 mL of 10% sodium thiosulfate per gallon of sample and cool to 4°C.

M.H.T. = 14 days for concentrated waste, soil, sediment or sludge.

M.H.T. = 7 days for liquid samples.

All extracts must be analyzed within 40 days.

QUALITY CONTROL: Internal, surrogate and five concentration level calibration standards are used. The calibration standards must be used with the analytical method blank. A quality control check sample concentrate containing benzo(b)fluoranthene at 10 μg/mL is required. The QC check sample concentrate may be prepared from pure standard materials or purchased as certified solutions. Use appropriate trip, matrix, control site, method, reagent and solvent blanks.

REFERENCE: Method 8310, SW-846, 3rd ed., Nov 1986.

PRIMARY NAME: Benzo(ghi)perylene Method 8100

TITLE: Polynuclear Aromatic Hydrocarbons MATRIX: groundwater, soils, sludges water miscible liquid wastes, and non-water miscible wastes.

CAS # : 191-24-2

APPLICATION: This method is used for the analysis of various PAHs. Samples are extracted, concentrated and analyzed using direct injection of both neat and diluted organic liquids. The method provides two optional GC columns that are better than Column 1 and that may help resolve analytes from interferences. The recovery of this compound at 80 and 800 times the MDL is reported as low (35% and 45% respectively) although the MDL was not reported.

INTERFERENCES: Solvents, reagents and glassware may introduce artifacts. Other interferences may come from coextracted compounds from samples.

INSTRUMENTATION: GC capable of on-column injections and a flame ionization detector (FID). Column 1: a 1.8 meter by 2 mm 3% OV-17 on Chromosorb W-AW-DCMS column. Column 2: a 30 meter by 0.25 mm SE-54 fused silica capillary colunm. Column 3: a 30 meter by 0.32 mm SE-54 fused silica capillary column.

RANGE: 0.1 to 425 µg/L MDL: Not reported

PRACTICAL QUANTITATION LIMIT FACTORS FOR MULTIPLYING TIMES FID MDL VALUE

Not available

PRECISION: 0.58X + 0.10 µg/L (overall precision)

ACCURACY: 0.44C + 0.30 µg/L (as recovery)

SAMPLING METHOD: Use 8 oz. widemouth glass bottles with Teflon lined caps for concentrated waste samples, soils, sediments and sludges. Use 1 or 2 1/2 gallon amber glass bottles with Teflon lined caps for liquid (water) samples.

STABILITY: Cool soil, sediment, sludge and liquid samples to 4°C. If residual chlorine is present in liquid samples add 3 mL of 10% sodium thiosulfate per gallon of sample and cool to 4°C.

M.H.T. = 14 days for concentrated waste, soil, sediment or sludge.
M.H.T. = 7 days for liquid samples.
All extracts must be analyzed within 40 days.

QUALITY CONTROL: A quality control check sample concentrate containing each analyte of interest is required. The QC check sample concentrate may be prepared from pure standard materials or purchased as certified solutions. Use appropriate trip, matrix, control site, method, reagent and solvent blanks. Internal, surrogate and five concentration level calibration standards are used. The quality control check sample concentrate should contain benzo(ghi)perylene at 10 µg/mL in acetonitrile.

REFERENCE: Method 8100, SW-846, 3rd ed., Nov 1986.

PRIMARY NAME: Benzo(ghi)perylene Method 8310

TITLE: Polynuclear Aromatic Hydrocarbons

MATRIX: groundwater, soils, sludges water miscible liquid wastes, and non-water miscible wastes.

CAS # : 191-24-2

APPLICATION: This method is used for the analysis of 16 polynuclear aromatic hydrocarbons (PAHs). Samples are extracted, concentrated and analyzed using HPLC with detection by UV and fluorescence detectors.

INTERFERENCES: Solvents, reagents and glassware may introduce artifacts. Other interferences may come from coextracted compounds from samples.

INSTRUMENTATION: HPLC with a gradient pumping system and a 250 mm by 2.6 mm reverse phase HC-ODS Sil-X 5-micron particle size column. The fluorescence detector uses an excitation wavelength of 280 nm and emission greater than 389 nm cutoff with dispersive optics.

RANGE: 0.1 to 425 μg/L

MDL: 0.076 μg/L (Fluorescence; reagent water)

PRACTICAL QUANTITATION LIMIT FACTORS FOR MULTIPLYING TIMES FID MDL VALUE

Matrix	Multiplication Factor
Groundwater	10
Low-level soil by sonication with GPC cleanup	670
High-level soil and sludge by sonication	10,000
Non-water miscible waste	100,000

PRECISION: 0.58X + 0.10 μg/L (overall precision)

ACCURACY: 0.44C + 0.30 μg/L (as recovery)

SAMPLING METHOD: Use 8 oz. widemouth glass bottles with Teflon lined caps for concentrated waste samples, soils, sediments and sludges. Use 1 or 2 1/2 gallon amber glass bottles with Teflon lined caps for liquid (water) samples.

STABILITY: Cool soil, sediment, sludge and liquid samples to 4°C. If residual chlorine is present in liquid samples add 3 mL of 10% sodium thiosulfate per gallon of sample and cool to 4°C.

M.H.T. = 14 days for concentrated waste, soil, sediment or sludge.

M.H.T. = 7 days for liquid samples.

All extracts must be analyzed within 40 days.

QUALITY CONTROL: Internal, surrogate and five concentration level calibration standards are used. The calibration standards must be used with the analytical method blank. A quality control check sample concentrate containing benzo(ghi)perylene at 10 μg/mL is required. The QC check sample concentrate may be prepared from pure standard materials or purchased as certified solutions. Use appropriate trip, matrix, control site, method, reagent and solvent blanks.

REFERENCE: Method 8310, SW-846, 3rd ed., Nov 1986.

PRIMARY NAME: Benzo(k)fluoranthene — Method 8100

TITLE: Polynuclear Aromatic Hydrocarbons

MATRIX: groundwater, soils, sludges water miscible liquid wastes, and non-water miscible wastes.

CAS # : 207-08-9

APPLICATION: This method is used for the analysis of various PAHs. Samples are extracted, concentrated and analyzed using direct injection of both neat and diluted organic liquids. The method provides two optional GC columns that are better than Column 1 and that may help resolve analytes from interferences.

INTERFERENCES: Solvents, reagents and glassware may introduce artifacts. Other interferences may come from coextracted compounds from samples.

INSTRUMENTATION: GC capable of on-column injections and a flame ionization detector (FID). Column 1: a 1.8 meter by 2 mm 3% OV-17 on Chromosorb W-AW-DCMS column. Column 2: a 30 meter by 0.25 mm SE-54 fused silica capillary colunm. Column 3: a 30 meter by 0.32 mm SE-54 fused silica capillary column.

RANGE: 0.1 to 425 µg/L

MDL: Not reported

PRACTICAL QUANTITATION LIMIT FACTORS FOR MULTIPLYING TIMES FID MDL VALUE

Not available

PRECISION: 0.69X + 0.10 µg/L (overall precision)

ACCURACY: 0.59C + 0.00 µg/L (as recovery)

SAMPLING METHOD: Use 8 oz. widemouth glass bottles with Teflon lined caps for concentrated waste samples, soils, sediments and sludges. Use 1 or 2 1/2 gallon amber glass bottles with Teflon lined caps for liquid (water) samples.

STABILITY: Cool soil, sediment, sludge and liquid samples to 4°C. If residual chlorine is present in liquid samples add 3 mL of 10% sodium thiosulfate per gallon of sample and cool to 4°C.

M.H.T. = 14 days for concentrated waste, soil, sediment or sludge.

M.H.T. = 7 days for liquid samples.

All extracts must be analyzed within 40 days.

QUALITY CONTROL: A quality control check sample concentrate containing each analyte of interest is required. The QC check sample concentrate may be prepared from pure standard materials or purchased as certified solutions. Use appropriate trip, matrix, control site, method, reagent and solvent blanks. Internal, surrogate and five concentration level calibration standards are used. The quality control check sample concentrate should contain benzo(k)fluoranthene at 5 μg/mL in acetonitrile.

REFERENCE: Method 8100, SW-846, 3rd ed., Nov 1986.

PRIMARY NAME: Benzo(k)fluoranthene Method 8310

TITLE: Polynuclear Aromatic Hydrocarbons MATRIX: groundwater, soils, sludges water miscible liquid wastes, and non-water miscible wastes.

CAS # : 207-08-9

APPLICATION: This method is used for the analysis of 16 polynuclear aromatic hydrocarbons (PAHs). Samples are extracted, concentrated and analyzed using HPLC with detection by UV and fluorescence detectors.

INTERFERENCES: Solvents, reagents and glassware may introduce artifacts. Other interferences may come from coextracted compounds from samples.

INSTRUMENTATION: HPLC with a gradient pumping system and a 250 mm by 2.6 mm reverse phase HC-ODS Sil-X 5-micron particle size column. The fluorescence detector uses an excitation wavelength of 280 nm and emission greater than 389 nm cutoff with dispersive optics.

RANGE: 0.1 to 425 μg/L

MDL: 0.017 μg/L (Fluorescence; reagent water)

PRACTICAL QUANTITATION LIMIT FACTORS FOR MULTIPLYING TIMES FID MDL VALUE

Matrix	Multiplication Factor
Groundwater	10
Low-level soil by sonication with GPC cleanup	670
High-level soil and sludge by sonication	10,000
Non-water miscible waste	100,000

PRECISION: 0.69X + 0.10 μg/L (overall precision)

ACCURACY: 0.59C + 0.00 μg/L (as recovery)

SAMPLING METHOD: Use 8 oz. widemouth glass bottles with Teflon lined caps for concentrated waste samples, soils, sediments and sludges. Use 1 or 2 1/2 gallon amber glass bottles with Teflon lined caps for liquid (water) samples.

STABILITY: Cool soil, sediment, sludge and liquid samples to 4°C. If residual chlorine is present in liquid samples add 3 mL of 10% sodium thiosulfate per gallon of sample and cool to 4°C.

M.H.T. = 14 days for concentrated waste, soil, sediment or sludge.

M.H.T. = 7 days for liquid samples.

All extracts must be analyzed within 40 days.

QUALITY CONTROL: Internal, surrogate and five concentration level calibration standards are used. The calibration standards must be used with the analytical method blank. A quality control check sample concentrate containing benzo(k)fluoranthene at 5 μg/mL is required. The QC check sample concentrate may be prepared from pure standard materials or purchased as certified solutions. Use appropriate trip, matrix, control site, method, reagent and solvent blanks.

REFERENCE: Method 8310, SW-846, 3rd ed., Nov 1986.

PRIMARY NAME: Butyl benzyl phthalate — Method 8060

TITLE: Phthalate Esters

MATRIX: groundwater, soils, sludges water miscible liquid wastes, and non-water miscible wastes.

CAS # : 85-68-7

APPLICATION: This method is used for the analysis of 6 phthalate esters. Samples are extracted, concentrated and analyzed using direct injection of both neat and diluted organic liquids into a gas chromatograph. Analytes are detected by a flame ionization detector (FID) or an electron capture detector (ECD). Groundwater samples should be determined by ECD. The method provides an optional GC column which is used for analyte confirmation and that may help resolve analytes from interferences.

INTERFERENCES: Solvents, reagents and glassware may introduce artifacts. Plastics, in particular, must be avoided. Other interferences may come from coextracted compounds from samples. There can be carry-over contamination with high and low level samples.

INSTRUMENTATION: GC capable of on-column injections and a flame ionization detector (FID) or electron capture detector (ECD). Column 1: 1.8 meter by 4 mm with 1.5% SP-2250 / 1.95% SP-2401 on Supelcoport. Column 2: 1.8 meter by 4 mm with 3% OV-1 on supelcoport.

RANGE: 0.7 to 106 µg/L

MDL: 15 µg/L (FID) and 0.34 µg/L (ECD)

PRACTICAL QUANTITATION LIMIT FACTORS FOR MULTIPLYING TIMES FID MDL VALUE

Matrix	Multiplication Factor
Groundwater	10
Low-level soil by sonication with GPC cleanup	670
High-level soil and sludge by sonication	10,000
Non-water miscible waste	100,000

PRECISION: 0.25X + 0.07 μg/L (overall precision using FID)

ACCURACY: 0.82C + 0.13 μg/L (as recovery using FID)

SAMPLING METHOD: Use 8 oz. widemouth glass bottles with Teflon lined caps for concentrated waste samples, soils, sediments and sludges. Use 1 or 2 1/2 gallon amber glass bottles with Teflon lined caps for liquid (water) samples.

STABILITY: Cool soil, sediment, sludge and liquid samples to 4°C. If residual chlorine is present in liquid samples add 3 mL of 10% sodium thiosulfate per gallon of sample and cool to 4°C.

M.H.T. = 14 days for concentrated waste, soil, sediment or sludge.

M.H.T. = 7 days for liquid samples.

All extracts must be analyzed within 40 days.

QUALITY CONTROL: A quality control check sample concentrate containing each analyte of interest is required. The QC check sample concentrate may be prepared from pure standard materials or purchased as certified solutions. Use appropriate trip, matrix, control site, method, reagent and solvent blanks. Internal, surrogate and five concentration level calibration standards are used. The QC check sample concentrate should contain this compound at 10 μg/mL in acetone.

REFERENCE: Method 8060, SW-846, 3rd ed., Nov 1986.

PRIMARY NAME: 4-Chloro-3-methylphenol — Method 8040

TITLE: Phenols

MATRIX: groundwater, soils, sludges water miscible liquid wastes, and non-water miscible wastes.

CAS # : 59-50-7

APPLICATION: This method is used for the analysis of 17 phenols. Samples are extracted, concentrated and analyzed using direct injection of both neat and diluted organic liquids. Pentafluorobenzylbromide (PFB) derivatives also may be made to increase sensitivity of the method.

INTERFERENCES: There can be carry-over contamination with high and low level samples. Solvents, reagents and glassware may introduce artifacts. Other interferences may come from coextracted compounds from samples.

INSTRUMENTATION: GC capable of on-column injections and a flame ionization detector (FID) or electron capture detector (ECD). Column for underivatized phenol: 1.8 meter by 2.0 mm with 1% SP-1240DA on Supelcoport. Column for derivatized phenols: 1.8 meter by 2.0 mm with 5% OV-17 on Chromosorb W-AW-DMCS.

RANGE: 12 to 450 µg/L

MDL: 0.36 µg/L (FID) and 1.8 µg/L (ECD)

PRACTICAL QUANTITATION LIMIT FACTORS FOR MULTIPLYING TIMES FID MDL VALUE

Matrix	Multiplication Factor
Groundwater	10
Low-level soil by sonication with GPC cleanup	670
High-level soil and sludge by sonication	10,000
Non-water miscible waste	100,000

PRECISION: 0.16X + 1.41 µg/L (overall precision using FID)

ACCURACY: 0.87C - 1.97 µg/L (as recovery using FID)

SAMPLING METHOD: Use 8 oz. widemouth glass bottles with Teflon lined caps for concentrated waste samples, soils, sediments and sludges. Use 1 or 2 1/2 gallon amber glass bottles with Teflon lined caps for liquid (water) samples.

STABILITY: Cool soil, sediment, sludge and liquid samples to 4°C. If residual chlorine is present in liquid samples add 3 mL of 10% sodium thiosulfate per gallon of sample and cool to 4°C.

M.H.T. = 14 days for concentrated waste, soil, sediment or sludge.
M.H.T. = 7 days for liquid samples.

All extracts must be analyzed within 40 days.

QUALITY CONTROL: A quality control check sample concentrate containing each analyte of interest is required. The QC check sample concentrate may be prepared from pure standard materials or purchased as certified solutions. Use appropriate trip, matrix, control site, method, reagent and solvent blanks. Internal, surrogate and five concentration level calibration standards are used. The QC check sample concentrate should contain this compound at 100 µg/mL in 2-propanol.

REFERENCE: Method 8040, SW-846, 3rd ed., Nov 1986.

PRIMARY NAME: 2-Chloronaphthalene

Method 8120

TITLE: Chlorinated Hydrocarbons

MATRIX: groundwater, soils, sludges water miscible liquid wastes, and non-water miscible wastes.

CAS # : 91-58-7

APPLICATION: This method is used for the analysis of various chlorinated hydrocarbons. Samples are extracted, concentrated and analyzed using direct injection of both neat and diluted organic liquid into a gas chromatograph (GC). The method provides an optional GC column which is used for analyte confirmation and that may help resolve analytes from interferences.

INTERFERENCES: Solvents, reagents and glassware may introduce artifacts. Other interferences may come from coextracted compounds from samples. Phthalate esters are common interferences when using an electron capture detector (ECD) so all plastics must be strictly avoided. Exhaustive cleanup of reagents and glassware may be required to eliminate phthalate contamination.

INSTRUMENTATION: GC capable of on-column injections and an ECD. Column 1: a 1.8 meter by 2 mm with 1% SP-1000 on Supelcoport column. Column 2: a 1.8 meter by 2 mm with 1.5% OV-1 / 2.4% OV-225 on Supelcoport column.

RANGE: 1 to 356 µg/L

MDL: 0.94 µg/L (in reagent water)

PRACTICAL QUANTITATION LIMIT FACTORS FOR MULTIPLYING TIMES FID MDL VALUE

Matrix	Multiplication Factor
Groundwater	10
Low-level soil by sonication with GPC cleanup	670
High-level soil and sludge by sonication	10,000
Non-water miscible waste	100,000

PRECISION: 0.38X - 1.39 μg/L (overall precision)

ACCURACY: 0.75C + 3.21 μg/L (as recovery)

SAMPLING METHOD: Use 8 oz. widemouth glass bottles with Teflon lined caps for concentrated waste samples, soils, sediments and sludges. Use 1 or 2 1/2 gallon amber glass bottles with Teflon lined caps for liquid (water) samples.

STABILITY: Cool soil, sediment, sludge and liquid samples to 4°C. If residual chlorine is present in liquid samples add 3 mL of 10% sodium thiosulfate per gallon of sample and cool to 4°C.

M.H.T. = 14 days for concentrated waste, soil, sediment or sludge.

M.H.T. = 7 days for liquid samples.

All extracts must be analyzed within 40 days.

QUALITY CONTROL: A quality control check sample concentrate containing each analyte of interest is required. The QC check sample concentrate may be prepared from pure standard materials or purchased as certified solutions. Use appropriate trip, matrix, control site, method, reagent and solvent blanks. Internal, surrogate and five concentration level calibration standards are used. The quality control check sample concentrate should contain 2-chloronaphthalene at 100 μg/mL in acetone.

REFERENCE: Method 8120, SW-846, 3rd ed., Nov 1986.

PRIMARY NAME: 2-Chlorophenol

Method 8040

TITLE: Phenols

MATRIX: groundwater, soils, sludges water miscible liquid wastes, and non-water miscible wastes.

CAS # : 95-57-8

APPLICATION: This method is used for the analysis of 17 phenols. Samples are extracted, concentrated and analyzed using direct injection of both neat and diluted organic liquids. Pentafluorobenzylbromide (PFB) derivatives also may be made to increase sensitivity of the method.

INTERFERENCES: There can be carry-over contamination with high and low level samples. Solvents, reagents and glassware may introduce artifacts. Other interferences may come from coextracted compounds from samples.

INSTRUMENTATION: GC capable of on-column injections and a flame ionization detector (FID) or electron capture detector (ECD). Column for underivatized phenol: 1.8 meter by 2.0 mm with 1% SP-1240DA on Supelcoport. Column for derivatized phenols: 1.8 meter by 2.0 mm with 5% OV-17 on Chromosorb W-AW-DMCS.

RANGE: 12 to 450 μg/L

MDL: 0.31 μg/L (FID) and 0.58 μg/L (ECD)

PRACTICAL QUANTITATION LIMIT FACTORS FOR MULTIPLYING TIMES FID MDL VALUE

Matrix	Multiplication Factor
Groundwater	10
Low-level soil by sonication with GPC cleanup	670
High-level soil and sludge by sonication	10,000
Non-water miscible waste	100,000

PRECISION: 0.21X + 0.75 μg/L (overall precision using FID)

ACCURACY: 0.83C - 0.84 μg/L (as recovery using FID)

SAMPLING METHOD: Use 8 oz. widemouth glass bottles with Teflon lined caps for concentrated waste samples, soils, sediments and sludges. Use 1 or 2 1/2 gallon amber glass bottles with Teflon lined caps for liquid (water) samples.

STABILITY: Cool soil, sediment, sludge and liquid samples to 4°C. If residual chlorine is present in liquid samples add 3 mL of 10% sodium thiosulfate per gallon of sample and cool to 4°C.

M.H.T. = 14 days for concentrated waste, soil, sediment or sludge.

M.H.T. = 7 days for liquid samples.

All extracts must be analyzed within 40 days.

QUALITY CONTROL: A quality control check sample concentrate containing each analyte of interest is required. The QC check sample concentrate may be prepared from pure standard materials or purchased as certified solutions. Use appropriate trip, matrix, control site, method, reagent and solvent blanks. Internal, surrogate and five concentration level calibration standards are used. The QC check sample concentrate should contain this compound at 100 µg/mL in 2-propanol.

REFERENCE: Method 8040, SW-846, 3rd ed., Nov 1986.

PRIMARY NAME: Chrysene Method 8100

TITLE: Polynuclear Aromatic Hydrocarbons

MATRIX: groundwater, soils, sludges water miscible liquid wastes, and non-water miscible wastes.

CAS # : 218-01-9

APPLICATION: This method is used for the analysis of various PAHs. Samples are extracted, concentrated and analyzed using direct injection of both neat and diluted organic liquids. The method provides two optional GC columns that are better than Column 1 and that may help resolve analytes from interferences.

INTERFERENCES: Solvents, reagents and glassware may introduce artifacts. Other interferences may come from coextracted compounds from samples.

INSTRUMENTATION: GC capable of on-column injections and a flame ionization detector (FID). Column 1: a 1.8 meter by 2 mm 3% OV-17 on Chromosorb W-AW-DCMS column. Column 2: a 30 meter by 0.25 mm SE-54 fused silica capillary colunm. Column 3: a 30 meter by 0.32 mm SE-54 fused silica capillary column.

RANGE: 0.1 to 425 µg/L MDL: Not reported

PRACTICAL QUANTITATION LIMIT FACTORS FOR MULTIPLYING TIMES FID MDL VALUE

Not available

PRECISION: 0.66X - 0.22 µg/L (overall precision)

ACCURACY: 0.77C - 0.18 µg/L (as recovery)

SAMPLING METHOD: Use 8 oz. widemouth glass bottles with Teflon lined caps for concentrated waste samples, soils, sediments and sludges. Use 1 or 2 1/2 gallon amber glass bottles with Teflon lined caps for liquid (water) samples.

STABILITY: Cool soil, sediment, sludge and liquid samples to 4°C. If residual chlorine is present in liquid samples add 3 mL of 10% sodium thiosulfate per gallon of sample and cool to 4°C.

M.H.T. = 14 days for concentrated waste, soil, sediment or sludge.

M.H.T. = 7 days for liquid samples.

All extracts must be analyzed within 40 days.

QUALITY CONTROL: A quality control check sample concentrate containing each analyte of interest is required. The QC check sample concentrate may be prepared from pure standard materials or purchased as certified solutions. Use appropriate trip, matrix, control site, method, reagent and solvent blanks. Internal, surrogate and five concentration level calibration standards are used. The quality control check sample concentrate should contain chrysene at 10 µg/mL in acetonitrile.

REFERENCE: Method 8100, SW-846, 3rd ed., Nov 1986.

PRIMARY NAME: Chrysene Method 8310

TITLE: Polynuclear Aromatic Hydrocarbons MATRIX: groundwater, soils, sludges water miscible liquid wastes, and non-water miscible wastes.

CAS # : 218-01-9

APPLICATION: This method is used for the analysis of 16 polynuclear aromatic hydrocarbons (PAHs). Samples are extracted, concentrated and analyzed using HPLC with detection by UV and fluorescence detectors.

INTERFERENCES: Solvents, reagents and glassware may introduce artifacts. Other interferences may come from coextracted compounds from samples.

INSTRUMENTATION: HPLC with a gradient pumping system and a 250 mm by 2.6 mm reverse phase HC-ODS Sil-X 5-micron particle size column. The fluorescence detector uses an excitation wavelength of 280 nm and emission greater than 389 nm cutoff with dispersive optics.

RANGE: 0.1 to 425 μg/L MDL: 0.15 μg/L (Fluorescence; reagent water)

PRACTICAL QUANTITATION LIMIT FACTORS FOR MULTIPLYING TIMES FID MDL VALUE

Matrix	Multiplication Factor
Groundwater	10
Low-level soil by sonication with GPC cleanup	670
High-level soil and sludge by sonication	10,000
Non-water miscible waste	100,000

PRECISION: 0.66X - 0.22 μg/L (overall precision)

ACCURACY: 0.77C - 0.18 μg/L (as recovery)

SAMPLING METHOD: Use 8 oz. widemouth glass bottles with Teflon lined caps for concentrated waste samples, soils, sediments and sludges. Use 1 or 2 1/2 gallon amber glass bottles with Teflon lined caps for liquid (water) samples.

STABILITY: Cool soil, sediment, sludge and liquid samples to 4°C. If residual chlorine is present in liquid samples add 3 mL of 10% sodium thiosulfate per gallon of sample and cool to 4°C.

M.H.T. = 14 days for concentrated waste, soil, sediment or sludge.

M.H.T. = 7 days for liquid samples.

All extracts must be analyzed within 40 days.

QUALITY CONTROL: Internal, surrogate and five concentration level calibration standards are used. The calibration standards must be used with the analytical method blank. A quality control check sample concentrate containing chrysene at 10 μg/mL is required. The QC check sample concentrate may be prepared from pure standard materials or purchased as certified solutions. Use appropriate trip, matrix, control site, method, reagent and solvent blanks.

REFERENCE: Method 8310, SW-846, 3rd ed., Nov 1986.

PRIMARY NAME: Dibenzo(a,h)anthracene — Method 8100

TITLE: Polynuclear Aromatic Hydrocarbons

MATRIX: groundwater, soils, sludges water miscible liquid wastes, and non-water miscible wastes.

CAS # : 53-70-3

APPLICATION: This method is used for the analysis of various PAHs. Samples are extracted, concentrated and analyzed using direct injection of both neat and diluted organic liquids. The method provides two optional GC columns that are better than Column 1 and that may help resolve analytes from interferences.

INTERFERENCES: Solvents, reagents and glassware may introduce artifacts. Other interferences may come from coextracted compounds from samples.

INSTRUMENTATION: GC capable of on-column injections and a flame ionization detector (FID). Column 1: a 1.8 meter by 2 mm 3% OV-17 on Chromosorb W-AW-DCMS column. Column 2: a 30 meter by 0.25 mm SE-54 fused silica capillary colunm. Column 3: a 30 meter by 0.32 mm SE-54 fused silica capillary column.

RANGE: 0.1 to 425 µg/L MDL: Not reported

PRACTICAL QUANTITATION LIMIT FACTORS FOR MULTIPLYING TIMES FID MDL VALUE

Not available

PRECISION: 0.45X + 0.03 µg/L (overall precision)

ACCURACY: 0.41C - 0.11 µg/L (as recovery)

SAMPLING METHOD: Use 8 oz. widemouth glass bottles with Teflon lined caps for concentrated waste samples, soils, sediments and sludges. Use 1 or 2 1/2 gallon amber glass bottles with Teflon lined caps for liquid (water) samples.

STABILITY: Cool soil, sediment, sludge and liquid samples to 4°C. If residual chlorine is present in liquid samples add 3 mL of 10% sodium thiosulfate per gallon of sample and cool to 4°C.

M.H.T. = 14 days for concentrated waste, soil, sediment or sludge.

M.H.T. = 7 days for liquid samples.

All extracts must be analyzed within 40 days.

QUALITY CONTROL: A quality control check sample concentrate containing each analyte of interest is required. The QC check sample concentrate may be prepared from pure standard materials or purchased as certified solutions. Use appropriate trip, matrix, control site, method, reagent and solvent blanks. Internal, surrogate and five concentration level calibration standards are used. The quality control check sample concentrate should contain dibenzo(a,h)anthracene at 10 µg/mL in acetonitrile.

REFERENCE: Method 8100, SW-846, 3rd ed., Nov 1986.

PRIMARY NAME: Dibenzo(a,h)anthracene Method 8310

TITLE: Polynuclear Aromatic Hydrocarbons MATRIX: groundwater, soils, sludges water miscible liquid wastes, and non-water miscible wastes.

CAS # : 53-70-3

APPLICATION: This method is used for the analysis of 16 polynuclear aromatic hydrocarbons (PAHs). Samples are extracted, concentrated and analyzed using HPLC with detection by UV and fluorescence detectors.

INTERFERENCES: Solvents, reagents and glassware may introduce artifacts. Other interferences may come from coextracted compounds from samples.

INSTRUMENTATION: HPLC with a gradient pumping system and a 250 mm by 2.6 mm reverse phase HC-ODS Sil-X 5-micron particle size column. The fluorescence detector uses an excitation wavelength of 280 nm and emission greater than 389 nm cutoff with dispersive optics.

RANGE: 0.1 to 425 μg/L MDL: 0.030 μg/L (Fluorescence; reagent water)

PRACTICAL QUANTITATION LIMIT FACTORS FOR MULTIPLYING TIMES FID MDL VALUE

Matrix	Multiplication Factor
Groundwater	10
Low-level soil by sonication with GPC cleanup	670
High-level soil and sludge by sonication	10,000
Non-water miscible waste	100,000

PRECISION: 0.45X + 0.03 μg/L (overall precision)

ACCURACY: 0.41C - 0.11 μg/L (as recovery)

SAMPLING METHOD: Use 8 oz. widemouth glass bottles with Teflon lined caps for concentrated waste samples, soils, sediments and sludges. Use 1 or

2 1/2 gallon amber glass bottles with Teflon lined caps for liquid (water) samples.

STABILITY: Cool soil, sediment, sludge and liquid samples to 4°C. If residual chlorine is present in liquid samples add 3 mL of 10% sodium thiosulfate per gallon of sample and cool to 4°C.

M.H.T. = 14 days for concentrated waste, soil, sediment or sludge.

M.H.T. = 7 days for liquid samples.

All extracts must be analyzed within 40 days.

QUALITY CONTROL: Internal, surrogate and five concentration level calibration standards are used. The calibration standards must be used with the analytical method blank. A quality control check sample concentrate containing dibenzo(a,h)anthracene at 10 µg/mL is required. The QC check sample concentrate may be prepared from pure standard materials or purchased as certified solutions. Use appropriate trip, matrix, control site, method, reagent and solvent blanks.

REFERENCE: Method 8310, SW-846, 3rd ed., Nov 1986.

PRIMARY NAME: Di-n-butyl phthalate — Method 8060

TITLE: Phthalate Esters

MATRIX: groundwater, soils, sludges water miscible liquid wastes, and non-water miscible wastes.

CAS # : 84-74-2

APPLICATION: This method is used for the analysis of 6 phthalate esters. Samples are extracted, concentrated and analyzed using direct injection of both neat and diluted organic liquids into a gas chromatograph. Analytes are detected by a flame ionization detector (FID) or an electron capture detector (ECD). Groundwater samples should be determined by ECD. The method provides an optional GC column which is used for analyte confirmation and that may help resolve analytes from interferences.

INTERFERENCES: Solvents, reagents and glassware may introduce artifacts. Plastics, in particular, must be avoided. Other interferences may come from coextracted compounds from samples. There can be carry-over contamination with high and low level samples.

INSTRUMENTATION: GC capable of on-column injections and a flame ionization detector (FID) or electron capture detector (ECD). Column 1: 1.8 meter by 4 mm with 1.5% SP-2250 / 1.95% SP-2401 on Supelcoport. Column 2: 1.8 meter by 4 mm with 3% OV-1 on supelcoport.

RANGE: 0.7 to 106 μg/L

MDL: 14 μg/L (FID) and 0.36 μg/L (ECD)

PRACTICAL QUANTITATION LIMIT FACTORS FOR MULTIPLYING TIMES FID MDL VALUE

Matrix	Multiplication Factor
Groundwater	10
Low-level soil by sonication with GPC cleanup	670
High-level soil and sludge by sonication	10,000
Non-water miscible waste	100,000

PRECISION: 0.29X + 0.06 μg/L (overall precision using FID)

ACCURACY: 0.79C + 0.17 μg/L (as recovery using FID)

SAMPLING METHOD: Use 8 oz. widemouth glass bottles with Teflon lined caps for concentrated waste samples, soils, sediments and sludges. Use 1 or 2 1/2 gallon amber glass bottles with Teflon lined caps for liquid (water) samples.

STABILITY: Cool soil, sediment, sludge and liquid samples to 4°C. If residual chlorine is present in liquid samples add 3 mL of 10% sodium thiosulfate per gallon of sample and cool to 4°C.

M.H.T. = 14 days for concentrated waste, soil, sediment or sludge.

M.H.T. = 7 days for liquid samples.

All extracts must be analyzed within 40 days.

QUALITY CONTROL: A quality control check sample concentrate containing each analyte of interest is required. The QC check sample concentrate may be prepared from pure standard materials or purchased as certified solutions. Use appropriate trip, matrix, control site, method, reagent and solvent blanks. Internal, surrogate and five concentration level calibration standards are used. The QC check sample concentrate should contain this compound at 25 μg/mL in acetone.

REFERENCE: Method 8060, SW-846, 3rd ed., Nov 1986.

PRIMARY NAME: 1,2-Dichlorobenzene

Method 8120

TITLE: Chlorinated Hydrocarbons

MATRIX: groundwater, soils, sludges water miscible liquid wastes, and non-water miscible wastes.

CAS # : 95-50-1

APPLICATION: This method is used for the analysis of various chlorinated hydrocarbons. Samples are extracted, concentrated and analyzed using direct injection of both neat and diluted organic liquid into a gas chromatograph (GC). The method provides an optional GC column which is used for analyte confirmation and that may help resolve analytes from interferences.

INTERFERENCES: Solvents, reagents and glassware may introduce artifacts. Other interferences may come from coextracted compounds from samples. Phthalate esters are common interferences when using an electron capture detector (ECD) so all plastics must be strictly avoided. Exhaustive cleanup of reagents and glassware may be required to eliminate phthalate contamination.

INSTRUMENTATION: GC capable of on-column injections and an ECD. Column 1: a 1.8 meter by 2 mm with 1% SP-1000 on Supelcoport column. Column 2: a 1.8 meter by 2 mm with 1.5% OV-1 / 2.4% OV-225 on Supelcoport column.

RANGE: 1 to 356 μg/L

MDL: 1.14 μg/L (in reagent water)

PRACTICAL QUANTITATION LIMIT FACTORS FOR MULTIPLYING TIMES FID MDL VALUE

Matrix	Multiplication Factor
Groundwater	10
Low-level soil by sonication with GPC cleanup	670
High-level soil and sludge by sonication	10,000
Non-water miscible waste	100,000

PRECISION: 0.41X - 3.92 μg/L (overall precision)

ACCURACY: 0.85C - 0.70 μg/L (as recovery)

SAMPLING METHOD: Use 8 oz. widemouth glass bottles with Teflon lined caps for concentrated waste samples, soils, sediments and sludges. Use 1 or 2 1/2 gallon amber glass bottles with Teflon lined caps for liquid (water) samples.

STABILITY: Cool soil, sediment, sludge and liquid samples to 4°C. If residual chlorine is present in liquid samples add 3 mL of 10% sodium thiosulfate per gallon of sample and cool to 4°C.

M.H.T. = 14 days for concentrated waste, soil, sediment or sludge.

M.H.T. = 7 days for liquid samples.

All extracts must be analyzed within 40 days.

QUALITY CONTROL: A quality control check sample concentrate containing each analyte of interest is required. The QC check sample concentrate may be prepared from pure standard materials or purchased as certified solutions. Use appropriate trip, matrix, control site, method, reagent and solvent blanks. Internal, surrogate and five concentration level calibration standards are used. The quality control check sample concentrate should contain 1,2-dichlorobenzene at 100 μg/mL in acetone.

REFERENCE: Method 8120, SW-846, 3rd ed., Nov 1986.

PRIMARY NAME: 1,3-Dichlorobenzene — Method 8120

TITLE: Chlorinated Hydrocarbons

MATRIX: groundwater, soils, sludges water miscible liquid wastes, and non-water miscible wastes.

CAS # : 541-73-1

APPLICATION: This method is used for the analysis of various chlorinated hydrocarbons. Samples are extracted, concentrated and analyzed using direct injection of both neat and diluted organic liquid into a gas chromatograph (GC). The method provides an optional GC column which is used for analyte confirmation and that may help resolve analytes from interferences.

INTERFERENCES: Solvents, reagents and glassware may introduce artifacts. Other interferences may come from coextracted compounds from samples. Phthalate esters are common interferences when using an electron capture detector (ECD) so all plastics must be strictly avoided. Exhaustive cleanup of reagents and glassware may be required to eliminate phthalate contamination.

INSTRUMENTATION: GC capable of on-column injections and an ECD. Column 1: a 1.8 meter by 2 mm with 1% SP-1000 on Supelcoport column. Column 2: a 1.8 meter by 2 mm with 1.5% OV-1 / 2.4% OV-225 on Supelcoport column.

RANGE: 1 to 356 μg/L

MDL: 1.19 μg/L (in reagent water)

PRACTICAL QUANTITATION LIMIT FACTORS FOR MULTIPLYING TIMES FID MDL VALUE

Matrix	Multiplication Factor
Groundwater	10
Low-level soil by sonication with GPC cleanup	670
High-level soil and sludge by sonication	10,000
Non-water miscible waste	100,000

PRECISION: 0.49X - 3.98 μg/L (overall precision)

ACCURACY: 0.72C + 0.87 μg/L (as recovery)

SAMPLING METHOD: Use 8 oz. widemouth glass bottles with Teflon lined caps for concentrated waste samples, soils, sediments and sludges. Use 1 or 2 1/2 gallon amber glass bottles with Teflon lined caps for liquid (water) samples.

STABILITY: Cool soil, sediment, sludge and liquid samples to 4°C. If residual chlorine is present in liquid samples add 3 mL of 10% sodium thiosulfate per gallon of sample and cool to 4°C.

M.H.T. = 14 days for concentrated waste, soil, sediment or sludge.

M.H.T. = 7 days for liquid samples.

All extracts must be analyzed within 40 days.

QUALITY CONTROL: A quality control check sample concentrate containing each analyte of interest is required. The QC check sample concentrate may be prepared from pure standard materials or purchased as certified solutions. Use appropriate trip, matrix, control site, method, reagent and solvent blanks. Internal, surrogate and five concentration level calibration standards are used. The quality control check sample concentrate should contain 1,3-dichlorobenzene at 100 µg/mL in acetone.

REFERENCE: Method 8120, SW-846, 3rd ed., Nov 1986.

PRIMARY NAME: 1,4-Dichlorobenzene Method 8120

TITLE: Chlorinated Hydrocarbons

MATRIX: groundwater, soils, sludges water miscible liquid wastes, and non-water miscible wastes.

CAS # : 106-46-7

APPLICATION: This method is used for the analysis of various chlorinated hydrocarbons. Samples are extracted, concentrated and analyzed using direct injection of both neat and diluted organic liquid into a gas chromatograph (GC). The method provides an optional GC column which is used for analyte confirmation and that may help resolve analytes from interferences.

INTERFERENCES: Solvents, reagents and glassware may introduce artifacts. Other interferences may come from coextracted compounds from samples. Phthalate esters are common interferences when using an electron capture detector (ECD) so all plastics must be strictly avoided. Exhaustive cleanup of reagents and glassware may be required to eliminate phthalate contamination.

INSTRUMENTATION: GC capable of on-column injections and an ECD. Column 1: a 1.8 meter by 2 mm with 1% SP-1000 on Supelcoport column. Column 2: a 1.8 meter by 2 mm with 1.5% OV-1 / 2.4% OV-225 on Supelcoport column.

RANGE: 1 to 356 μg/L

MDL: 1.34 μg/L (in reagent water)

PRACTICAL QUANTITATION LIMIT FACTORS FOR MULTIPLYING TIMES FID MDL VALUE

Matrix	Multiplication Factor
Groundwater	10
Low-level soil by sonication with GPC cleanup	670
High-level soil and sludge by sonication	10,000
Non-water miscible waste	100,000

PRECISION: 0.35X - 0.57 μg/L (overall precision)

ACCURACY: 0.72C + 2.80 μg/L (as recovery)

SAMPLING METHOD: Use 8 oz. widemouth glass bottles with Teflon lined caps for concentrated waste samples, soils, sediments and sludges. Use 1 or 2 1/2 gallon amber glass bottles with Teflon lined caps for liquid (water) samples.

STABILITY: Cool soil, sediment, sludge and liquid samples to 4°C. If residual chlorine is present in liquid samples add 3 mL of 10% sodium thiosulfate per gallon of sample and cool to 4°C.

M.H.T. = 14 days for concentrated waste, soil, sediment or sludge.

M.H.T. = 7 days for liquid samples.

All extracts must be analyzed within 40 days.

QUALITY CONTROL: A quality control check sample concentrate containing each analyte of interest is required. The QC check sample concentrate may be prepared from pure standard materials or purchased as certified solutions. Use appropriate trip, matrix, control site, method, reagent and solvent blanks. Internal, surrogate and five concentration level calibration standards are used. The quality control check sample concentrate should contain 1,4-dichlorobenzene at 100 μg/mL in acetone.

REFERENCE: Method 8120, SW-846, 3rd ed., Nov 1986.

PRIMARY NAME: 2,4-Dichlorophenol

Method 8040

TITLE: Phenols

MATRIX: groundwater, soils, sludges water miscible liquid wastes, and non-water miscible wastes.

CAS # : 120-83-2

APPLICATION: This method is used for the analysis of 17 phenols. Samples are extracted, concentrated and analyzed using direct injection of both neat and diluted organic liquids. Pentafluorobenzylbromide (PFB) derivatives also may be made to increase sensitivity of the method.

INTERFERENCES: There can be carry-over contamination with high and low level samples. Solvents, reagents and glassware may introduce artifacts. Other interferences may come from coextracted compounds from samples.

INSTRUMENTATION: GC capable of on-column injections and a flame ionization detector (FID) or electron capture detector (ECD). Column for underivatized phenol: 1.8 meter by 2.0 mm with 1% SP-1240DA on Supelcoport. Column for derivatized phenols: 1.8 meter by 2.0 mm with 5% OV-17 on Chromosorb W-AW-DMCS.

RANGE: 12 to 450 μg/L

MDL: 0.39 μg/L (FID) and 0.68 μg/L (ECD)

PRACTICAL QUANTITATION LIMIT FACTORS FOR MULTIPLYING TIMES FID MDL VALUE

Matrix	Multiplication Factor
Groundwater	10
Low-level soil by sonication with GPC cleanup	670
High-level soil and sludge by sonication	10,000
Non-water miscible waste	100,000

PRECISION: 0.18X + 0.62 μg/L (overall precision using FID)

ACCURACY: 0.81C + 0.48 μg/L (as recovery using FID)

SAMPLING METHOD: Use 8 oz. widemouth glass bottles with Teflon lined

caps for concentrated waste samples, soils, sediments and sludges. Use 1 or 2 1/2 gallon amber glass bottles with Teflon lined caps for liquid (water) samples.

STABILITY: Cool soil, sediment, sludge and liquid samples to 4°C. If residual chlorine is present in liquid samples add 3 mL of 10% sodium thiosulfate per gallon of sample and cool to 4°C.

M.H.T. = 14 days for concentrated waste, soil, sediment or sludge.

M.H.T. = 7 days for liquid samples.

All extracts must be analyzed within 40 days.

QUALITY CONTROL: A quality control check sample concentrate containing each analyte of interest is required. The QC check sample concentrate may be prepared from pure standard materials or purchased as certified solutions. Use appropriate trip, matrix, control site, method, reagent and solvent blanks. Internal, surrogate and five concentration level calibration standards are used. The QC check sample concentrate should contain this compound at 100 μg/mL in 2-propanol.

REFERENCE: Method 8040, SW-846, 3rd ed., Nov 1986.

PRIMARY NAME: Diethyl phthalate

Method 8060

TITLE: Phthalate Esters

MATRIX: groundwater, soils, sludges water miscible liquid wastes, and non-water miscible wastes.

CAS # : 84-66-2

APPLICATION: This method is used for the analysis of 6 phthalate esters. Samples are extracted, concentrated and analyzed using direct injection of both neat and diluted organic liquids into a gas chromatograph. Analytes are detected by a flame ionization detector (FID) or an electron capture detector (ECD). Groundwater samples should be determined by ECD. The method provides an optional GC column which is used for analyte confirmation and that may help resolve analytes from interferences.

INTERFERENCES: Solvents, reagents and glassware may introduce artifacts. Plastics, in particular, must be avoided. Other interferences may come from coextracted compounds from samples. There can be carry-over contamination with high and low level samples.

INSTRUMENTATION: GC capable of on-column injections and a flame ionization detector (FID) or electron capture detector (ECD). Column 1: 1.8 meter by 4 mm with 1.5% SP-2250 / 1.95% SP-2401 on Supelcoport. Column 2: 1.8 meter by 4 mm with 3% OV-1 on supelcoport.

RANGE: 0.7 to 106 µg/L

MDL: 31 µg/L (FID) and 0.49 µg/L (ECD)

PRACTICAL QUANTITATION LIMIT FACTORS FOR MULTIPLYING TIMES FID MDL VALUE

Matrix	Multiplication Factor
Groundwater	10
Low-level soil by sonication with GPC cleanup	670
High-level soil and sludge by sonication	10,000
Non-water miscible waste	100,000

PRECISION: 0.45X + 0.11 µg/L (overall precision using FID)

ACCURACY: 0.70C + 0.13 µg/L (as recovery using FID)

SAMPLING METHOD: Use 8 oz. widemouth glass bottles with Teflon lined caps for concentrated waste samples, soils, sediments and sludges. Use 1 or 2 1/2 gallon amber glass bottles with Teflon lined caps for liquid (water) samples.

STABILITY: Cool soil, sediment, sludge and liquid samples to 4°C. If residual chlorine is present in liquid samples add 3 mL of 10% sodium thiosulfate per gallon of sample and cool to 4°C.

M.H.T. = 14 days for concentrated waste, soil, sediment or sludge.

M.H.T. = 7 days for liquid samples.

All extracts must be analyzed within 40 days.

QUALITY CONTROL: A quality control check sample concentrate containing each analyte of interest is required. The QC check sample concentrate may be prepared from pure standard materials or purchased as certified solutions. Use appropriate trip, matrix, control site, method, reagent and solvent blanks. Internal, surrogate and five concentration level calibration standards are used. The QC check sample concentrate should contain this compound at 25 µg/mL in acetone.

REFERENCE: Method 8060, SW-846, 3rd ed., Nov 1986.

PRIMARY NAME: 2,4-Dimethylphenol

Method 8040

TITLE: Phenols

MATRIX: groundwater, soils, sludges water miscible liquid wastes, and non-water miscible wastes.

CAS # : 105-67-9

APPLICATION: This method is used for the analysis of 17 phenols. Samples are extracted, concentrated and analyzed using direct injection of both neat and diluted organic liquids. Pentafluorobenzylbromide (PFB) derivatives also may be made to increase sensitivity of the method.

INTERFERENCES: There can be carry-over contamination with high and low level samples. Solvents, reagents and glassware may introduce artifacts. Other interferences may come from coextracted compounds from samples.

INSTRUMENTATION: GC capable of on-column injections and a flame ionization detector (FID) or electron capture detector (ECD). Column for underivatized phenol: 1.8 meter by 2.0 mm with 1% SP-1240DA on Supelcoport. Column for derivatized phenols: 1.8 meter by 2.0 mm with 5% OV-17 on Chromosorb W-AW-DMCS.

RANGE: 12 to 450 μg/L

MDL: 0.32 μg/L (FID) and 0.63 μg/L (ECD)

PRACTICAL QUANTITATION LIMIT FACTORS FOR MULTIPLYING TIMES FID MDL VALUE

Matrix	Multiplication Factor
Groundwater	10
Low-level soil by sonication with GPC cleanup	670
High-level soil and sludge by sonication	10,000
Non-water miscible waste	100,000

PRECISION: 0.25X + 0.48 μg/L (overall precision using FID)

ACCURACY: 0.62C - 1.64 μg/L (as recovery using FID)

SAMPLING METHOD: Use 8 oz. widemouth glass bottles with Teflon lined caps for concentrated waste samples, soils, sediments and sludges. Use 1 or 2 1/2 gallon amber glass bottles with Teflon lined caps for liquid (water) samples.

STABILITY: Cool soil, sediment, sludge and liquid samples to 4°C. If residual chlorine is present in liquid samples add 3 mL of 10% sodium thiosulfate per gallon of sample and cool to 4°C.
M.H.T. = 14 days for concentrated waste, soil, sediment or sludge
M.H.T. = 7 days for liquid samples.
All extracts must be analyzed within 40 days.

QUALITY CONTROL: A quality control check sample concentrate containing each analyte of interest is required. The QC check sample concentrate may be prepared from pure standard materials or purchased as certified solutions. Use appropriate trip, matrix, control site, method, reagent and solvent blanks. Internal, surrogate and five concentration level calibration standards are used. The QC check sample concentrate should contain this compound at 100 μg/mL in 2-propanol.

REFERENCE: Method 8040, SW-846, 3rd ed., Nov 1986.

PRIMARY NAME: Dimethyl phthalate

Method 8060

TITLE: Phthalate Esters

MATRIX: groundwater, soils, sludges water miscible liquid wastes, and non-water miscible wastes.

CAS # : 131-11-13

APPLICATION: This method is used for the analysis of 6 phthalate esters. Samples are extracted, concentrated and analyzed using direct injection of both neat and diluted organic liquids into a gas chromatograph. Analytes are detected by a flame ionization detector (FID) or an electron capture detector (ECD). Groundwater samples should be determined by ECD. The method provides an optional GC column which is used for analyte confirmation and that may help resolve analytes from interferences.

INTERFERENCES: Solvents, reagents and glassware may introduce artifacts. Plastics, in particular, must be avoided. Other interferences may come from coextracted compounds from samples. There can be carry-over contamination with high and low level samples.

INSTRUMENTATION: GC capable of on-column injections and a flame ionization detector (FID) or electron capture detector (ECD). Column 1: 1.8 meter by 4 mm with 1.5% SP-2250 / 1.95% SP-2401 on Supelcoport. Column 2: 1.8 meter by 4 mm with 3% OV-1 on supelcoport.

RANGE: 0.7 to 106 μg/L

MDL: 19 mg/L (FID) and 0.29 μg/L (ECD)

PRACTICAL QUANTITATION LIMIT FACTORS FOR MULTIPLYING TIMES FID MDL VALUE

Matrix	Multiplication Factor
Groundwater	10
Low-level soil by sonication with GPC cleanup	670
High-level soil and sludge by sonication	10,000
Non-water miscible waste	100,000

PRECISION: 0.44X + 0.31 μg/L (overall precision using FID)

ACCURACY: 0.73C + 0.17 μg/L (as recovery using FID)

SAMPLING METHOD: Use 8 oz. widemouth glass bottles with Teflon lined caps for concentrated waste samples, soils, sediments and sludges. Use 1 or 2 1/2 gallon amber glass bottles with Teflon lined caps for liquid (water) samples.

STABILITY: Cool soil, sediment, sludge and liquid samples to 4°C. If residual chlorine is present in liquid samples add 3 mL of 10% sodium thiosulfate per gallon of sample and cool to 4°C.

M.H.T. = 14 days for concentrated waste, soil, sediment or sludge.

M.H.T. = 7 days for liquid samples.

All extracts must be analyzed within 40 days.

QUALITY CONTROL: A quality control check sample concentrate containing each analyte of interest is required. The QC check sample concentrate may be prepared from pure standard materials or purchased as certified solutions. Use appropriate trip, matrix, control site, method, reagent and solvent blanks. Internal, surrogate and five concentration level calibration

standards are used. The QC check sample concentrate should contain this compound at 25 µg/mL in acetone.

REFERENCE: Method 8060, SW-846, 3rd ed., Nov 1986.

PRIMARY NAME: 2,4-Dinitrophenol

Method 8040

TITLE: Phenols

MATRIX: groundwater, soils, sludges water miscible liquid wastes, and non-water miscible wastes.

CAS # : 51-28-5

APPLICATION: This method is used for the analysis of 17 phenols. Samples are extracted, concentrated and analyzed using direct injection of both neat and diluted organic liquids. Pentafluorobenzylbromide (PFB) derivatives also may be made to increase sensitivity of the method.

INTERFERENCES: There can be carry-over contamination with high and low level samples. Solvents, reagents and glassware may introduce artifacts. Other interferences may come from coextracted compounds from samples.

INSTRUMENTATION: GC capable of on-column injections and a flame ionization detector (FID) or electron capture detector (ECD). Column for underivatized phenol: 1.8 meter by 2.0 mm with 1% SP-1240DA on Supelcoport. Column for derivatized phenols: 1.8 meter by 2.0 mm with 5% OV-17 on Chromosorb W-AW-DMCS.

RANGE: 12 to 450 µg/L

MDL: 13.0 µg/L (FID)

PRACTICAL QUANTITATION LIMIT FACTORS FOR MULTIPLYING TIMES FID MDL VALUE

Matrix	Multiplication Factor
Groundwater	10
Low-level soil by sonication with GPC cleanup	670
High-level soil and sludge by sonication	10,000
Non-water miscible waste	100,000

PRECISION: 0.29X + 4.51 µg/L (overall precision using FID)

ACCURACY: 0.80C - 1.58 µg/L (as recovery using FID)

SAMPLING METHOD: Use 8 oz. widemouth glass bottles with Teflon lined caps for concentrated waste samples, soils, sediments and sludges. Use 1 or 2 1/2 gallon amber glass bottles with Teflon lined caps for liquid (water) samples.

STABILITY: Cool soil, sediment, sludge and liquid samples to 4°C. If residual chlorine is present in liquid samples add 3 mL of 10% sodium thiosulfate per gallon of sample and cool to 4°C.

M.H.T. = 14 days for concentrated waste, soil, sediment or sludge.

M.H.T. = 7 days for liquid samples.

All extracts must be analyzed within 40 days.

QUALITY CONTROL: A quality control check sample concentrate containing each analyte of interest is required. The QC check sample concentrate may be prepared from pure standard materials or purchased as certified solutions. Use appropriate trip, matrix, control site, method, reagent and solvent blanks. Internal, surrogate and five concentration level calibration standards are used. The QC check sample concentrate should contain this compound at 100 µg/mL in 2-propanol.

REFERENCE: Method 8040, SW-846, 3rd ed., Nov 1986.

PRIMARY NAME: 2,4-Dinitrotoluene — Method 8090

TITLE: Nitroaromatics and Cyclic Ketones

MATRIX: groundwater, soils, sludges water miscible liquid wastes, and non-water miscible wastes.

CAS # : 121-14-2

APPLICATION: This method is used for the analysis of various nitroaromatic and cyclic ketone compounds. Samples are extracted, concentrated and analyzed using direct injection of both neat and diluted organic liquids. Dinitrotoluenes are determined using ECD and the other compounds amenable to this method are determined using FID. The method provides an optional GC column which is used for analyte confirmation and that may help resolve analytes from interferences.

INTERFERENCES: Solvents, reagents and glassware may introduce artifacts. Other interferences may come from coextracted compounds from samples.

INSTRUMENTATION: GC capable of on-column injections and a flame ionization detector (FID) or electron capture detector (ECD). Column 1: a 1.2 meter by 2 mm or 4 mm with 1.95% QF-1/1.5% OV-17 on Gas-Chrom Q. Column 2: a 3 meter by 2 mm or 4 mm with 3% OV-101 on Gas-Chrom Q.

RANGE: 1 to 515 µg/L MDL: 0.02 µg/L (ECD)

PRACTICAL QUANTITATION LIMIT FACTORS FOR MULTIPLYING TIMES FID MDL VALUE:

Matrix	Multiplication Factor
Groundwater	10
Low-level soil by sonication with GPC cleanup	670
High-level soil and sludge by sonication	10,000
Non-water miscible waste	100,000

PRECISION: 0.37X - 0.07 µg/L (overall precision)

ACCURACY: 0.65C + 0.22 µg/L (as recovery)

SAMPLING METHOD: Use 8 oz. widemouth glass bottles with Teflon lined caps for concentrated waste samples, soils, sediments and sludges. Use 1 or 2 1/2 gallon amber glass bottles with Teflon lined caps for liquid (water) samples.

STABILITY: Cool soil, sediment, sludge and liquid samples to 4°C. If residual chlorine is present in liquid samples add 3 mL of 10% sodium thiosulfate per gallon of sample and cool to 4°C.

M.H.T. = 14 days for concentrated waste, soil, sediment or sludge.

M.H.T. = 7 days for liquid samples.

All extracts must be analyzed within 40 days.

QUALITY CONTROL: A quality control check sample concentrate containing each analyte of interest is required. The QC check sample concentrate may be prepared from pure standard materials or purchased as certified solutions. Use appropriate trip, matrix, control site, method, reagent and solvent blanks. Internal, surrogate and five concentration level calibration standards are used. The QC check sample concentrate should contain this compound at 20 µg/mL in acetone.

REFERENCE: Method 8090, SW-846, 3rd ed., Nov 1986.

PRIMARY NAME: 2,6-Dinitrotoluene

Method 8090

TITLE: Nitroaromatics and Cyclic Ketones

MATRIX: groundwater, soils, sludges water miscible liquid wastes, and non-water miscible wastes.

CAS # : 606-20-2

APPLICATION: This method is used for the analysis of various nitroaromatic and cyclic ketone compounds. Samples are extracted, concentrated and analyzed using direct injection of both neat and diluted organic liquids. Dinitrotoluenes are determined using ECD and the other compounds amenable to this method are determined using FID. The method provides an optional GC column which is used for analyte confirmation and that may help resolve analytes from interferences.

INTERFERENCES: Solvents, reagents and glassware may introduce artifacts. Other interferences may come from coextracted compounds from samples.

INSTRUMENTATION: GC capable of on-column injections and a flame ionization detector (FID) or electron capture detector (ECD). Column 1: a 1.2 meter by 2 mm or 4 mm with 1.95% QF-1/1.5% OV-17 on Gas-Chrom Q. Column 2: a 3 meter by 2 mm or 4 mm with 3% OV-101 on Gas-Chrom Q.

RANGE: 1 to 515 μg/L

MDL: 0.01 μg/L (ECD)

PRACTICAL QUANTITATION LIMIT FACTORS FOR MULTIPLYING TIMES FID MDL VALUE

Matrix	Multiplication Factor
Groundwater	10
Low-level soil by sonication with GPC cleanup	670
High-level soil and sludge by sonication	10,000
Non-water miscible waste	100,000

PRECISION: 0.36X - 0.00 μg/L (overall precision)

ACCURACY: 0.66C + 0.20 μg/L (as recovery)

SAMPLING METHOD: Use 8 oz. widemouth glass bottles with Teflon lined caps for concentrated waste samples, soils, sediments and sludges. Use 1 or 2 1/2 gallon amber glass bottles with Teflon lined caps for liquid (water) samples.

STABILITY: Cool soil, sediment, sludge and liquid samples to 4°C. If residual chlorine is present in liquid samples add 3 mL of 10% sodium thiosulfate per gallon of sample and cool to 4°C.

M.H.T. = 14 days for concentrated waste, soil, sediment or sludge.

M.H.T. = 7 days for liquid samples.

All extracts must be analyzed within 40 days.

QUALITY CONTROL: A quality control check sample concentrate containing each analyte of interest is required. The QC check sample concentrate may be prepared from pure standard materials or purchased as certified solutions. Use appropriate trip, matrix, control site, method, reagent and solvent blanks. Internal, surrogate and five concentration level calibration standards are used. The QC check sample concentrate should contain this compound at 20 μg/mL in acetone.

REFERENCE: Method 8090, SW-846, 3rd ed., Nov 1986.

PRIMARY NAME: Di-n-octyl phthalate — Method 8060

TITLE: Phthalate Esters

MATRIX: groundwater, soils, sludges water miscible liquid wastes, and non-water miscible wastes.

CAS # : 117-84-0

APPLICATION: This method is used for the analysis of 6 phthalate esters. Samples are extracted, concentrated and analyzed using direct injection of both neat and diluted organic liquids into a gas chromatograph. Analytes are detected by a flame ionization detector (FID) or an electron capture detector (ECD). Groundwater samples should be determined by ECD. The method provides an optional GC column which is used for analyte confirmation and that may help resolve analytes from interferences.

INTERFERENCES: Solvents, reagents and glassware may introduce artifacts. Plastics, in particular, must be avoided. Other interferences may come from coextracted compounds from samples. There can be carry-over contamination with high and low level samples.

INSTRUMENTATION: GC capable of on-column injections and a flame ionization detector (FID) or electron capture detector (ECD). Column 1: 1.8 meter by 4 mm with 1.5% SP-2250 / 1.95% SP-2401 on Supelcoport. Column 2: 1.8 meter by 4 mm with 3% OV-1 on supelcoport.

RANGE: 0.7 to 106 μg/L

MDL: 31 μg/L (FID) and 3.0 μg/L (ECD)

PRACTICAL QUANTITATION LIMIT FACTORS FOR MULTIPLYING TIMES FID MDL VALUE

Matrix	Multiplication Factor
Groundwater	10
Low-level soil by sonication with GPC cleanup	670
High-level soil and sludge by sonication	10,000
Non-water miscible waste	100,000

PRECISION: 0.62X + 0.34 μg/L (overall precision using FID)

ACCURACY: 0.35C - 0.71 μg/L (as recovery using FID)

SAMPLING METHOD: Use 8 oz. widemouth glass bottles with Teflon lined caps for concentrated waste samples, soils, sediments and sludges. Use 1 or 2 1/2 gallon amber glass bottles with Teflon lined caps for liquid (water) samples.

STABILITY: Cool soil, sediment, sludge and liquid samples to 4°C. If residual chlorine is present in liquid samples add 3 mL of 10% sodium thiosulfate per gallon of sample and cool to 4°C.

M.H.T. = 14 days for concentrated waste, soil, sediment or sludge.

M.H.T. = 7 days for liquid samples.

All extracts must be analyzed within 40 days.

QUALITY CONTROL: A quality control check sample concentrate containing each analyte of interest is required. The QC check sample concentrate may be prepared from pure standard materials or purchased as certified solutions. Use appropriate trip, matrix, control site, method, reagent and solvent blanks. Internal, surrogate and five concentration level calibration standards are used. The QC check sample concentrate should contain this compound at 50 µg/mL in acetone.

REFERENCE: Method 8060, SW-846, 3rd ed., Nov 1986.

PRIMARY NAME: bis(2-Ethylhexyl) phthalate Method 8060

TITLE: Phthalate Esters

MATRIX: groundwater, soils, sludges water miscible liquid wastes, and non-water miscible wastes.

CAS # : 117-81-7

APPLICATION: This method is used for the analysis of 6 phthalate esters. Samples are extracted, concentrated and analyzed using direct injection of both neat and diluted organic liquids into a gas chromatograph. Analytes are detected by a flame ionization detector (FID) or an electron capture detector (ECD). Groundwater samples should be determined by ECD. The method provides an optional GC column which is used for analyte confirmation and that may help resolve analytes from interferences.

INTERFERENCES: Solvents, reagents and glassware may introduce artifacts. Plastics, in particular, must be avoided. Other interferences may come from coextracted compounds from samples. There can be carry-over contamination with high and low level samples.

INSTRUMENTATION: GC capable of on-column injections and a flame ionization detector (FID) or electron capture detector (ECD). Column 1: 1.8 meter by 4 mm with 1.5% SP-2250 / 1.95% SP-2401 on Supelcoport. Column 2: 1.8 meter by 4 mm with 3% OV-1 on supelcoport.

RANGE: 0.7 to 106 μg/L

MDL: 20 μg/L (FID) and 2.0 μg/L (ECD)

PRACTICAL QUANTITATION LIMIT FACTORS FOR MULTIPLYING TIMES FID MDL VALUE

Matrix	Multiplication Factor
Groundwater	10
Low-level soil by sonication with GPC cleanup	670
High-level soil and sludge by sonication	10,000
Non-water miscible waste	100,000

PRECISION: 0.73X - 0.17 μg/L (overall precision using FID)

ACCURACY: 0.53C + 2.02 μg/L (as recovery using FID)

SAMPLING METHOD: Use 8 oz. widemouth glass bottles with Teflon lined caps for concentrated waste samples, soils, sediments and sludges. Use 1 or 2 1/2 gallon amber glass bottles with Teflon lined caps for liquid (water) samples.

STABILITY: Cool soil, sediment, sludge and liquid samples to 4°C. If residual chlorine is present in liquid samples add 3 mL of 10% sodium thiosulfate per gallon of sample and cool to 4°C.

M.H.T. = 14 days for concentrated waste, soil, sediment or sludge.
M.H.T. = 7 days for liquid samples.

All extracts must be analyzed within 40 days.

QUALITY CONTROL: A quality control check sample concentrate containing each analyte of interest is required. The QC check sample concentrate may be prepared from pure standard materials or purchased as certified solutions. Use appropriate trip, matrix, control site, method, reagent and solvent blanks. Internal, surrogate and five concentration level calibration standards are used. The QC check sample concentrate should contain this compound at 50 μg/mL in acetone.

REFERENCE: Method 8060, SW-846, 3rd ed., Nov 1986.

PRIMARY NAME: Fluoranthene Method 8100

TITLE: Polynuclear Aromatic Hydrocarbons

MATRIX: groundwater, soils, sludges water miscible liquid wastes, and non-water miscible wastes.

CAS # : 206-44-0

APPLICATION: This method is used for the analysis of various PAHs. Samples are extracted, concentrated and analyzed using direct injection of both neat and diluted organic liquids. The method provides two optional GC columns that are better than Column 1 and that may help resolve analytes from interferences.

INTERFERENCES: Solvents, reagents and glassware may introduce artifacts. Other interferences may come from coextracted compounds from samples.

INSTRUMENTATION: GC capable of on-column injections and a flame ionization detector (FID). Column 1: a 1.8 meter by 2 mm 3% OV-17 on Chromosorb W-AW-DCMS column. Column 2: a 30 meter by 0.25 mm SE-54 fused silica capillary colunm. Column 3: a 30 meter by 0.32 mm SE-54 fused silica capillary column.

RANGE: 0.1 to 425 µg/L

MDL: Not reported

PRACTICAL QUANTITATION LIMIT FACTORS FOR MULTIPLYING TIMES FID MDL VALUE

Not available

PRECISION: 0.32X + 0.03 µg/L (overall precision)

ACCURACY: 0.68C + 0.07 µg/L (as recovery)

SAMPLING METHOD: Use 8 oz. widemouth glass bottles with Teflon lined caps for concentrated waste samples, soils, sediments and sludges. Use 1 or 2 1/2 gallon amber glass bottles with Teflon lined caps for liquid (water) samples.

STABILITY: Cool soil, sediment, sludge and liquid samples to 4°C. If residual chlorine is present in liquid samples add 3 mL of 10% sodium thiosulfate per gallon of sample and cool to 4°C.

M.H.T. = 14 days for concentrated waste, soil, sediment or sludge.

M.H.T. = 7 days for liquid samples.

All extracts must be analyzed within 40 days.

QUALITY CONTROL: A quality control check sample concentrate containing each analyte of interest is required. The QC check sample concentrate may be prepared from pure standard materials or purchased as certified solutions. Use appropriate trip, matrix, control site, method, reagent and solvent blanks. Internal, surrogate and five concentration level calibration standards are used. The quality control check sample concentrate should contain fluoranthene at 10 µg/mL in acetonitrile.

REFERENCE: Method 8100, SW-846, 3rd ed., Nov 1986.

PRIMARY NAME: Fluoranthene

Method 8310

TITLE: Polynuclear Aromatic Hydrocarbons

MATRIX: groundwater, soils, sludges water miscible liquid wastes, and non-water miscible wastes.

CAS # : 206-44-0

APPLICATION: This method is used for the analysis of 16 polynuclear aromatic hydrocarbons (PAHs). Samples are extracted, concentrated and analyzed using HPLC with detection by UV and fluorescence detectors.

INTERFERENCES: Solvents, reagents and glassware may introduce artifacts. Other interferences may come from coextracted compounds from samples.

INSTRUMENTATION: HPLC with a gradient pumping system and a 250 mm by 2.6 mm reverse phase HC-ODS Sil-X 5-micron particle size column. The fluorescence detector uses an excitation wavelength of 280 nm and emission greater than 389 nm cutoff with dispersive optics.

RANGE: 0.1 to 425 µg/L

MDL: 0.21 µg/L (Fluorescence; reagent water)

PRACTICAL QUANTITATION LIMIT FACTORS FOR MULTIPLYING TIMES FID MDL VALUE:

Matrix	Multiplication Factor
Groundwater	10
Low-level soil by sonication with GPC cleanup	670
High-level soil and sludge by sonication	10,000
Non-water miscible waste	100,000

PRECISION: 0.32X + 0.03 μg/L (overall precision)

ACCURACY: 0.68C + 0.07 μg/L (as recovery)

SAMPLING METHOD: Use 8 oz. widemouth glass bottles with Teflon lined caps for concentrated waste samples, soils, sediments and sludges. Use 1 or 2 1/2 gallon amber glass bottles with Teflon lined caps for liquid (water) samples.

STABILITY: Cool soil, sediment, sludge and liquid samples to 4°C. If residual chlorine is present in liquid samples add 3 mL of 10% sodium thiosulfate per gallon of sample and cool to 4°C.

M.H.T. = 14 days for concentrated waste, soil, sediment or sludge.

M.H.T. = 7 days for liquid samples.

All extracts must be analyzed within 40 days.

QUALITY CONTROL: Internal, surrogate and five concentration level calibration standards are used. The calibration standards must be used with the analytical method blank. A quality control check sample concentrate containing fluoranthene at 10 μg/mL is required. The QC check sample concentrate may be prepared from pure standard materials or purchased as certified solutions. Use appropriate trip, matrix, control site, method, reagent and solvent blanks.

REFERENCE: Method 8310, SW-846, 3rd ed., Nov 1986.

PRIMARY NAME: Fluorene — Method 8100

TITLE: Polynuclear Aromatic Hydrocarbons

MATRIX: groundwater, soils, sludges water miscible liquid wastes, and non-water miscible wastes.

CAS # : 86-73-7

APPLICATION: This method is used for the analysis of various PAHs. Samples are extracted, concentrated and analyzed using direct injection of both neat and diluted organic liquids. The method provides two optional GC columns that are better than Column 1 and that may help resolve analytes from interferences.

INTERFERENCES: Solvents, reagents and glassware may introduce artifacts. Other interferences may come from coextracted compounds from samples.

INSTRUMENTATION: GC capable of on-column injections and a flame ionization detector (FID). Column 1: a 1.8 meter by 2 mm 3% OV-17 on Chromosorb W-AW-DCMS column. Column 2: a 30 meter by 0.25 mm SE-54 fused silica capillary colunm. Column 3: a 30 meter by 0.32 mm SE-54 fused silica capillary column.

RANGE: 0.1 to 425 mg/L MDL: Not reported

PRACTICAL QUANTITATION LIMIT FACTORS FOR MULTIPLYING TIMES FID MDL VALUE

Not available

PRECISION: 0.63X - 0.65 μg/L (overall precision)

ACCURACY: 0.56C - 0.52 μg/L (as recovery)

SAMPLING METHOD: Use 8 oz. widemouth glass bottles with Teflon lined caps for concentrated waste samples, soils, sediments and sludges. Use 1 or 2 1/2 gallon amber glass bottles with Teflon lined caps for liquid (water) samples.

STABILITY: Cool soil, sediment, sludge and liquid samples to 4°C. If residual chlorine is present in liquid samples add 3 mL of 10% sodium thiosulfate per gallon of sample and cool to 4°C.

M.H.T. = 14 days for concentrated waste, soil, sediment or sludge.

M.H.T. = 7 days for liquid samples.

All extracts must be analyzed within 40 days.

QUALITY CONTROL: A quality control check sample concentrate containing each analyte of interest is required. The QC check sample concentrate may be prepared from pure standard materials or purchased as certified solutions. Use appropriate trip, matrix, control site, method, reagent and solvent blanks. Internal, surrogate and five concentration level calibration

standards are used. The quality control check sample concentrate should contain fluorene at 100 μg/mL in acetonitrile.

REFERENCE: Method 8100, SW-846, 3rd ed., Nov 1986.

PRIMARY NAME: Fluorene — Method 8310

TITLE: Polynuclear Aromatic Hydrocarbons — MATRIX: groundwater, soils, sludges water miscible liquid wastes, and non-water miscible wastes.

CAS # : 86-73-7

APPLICATION: This method is used for the analysis of 16 polynuclear aromatic hydrocarbons (PAHs). Samples are extracted, concentrated and analyzed using HPLC with detection by UV and fluorescence detectors.

INTERFERENCES: Solvents, reagents and glassware may introduce artifacts. Other interferences may come from coextracted compounds from samples.

INSTRUMENTATION: HPLC with a gradient pumping system and a 250 mm by 2.6 mm reverse phase HC-ODS Sil-X 5-micron particle size column. The UV detector uses an excitation wavelength of 254 nm coupled to the fluorescence detector. The fluorescence detector uses an excitation wavelength of 280 nm and emission greater than 389 nm cutoff with dispersive optics.

RANGE: 0.1 to 425 μg/L — MDL: 0.21 μg/L (UV detector; reagent water)

PRACTICAL QUANTITATION LIMIT FACTORS FOR MULTIPLYING TIMES FID MDL VALUE

Matrix	Multiplication Factor
Groundwater	10
Low-level soil by sonication with GPC cleanup	670
High-level soil and sludge by sonication	10,000
Non-water miscible waste	100,000

PRECISION: 0.63X - 0.65 μg/L (overall precision)

ACCURACY: 0.56C - 0.52 μg/L (as recovery)

SAMPLING METHOD: Use 8 oz. widemouth glass bottles with Teflon lined caps for concentrated waste samples, soils, sediments and sludges. Use 1 or 2 1/2 gallon amber glass bottles with Teflon lined caps for liquid (water) samples.

STABILITY: Cool soil, sediment, sludge and liquid samples to 4°C. If residual chlorine is present in liquid samples add 3 mL of 10% sodium thiosulfate per gallon of sample and cool to 4°C.

M.H.T. = 14 days for concentrated waste, soil, sediment or sludge.

M.H.T. = 7 days for liquid samples.

All extracts must be analyzed within 40 days.

QUALITY CONTROL: Internal, surrogate and five concentration level calibration standards are used. The calibration standards must be used with the analytical method blank. A quality control check sample concentrate containing fluorene at 100 μg/mL is required. The QC check sample concentrate may be prepared from pure standard materials or purchased as certified solutions. Use appropriate trip, matrix, control site, method, reagent and solvent blanks.

REFERENCE: Method 8310, SW-846, 3rd ed., Nov 1986.

PRIMARY NAME: Hexachlorobenzene — Method 8120

TITLE: Chlorinated Hydrocarbons

MATRIX: groundwater, soils, sludges water miscible liquid wastes, and non-water miscible wastes.

CAS # : 118-74-1

APPLICATION: This method is used for the analysis of various chlorinated hydrocarbons. Samples are extracted, concentrated and analyzed using direct injection of both neat and diluted organic liquid into a gas chromatograph (GC). The method provides an optional GC column which is used for analyte confirmation and that may help resolve analytes from interferences.

INTERFERENCES: Solvents, reagents and glassware may introduce artifacts. Other interferences may come from coextracted compounds from samples. Phthalate esters are common interferences when using an electron capture detector (ECD) so all plastics must be strictly avoided. Exhaustive cleanup of reagents and glassware may be required to eliminate phthalate contamination.

INSTRUMENTATION: GC capable of on-column injections and an ECD. Column 1: a 1.8 meter by 2 mm with 1% SP-1000 on Supelcoport column. Column 2: a 1.8 meter by 2 mm with 1.5% OV-1 / 2.4% OV-225 on Supelcoport column.

RANGE: 1 to 356 μg/L

MDL: 0.05 μg/L (in reagent water)

PRACTICAL QUANTITATION LIMIT FACTORS FOR MULTIPLYING TIMES FID MDL VALUE

Matrix	Multiplication Factor
Groundwater	10
Low-level soil by sonication with GPC cleanup	670
High-level soil and sludge by sonication	10,000
Non-water miscible waste	100,000

PRECISION: 0.36X - 0.19 μg/L (overall precision)

ACCURACY: 0.87C - 0.02 μg/L (as recovery)

SAMPLING METHOD: Use 8 oz. widemouth glass bottles with Teflon lined caps for concentrated waste samples, soils, sediments and sludges. Use 1 or 2 1/2 gallon amber glass bottles with Teflon lined caps for liquid (water) samples.

STABILITY: Cool soil, sediment, sludge and liquid samples to 4°C. If residual chlorine is present in liquid samples add 3 mL of 10% sodium thiosulfate per gallon of sample and cool to 4°C.

M.H.T. = 14 days for concentrated waste, soil, sediment or sludge.

M.H.T. = 7 days for liquid samples.

All extracts must be analyzed within 40 days.

QUALITY CONTROL: A quality control check sample concentrate containing each analyte of interest is required. The QC check sample concentrate may be prepared from pure standard materials or purchased as certified solutions. Use appropriate trip, matrix, control site, method, reagent and solvent blanks. Internal, surrogate and five concentration level calibration standards are used. The quality control check sample concentrate should contain hexachlorobenzene at 10 μg/mL in acetone.

REFERENCE: Method 8120, SW-846, 3rd ed., Nov 1986.

PRIMARY NAME: Hexachlorobutadiene

Method 8120

TITLE: Chlorinated Hydrocarbons

MATRIX: groundwater, soils, sludges water miscible liquid wastes, and non-water miscible wastes

CAS # : 87-68-3

APPLICATION: This method is used for the analysis of various chlorinated hydrocarbons. Samples are extracted, concentrated and analyzed using direct injection of both neat and diluted organic liquid into a gas chromatograph (GC). The method provides an optional GC column which is used for analyte confirmation and that may help resolve analytes from interferences.

INTERFERENCES: Solvents, reagents and glassware may introduce artifacts. Other interferences may come from coextracted compounds from samples. Phthalate esters are common interferences when using an electron capture detector (ECD) so all plastics must be strictly avoided. Exhaustive cleanup of reagents and glassware may be required to eliminate phthalate contamination.

INSTRUMENTATION: GC capable of on-column injections and an ECD. Column 1: a 1.8 meter by 2 mm with 1% SP-1000 on Supelcoport column. Column 2: a 1.8 meter by 2 mm with 1.5% OV-1 / 2.4% OV-225 on Supelcoport column.

RANGE: 1 to 356 μg/L

MDL: 0.34 μg/L (in reagent water)

PRACTICAL QUANTITATION LIMIT FACTORS FOR MULTIPLYING TIMES FID MDL VALUE

Matrix	Multiplication Factor
Groundwater	10
Low-level soil by sonication with GPC cleanup	670
High-level soil and sludge by sonication	10,000
Non-water miscible waste	100,000

PRECISION: 0.53X - 0.12 μg/L (overall precision)

ACCURACY: 0.61C + 0.03 μg/L (as recovery)

SAMPLING METHOD: Use 8 oz. widemouth glass bottles with Teflon lined caps for concentrated waste samples, soils, sediments and sludges. Use 1 or 2 1/2 gallon amber glass bottles with Teflon lined caps for liquid (water) samples.

STABILITY: Cool soil, sediment, sludge and liquid samples to 4°C. If residual chlorine is present in liquid samples add 3 mL of 10% sodium thiosulfate per gallon of sample and cool to 4°C.

M.H.T. = 14 days for concentrated waste, soil, sediment or sludge.
M.H.T. = 7 days for liquid samples.

All extracts must be analyzed within 40 days.

QUALITY CONTROL: A quality control check sample concentrate containing each analyte of interest is required. The QC check sample concentrate may be prepared from pure standard materials or purchased as certified solutions. Use appropriate trip, matrix, control site, method, reagent and solvent blanks. Internal, surrogate and five concentration level calibration standards are used. The quality control check sample concentrate should contain hexachlorobutadiene at 10 μg/mL in acetone.

REFERENCE: Method 8120, SW-846, 3rd ed., Nov 1986.

PRIMARY NAME: Hexachlorocyclopentadiene Method 8120

TITLE: Chlorinated Hydrocarbons

MATRIX: groundwater, soils, sludges water miscible liquid wastes, and non-water miscible wastes.

CAS # : 77-47-4

APPLICATION: This method is used for the analysis of various chlorinated hydrocarbons. Samples are extracted, concentrated and analyzed using direct injection of both neat and diluted organic liquid into a gas chromatograph (GC). The method provides an optional GC column which is used for analyte confirmation and that may help resolve analytes from interferences.

INTERFERENCES: Solvents, reagents and glassware may introduce artifacts. Other interferences may come from coextracted compounds from samples. Phthalate esters are common interferences when using an electron capture detector (ECD) so all plastics must be strictly avoided. Exhaustive cleanup of reagents and glassware may be required to eliminate phthalate contamination.

INSTRUMENTATION: GC capable of on-column injections and an ECD. Column 1: a 1.8 meter by 2 mm with 1% SP-1000 on Supelcoport column. Column 2: a 1.8 meter by 2 mm with 1.5% OV-1 / 2.4% OV-225 on Supelcoport column.

RANGE: 1 to 356 μg/L

MDL: 0.40 μg/L (in reagent water)

PRACTICAL QUANTITATION LIMIT FACTORS FOR MULTIPLYING TIMES FID MDL VALUE

Matrix	Multiplication Factor
Groundwater	10
Low-level soil by sonication with GPC cleanup	670
High-level soil and sludge by sonication	10,000
Non-water miscible waste	100,000

PRECISION: 0.50X μg/L (overall precision)

ACCURACY: 0.47C μg/L (as recovery)

SAMPLING METHOD: Use 8 oz. widemouth glass bottles with Teflon lined caps for concentrated waste samples, soils, sediments and sludges. Use 1 or 2 1/2 gallon amber glass bottles with Teflon lined caps for liquid (water) samples.

STABILITY: Cool soil, sediment, sludge and liquid samples to 4°C. If residual chlorine is present in liquid samples add 3 mL of 10% sodium thiosulfate per gallon of sample and cool to 4°C.

M.H.T. = 14 days for concentrated waste, soil, sediment or sludge.

M.H.T. = 7 days for liquid samples.

All extracts must be analyzed within 40 days.

QUALITY CONTROL: A quality control check sample concentrate containing each analyte of interest is required. The QC check sample concentrate may be prepared from pure standard materials or purchased as certified solutions. Use appropriate trip, matrix, control site, method, reagent and solvent blanks. Internal, surrogate and five concentration level calibration standards are used. The quality control check sample concentrate should contain hexachlorocyclopentadiene at 10 μg/mL in acetone.

REFERENCE: Method 8120, SW-846, 3rd ed., Nov 1986.

PRIMARY NAME: Hexachloroethane — Method 8120

TITLE: Chlorinated Hydrocarbons

MATRIX: groundwater, soils, sludges water miscible liquid wastes, and non-water miscible wastes.

CAS # : 67-72-1

APPLICATION: This method is used for the analysis of various chlorinated hydrocarbons. Samples are extracted, concentrated and analyzed using direct injection of both neat and diluted organic liquid into a gas chromatograph (GC). The method provides an optional GC column which is used for analyte confirmation and that may help resolve analytes from interferences.

INTERFERENCES: Solvents, reagents and glassware may introduce artifacts. Other interferences may come from coextracted compounds from samples. Phthalate esters are common interferences when using an electron capture detector (ECD) so all plastics must be strictly avoided. Exhaustive cleanup of reagents and glassware may be required to eliminate phthalate contamination.

INSTRUMENTATION: GC capable of on-column injections and an ECD. Column 1: a 1.8 meter by 2 mm with 1% SP-1000 on Supelcoport column. Column 2: a 1.8 meter by 2 mm with 1.5% OV-1 / 2.4% OV-225 on Supelcoport column.

RANGE: 1 to 356 μg/L

MDL: 0.03 μg/L (in reagent water)

PRACTICAL QUANTITATION LIMIT FACTORS FOR MULTIPLYING TIMES FID MDL VALUE

Matrix	Multiplication Factor
Groundwater	10
Low-level soil by sonication with GPC cleanup	670
High-level soil and sludge by sonication	10,000
Non-water miscible waste	100,000

PRECISION: 0.36X - 0.00 μg/L (overall precision)

ACCURACY: 0.74C - 0.02 μg/L (as recovery)

SAMPLING METHOD: Use 8 oz. widemouth glass bottles with Teflon lined caps for concentrated waste samples, soils, sediments and sludges. Use 1 or 2 1/2 gallon amber glass bottles with Teflon lined caps for liquid (water) samples.

STABILITY: Cool soil, sediment, sludge and liquid samples to 4°C. If residual chlorine is present in liquid samples add 3 mL of 10% sodium thiosulfate per gallon of sample and cool to 4°C.

M.H.T. = 14 days for concentrated waste, soil, sediment or sludge.

M.H.T. = 7 days for liquid samples.

All extracts must be analyzed within 40 days.

QUALITY CONTROL: A quality control check sample concentrate containing each analyte of interest is required. The QC check sample concentrate may be prepared from pure standard materials or purchased as certified solutions. Use appropriate trip, matrix, control site, method, reagent and solvent blanks. Internal, surrogate and five concentration level calibration standards are used. The quality control check sample concentrate should contain hexachloroethane at 10 µg/mL in acetone.

REFERENCE: Method 8120, SW-846, 3rd ed., Nov 1986.

PRIMARY NAME: Indeno(1,2,3-cd)pyrene Method 8100

TITLE: Polynuclear Aromatic Hydrocarbons MATRIX: groundwater, soils, sludges water miscible liquid wastes, and non-water miscible wastes.

CAS # : 193-39-5

APPLICATION: This method is used for the analysis of various PAHs. Samples are extracted, concentrated and analyzed using direct injection of both neat and diluted organic liquids. The method provides two optional GC columns that are better than Column 1 and that may help resolve analytes from interferences.

INTERFERENCES: Solvents, reagents and glassware may introduce artifacts. Other interferences may come from coextracted compounds from samples.

INSTRUMENTATION: GC capable of on-column injections and a flame ionization detector (FID). Column 1: a 1.8 meter by 2 mm 3% OV-17 on Chromosorb W-AW-DCMS column. Column 2: a 30 meter by 0.25 mm SE-54 fused silica capillary colunm. Column 3: a 30 meter by 0.32 mm SE-54 fused silica capillary column.

RANGE: 0.1 to 425 µg/L MDL: Not reported

PRACTICAL QUANTITATION LIMIT FACTORS FOR MULTIPLYING TIMES FID MDL VALUE

Not available

PRECISION: 0.42X + 0.01 µg/L (overall precision)

ACCURACY: 0.54C + 0.06 µg/L (as recovery)

SAMPLING METHOD: Use 8 oz. widemouth glass bottles with Teflon lined caps for concentrated waste samples, soils, sediments and sludges. Use 1 or 2 1/2 gallon amber glass bottles with Teflon lined caps for liquid (water) samples.

STABILITY: Cool soil, sediment, sludge and liquid samples to 4°C. If residual chlorine is present in liquid samples add 3 mL of 10% sodium thiosulfate per gallon of sample and cool to 4°C.

M.H.T. = 14 days for concentrated waste, soil, sediment or sludge.

M.H.T. = 7 days for liquid samples.

All extracts must be analyzed within 40 days.

QUALITY CONTROL: A quality control check sample concentrate containing each analyte of interest is required. The QC check sample concentrate may be prepared from pure standard materials or purchased as certified solutions. Use appropriate trip, matrix, control site, method, reagent and solvent blanks. Internal, surrogate and five concentration level calibration standards are used. The quality control check sample concentrate should contain indeno(1,2,3-cd)pyrene at 10 µg/mL in acetonitrile.

REFERENCE: Method 8100, SW-846, 3rd ed., Nov 1986.

PRIMARY NAME: Indeno(1,2,3-cd)pyrene Method 8310

TITLE: Polynuclear Aromatic Hydrocarbons MATRIX: groundwater, soils, sludges water miscible liquid wastes, and non-water miscible wastes.

CAS # : 193-39-5

APPLICATION: This method is used for the analysis of 16 polynuclear aromatic hydrocarbons (PAHs). Samples are extracted, concentrated and analyzed using HPLC with detection by UV and fluorescence detectors.

INTERFERENCES: Solvents, reagents and glassware may introduce artifacts. Other interferences may come from coextracted compounds from samples.

INSTRUMENTATION: HPLC with a gradient pumping system and a 250 mm by 2.6 mm reverse phase HC-ODS Sil-X 5-micron particle size column. The fluorescence detector uses an excitation wavelength of 280 nm and emission greater than 389 nm cutoff with dispersive optics.

RANGE: 0.1 to 425 µg/L

MDL: 0.043 µg/L (Fluorescence; reagent water)

PRACTICAL QUANTITATION LIMIT FACTORS FOR MULTIPLYING TIMES FID MDL VALUE

Matrix	Multiplication Factor
Groundwater	10
Low-level soil by sonication with GPC cleanup	670
High-level soil and sludge by sonication	10,000
Non-water miscible waste	100,000

PRECISION: 0.42X + 0.01 µg/L (overall precision)

ACCURACY: 0.54C + 0.06 µg/L (as recovery)

SAMPLING METHOD: Use 8 oz. widemouth glass bottles with Teflon lined caps for concentrated waste samples, soils, sediments and sludges. Use 1 or 2 1/2 gallon amber glass bottles with Teflon lined caps for liquid (water) samples.

STABILITY: Cool soil, sediment, sludge and liquid samples to 4°C. If residual chlorine is present in liquid samples add 3 mL of 10% sodium thiosulfate per gallon of sample and cool to 4°C.

M.H.T. = 14 days for concentrated waste, soil, sediment or sludge.

M.H.T. = 7 days for liquid samples.

All extracts must be analyzed within 40 days.

QUALITY CONTROL: Internal, surrogate and five concentration level calibration standards are used. The calibration standards must be used with the analytical method blank. A quality control check sample concentrate containing indeno(1,2,3-cd)pyrene at 10 µg/mL is required. The QC check sample concentrate may be prepared from pure standard materials or purchased as certified solutions. Use appropriate trip, matrix, control site, method, reagent and solvent blanks.

REFERENCE: Method 8310, SW-846, 3rd ed., Nov 1986.

PRIMARY NAME: Isophorone

Method 8090

TITLE: Nitroaromatics and Cyclic Ketones

MATRIX: groundwater, soils, sludges water miscible liquid wastes, and non-water miscible wastes.

CAS # : 78-59-1

APPLICATION: This method is used for the analysis of various nitroaromatic and cyclic ketone compounds. Samples are extracted, concentrated and analyzed using direct injection of both neat and diluted organic liquids. Dinitrotoluenes are determined using ECD and the other compounds amenable to this method are determined using FID. The method provides an optional GC column which is used for analyte confirmation and that may help resolve analytes from interferences.

INTERFERENCES: Solvents, reagents and glassware may introduce artifacts. Other interferences may come from coextracted compounds from samples.

INSTRUMENTATION: GC capable of on-column injections and a flame ionization detector (FID) or electron capture detector (ECD). Column 1: a 1.2 meter by 2 mm or 4 mm with 1.95% QF-1/1.5% OV-17 on Gas-Chrom Q. Column 2: a 3 meter by 2 mm or 4 mm with 3% OV-101 on Gas-Chrom Q.

RANGE: 1 to 515 μg/L

MDL: 5.7 μg/L (FID) and 15.7 μg/L (ECD)

PRACTICAL QUANTITATION LIMIT FACTORS FOR MULTIPLYING TIMES FID MDL VALUE

Matrix	Multiplication Factor
Groundwater	10
Low-level soil by sonication with GPC cleanup	670
High-level soil and sludge by sonication	10,000
Non-water miscible waste	100,000

PRECISION: 0.46X + 0.31 μg/L (overall precision)

ACCURACY: 0.49C + 2.93 μg/L (as recovery)

SAMPLING METHOD: Use 8 oz. widemouth glass bottles with Teflon lined caps for concentrated waste samples, soils, sediments and sludges. Use 1 or 2 1/2 gallon amber glass bottles with Teflon lined caps for liquid (water) samples.

STABILITY: Cool soil, sediment, sludge and liquid samples to 4°C. If residual chlorine is present in liquid samples add 3 mL of 10% sodium thiosulfate per gallon of sample and cool to 4°C.

M.H.T. = 14 days for concentrated waste, soil, sediment or sludge.

M.H.T. = 7 days for liquid samples.

All extracts must be analyzed within 40 days.

QUALITY CONTROL: A quality control check sample concentrate containing each analyte of interest is required. The QC check sample concentrate may be prepared from pure standard materials or purchased as certified solutions. Use appropriate trip, matrix, control site, method, reagent and solvent blanks. Internal, surrogate and five concentration level calibration standards are used. The QC check sample concentrate should contain this compound at 100 μg/mL in acetone.

REFERENCE: Method 8090, SW-846, 3rd ed., Nov 1986.

PRIMARY NAME: 2-Methyl-4,6-dinitrophenol — Method 8040

TITLE: Phenols

MATRIX: groundwater, soils, sludges water miscible liquid wastes, and non-water miscible wastes.

CAS # : 534-52-1

APPLICATION: This method is used for the analysis of 17 phenols. Samples are extracted, concentrated and analyzed using direct injection of both neat and diluted organic liquids. Pentafluorobenzylbromide (PFB) derivatives also may be made to increase sensitivity of the method.

INTERFERENCES: There can be carry-over contamination with high and low level samples. Solvents, reagents and glassware may introduce artifacts. Other interferences may come from coextracted compounds from samples.

INSTRUMENTATION: GC capable of on-column injections and a flame

ionization detector (FID) or electron capture detector (ECD). Column for underivatized phenol: 1.8 meter by 2.0 mm with 1% SP-1240DA on Supelcoport. Column for derivatized phenols: 1.8 meter by 2.0 mm with 5% OV-17 on Chromosorb W-AW-DMCS.

RANGE: 12 to 450 μg/L MDL: 16.0 μg/L (FID)

PRACTICAL QUANTITATION LIMIT FACTORS FOR MULTIPLYING TIMES FID MDL VALUE

Matrix	Multiplication Factor
Groundwater	10
Low-level soil by sonication with GPC cleanup	670
High-level soil and sludge by sonication	10,000
Non-water miscible waste	100,000

PRECISION: 0.19X + 5.85 μg/L (overall precision using FID)

ACCURACY: 0.84C - 1.01 μg/L (as recovery using FID)

SAMPLING METHOD: Use 8 oz. widemouth glass bottles with Teflon lined caps for concentrated waste samples, soils, sediments and sludges. Use 1 or 2 1/2 gallon amber glass bottles with Teflon lined caps for liquid (water) samples.

STABILITY: Cool soil, sediment, sludge and liquid samples to 4°C. If residual chlorine is present in liquid samples add 3 mL of 10% sodium thiosulfate per gallon of sample and cool to 4°C.

M.H.T. = 14 days for concentrated waste, soil, sediment or sludge.

M.H.T. = 7 days for liquid samples.

All extracts must be analyzed within 40 days.

QUALITY CONTROL: A quality control check sample concentrate containing each analyte of interest is required. The QC check sample concentrate may be prepared from pure standard materials or purchased as certified solutions. Use appropriate trip, matrix, control site, method, reagent and solvent blanks. Internal, surrogate and five concentration level calibration standards are used. The QC check sample concentrate should contain this compound at 100 μg/mL in 2-propanol.

REFERENCE: Method 8040, SW-846, 3rd ed., Nov 1986.

PRIMARY NAME: Naphthalene

Method 8100

TITLE: Polynuclear Aromatic Hydrocarbons

MATRIX: groundwater, soils, sludges water miscible liquid wastes, and non-water miscible wastes.

CAS # : 91-20-3

APPLICATION: This method is used for the analysis of various PAHs. Samples are extracted, concentrated and analyzed using direct injection of both neat and diluted organic liquids. The method provides two optional GC columns that are better than Column 1 and that may help resolve analytes from interferences.

INTERFERENCES: Solvents, reagents and glassware may introduce artifacts. Other interferences may come from coextracted compounds from samples.

INSTRUMENTATION: GC capable of on-column injections and a flame ionization detector (FID). Column 1: a 1.8 meter by 2 mm 3% OV-17 on Chromosorb W-AW-DCMS column. Column 2: a 30 meter by 0.25 mm SE-54 fused silica capillary colunm. Column 3: a 30 meter by 0.32 mm SE-54 fused silica capillary column.

RANGE: 0.1 to 425 μg/L

MDL: Not reported

PRACTICAL QUANTITATION LIMIT FACTORS FOR MULTIPLYING TIMES FID MDL VALUE

Not available

PRECISION: 0.41X + 0.74 μg/L (overall precision)

ACCURACY: 0.57C - 0.70 μg/L (as recovery)

SAMPLING METHOD: Use 8 oz. widemouth glass bottles with Teflon lined caps for concentrated waste samples, soils, sediments and sludges. Use 1 or 2 1/2 gallon amber glass bottles with Teflon lined caps for liquid (water) samples.

STABILITY: Cool soil, sediment, sludge and liquid samples to 4°C. If residual chlorine is present in liquid samples add 3 mL of 10% sodium thiosulfate per gallon of sample and cool to 4°C.

M.H.T. = 14 days for concentrated waste, soil, sediment or sludge.
M.H.T. = 7 days for liquid samples.
All extracts must be analyzed within 40 days.

QUALITY CONTROL: A quality control check sample concentrate containing each analyte of interest is required. The QC check sample concentrate may be prepared from pure standard materials or purchased as certified solutions. Use appropriate trip, matrix, control site, method, reagent and solvent blanks. Internal, surrogate and five concentration level calibration standards are used. The quality control check sample concentrate should contain naphthalene at 100 μg/mL in acetonitrile.

REFERENCE: Method 8100, SW-846, 3rd ed., Nov 1986.

PRIMARY NAME: Naphthalene

Method 8310

TITLE: Polynuclear Aromatic Hydrocarbons

MATRIX: groundwater, soils, sludges water miscible liquid wastes, and non-water miscible wastes.

CAS # : 91-20-3

APPLICATION: This method is used for the analysis of 16 polynuclear aromatic hydrocarbons (PAHs). Samples are extracted, concentrated and analyzed using HPLC with detection by UV and fluorescence detectors.

INTERFERENCES: Solvents, reagents and glassware may introduce artifacts. Other interferences may come from coextracted compounds from samples.

INSTRUMENTATION: HPLC with a gradient pumping system and a 250 mm by 2.6 mm reverse phase HC-ODS Sil-X 5-micron particle size column. The UV detector uses an excitation wavelength of 254 nm coupled to the fluorescence detector. The fluorescence detector uses an excitation wavelength of 280 nm and emission greater than 389 nm cutoff with dispersive optics.

RANGE: 0.1 to 425 μg/L

MDL: 1.8 μg/L (UV detector; reagent water)

PRACTICAL QUANTITATION LIMIT FACTORS FOR MULTIPLYING TIMES FID MDL VALUE

Matrix	Multiplication Factor
Groundwater	10
Low-level soil by sonication with GPC cleanup	670
High-level soil and sludge by sonication	10,000
Non-water miscible waste	100,000

PRECISION: 0.41X + 0.74 μg/L (overall precision)

ACCURACY: 0.57C - 0.70 μg/L (as recovery)

SAMPLING METHOD: Use 8 oz. widemouth glass bottles with Teflon lined caps for concentrated waste samples, soils, sediments and sludges. Use 1 or 2 1/2 gallon amber glass bottles with Teflon lined caps for liquid (water) samples.

STABILITY: Cool soil, sediment, sludge and liquid samples to 4°C. If residual chlorine is present in liquid samples add 3 mL of 10% sodium thiosulfate per gallon of sample and cool to 4°C.

M.H.T. = 14 days for concentrated waste, soil, sediment or sludge.

M.H.T. = 7 days for liquid samples.

All extracts must be analyzed within 40 days.

QUALITY CONTROL: Internal, surrogate and five concentration level calibration standards are used. The calibration standards must be used with the analytical method blank. A quality control check sample concentrate containing naphthalene at 100 μg/mL is required. The QC check sample concentrate may be prepared from pure standard materials or purchased as certified solutions. Use appropriate trip, matrix, control site, method, reagent and solvent blanks.

REFERENCE: Method 8310, SW-846, 3rd ed., Nov 1986.

PRIMARY NAME: Nitrobenzene — Method 8090

TITLE: Nitroaromatics and Cyclic Ketones

MATRIX: groundwater, soils, sludges water miscible liquid wastes, and non-water miscible wastes.

CAS # : 98-95-3

APPLICATION: This method is used for the analysis of various nitroaromatic and cyclic ketone compounds. Samples are extracted, concentrated and analyzed using direct injection of both neat and diluted organic liquids. Dinitrotoluenes are determined using ECD and the other compounds amenable to this method are determined using FID. The method provides an optional GC column which is used for analyte confirmation and that may help resolve analytes from interferences.

INTERFERENCES: Solvents, reagents and glassware may introduce artifacts. Other interferences may come from coextracted compounds from samples.

INSTRUMENTATION: GC capable of on-column injections and a flame ionization detector (FID) or electron capture detector (ECD). Column 1: a 1.2 meter by 2 mm or 4 mm with 1.95% QF-1/1.5% OV-17 on Gas-Chrom Q. Column 2: a 3 meter by 2 mm or 4 mm with 3% OV-101 on Gas-Chrom Q.

RANGE: 1 to 515 μg/L

MDL: 3.6 μg/L (FID) and 13.7 μg/L (ECD)

PRACTICAL QUANTITATION LIMIT FACTORS FOR MULTIPLYING TIMES FID MDL VALUE

Matrix	Multiplication Factor
Groundwater	10
Low-level soil by sonication with GPC cleanup	670
High-level soil and sludge by sonication	10,000
Non-water miscible waste	100,000

PRECISION: 0.37X - 0.78 µg/L (overall precision)

ACCURACY: 0.60C + 2.00 µg/L (as recovery)

SAMPLING METHOD: Use 8 oz. widemouth glass bottles with Teflon lined caps for concentrated waste samples, soils, sediments and sludges. Use 1 or 2 1/2 gallon amber glass bottles with Teflon lined caps for liquid (water) samples.

STABILITY: Cool soil, sediment, sludge and liquid samples to 4°C. If residual chlorine is present in liquid samples add 3 mL of 10% sodium thiosulfate per gallon of sample and cool to 4°C.

M.H.T. = 14 days for concentrated waste, soil, sediment or sludge.

M.H.T. = 7 days for liquid samples.

All extracts must be analyzed within 40 days.

QUALITY CONTROL: A quality control check sample concentrate containing each analyte of interest is required. The QC check sample concentrate may be prepared from pure standard materials or purchased as certified solutions. Use appropriate trip, matrix, control site, method, reagent and solvent blanks. Internal, surrogate and five concentration level calibration standards are used. The QC check sample concentrate should contain this compound at 100 µg/mL in acetone.

REFERENCE: Method 8090, SW-846, 3rd ed., Nov 1986.

PRIMARY NAME: 2-Nitrophenol

Method 8040

TITLE: Phenols

MATRIX: groundwater, soils, sludges water miscible liquid wastes, and non-water miscible wastes.

CAS # : 88-75-5

APPLICATION: This method is used for the analysis of 17 phenols. Samples are extracted, concentrated and analyzed using direct injection of both neat and diluted organic liquids. Pentafluorobenzylbromide (PFB) derivatives also may be made to increase sensitivity of the method.

INTERFERENCES: There can be carry-over contamination with high and low level samples. Solvents, reagents and glassware may introduce artifacts. Other interferences may come from coextracted compounds from samples.

INSTRUMENTATION: GC capable of on-column injections and a flame ionization detector (FID) or electron capture detector (ECD). Column for underivatized phenol: 1.8 meter by 2.0 mm with 1% SP-1240DA on Supelcoport. Column for derivatized phenols: 1.8 meter by 2.0 mm with 5% OV-17 on Chromosorb W-AW-DMCS.

RANGE: 12 to 450 μg/L

MDL: 0.45 μg/L (FID) and 0.77 μg/L (ECD)

PRACTICAL QUANTITATION LIMIT FACTORS FOR MULTIPLYING TIMES FID MDL VALUE

Matrix	Multiplication Factor
Groundwater	10
Low-level soil by sonication with GPC cleanup	670
High-level soil and sludge by sonication	10,000
Non-water miscible waste	100,000

PRECISION: 0.14X + 3.84 μg/L (overall precision using FID)

ACCURACY: 0.81C - 0.76 μg/L (as recovery using FID)

SAMPLING METHOD: Use 8 oz. widemouth glass bottles with Teflon lined caps for concentrated waste samples, soils, sediments and sludges. Use 1 or 2 1/2 gallon amber glass bottles with Teflon lined caps for liquid (water) samples.

STABILITY: Cool soil, sediment, sludge and liquid samples to 4°C. If residual chlorine is present in liquid samples add 3 mL of 10% sodium thiosulfate per gallon of sample and cool to 4°C.

M.H.T. = 14 days for concentrated waste, soil, sediment or sludge.

M.H.T. = 7 days for liquid samples.

All extracts must be analyzed within 40 days.

QUALITY CONTROL: A quality control check sample concentrate containing each analyte of interest is required. The QC check sample concentrate may be prepared from pure standard materials or purchased as certified solutions. Use appropriate trip, matrix, control site, method, reagent and solvent blanks. Internal, surrogate and five concentration level calibration standards are used. The QC check sample concentrate should contain this compound at 100 μg/mL in 2-propanol.

REFERENCE: Method 8040, SW-846, 3rd ed., Nov 1986.

PRIMARY NAME: 4-Nitrophenol

Method 8040

TITLE: Phenols

MATRIX: groundwater, soils, sludges water miscible liquid wastes, and non-water miscible wastes.

CAS # : 100-02-7

APPLICATION: This method is used for the analysis of 17 phenols. Samples are extracted, concentrated and analyzed using direct injection of both neat and diluted organic liquids. Pentafluorobenzylbromide (PFB) derivatives also may be made to increase sensitivity of the method.

INTERFERENCES: There can be carry-over contamination with high and low level samples. Solvents, reagents and glassware may introduce artifacts. Other interferences may come from coextracted compounds from samples.

INSTRUMENTATION: GC capable of on-column injections and a flame ionization detector (FID) or electron capture detector (ECD). Column for underivatized phenol: 1.8 meter by 2.0 mm with 1% SP-1240DA on Supelcoport. Column for derivatized phenols: 1.8 meter by 2.0 mm with 5% OV-17 on Chromosorb W-AW-DMCS.

RANGE: 12 to 450 μg/L

MDL: 2.8 μg/L (FID) and 0.70 μg/L (ECD)

PRACTICAL QUANTITATION LIMIT FACTORS FOR MULTIPLYING TIMES FID MDL VALUE

Matrix	Multiplication Factor
Groundwater	10
Low-level soil by sonication with GPC cleanup	670
High-level soil and sludge by sonication	10,000
Non-water miscible waste	100,000

PRECISION: 0.19X + 4.79 μg/L (overall precision using FID)

ACCURACY: 0.46C + 0.18 µg/L (as recovery using FID)

SAMPLING METHOD: Use 8 oz. widemouth glass bottles with Teflon lined caps for concentrated waste samples, soils, sediments and sludges. Use 1 or 2 1/2 gallon amber glass bottles with Teflon lined caps for liquid (water) samples.

STABILITY: Cool soil, sediment, sludge and liquid samples to 4°C. If residual chlorine is present in liquid samples add 3 mL of 10% sodium thiosulfate per gallon of sample and cool to 4°C.
M.H.T. = 14 days for concentrated waste, soil, sediment or sludge.
M.H.T. = 7 days for liquid samples.
All extracts must be analyzed within 40 days.

QUALITY CONTROL: A quality control check sample concentrate containing each analyte of interest is required. The QC check sample concentrate may be prepared from pure standard materials or purchased as certified solutions. Use appropriate trip, matrix, control site, method, reagent and solvent blanks. Internal, surrogate and five concentration level calibration standards are used. The QC check sample concentrate should contain this compound at 100 µg/mL in 2-propanol.

REFERENCE: Method 8040, SW-846, 3rd ed., Nov 1986.

PRIMARY NAME: Pentachlorophenol Method 8040

TITLE: Phenols MATRIX: groundwater, soils, sludges water miscible liquid wastes, and non-water miscible wastes.

CAS # : 87-86-5

APPLICATION: This method is used for the analysis of 17 phenols. Samples are extracted, concentrated and analyzed using direct injection of both neat and diluted organic liquids. Pentafluorobenzylbromide (PFB) derivatives also may be made to increase sensitivity of the method.

INTERFERENCES: There can be carry-over contamination with high and low level samples. Solvents, reagents and glassware may introduce artifacts.

Other interferences may come from coextracted compounds from samples.

INSTRUMENTATION: GC capable of on-column injections and a flame ionization detector (FID) or electron capture detector (ECD). Column for underivatized phenol: 1.8 meter by 2.0 mm with 1% SP-1240DA on Supelcoport. Column for derivatized phenols: 1.8 meter by 2.0 mm with 5% OV-17 on Chromosorb W-AW-DMCS.

RANGE: 12 to 450 μg/L

MDL: 7.4 μg/L (FID) and 0.59 μg/L (ECD)

PRACTICAL QUANTITATION LIMIT FACTORS FOR MULTIPLYING TIMES FID MDL VALUE

Matrix	Multiplication Factor
Groundwater	10
Low-level soil by sonication with GPC cleanup	670
High-level soil and sludge by sonication	10,000
Non-water miscible waste	100,000

PRECISION: 0.23X + 0.57 μg/L (overall precision using FID)

ACCURACY: 0.83C + 2.07 μg/L (as recovery using FID)

SAMPLING METHOD: Use 8 oz. widemouth glass bottles with Teflon lined caps for concentrated waste samples, soils, sediments and sludges. Use 1 or 2 1/2 gallon amber glass bottles with Teflon lined caps for liquid (water) samples.

STABILITY: Cool soil, sediment, sludge and liquid samples to 4°C. If residual chlorine is present in liquid samples add 3 mL of 10% sodium thiosulfate per gallon of sample and cool to 4°C.

M.H.T. = 14 days for concentrated waste, soil, sediment or sludge.

M.H.T. = 7 days for liquid samples.

All extracts must be analyzed within 40 days.

QUALITY CONTROL: A quality control check sample concentrate containing each analyte of interest is required. The QC check sample concentrate may be prepared from pure standard materials or purchased as certified solutions. Use appropriate trip, matrix, control site, method, reagent and solvent blanks. Internal, surrogate and five concentration level calibration

standards are used. The QC check sample concentrate should contain this compound at 100 μg/mL in 2-propanol.

REFERENCE: Method 8040, SW-846, 3rd ed., Nov 1986.

PRIMARY NAME: Phenanthrene Method 8100

TITLE: Polynuclear Aromatic Hydrocarbons MATRIX: groundwater, soils, sludges water miscible liquid wastes, and non-water miscible wastes.

CAS # : 85-01-8

APPLICATION: This method is used for the analysis of various PAHs. Samples are extracted, concentrated and analyzed using direct injection of both neat and diluted organic liquids. The method provides two optional GC columns that are better than Column 1 and that may help resolve analytes from interferences.

INTERFERENCES: Solvents, reagents and glassware may introduce artifacts. Other interferences may come from coextracted compounds from samples.

INSTRUMENTATION: GC capable of on-column injections and a flame ionization detector (FID). Column 1: a 1.8 meter by 2 mm 3% OV-17 on Chromosorb W-AW-DCMS column. Column 2: a 30 meter by 0.25 mm SE-54 fused silica capillary colunm. Column 3: a 30 meter by 0.32 mm SE-54 fused silica capillary column.

RANGE: 0.1 to 425 μg/L MDL: Not reported

PRACTICAL QUANTITATION LIMIT FACTORS FOR MULTIPLYING TIMES FID MDL VALUE

Not available

PRECISION: 0.47X - 0.25 μg/L (overall precision)

ACCURACY: 0.72C - 0.95 μg/L (as recovery)

SAMPLING METHOD: Use 8 oz. widemouth glass bottles with Teflon lined

caps for concentrated waste samples, soils, sediments and sludges. Use 1 or 2 1/2 gallon amber glass bottles with Teflon lined caps for liquid (water) samples.

STABILITY: Cool soil, sediment, sludge and liquid samples to 4°C. If residual chlorine is present in liquid samples add 3 mL of 10% sodium thiosulfate per gallon of sample and cool to 4°C.

M.H.T. = 14 days for concentrated waste, soil, sediment or sludge.

M.H.T. = 7 days for liquid samples.

All extracts must be analyzed within 40 days.

QUALITY CONTROL: A quality control check sample concentrate containing each analyte of interest is required. The QC check sample concentrate may be prepared from pure standard materials or purchased as certified solutions. Use appropriate trip, matrix, control site, method, reagent and solvent blanks. Internal, surrogate and five concentration level calibration standards are used. The quality control check sample concentrate should contain phenanthrene at 100 μg/mL in acetonitrile.

REFERENCE: Method 8100, SW-846, 3rd ed., Nov 1986.

PRIMARY NAME: Phenanthrene Method 8310

TITLE: Polynuclear Aromatic Hydrocarbons

MATRIX: groundwater, soils, sludges water miscible liquid wastes, and non-water miscible wastes.

CAS # : 85-01-8

APPLICATION: This method is used for the analysis of 16 polynuclear aromatic hydrocarbons (PAHs). Samples are extracted, concentrated and analyzed using HPLC with detection by UV and fluorescence detectors.

INTERFERENCES: Solvents, reagents and glassware may introduce artifacts. Other interferences may come from coextracted compounds from samples.

INSTRUMENTATION: HPLC with a gradient pumping system and a 250 mm by 2.6 mm reverse phase HC-ODS Sil-X 5-micron particle size column.

The fluorescence detector uses an excitation wavelength of 280 nm and emission greater than 389 nm cutoff with dispersive optics.

RANGE: 0.1 to 425 µg/L

MDL: 0.64 µg/L (Fluorescence; reagent water)

PRACTICAL QUANTITATION LIMIT FACTORS FOR MULTIPLYING TIMES FID MDL VALUE

Matrix	Multiplication Factor
Groundwater	10
Low-level soil by sonication with GPC cleanup	670
High-level soil and sludge by sonication	10,000
Non-water miscible waste	100,000

PRECISION: 0.47X - 0.25 µg/L (overall precision)

ACCURACY: 0.72C - 0.95 µg/L (as recovery)

SAMPLING METHOD: Use 8 oz. widemouth glass bottles with Teflon lined caps for concentrated waste samples, soils, sediments and sludges. Use 1 or 2 1/2 gallon amber glass bottles with Teflon lined caps for liquid (water) samples.

STABILITY: Cool soil, sediment, sludge and liquid samples to 4°C. If residual chlorine is present in liquid samples add 3 mL of 10% sodium thiosulfate per gallon of sample and cool to 4°C.

M.H.T. = 14 days for concentrated waste, soil, sediment or sludge.

M.H.T. = 7 days for liquid samples.

All extracts must be analyzed within 40 days.

QUALITY CONTROL: Internal, surrogate and five concentration level calibration standards are used. The calibration standards must be used with the analytical method blank. A quality control check sample concentrate containing phenanthrene at 100 µg/mL is required. The QC check sample concentrate may be prepared from pure standard materials or purchased as certified solutions. Use appropriate trip, matrix, control site, method, reagent and solvent blanks.

REFERENCE: Method 8310, SW-846, 3rd ed., Nov 1986.

PRIMARY NAME: Phenol

Method 8040

TITLE: Phenols

MATRIX: groundwater, soils, sludges water miscible liquid wastes, and non-water miscible wastes.

CAS # : 108-95-2

APPLICATION: This method is used for the analysis of 17 phenols. Samples are extracted, concentrated and analyzed using direct injection of both neat and diluted organic liquids. Pentafluorobenzylbromide (PFB) derivatives also may be made to increase sensitivity of the method.

INTERFERENCES: There can be carry-over contamination with high and low level samples. Solvents, reagents and glassware may introduce artifacts. Other interferences may come from coextracted compounds from samples.

INSTRUMENTATION: GC capable of on-column injections and a flame ionization detector (FID) or electron capture detector (ECD). Column for underivatized phenol: 1.8 meter by 2.0 mm with 1% SP-1240DA on Supelcoport. Column for derivatized phenols: 1.8 meter by 2.0 mm with 5% OV-17 on Chromosorb W-AW-DMCS.

RANGE: 12 to 450 μg/L

MDL: 0.14 μg/L (FID) and 2.2 μg/L (ECD)

PRACTICAL QUANTITATION LIMIT FACTORS FOR MULTIPLYING TIMES FID MDL VALUE

Matrix	Multiplication Factor
Groundwater	10
Low-level soil by sonication with GPC cleanup	670
High-level soil and sludge by sonication	10,000
Non-water miscible waste	100,000

PRECISION: 0.17X + 0.77 μg/L (overall precision using FID)

ACCURACY: 0.43C + 0.11 μg/L (as recovery using FID)

SAMPLING METHOD: Use 8 oz. widemouth glass bottles with Teflon lined caps for concentrated waste samples, soils, sediments and sludges. Use 1 or 2 1/2 gallon amber glass bottles with Teflon lined caps for liquid (water) samples.

STABILITY: Cool soil, sediment, sludge and liquid samples to 4°C. If residual chlorine is present in liquid samples add 3 mL of 10% sodium thiosulfate per gallon of sample and cool to 4°C.

M.H.T. = 14 days for concentrated waste, soil, sediment or sludge.

M.H.T. = 7 days for liquid samples.

All extracts must be analyzed within 40 days.

QUALITY CONTROL: A quality control check sample concentrate containing each analyte of interest is required. The QC check sample concentrate may be prepared from pure standard materials or purchased as certified solutions. Use appropriate trip, matrix, control site, method, reagent and solvent blanks. Internal, surrogate and five concentration level calibration standards are used. The QC check sample concentrate should contain this compound at 100 μg/mL in 2-propanol.

REFERENCE: Method 8040, SW-846, 3rd ed., Nov 1986.

PRIMARY NAME: Pyrene Method 8100

TITLE: Polynuclear Aromatic Hydrocarbons

MATRIX: groundwater, soils, sludges water miscible liquid wastes, and non-water miscible wastes.

CAS # : 129-00-0

APPLICATION: This method is used for the analysis of various PAHs. Samples are extracted, concentrated and analyzed using direct injection of both neat and diluted organic liquids. The method provides two optional GC columns that are better than Column 1 and that may help resolve analytes from interferences.

INTERFERENCES: Solvents, reagents and glassware may introduce artifacts. Other interferences may come from coextracted compounds from samples.

INSTRUMENTATION: GC capable of on-column injections and a flame

ionization detector (FID). Column 1: a 1.8 meter by 2 mm 3% OV-17 on Chromosorb W-AW-DCMS column. Column 2: a 30 meter by 0.25 mm SE-54 fused silica capillary colunm. Column 3: a 30 meter by 0.32 mm SE-54 fused silica capillary column.

RANGE: 0.1 to 425 μg/L MDL: Not reported

PRACTICAL QUANTITATION LIMIT FACTORS FOR MULTIPLYING TIMES FID MDL VALUE

Not available

PRECISION: 0.42X - 0.00 μg/L (overall precision)

ACCURACY: 0.69C - 0.12 μg/L (as recovery)

SAMPLING METHOD: Use 8 oz. widemouth glass bottles with Teflon lined caps for concentrated waste samples, soils, sediments and sludges. Use 1 or 2 1/2 gallon amber glass bottles with Teflon lined caps for liquid (water) samples.

STABILITY: Cool soil, sediment, sludge and liquid samples to 4°C. If residual chlorine is present in liquid samples add 3 mL of 10% sodium thiosulfate per gallon of sample and cool to 4°C.

M.H.T. = 14 days for concentrated waste, soil, sediment or sludge.

M.H.T. = 7 days for liquid samples.

All extracts must be analyzed within 40 days.

QUALITY CONTROL: A quality control check sample concentrate containing each analyte of interest is required. The QC check sample concentrate may be prepared from pure standard materials or purchased as certified solutions. Use appropriate trip, matrix, control site, method, reagent and solvent blanks. Internal, surrogate and five concentration level calibration standards are used. The quality control check sample concentrate should contain pyrene at 10 μg/mL in acetonitrile.

REFERENCE: Method 8100, SW-846, 3rd ed., Nov 1986.

PRIMARY NAME: Pyrene

Method 8310

TITLE: Polynuclear Aromatic Hydrocarbons

MATRIX: groundwater, soils, sludges water miscible liquid wastes, and non-water miscible wastes.

CAS # : 129-00-0

APPLICATION: This method is used for the analysis of 16 polynuclear aromatic hydrocarbons (PAHs). Samples are extracted, concentrated and analyzed using HPLC with detection by UV and fluorescence detectors.

INTERFERENCES: Solvents, reagents and glassware may introduce artifacts. Other interferences may come from coextracted compounds from samples.

INSTRUMENTATION: HPLC with a gradient pumping system and a 250 mm by 2.6 mm reverse phase HC-ODS Sil-X 5-micron particle size column. The fluorescence detector uses an excitation wavelength of 280 nm and emission greater than 389 nm cutoff with dispersive optics.

RANGE: 0.1 to 425 µg/L

MDL: 0.27 µg/L (Fluorescence; reagent water)

PRACTICAL QUANTITATION LIMIT FACTORS FOR MULTIPLYING TIMES FID MDL VALUE

Matrix	Multiplication Factor
Groundwater	10
Low-level soil by sonication with GPC cleanup	670
High-level soil and sludge by sonication	10,000
Non-water miscible waste	100,000

PRECISION: 0.42X - 0.00 µg/L (overall precision)

ACCURACY: 0.69C - 0.12 μg/L (as recovery)

SAMPLING METHOD: Use 8 oz. widemouth glass bottles with Teflon lined caps for concentrated waste samples, soils, sediments and sludges. Use 1 or 2 1/2 gallon amber glass bottles with Teflon lined caps for liquid (water) samples.

STABILITY: Cool soil, sediment, sludge and liquid samples to 4°C. If residual chlorine is present in liquid samples add 3 mL of 10% sodium thiosulfate per gallon of sample and cool to 4°C.
M.H.T. = 14 days for concentrated waste, soil, sediment or sludge.
M.H.T. = 7 days for liquid samples.
All extracts must be analyzed within 40 days.

QUALITY CONTROL: Internal, surrogate and five concentration level calibration standards are used. The calibration standards must be used with the analytical method blank. A quality control check sample concentrate containing pyrene at 10 μg/mL is required. The QC check sample concentrate may be prepared from pure standard materials or purchased as certified solutions. Use appropriate trip, matrix, control site, method, reagent and solvent blanks.

REFERENCE: Method 8310, SW-846, 3rd ed., Nov 1986.

PRIMARY NAME: 1,2,4-Trichlorobenzene

Method 8120

TITLE: Chlorinated Hydrocarbons

MATRIX: groundwater, soils, sludges water miscible liquid wastes, and non-water miscible wastes.

CAS # : 120-82-1

APPLICATION: This method is used for the analysis of various chlorinated hydrocarbons. Samples are extracted, concentrated and analyzed using direct injection of both neat and diluted organic liquid into a gas chromatograph (GC). The method provides an optional GC column which is used for analyte confirmation and that may help resolve analytes from interferences.

INTERFERENCES: Solvents, reagents and glassware may introduce artifacts.

Other interferences may come from coextracted compounds from samples. Phthalate esters are common interferences when using an electron capture detector (ECD) so all plastics must be strictly avoided. Exhaustive cleanup of reagents and glassware may be required to eliminate phthalate contamination.

INSTRUMENTATION: GC capable of on-column injections and an ECD. Column 1: a 1.8 meter by 2 mm with 1% SP-1000 on Supelcoport column. Column 2: a 1.8 meter by 2 mm with 1.5% OV-1 / 2.4% OV-225 on Supelcoport column.

RANGE: 1 to 356 μg/L

MDL: 0.05 μg/L (in reagent water)

PRACTICAL QUANTITATION LIMIT FACTORS FOR MULTIPLYING TIMES FID MDL VALUE

Matrix	Multiplication Factor
Groundwater	10
Low-level soil by sonication with GPC cleanup	670
High-level soil and sludge by sonication	10,000
Non-water miscible waste	100,000

PRECISION: 0.40X - 1.37 μg/L (overall precision)

ACCURACY: 0.76C + 0.98 μg/L (as recovery)

SAMPLING METHOD: Use 8 oz. widemouth glass bottles with Teflon lined caps for concentrated waste samples, soils, sediments and sludges. Use 1 or 2 1/2 gallon amber glass bottles with Teflon lined caps for liquid (water) samples.

STABILITY: Cool soil, sediment, sludge and liquid samples to 4°C. If residual chlorine is present in liquid samples add 3 mL of 10% sodium thiosulfate per gallon of sample and cool to 4°C.

M.H.T. = 14 days for concentrated waste, soil, sediment or sludge.

M.H.T. = 7 days for liquid samples.

All extracts must be analyzed within 40 days.

QUALITY CONTROL: A quality control check sample concentrate containing each analyte of interest is required. The QC check sample concentrate may be prepared from pure standard materials or purchased as certified solutions. Use appropriate trip, matrix, control site, method, reagent and solvent blanks. Internal, surrogate and five concentration level calibration standards are used. The quality control check sample concentrate should contain 1,2,4-trichlorobenzene at 100 μg/mL in acetone.

REFERENCE: Method 8120, SW-846, 3rd ed., Nov 1986.

PRIMARY NAME: 2,4,6-Trichlorophenol — Method 8040

TITLE: Phenols

MATRIX: groundwater, soils, sludges water miscible liquid wastes, and non-water miscible wastes.

CAS # : 88-06-2

APPLICATION: This method is used for the analysis of 17 phenols. Samples are extracted, concentrated and analyzed using direct injection of both neat and diluted organic liquids. Pentafluorobenzylbromide (PFB) derivatives also may be made to increase sensitivity of the method.

INTERFERENCES: There can be carry-over contamination with high and low level samples. Solvents, reagents and glassware may introduce artifacts. Other interferences may come from coextracted compounds from samples.

INSTRUMENTATION: GC capable of on-column injections and a flame ionization detector (FID) or electron capture detector (ECD). Column for underivatized phenol: 1.8 meter by 2.0 mm with 1% SP-1240DA on Supelcoport. Column for derivatized phenols: 1.8 meter by 2.0 mm with 5% OV-17 on Chromosorb W-AW-DMCS.

RANGE: 12 to 450 μg/L

MDL: 0.64 μg/L (FID) and 0.58 μg/L (ECD)

PRACTICAL QUANTITATION LIMIT FACTORS FOR MULTIPLYING TIMES FID MDL VALUE

Matrix	Multiplication Factor
Groundwater	10
Low-level soil by sonication with GPC cleanup	670
High-level soil and sludge by sonication	10,000
Non-water miscible waste	100,000

PRECISION: 0.13X + 2.40 μg/L (overall precision using FID)

ACCURACY: 0.86C - 0.40 μg/L (as recovery using FID)

SAMPLING METHOD: Use 8 oz. widemouth glass bottles with Teflon lined caps for concentrated waste samples, soils, sediments and sludges. Use 1 or 2 1/2 gallon amber glass bottles with Teflon lined caps for liquid (water) samples.

STABILITY: Cool soil, sediment, sludge and liquid samples to 4°C. If residual chlorine is present in liquid samples add 3 mL of 10% sodium thiosulfate per gallon of sample and cool to 4°C.

M.H.T. = 14 days for concentrated waste, soil, sediment or sludge.
M.H.T. = 7 days for liquid samples.

All extracts must be analyzed within 40 days.

QUALITY CONTROL: A quality control check sample concentrate containing each analyte of interest is required. The QC check sample concentrate may be prepared from pure standard materials or purchased as certified solutions. Use appropriate trip, matrix, control site, method, reagent and solvent blanks. Internal, surrogate and five concentration level calibration standards are used. The QC check sample concentrate should contain this compound at 100 μg/mL in 2-propanol.

REFERENCE: Method 8040, SW-846, 3rd ed., Nov 1986.

Chapter 5

Pesticides, Herbicides, PCBs, Dioxins, and Furans

PRIMARY NAME: Aldrin — Method 8080

TITLE: Organochlorine Pesticides and PCBs

MATRIX: groundwater, soils, sludges water miscible liquid wastes, and non-water miscible wastes.

CAS # : 309-00-2

APPLICATION: This method is used for the analysis of 19 pesticides and 7 Aroclor (PCB) mixtures. Samples are extracted, concentrated and analyzed using direct injection of both neat and diluted organic liquid into a gas chromatograph (GC).

INTERFERENCES: Solvents, reagents and glassware may introduce artifacts. Other interferences may come from coextracted compounds from samples. Phthalate esters are common interferences when using an electron capture detector (ECD) so all plastics must be strictly avoided. Exhaustive cleanup of reagents and glassware may be required to eliminate phthalate contamination. Use of a halogen specific microcoulometric or electrolytic conductivity detector will eliminate phthalate interference.

INSTRUMENTATION: GC capable of on-column injections and an ECD or a halogen specific detector (HSD). Column 1: 1.8 meter by 4 mm with 1.5%

SP-2250 / 1.95% SP-2401 on Supelcoport. Column 2: 1.8 meter by 4 mm with 3% OV-1 on Supelcoport.

RANGE: 8.5 to 400 μg/L

MDL: 0.004 μg/L (in reagent water)

PRACTICAL QUANTITATION LIMIT FACTORS FOR MULTIPLYING TIMES FID MDL VALUE

Matrix	Multiplication Factor
Groundwater	10
Low-level soil by sonication with GPC cleanup	670
High-level soil and sludge by sonication	10,000
Non-water miscible waste	100,000

PRECISION: 0.20X - 0.01 μg/L (overall precision)

ACCURACY: 0.81C + 0.04 μg/L (as recovery)

SAMPLING METHOD: Use 8 oz. widemouth glass bottles with Teflon lined caps for concentrated waste samples, soils, sediments and sludges. Use 1 or 2 1/2 gallon amber glass bottles with Teflon lined caps for liquid (water) samples.

STABILITY: Cool soil, sediment, sludge and liquid samples to 4°C. If residual chlorine is present in liquid samples add 3 mL of 10% sodium thiosulfate per gallon of sample and cool to 4°C.

M.H.T. = 14 days for concentrated waste, soil, sediment or sludge.

M.H.T. = 7 days for liquid samples.

All extracts must be analyzed within 40 days.

QUALITY CONTROL: A quality control check sample concentrate containing each analyte of interest is required. The QC check sample concentrate may be prepared from pure standard materials or purchased as certified solutions. Use appropriate trip, matrix, control site, method, reagent and solvent blanks. Internal, surrogate and five concentration level calibration standards are used. The quality control check sample concentrate should contain aldrin at 2 μg/mL in acetone.

REFERENCE: Method 8080, SW-846, 3rd ed., Nov 1986.

PRIMARY NAME: Aroclor 1016 (PCB-1016) Method 8080

TITLE: Organochlorine Pesticides and PCBs

MATRIX: groundwater, soils, sludges water miscible liquid wastes, and non-water miscible wastes.

CAS # : 12674-11-2

APPLICATION: This method is used for the analysis of 19 pesticides and 7 Aroclor (PCB) mixtures. Samples are extracted, concentrated and analyzed using direct injection of both neat and diluted organic liquid into a gas chromatograph (GC).

INTERFERENCES: Solvents, reagents and glassware may introduce artifacts. Other interferences may come from coextracted compounds from samples. Phthalate esters are common interferences when using an electron capture detector (ECD) so all plastics must be strictly avoided. Exhaustive cleanup of reagents and glassware may be required to eliminate phthalate contamination. Use of a halogen specific microcoulometric or electrolytic conductivity detector will eliminate phthalate interference.

INSTRUMENTATION: GC capable of on-column injections and an ECD or a halogen specific detector (HSD). Column 1: 1.8 meter by 4 mm with 1.5% SP-2250 / 1.95% SP-2401 on Supelcoport. Column 2: 1.8 meter by 4 mm with 3% OV-1 on Supelcoport.

RANGE: 8.5 to 400 μg/L MDL: Not determined

PRACTICAL QUANTITATION LIMIT FACTORS FOR MULTIPLYING TIMES FID MDL VALUE

Matrix	Multiplication Factor
Groundwater	10
Low-level soil by sonication with GPC cleanup	670
High-level soil and sludge by sonication	10,000
Non-water miscible waste	100,000

PRECISION: 0.15X + 0.45 µg/L (overall precision)

ACCURACY: 0.81C + 0.50 µg/L (as recovery)

SAMPLING METHOD: Use 8 oz. widemouth glass bottles with Teflon lined caps for concentrated waste samples, soils, sediments and sludges. Use 1 or 2 1/2 gallon amber glass bottles with Teflon lined caps for liquid (water) samples.

STABILITY: Cool soil, sediment, sludge and liquid samples to 4°C. If residual chlorine is present in liquid samples add 3 mL of 10% sodium thiosulfate per gallon of sample and cool to 4°C.

M.H.T. = 14 days for concentrated waste, soil, sediment or sludge.

M.H.T. = 7 days for liquid samples.

All extracts must be analyzed within 40 days.

QUALITY CONTROL: A quality control check sample concentrate containing each analyte of interest is required. The QC check sample concentrate may be prepared from pure standard materials or purchased as certified solutions. Use appropriate trip, matrix, control site, method, reagent and solvent blanks. Internal, surrogate and five concentration level calibration standards are used. The quality control check sample concentrate should contain Aroclor 1016 at 50 µg/mL in acetone.

REFERENCE: Method 8080, SW-846, 3rd ed., Nov 1986.

PRIMARY NAME: Aroclor 1221 (PCB-1221) Method 8080

TITLE: Organochlorine Pesticides and PCBs

MATRIX: groundwater, soils, sludges water miscible liquid wastes, and non-water miscible wastes.

CAS # : 11104-28-2

APPLICATION: This method is used for the analysis of 19 pesticides and 7 Aroclor (PCB) mixtures. Samples are extracted, concentrated and analyzed using direct injection of both neat and diluted organic liquid into a gas chromatograph (GC).

INTERFERENCES: Solvents, reagents and glassware may introduce artifacts. Other interferences may come from coextracted compounds from samples. Phthalate esters are common interferences when using an electron capture detector (ECD) so all plastics must be strictly avoided. Exhaustive cleanup of reagents and glassware may be required to eliminate phthalate contamination. Use of a halogen specific microcoulometric or electrolytic conductivity detector will eliminate phthalate interference.

INSTRUMENTATION: GC capable of on-column injections and an ECD or a halogen specific detector (HSD). Column 1: 1.8 meter by 4 mm with 1.5% SP-2250 / 1.95% SP-2401 on Supelcoport. Column 2: 1.8 meter by 4 mm with 3% OV-1 on Supelcoport.

RANGE: 8.5 to 400 µg/L MDL: Not determined

PRACTICAL QUANTITATION LIMIT FACTORS FOR MULTIPLYING TIMES FID MDL VALUE

Matrix	Multiplication Factor
Groundwater	10
Low-level soil by sonication with GPC cleanup	670
High-level soil and sludge by sonication	10,000
Non-water miscible waste	100,000

PRECISION: 0.35X - 0.62 µg/L (overall precision)

ACCURACY: 0.96C + 0.65 µg/L (as recovery)

SAMPLING METHOD: Use 8 oz. widemouth glass bottles with Teflon lined caps for concentrated waste samples, soils, sediments and sludges. Use 1 or 2 1/2 gallon amber glass bottles with Teflon lined caps for liquid (water) samples.

STABILITY: Cool soil, sediment, sludge and liquid samples to 4°C. If residual chlorine is present in liquid samples add 3 mL of 10% sodium thiosulfate per gallon of sample and cool to 4°C.

M.H.T. = 14 days for concentrated waste, soil, sediment or sludge.

M.H.T. = 7 days for liquid samples.

All extracts must be analyzed within 40 days.

QUALITY CONTROL: A quality control check sample concentrate containing each analyte of interest is required. The QC check sample concentrate may be prepared from pure standard materials or purchased as certified solutions. Use appropriate trip, matrix, control site, method, reagent and solvent blanks. Internal, surrogate and five concentration level calibration standards are used. The quality control check sample concentrate should contain Aroclor 1221 at 50 μg/mL in acetone.

REFERENCE: Method 8080, SW-846, 3rd ed., Nov 1986.

PRIMARY NAME: Aroclor 1232 (PCB-1232) Method 8080

TITLE: Organochlorine Pesticides and PCBs

MATRIX: groundwater, soils, sludges water miscible liquid wastes, and non-water miscible wastes.

CAS # : 11141-16-5

APPLICATION: This method is used for the analysis of 19 pesticides and 7 Aroclor (PCB) mixtures. Samples are extracted, concentrated and analyzed using direct injection of both neat and diluted organic liquid into a gas chromatograph (GC).

INTERFERENCES: Solvents, reagents and glassware may introduce artifacts. Other interferences may come from coextracted compounds from samples. Phthalate esters are common interferences when using an electron capture detector (ECD) so all plastics must be strictly avoided. Exhaustive cleanup of reagents and glassware may be required to eliminate phthalate contamination. Use of a halogen specific microcoulometric or electrolytic conductivity detector will eliminate phthalate interference.

INSTRUMENTATION: GC capable of on-column injections and an ECD or a halogen specific detector (HSD). Column 1: 1.8 meter by 4 mm with 1.5% SP-2250 / 1.95% SP-2401 on Supelcoport. Column 2: 1.8 meter by 4 mm with 3% OV-1 on Supelcoport.

RANGE: 8.5 to 400 μg/L

MDL: Not determined

PRACTICAL QUANTITATION LIMIT FACTORS FOR MULTIPLYING TIMES FID MDL VALUE

Matrix	Multiplication Factor
Groundwater	10
Low-level soil by sonication with GPC cleanup	670
High-level soil and sludge by sonication	10,000
Non-water miscible waste	100,000

PRECISION: 0.31X + 3.50 µg/L (overall precision)

ACCURACY: 0.91C + 10.79 µg/L (as recovery)

SAMPLING METHOD: Use 8 oz. widemouth glass bottles with Teflon lined caps for concentrated waste samples, soils, sediments and sludges. Use 1 or 2 1/2 gallon amber glass bottles with Teflon lined caps for liquid (water) samples.

STABILITY: Cool soil, sediment, sludge and liquid samples to 4°C. If residual chlorine is present in liquid samples add 3 mL of 10% sodium thiosulfate per gallon of sample and cool to 4°C.

M.H.T. = 14 days for concentrated waste, soil, sediment or sludge.

M.H.T. = 7 days for liquid samples.

All extracts must be analyzed within 40 days.

QUALITY CONTROL: A quality control check sample concentrate containing each analyte of interest is required. The QC check sample concentrate may be prepared from pure standard materials or purchased as certified solutions. Use appropriate trip, matrix, control site, method, reagent and solvent blanks. Internal, surrogate and five concentration level calibration standards are used. The quality control check sample concentrate should contain Aroclor 1232 at 50 µg/mL in acetone.

REFERENCE: Method 8080, SW-846, 3rd ed., Nov 1986.

PRIMARY NAME: Aroclor 1242 (PCB-1242) Method 8080

TITLE: Organochlorine Pesticides and PCBs MATRIX: groundwater, soils, sludges water miscible liquid wastes, and non-water miscible wastes.

CAS # : 53469-21-9

APPLICATION: This method is used for the analysis of 19 pesticides and 7 Aroclor (PCB) mixtures. Samples are extracted, concentrated and analyzed using direct injection of both neat and diluted organic liquid into a gas chromatograph (GC).

INTERFERENCES: Solvents, reagents and glassware may introduce artifacts. Other interferences may come from coextracted compounds from samples. Phthalate esters are common interferences when using an electron capture detector (ECD) so all plastics must be strictly avoided. Exhaustive cleanup of reagents and glassware may be required to eliminate phthalate contamination. Use of a halogen specific microcoulometric or electrolytic conductivity detector will eliminate phthalate interference.

INSTRUMENTATION: GC capable of on-column injections and an ECD or a halogen specific detector (HSD). Column 1: 1.8 meter by 4 mm with 1.5% SP-2250 / 1.95% SP-2401 on Supelcoport. Column 2: 1.8 meter by 4 mm with 3% OV-1 on Supelcoport.

RANGE: 8.5 to 400 μg/L MDL: 0.065 μg/L (in reagent water)

PRACTICAL QUANTITATION LIMIT FACTORS FOR MULTIPLYING TIMES FID MDL VALUE

Matrix	Multiplication Factor
Groundwater	10
Low-level soil by sonication with GPC cleanup	670
High-level soil and sludge by sonication	10,000
Non-water miscible waste	100,000

PRECISION: 0.21X + 1.52 μg/L (overall precision)

ACCURACY: 0.93C + 0.70 μg/L (as recovery)

SAMPLING METHOD: Use 8 oz. widemouth glass bottles with Teflon lined caps for concentrated waste samples, soils, sediments and sludges. Use 1 or 2 1/2 gallon amber glass bottles with Teflon lined caps for liquid (water) samples.

STABILITY: Cool soil, sediment, sludge and liquid samples to 4°C. If residual chlorine is present in liquid samples add 3 mL of 10% sodium thiosulfate per gallon of sample and cool to 4°C.
M.H.T. = 14 days for concentrated waste, soil, sediment or sludge.
M.H.T. = 7 days for liquid samples.
All extracts must be analyzed within 40 days.

QUALITY CONTROL: A quality control check sample concentrate containing each analyte of interest is required. The QC check sample concentrate may be prepared from pure standard materials or purchased as certified solutions. Use appropriate trip, matrix, control site, method, reagent and solvent blanks. Internal, surrogate and five concentration level calibration standards are used. The quality control check sample concentrate should contain Aroclor 1242 at 50 μg/mL in acetone.

REFERENCE: Method 8080, SW-846, 3rd ed., Nov 1986.

PRIMARY NAME: Aroclor 1248 (PCB-1248) Method 8080

TITLE: Organochlorine Pesticides and PCBs

MATRIX: groundwater, soils, sludges water miscible liquid wastes, and non-water miscible wastes.

CAS # : 12672-29-6

APPLICATION: This method is used for the analysis of 19 pesticides and 7 Aroclor (PCB) mixtures. Samples are extracted, concentrated and analyzed using direct injection of both neat and diluted organic liquid into a gas chromatograph (GC).

INTERFERENCES: Solvents, reagents and glassware may introduce artifacts. Other interferences may come from coextracted compounds from samples. Phthalate esters are common interferences when using an electron capture detector (ECD) so all plastics must be strictly avoided. Exhaustive cleanup of reagents and glassware may be required to eliminate phthalate contamination. Use of a halogen specific microcoulometric or electrolytic conductivity detector will eliminate phthalate interference.

INSTRUMENTATION: GC capable of on-column injections and an ECD or a halogen specific detector (HSD). Column 1: 1.8 meter by 4 mm with 1.5% SP-2250 / 1.95% SP-2401 on Supelcoport. Column 2: 1.8 meter by 4 mm with 3% OV-1 on Supelcoport.

RANGE: 8.5 to 400 μg/L MDL: Not determined

PRACTICAL QUANTITATION LIMIT FACTORS FOR MULTIPLYING TIMES FID MDL VALUE

Matrix	Multiplication Factor
Groundwater	10
Low-level soil by sonication with GPC cleanup	670
High-level soil and sludge by sonication	10,000
Non-water miscible waste	100,000

PRECISION: 0.25X - 0.37 μg/L (overall precision)

ACCURACY: 0.97C + 1.06 μg/L (as recovery)

SAMPLING METHOD: Use 8 oz. widemouth glass bottles with Teflon lined caps for concentrated waste samples, soils, sediments and sludges. Use 1 or 2 1/2 gallon amber glass bottles with Teflon lined caps for liquid (water) samples.

STABILITY: Cool soil, sediment, sludge and liquid samples to 4°C. If residual chlorine is present in liquid samples add 3 mL of 10% sodium thiosulfate per gallon of sample and cool to 4°C.

M.H.T. = 14 days for concentrated waste, soil, sediment or sludge.

M.H.T. = 7 days for liquid samples.

All extracts must be analyzed within 40 days.

QUALITY CONTROL: A quality control check sample concentrate containing each analyte of interest is required. The QC check sample concentrate may be prepared from pure standard materials or purchased as certified solutions. Use appropriate trip, matrix, control site, method, reagent and solvent blanks. Internal, surrogate and five concentration level calibration standards are used. The quality control check sample concentrate should contain Aroclor 1248 at 50 µg/mL in acetone.

REFERENCE: Method 8080, SW-846, 3rd ed., Nov 1986.

PRIMARY NAME: Aroclor 1254 (PCB-1254) Method 8080

TITLE: Organochlorine Pesticides and PCBs

MATRIX: groundwater, soils, sludges water miscible liquid wastes, and non-water miscible wastes.

CAS # : 11097-69-1

APPLICATION: This method is used for the analysis of 19 pesticides and 7 Aroclor (PCB) mixtures. Samples are extracted, concentrated and analyzed using direct injection of both neat and diluted organic liquid into a gas chromatograph (GC).

INTERFERENCES: Solvents, reagents and glassware may introduce artifacts. Other interferences may come from coextracted compounds from samples. Phthalate esters are common interferences when using an electron capture detector (ECD) so all plastics must be strictly avoided. Exhaustive cleanup of reagents and glassware may be required to eliminate phthalate contamination. Use of a halogen specific microcoulometric or electrolytic conductivity detector will eliminate phthalate interference.

INSTRUMENTATION: GC capable of on-column injections and an ECD or a halogen specific detector (HSD). Column 1: 1.8 meter by 4 mm with 1.5% SP-2250 / 1.95% SP-2401 on Supelcoport. Column 2: 1.8 meter by 4 mm with 3% OV-1 on Supelcoport.

RANGE: 8.5 to 400 µg/L

MDL: Not determined

PRACTICAL QUANTITATION LIMIT FACTORS FOR MULTIPLYING TIMES FID MDL VALUE

Matrix	Multiplication Factor
Groundwater	10
Low-level soil by sonication with GPC cleanup	670
High-level soil and sludge by sonication	10,000
Non-water miscible waste	100,000

PRECISION: 0.17X + 3.62 μg/L (overall precision)

ACCURACY: 0.76C + 2.07 μg/L (as recovery)

SAMPLING METHOD: Use 8 oz. widemouth glass bottles with Teflon lined caps for concentrated waste samples, soils, sediments and sludges. Use 1 or 2 1/2 gallon amber glass bottles with Teflon lined caps for liquid (water) samples.

STABILITY: Cool soil, sediment, sludge and liquid samples to 4°C. If residual chlorine is present in liquid samples add 3 mL of 10% sodium thiosulfate per gallon of sample and cool to 4°C.

M.H.T. = 14 days for concentrated waste, soil, sediment or sludge.

M.H.T. = 7 days for liquid samples.

All extracts must be analyzed within 40 days.

QUALITY CONTROL: A quality control check sample concentrate containing each analyte of interest is required. The QC check sample concentrate may be prepared from pure standard materials or purchased as certified solutions. Use appropriate trip, matrix, control site, method, reagent and solvent blanks. Internal, surrogate and five concentration level calibration standards are used. The quality control check sample concentrate should contain Aroclor 1254 at 50 μg/mL in acetone.

REFERENCE: Method 8080, SW-846, 3rd ed., Nov 1986.

PRIMARY NAME: Aroclor 1260 (PCB-1260) Method 8080

TITLE: Organochlorine Pesticides and PCBs

MATRIX: groundwater, soils, sludges water miscible liquid wastes, and non-water miscible wastes.

CAS # : 11096-82-5

APPLICATION: This method is used for the analysis of 19 pesticides and 7 Aroclor (PCB) mixtures. Samples are extracted, concentrated and analyzed using direct injection of both neat and diluted organic liquid into a gas chromatograph (GC).

INTERFERENCES: Solvents, reagents and glassware may introduce artifacts. Other interferences may come from coextracted compounds from samples. Phthalate esters are common interferences when using an electron capture detector (ECD) so all plastics must be strictly avoided. Exhaustive cleanup of reagents and glassware may be required to eliminate phthalate contamination. Use of a halogen specific microcoulometric or electrolytic conductivity detector will eliminate phthalate interference.

INSTRUMENTATION: GC capable of on-column injections and an ECD or a halogen specific detector (HSD). Column 1: 1.8 meter by 4 mm with 1.5% SP-2250 / 1.95% SP-2401 on Supelcoport. Column 2: 1.8 meter by 4 mm with 3% OV-1 on Supelcoport.

RANGE: 8.5 to 400 µg/L MDL: Not determined

PRACTICAL QUANTITATION LIMIT FACTORS FOR MULTIPLYING TIMES FID MDL VALUE

Matrix	Multiplication Factor
Groundwater	10
Low-level soil by sonication with GPC cleanup	670
High-level soil and sludge by sonication	10,000
Non-water miscible waste	100,000

PRECISION: 0.39X - 4.86 μg/L (overall precision)

ACCURACY: 0.66C + 3.76 μg/L (as recovery)

SAMPLING METHOD: Use 8 oz. widemouth glass bottles with Teflon lined caps for concentrated waste samples, soils, sediments and sludges. Use 1 or 2 1/2 gallon amber glass bottles with Teflon lined caps for liquid (water) samples.

STABILITY: Cool soil, sediment, sludge and liquid samples to 4°C. If residual chlorine is present in liquid samples add 3 mL of 10% sodium thiosulfate per gallon of sample and cool to 4°C.

M.H.T. = 14 days for concentrated waste, soil, sediment or sludge.

M.H.T. = 7 days for liquid samples.

All extracts must be analyzed within 40 days.

QUALITY CONTROL: A quality control check sample concentrate containing each analyte of interest is required. The QC check sample concentrate may be prepared from pure standard materials or purchased as certified solutions. Use appropriate trip, matrix, control site, method, reagent and solvent blanks. Internal, surrogate and five concentration level calibration standards are used. The quality control check sample concentrate should contain Aroclor 1260 at 50 μg/mL in acetone.

REFERENCE: Method 8080, SW-846, 3rd ed., Nov 1986.

PRIMARY NAME: Azinophos, methyl — Method 8140

TITLE: Organophosphorus Pesticides

MATRIX: groundwater, soils, sludges water miscible liquid wastes, and non-water miscible wastes.

CAS # : 86-50-0

APPLICATION: This method is used for the analysis of 21 organophosphorus pesticides. Samples are extracted, concentrated and analyzed using direct injection of both neat and diluted organic liquid into a gas chromatograph (GC).

INTERFERENCES: Solvents, reagents and glassware may introduce artifacts. Other interferences may come from coextracted compounds from samples. The use of Florisil cleanup materials may produce low recoveries. Elemental sulfur may interfere with some compounds when using a flame photometric detector. Sulfur cleanup (Method 3660) may alleviate sulfur interference.

INSTRUMENTATION: GC capable of on-column injections and a flame photometric detector (FPD) or a thermionic detector. Column 1: 1.8 meter by 2 mm with 5% SP-2401 on Supelcoport. Column 2: 1.8 meter by 2 mm with 3% SP-2401 on Supelcoport. Column 3: 50 cm by 1/8 inch Teflon with 15% SE-54 on Gas Chrom Q. The preferred column is Column Number 1.

RANGE: 21 to 250 μg/L

MDL: 1.5 μg/L (in reagent water)

PRACTICAL QUANTITATION LIMIT FACTORS FOR MULTIPLYING TIMES FID MDL VALUE

Matrix	Multiplication Factor
Groundwater	10
Low-level soil by sonication with GPC cleanup	670
High-level soil and sludge by sonication	10,000
Non-water miscible waste	100,000

PRECISION: 18.8% (single operator standard deviation)

ACCURACY: 72.7% (single operator average recovery)

SAMPLING METHOD: Use 8 oz. widemouth glass bottles with Teflon lined caps for concentrated waste samples, soils, sediments and sludges. Use 1 or 2 1/2 gallon amber glass bottles with Teflon lined caps for liquid (water) samples.

STABILITY: Cool soil, sediment, sludge and liquid samples to 4°C. If residual chlorine is present in liquid samples add 3 mL of 10% sodium thiosulfate per gallon of sample and cool to 4°C.

M.H.T. = 14 days for concentrated waste, soil, sediment or sludge.

M.H.T. = 7 days for liquid samples.

All extracts must be analyzed within 40 days.

QUALITY CONTROL: A quality control check sample concentrate containing this compound in acetone at a concentration 1,000 times more concentrated than the selected spike concentration is required. The QC check sample concentrate may be prepared from pure standard materials or purchased as certified solutions. Use appropriate trip, matrix, control site, method, reagent and solvent blanks. Internal, surrogate and five concentration level calibration standards are used.

REFERENCE: Method 8140, SW-846, 3rd ed., Sep 1986.

PRIMARY NAME: alpha-BHC Method 8080

TITLE: Organochlorine Pesticides and PCBs

MATRIX: groundwater, soils, sludges water miscible liquid wastes, and non-water miscible wastes.

CAS # : 319-84-6

APPLICATION: This method is used for the analysis of 19 pesticides and 7 Aroclor (PCB) mixtures. Samples are extracted, concentrated and analyzed using direct injection of both neat and diluted organic liquid into a gas chromatograph (GC).

INTERFERENCES: Solvents, reagents and glassware may introduce artifacts. Other interferences may come from coextracted compounds from samples. Phthalate esters are common interferences when using an electron capture detector (ECD) so all plastics must be strictly avoided. Exhaustive cleanup of reagents and glassware may be required to eliminate phthalate contamination. Use of a halogen specific microcoulometric or electrolytic conductivity detector will eliminate phthalate interference.

INSTRUMENTATION: GC capable of on-column injections and an ECD or a halogen specific detector (HSD). Column 1: 1.8 meter by 4 mm with 1.5% SP-2250 / 1.95% SP-2401 on Supelcoport. Column 2: 1.8 meter by 4 mm with 3% OV-1 on Supelcoport.

RANGE: 8.5 to 400 µg/L

MDL: 0.003 µg/L (in reagent water)

PRACTICAL QUANTITATION LIMIT FACTORS FOR MULTIPLYING TIMES FID MDL VALUE

Matrix	Multiplication Factor
Groundwater	10
Low-level soil by sonication with GPC cleanup	670
High-level soil and sludge by sonication	10,000
Non-water miscible waste	100,000

PRECISION: 0.23X - 0.00 μg/L (overall precision)

ACCURACY: 0.84C + 0.03 μg/L (as recovery)

SAMPLING METHOD: Use 8 oz. widemouth glass bottles with Teflon lined caps for concentrated waste samples, soils, sediments and sludges. Use 1 or 2 1/2 gallon amber glass bottles with Teflon lined caps for liquid (water) samples.

STABILITY: Cool soil, sediment, sludge and liquid samples to 4°C. If residual chlorine is present in liquid samples add 3 mL of 10% sodium thiosulfate per gallon of sample and cool to 4°C.

M.H.T. = 14 days for concentrated waste, soil, sediment or sludge.

M.H.T. = 7 days for liquid samples.

All extracts must be analyzed within 40 days.

QUALITY CONTROL: A quality control check sample concentrate containing each analyte of interest is required. The QC check sample concentrate may be prepared from pure standard materials or purchased as certified solutions. Use appropriate trip, matrix, control site, method, reagent and solvent blanks. Internal, surrogate and five concentration level calibration standards are used. The quality control check sample concentrate should contain alpha-BHC at 2 μg/mL in acetone.

REFERENCE: Method 8080, SW-846, 3rd ed., Nov 1986.

PRIMARY NAME: beta-BHC

Method 8080

TITLE: Organochlorine Pesticides and PCBs

MATRIX: groundwater, soils, sludges water miscible liquid wastes, and non-water miscible wastes.

CAS # : 319-85-7

APPLICATION: This method is used for the analysis of 19 pesticides and 7 Aroclor (PCB) mixtures. Samples are extracted, concentrated and analyzed using direct injection of both neat and diluted organic liquid into a gas chromatograph (GC).

INTERFERENCES: Solvents, reagents and glassware may introduce artifacts. Other interferences may come from coextracted compounds from samples. Phthalate esters are common interferences when using an electron capture detector (ECD) so all plastics must be strictly avoided. Exhaustive cleanup of reagents and glassware may be required to eliminate phthalate contamination. Use of a halogen specific microcoulometric or electrolytic conductivity detector will eliminate phthalate interference.

INSTRUMENTATION: GC capable of on-column injections and an ECD or a halogen specific detector (HSD). Column 1: 1.8 meter by 4 mm with 1.5% SP-2250 / 1.95% SP-2401 on Supelcoport. Column 2: 1.8 meter by 4 mm with 3% OV-1 on Supelcoport.

RANGE: 8.5 to 400 μg/L

MDL: 0.006 μg/L (in reagent water)

PRACTICAL QUANTITATION LIMIT FACTORS FOR MULTIPLYING TIMES FID MDL VALUE

Matrix	Multiplication Factor
Groundwater	10
Low-level soil by sonication with GPC cleanup	670
High-level soil and sludge by sonication	10,000
Non-water miscible waste	100,000

PRECISION: 0.33X - 0.95 μg/L (overall precision)

ACCURACY: 0.81C + 0.07 μg/L (as recovery)

SAMPLING METHOD: Use 8 oz. widemouth glass bottles with Teflon lined caps for concentrated waste samples, soils, sediments and sludges. Use 1 or 2 1/2 gallon amber glass bottles with Teflon lined caps for liquid (water) samples.

STABILITY: Cool soil, sediment, sludge and liquid samples to 4°C. If residual chlorine is present in liquid samples add 3 mL of 10% sodium thiosulfate per gallon of sample and cool to 4°C.
M.H.T. = 14 days for concentrated waste, soil, sediment or sludge.
M.H.T. = 7 days for liquid samples.
All extracts must be analyzed within 40 days.

QUALITY CONTROL: A quality control check sample concentrate containing each analyte of interest is required. The QC check sample concentrate may be prepared from pure standard materials or purchased as certified solutions. Use appropriate trip, matrix, control site, method, reagent and solvent blanks. Internal, surrogate and five concentration level calibration standards are used. The quality control check sample concentrate should contain beta-BHC at 2 μg/mL in acetone.

REFERENCE: Method 8080, SW-846, 3rd ed., Nov 1986.

PRIMARY NAME: delta-BHC Method 8080

TITLE: Organochlorine Pesticides and PCBs MATRIX: groundwater, soils, sludges water miscible liquid wastes, and non-water miscible wastes.

CAS # : 319-86-8

APPLICATION: This method is used for the analysis of 19 pesticides and 7 Aroclor (PCB) mixtures. Samples are extracted, concentrated and analyzed using direct injection of both neat and diluted organic liquid into a gas chromatograph (GC).

INTERFERENCES: Solvents, reagents and glassware may introduce artifacts. Other interferences may come from coextracted compounds from samples. Phthalate esters are common interferences when using an electron capture detector (ECD) so all plastics must be strictly avoided. Exhaustive cleanup of reagents and glassware may be required to eliminate phthalate contamination. Use of a halogen specific microcoulometric or electrolytic conductivity detector will eliminate phthalate interference.

INSTRUMENTATION: GC capable of on-column injections and an ECD or a halogen specific detector (HSD). Column 1: 1.8 meter by 4 mm with 1.5% SP-2250 / 1.95% SP-2401 on Supelcoport. Column 2: 1.8 meter by 4 mm with 3% OV-1 on Supelcoport.

RANGE: 8.5 to 400 μg/L

MDL: 0.009 μg/L (in reagent water)

PRACTICAL QUANTITATION LIMIT FACTORS FOR MULTIPLYING TIMES FID MDL VALUE

Matrix	Multiplication Factor
Groundwater	10
Low-level soil by sonication with GPC cleanup	670
High-level soil and sludge by sonication	10,000
Non-water miscible waste	100,000

PRECISION: 0.25X + 0.03 μg/L (overall precision)

ACCURACY: 0.81C + 0.07 μg/L (as recovery)

SAMPLING METHOD: Use 8 oz. widemouth glass bottles with Teflon lined caps for concentrated waste samples, soils, sediments and sludges. Use 1 or 2 1/2 gallon amber glass bottles with Teflon lined caps for liquid (water) samples.

STABILITY: Cool soil, sediment, sludge and liquid samples to 4°C. If residual chlorine is present in liquid samples add 3 mL of 10% sodium thiosulfate per gallon of sample and cool to 4°C.

M.H.T. = 14 days for concentrated waste, soil, sediment or sludge.
M.H.T. = 7 days for liquid samples.
All extracts must be analyzed within 40 days.

QUALITY CONTROL: A quality control check sample concentrate containing each analyte of interest is required. The QC check sample concentrate may be prepared from pure standard materials or purchased as certified solutions. Use appropriate trip, matrix, control site, method, reagent and solvent blanks. Internal, surrogate and five concentration level calibration standards are used. The quality control check sample concentrate should contain delta-BHC at 2 μg/mL in acetone.

REFERENCE: Method 8080, SW-846, 3rd ed., Nov 1986.

PRIMARY NAME: gamma-BHC (Lindane) Method 8080

TITLE: Organochlorine Pesticides and PCBs

MATRIX: groundwater, soils, sludges water miscible liquid wastes, and non-water miscible wastes.

CAS # : 58-89-9

APPLICATION: This method is used for the analysis of 19 pesticides and 7 Aroclor (PCB) mixtures. Samples are extracted, concentrated and analyzed using direct injection of both neat and diluted organic liquid into a gas chromatograph (GC).

INTERFERENCES: Solvents, reagents and glassware may introduce artifacts. Other interferences may come from coextracted compounds from samples. Phthalate esters are common interferences when using an electron capture detector (ECD) so all plastics must be strictly avoided. Exhaustive cleanup of reagents and glassware may be required to eliminate phthalate contamination. Use of a halogen specific microcoulometric or electrolytic conductivity detector will eliminate phthalate interference.

INSTRUMENTATION: GC capable of on-column injections and an ECD or a halogen specific detector (HSD). Column 1: 1.8 meter by 4 mm with 1.5% SP-2250 / 1.95% SP-2401 on Supelcoport. Column 2: 1.8 meter by 4 mm with 3% OV-1 on Supelcoport.

RANGE: 8.5 to 400 µg/L

MDL: 0.004 µg/L (in reagent water)

PRACTICAL QUANTITATION LIMIT FACTORS FOR MULTIPLYING TIMES FID MDL VALUE

Matrix	Multiplication Factor
Groundwater	10
Low-level soil by sonication with GPC cleanup	670
High-level soil and sludge by sonication	10,000
Non-water miscible waste	100,000

PRECISION: 0.22X + 0.04 µg/L (overall precision)

ACCURACY: 0.82C - 0.05 µg/L (as recovery)

SAMPLING METHOD: Use 8 oz. widemouth glass bottles with Teflon lined caps for concentrated waste samples, soils, sediments and sludges. Use 1 or 2 1/2 gallon amber glass bottles with Teflon lined caps for liquid (water) samples.

STABILITY: Cool soil, sediment, sludge and liquid samples to 4°C. If residual chlorine is present in liquid samples add 3 mL of 10% sodium thiosulfate per gallon of sample and cool to 4°C.

M.H.T. = 14 days for concentrated waste, soil, sediment or sludge.

M.H.T. = 7 days for liquid samples.

All extracts must be analyzed within 40 days.

QUALITY CONTROL: A quality control check sample concentrate containing each analyte of interest is required. The QC check sample concentrate may be prepared from pure standard materials or purchased as certified solutions. Use appropriate trip, matrix, control site, method, reagent and solvent blanks. Internal, surrogate and five concentration level calibration standards are used. The quality control check sample concentrate should contain gamma-BHC at 2 µg/mL in acetone.

REFERENCE: Method 8080, SW-846, 3rd ed., Nov 1986.

PRIMARY NAME: Bolstar

Method 8140

TITLE: Organophosphorus Pesticides

MATRIX: groundwater, soils, sludges water miscible liquid wastes, and non-water miscible wastes.

CAS # : 35400-43-2

APPLICATION: This method is used for the analysis of 21 organophosphorus pesticides. Samples are extracted, concentrated and analyzed using direct injection of both neat and diluted organic liquid into a gas chromatograph (GC).

INTERFERENCES: Solvents, reagents and glassware may introduce artifacts. Other interferences may come from coextracted compounds from samples. The use of Florisil cleanup materials may produce low recoveries. Elemental sulfur may interfere with some compounds when using a flame photometric detector. Sulfur cleanup (Method 3660) may alleviate sulfur interference.

INSTRUMENTATION: GC capable of on-column injections and a flame photometric detector (FPD) or a thermionic detector. Column 1: 1.8 meter by 2 mm with 5% SP-2401 on Supelcoport. Column 2: 1.8 meter by 2 mm with 3% SP-2401 on Supelcoport. Column 3: 50 cm by 1/8 inch Teflon with 15% SE-54 on Gas Chrom Q. The preferred column is Column Number 1.

RANGE: 4.9 to 46 μg/L

MDL: 0.15 μg/L (in reagent water)

PRACTICAL QUANTITATION LIMIT FACTORS FOR MULTIPLYING TIMES FID MDL VALUE

Matrix	Multiplication Factor
Groundwater	10
Low-level soil by sonication with GPC cleanup	670
High-level soil and sludge by sonication	10,000
Non-water miscible waste	100,000

PRECISION: 6.3% (single operator standard deviation)

ACCURACY: 64.6% (single operator average recovery)

SAMPLING METHOD: Use 8 oz. widemouth glass bottles with Teflon lined caps for concentrated waste samples, soils, sediments and sludges. Use 1 or 2 1/2 gallon amber glass bottles with Teflon lined caps for liquid (water) samples.

STABILITY: Cool soil, sediment, sludge and liquid samples to 4°C. If residual chlorine is present in liquid samples add 3 mL of 10% sodium thiosulfate per gallon of sample and cool to 4°C.

M.H.T. = 14 days for concentrated waste, soil, sediment or sludge.

M.H.T. = 7 days for liquid samples.

All extracts must be analyzed within 40 days.

QUALITY CONTROL: A quality control check sample concentrate containing this compound in acetone at a concentration 1,000 times more concentrated than the selected spike concentration is required. The QC check sample concentrate may be prepared from pure standard materials or purchased as certified solutions. Use appropriate trip, matrix, control site, method, reagent and solvent blanks. Internal, surrogate and five concentration level calibration standards are used.

REFERENCE: Method 8140, SW-846, 3rd ed., Sep 1986.

PRIMARY NAME: Chlordane (technical) Method 8080

TITLE: Organochlorine Pesticides and PCBs

MATRIX: groundwater, soils, sludges water miscible liquid wastes, and non-water miscible wastes.

CAS # : 57-74-9

APPLICATION: This method is used for the analysis of 19 pesticides and 7 Aroclor (PCB) mixtures. Samples are extracted, concentrated and analyzed using direct injection of both neat and diluted organic liquid into a gas chromatograph (GC).

INTERFERENCES: Solvents, reagents and glassware may introduce artifacts. Other interferences may come from coextracted compounds from samples. Phthalate esters are common interferences when using an electron capture detector (ECD) so all plastics must be strictly avoided. Exhaustive cleanup of reagents and glassware may be required to eliminate phthalate contamination. Use of a halogen specific microcoulometric or electrolytic conductivity detector will eliminate phthalate interference.

INSTRUMENTATION: GC capable of on-column injections and an ECD or a halogen specific detector (HSD). Column 1: 1.8 meter by 4 mm with 1.5% SP-2250 / 1.95% SP-2401 on Supelcoport. Column 2: 1.8 meter by 4 mm with 3% OV-1 on Supelcoport.

RANGE: 8.5 to 400 μg/L

MDL: 0.014 μg/L (in reagent water)

PRACTICAL QUANTITATION LIMIT FACTORS FOR MULTIPLYING TIMES FID MDL VALUE

Matrix	Multiplication Factor
Groundwater	10
Low-level soil by sonication with GPC cleanup	670
High-level soil and sludge by sonication	10,000
Non-water miscible waste	100,000

PRECISION: 0.18 + 0.18 μg/L (overall precision)

ACCURACY: 0.82C - 0.04 μg/L (as recovery)

SAMPLING METHOD: Use 8 oz. widemouth glass bottles with Teflon lined caps for concentrated waste samples, soils, sediments and sludges. Use 1 or 2 1/2 gallon amber glass bottles with Teflon lined caps for liquid (water) samples.

STABILITY: Cool soil, sediment, sludge and liquid samples to 4°C. If residual chlorine is present in liquid samples add 3 mL of 10% sodium thiosulfate per gallon of sample and cool to 4°C.

M.H.T. = 14 days for concentrated waste, soil, sediment or sludge.
M.H.T. = 7 days for liquid samples.

All extracts must be analyzed within 40 days.

QUALITY CONTROL: A quality control check sample concentrate containing each analyte of interest is required. The QC check sample concentrate may be prepared from pure standard materials or purchased as certified solutions. Use appropriate trip, matrix, control site, method, reagent and solvent blanks. Internal, surrogate and five concentration level calibration standards are used. The quality control check sample concentrate should contain chlordane at 50 µg/mL in acetone.

REFERENCE: Method 8080, SW-846, 3rd ed., Nov 1986.

PRIMARY NAME: Chlorpyrifos

Method 8140

TITLE: Organophosphorus Pesticides

MATRIX: groundwater, soils, sludges water miscible liquid wastes, and non-water miscible wastes.

CAS # : 2921-88-2

APPLICATION: This method is used for the analysis of 21 organophosphorus pesticides. Samples are extracted, concentrated and analyzed using direct injection of both neat and diluted organic liquid into a gas chromatograph (GC).

INTERFERENCES: Solvents, reagents and glassware may introduce artifacts. Other interferences may come from coextracted compounds from samples. The use of Florisil cleanup materials may produce low recoveries. Elemental sulfur may interfere with some compounds when using a flame photometric detector. Sulfur cleanup (Method 3660) may alleviate sulfur interference.

INSTRUMENTATION: GC capable of on-column injections and a flame photometric detector (FPD) or a thermionic detector. Column 1: 1.8 meter by 2 mm with 5% SP-2401 on Supelcoport. Column 2: 1.8 meter by 2 mm with 3% SP-2401 on Supelcoport. Column 3: 50 cm by 1/8 inch Teflon with 15% SE-54 on Gas Chrom Q. The preferred column is Column Number 2.

RANGE: 1.0 to 50.5 µg/L

MDL: 0.3 µg/L (in reagent water)

PRACTICAL QUANTITATION LIMIT FACTORS FOR MULTIPLYING TIMES FID MDL VALUE

Matrix	Multiplication Factor
Groundwater	10
Low-level soil by sonication with GPC cleanup	670
High-level soil and sludge by sonication	10,000
Non-water miscible waste	100,000

PRECISION: 5.5% (single operator standard deviation)

ACCURACY: 98.3% (single operator average recovery)

SAMPLING METHOD: Use 8 oz. widemouth glass bottles with Teflon lined caps for concentrated waste samples, soils, sediments and sludges. Use 1 or 2 1/2 gallon amber glass bottles with Teflon lined caps for liquid (water) samples.

STABILITY: Cool soil, sediment, sludge and liquid samples to 4°C. If residual chlorine is present in liquid samples add 3 mL of 10% sodium thiosulfate per gallon of sample and cool to 4°C.

M.H.T. = 14 days for concentrated waste, soil, sediment or sludge.

M.H.T. = 7 days for liquid samples.

All extracts must be analyzed within 40 days.

QUALITY CONTROL: A quality control check sample concentrate containing this compound in acetone at a concentration 1,000 times more concentrated than the selected spike concentration is required. The QC check sample concentrate may be prepared from pure standard materials or purchased as certified solutions. Use appropriate trip, matrix, control site, method, reagent and solvent blanks. Internal, surrogate and five concentration level calibration standards are used.

REFERENCE: Method 8140, SW-846, 3rd ed., Sep 1986.

PRIMARY NAME: Coumaphos

Method 8140

TITLE: Organophosphorus Pesticides

MATRIX: groundwater, soils, sludges water miscible liquid wastes, and non-water miscible wastes.

CAS # : 56-72-4

APPLICATION: This method is used for the analysis of 21 organophosphorus pesticides. Samples are extracted, concentrated and analyzed using direct injection of both neat and diluted organic liquid into a gas chromatograph (GC).

INTERFERENCES: Solvents, reagents and glassware may introduce artifacts. Other interferences may come from coextracted compounds from samples. The use of Florisil cleanup materials may produce low recoveries. Elemental sulfur may interfere with some compounds when using a flame photometric detector. Sulfur cleanup (Method 3660) may alleviate sulfur interference.

INSTRUMENTATION: GC capable of on-column injections and a flame photometric detector (FPD) or a thermionic detector. Column 1: 1.8 meter by 2 mm with 5% SP-2401 on Supelcoport. Column 2: 1.8 meter by 2 mm with 3% SP-2401 on Supelcoport. Column 3: 50 cm by 1/8 inch Teflon with 15% SE-54 on Gas Chrom Q. The preferred column is Column Number 1.

RANGE: 25 to 225 μg/L

MDL: 1.5 μg/L (in reagent water)

PRACTICAL QUANTITATION LIMIT FACTORS FOR MULTIPLYING TIMES FID MDL VALUE

Matrix	Multiplication Factor
Groundwater	10
Low-level soil by sonication with GPC cleanup	670
High-level soil and sludge by sonication	10,000
Non-water miscible waste	100,000

PRECISION: 12.7% (single operator standard deviation)

ACCURACY: 109% (single operator average recovery)

SAMPLING METHOD: Use 8 oz. widemouth glass bottles with Teflon lined caps for concentrated waste samples, soils, sediments and sludges. Use 1 or 2 1/2 gallon amber glass bottles with Teflon lined caps for liquid (water) samples.

STABILITY: Cool soil, sediment, sludge and liquid samples to 4°C. If residual chlorine is present in liquid samples add 3 mL of 10% sodium thiosulfate per gallon of sample and cool to 4°C.

M.H.T. = 14 days for concentrated waste, soil, sediment or sludge.

M.H.T. = 7 days for liquid samples.

All extracts must be analyzed within 40 days.

QUALITY CONTROL: A quality control check sample concentrate containing this compound in acetone at a concentration 1,000 times more concentrated than the selected spike concentration is required. The QC check sample concentrate may be prepared from pure standard materials or purchased as certified solutions. Use appropriate trip, matrix, control site, method, reagent and solvent blanks. Internal, surrogate and five concentration level calibration standards are used.

REFERENCE: Method 8140, SW-846, 3rd ed., Sep 1986.

PRIMARY NAME: 2,4-D

Method 8150

TITLE: Chlorinated Herbicides

MATRIX: groundwater, soils, sludges water miscible liquid wastes, and non-water miscible wastes.

CAS # : 94-75-7

APPLICATION: This method is used for the analysis of 10 chlorinated herbicides. Samples are extracted, hydrolyzed with potassium hydroxide, and extraneous organics are removed by a solvent wash. After acidification, the acids are extracted, concentrated and converted to their methyl esters using diazomethane. They are then analyzed using direct injection into a gas chromatograph (GC). Be very careful because diazomethane can explode under certain conditions and it is also a carcinogen.

INTERFERENCES: Organic acids and phenols (especially chlorinated acids and phenols) may cause interferences. Phthalate esters are not as significant an interference as with other GC-ECD methods if an electron capture detector is used. The herbicides may react readily with alkaline substances and be lost during analysis so all glassware and glass wool must be acid rinsed and sodium sulfate must be acidified with sulfuric acid prior to use. Sensitivity usually depends on the level of interferences rather than on instrumentation.

INSTRUMENTATION: GC capable of on-column injections and an electron capture detector (ECD)or a halogen specific detector. Column 1: 1.8 meter by 4 mm with 1.5% SP-2250 / 1.95% SP-2401 on Supelcoport. Column 2: 1.8 meter by 4 mm with 5% OV-210 on Gas Chrom Q. Column 3: 1.98 meter by 2 mm with 0.1% SP-1000 on Carbopack C. The preferred column is Column Number 1 or 2.

RANGE: Not listed

MDL: 1.2 μg/L (in reagent water; ECD)

PRACTICAL QUANTITATION LIMIT FACTORS FOR MULTIPLYING TIMES FID MDL VALUE

Matrix	Multiplication Factor
Groundwater	10
Low-level soil by sonication with GPC cleanup	670
High-level soil and sludge by sonication	10,000
Non-water miscible waste	100,000

PRECISION: (as standard deviation) 4% with 10.9 μg/L spike in drinking water; 4% with 10.1 μg/L in municipal water.

ACCURACY: (as mean recovery) 75% with 10.9 μg/L spike in drinking water; 77% with 10.1 μg/L in municipal water.

SAMPLING METHOD: Use 8 oz. widemouth glass bottles with Teflon lined caps for concentrated waste samples, soils, sediments and sludges. Use 1 or 2 1/2 gallon amber glass bottles with Teflon lined caps for liquid (water) samples.

STABILITY: Cool soil, sediment, sludge and liquid samples to 4°C. If residual chlorine is present in liquid samples add 3 mL of 10% sodium thiosulfate per gallon of sample and cool to 4°C.

M.H.T. = 14 days for concentrated waste, soil, sediment or sludge.
M.H.T. = 7 days for liquid samples.
All extracts must be analyzed within 40 days.

QUALITY CONTROL: A quality control check sample concentrate containing this compound in acetone at a concentration 1,000 times more concentrated than the selected spike concentration is required. The QC check sample concentrate may be prepared from pure standard materials or purchased as certified solutions. Use appropriate trip, matrix, control site, method, reagent and solvent blanks. Internal, surrogate and five concentration level calibration standards are used.

REFERENCE: Method 8150, SW-846, 3rd ed., Sep 1986.

PRIMARY NAME: Dalapon Method 8150

TITLE: Chlorinated Herbicides

MATRIX: groundwater, soils, sludges water miscible liquid wastes, and non-water miscible wastes.

CAS # : 75-99-0

APPLICATION: This method is used for the analysis of 10 chlorinated herbicides. Samples are extracted, hydrolyzed with potassium hydroxide, and extraneous organics are removed by a solvent wash. After acidification, the acids are extracted, concentrated and converted to their methyl esters using diazomethane. They are then analyzed using direct injection into a gas chromatograph (GC). Be very careful because diazomethane can explode under certain conditions and it is also a carcinogen.

INTERFERENCES: Organic acids and phenols (especially chlorinated acids and phenols) may cause interferences. Phthalate esters are not as significant an interference as with other GC-ECD methods if an electron capture detector is used. The herbicides may react readily with alkaline substances and be lost during analysis so all glassware and glass wool must be acid rinsed and sodium sulfate must be acidified with sulfuric acid prior to use. Sensitivity usually depends on the level of interferences rather than on instrumentation.

INSTRUMENTATION: GC capable of on-column injections and an electron capture detector (ECD)or a halogen specific detector. Column 1: 1.8 meter by 4 mm with 1.5% SP-2250 / 1.95% SP-2401 on Supelcoport. Column 2: 1.8 meter by 4 mm with 5% OV-210 on Gas Chrom Q. Column 3: 1.98 meter by 2 mm with 0.1% SP-1000 on Carbopack C. The preferred column is Column Number 3.

RANGE: Not listed

MDL: 5.8 μg/L (in reagent water; ECD)

PRACTICAL QUANTITATION LIMIT FACTORS FOR MULTIPLYING TIMES FID MDL VALUE

Matrix	Multiplication Factor
Groundwater	10
Low-level soil by sonication with GPC cleanup	670
High-level soil and sludge by sonication	10,000
Non-water miscible waste	100,000

PRECISION: (as standard deviation) 8% with 23.4 μg/L spike in drinking water; 5% with 200 μg/L in municipal water.

ACCURACY: (as mean recovery) 66% with 23.4 μg/L spike in drinking water; 65% with 200 μg/L in municipal water.

SAMPLING METHOD: Use 8 oz. widemouth glass bottles with Teflon lined caps for concentrated waste samples, soils, sediments and sludges. Use 1 or 2 1/2 gallon amber glass bottles with Teflon lined caps for liquid (water) samples.

STABILITY: Cool soil, sediment, sludge and liquid samples to 4°C. If residual chlorine is present in liquid samples add 3 mL of 10% sodium thiosulfate per gallon of sample and cool to 4°C.

M.H.T. = 14 days for concentrated waste, soil, sediment or sludge.

M.H.T. = 7 days for liquid samples.

All extracts must be analyzed within 40 days.

QUALITY CONTROL: A quality control check sample concentrate containing this compound in acetone at a concentration 1,000 times more concentrated than the selected spike concentration is required. The QC check sample concentrate may be prepared from pure standard materials or purchased as certified solutions. Use appropriate trip, matrix, control site, method, reagent and solvent blanks. Internal, surrogate and five concentration level calibration standards are used.

REFERENCE: Method 8150, SW-846, 3rd ed., Sep 1986.

PRIMARY NAME: 2,4-DB

Method 8150

TITLE: Chlorinated Herbicides

MATRIX: groundwater, soils, sludges water miscible liquid wastes, and non-water miscible wastes.

CAS # : 94-82-6

APPLICATION: This method is used for the analysis of 10 chlorinated herbicides. Samples are extracted, hydrolyzed with potassium hydroxide, and extraneous organics are removed by a solvent wash. After acidification, the acids are extracted, concentrated and converted to their methyl esters using diazomethane. They are then analyzed using direct injection into a gas chromatograph (GC). Be very careful because diazomethane can explode under certain conditions and it is also a carcinogen.

INTERFERENCES: Organic acids and phenols (especially chlorinated acids and phenols) may cause interferences. Phthalate esters are not as significant an interference as with other GC-ECD methods if an electron capture detector is used. The herbicides may react readily with alkaline substances and be lost during analysis so all glassware and glass wool must be acid rinsed and sodium sulfate must be acidified with sulfuric acid prior to use. Sensitivity usually depends on the level of interferences rather than on instrumentation.

INSTRUMENTATION: GC capable of on-column injections and an electron capture detector (ECD)or a halogen specific detector. Column 1: 1.8 meter by 4 mm with 1.5% SP-2250 / 1.95% SP-2401 on Supelcoport. Column 2: 1.8 meter by 4 mm with 5% OV-210 on Gas Chrom Q. Column 3: 1.98 meter by 2 mm with 0.1% SP-1000 on Carbopack C. The preferred column is Column Number 1.

RANGE: Not listed

MDL: 0.91 µg/L (in reagent water; ECD)

PRACTICAL QUANTITATION LIMIT FACTORS FOR MULTIPLYING TIMES FID MDL VALUE

Matrix	Multiplication Factor
Groundwater	10
Low-level soil by sonication with GPC cleanup	670
High-level soil and sludge by sonication	10,000
Non-water miscible waste	100,000

PRECISION: (as standard deviation) 3% with 10.3 µg/L spike in drinking water; 3% with 10.4 µg/L in municipal water.

ACCURACY: (as mean recovery) 93% with 10.3 µg/L spike in drinking water; 93% with 10.4 µg/L in municipal water.

SAMPLING METHOD: Use 8 oz. widemouth glass bottles with Teflon lined caps for concentrated waste samples, soils, sediments and sludges. Use 1 or 2 1/2 gallon amber glass bottles with Teflon lined caps for liquid (water) samples.

STABILITY: Cool soil, sediment, sludge and liquid samples to 4°C. If residual chlorine is present in liquid samples add 3 mL of 10% sodium thiosulfate per gallon of sample and cool to 4°C.

M.H.T. = 14 days for concentrated waste, soil, sediment or sludge.

M.H.T. = 7 days for liquid samples.

All extracts must be analyzed within 40 days.

QUALITY CONTROL: A quality control check sample concentrate containing this compound in acetone at a concentration 1,000 times more concentrated than the selected spike concentration is required. The QC check sample concentrate may be prepared from pure standard materials or purchased as certified solutions. Use appropriate trip, matrix, control site, method, reagent and solvent blanks. Internal, surrogate and five concentration level calibration standards are used.

REFERENCE: Method 8150, SW-846, 3rd ed., Sep 1986.

PRIMARY NAME: 4,4'-DDD

Method 8080

TITLE: Organochlorine Pesticides and PCBs

MATRIX: groundwater, soils, sludges water miscible liquid wastes, and non-water miscible wastes.

CAS # : 72-54-8

APPLICATION: This method is used for the analysis of 19 pesticides and 7 Aroclor (PCB) mixtures. Samples are extracted, concentrated and analyzed using direct injection of both neat and diluted organic liquid into a gas chromatograph (GC).

INTERFERENCES: Solvents, reagents and glassware may introduce artifacts. Other interferences may come from coextracted compounds from samples. Phthalate esters are common interferences when using an electron capture detector (ECD) so all plastics must be strictly avoided. Exhaustive cleanup of reagents and glassware may be required to eliminate phthalate contamination. Use of a halogen specific microcoulometric or electrolytic conductivity detector will eliminate phthalate interference.

INSTRUMENTATION: GC capable of on-column injections and an ECD or a halogen specific detector (HSD). Column 1: 1.8 meter by 4 mm with 1.5% SP-2250 / 1.95% SP-2401 on Supelcoport. Column 2: 1.8 meter by 4 mm with 3% OV-1 on Supelcoport.

RANGE: 8.5 to 400 μg/L

MDL: 0.011 μg/L (in reagent water)

PRACTICAL QUANTITATION LIMIT FACTORS FOR MULTIPLYING TIMES FID MDL VALUE

Matrix	Multiplication Factor
Groundwater	10
Low-level soil by sonication with GPC cleanup	670
High-level soil and sludge by sonication	10,000
Non-water miscible waste	100,000

PRECISION: 0.27X - 0.14 μg/L (overall precision)

ACCURACY: 0.84C + 0.30 μg/L (as recovery)

SAMPLING METHOD: Use 8 oz. widemouth glass bottles with Teflon lined caps for concentrated waste samples, soils, sediments and sludges. Use 1 or 2 1/2 gallon amber glass bottles with Teflon lined caps for liquid (water) samples.

STABILITY: Cool soil, sediment, sludge and liquid samples to 4°C. If residual chlorine is present in liquid samples add 3 mL of 10% sodium thiosulfate per gallon of sample and cool to 4°C.

M.H.T. = 14 days for concentrated waste, soil, sediment or sludge.

M.H.T. = 7 days for liquid samples.

All extracts must be analyzed within 40 days.

QUALITY CONTROL: A quality control check sample concentrate containing each analyte of interest is required. The QC check sample concentrate may be prepared from pure standard materials or purchased as certified solutions. Use appropriate trip, matrix, control site, method, reagent and solvent blanks. Internal, surrogate and five concentration level calibration standards are used. The quality control check sample concentrate should contain 4,4'-DDD at 10 μg/mL in acetone.

REFERENCE: Method 8080, SW-846, 3rd ed., Nov 1986.

PRIMARY NAME: 4,4'-DDE

Method 8080

TITLE: Organochlorine Pesticides and PCBs

MATRIX: groundwater, soils, sludges water miscible liquid wastes, and non-water miscible wastes.

CAS # : 72-55-9

APPLICATION: This method is used for the analysis of 19 pesticides and 7 Aroclor (PCB) mixtures. Samples are extracted, concentrated and analyzed using direct injection of both neat and diluted organic liquid into a gas chromatograph (GC).

INTERFERENCES: Solvents, reagents and glassware may introduce artifacts. Other interferences may come from coextracted compounds from samples. Phthalate esters are common interferences when using an electron capture detector (ECD) so all plastics must be strictly avoided. Exhaustive cleanup of reagents and glassware may be required to eliminate phthalate contamination. Use of a halogen specific microcoulometric or electrolytic conductivity detector will eliminate phthalate interference.

INSTRUMENTATION: GC capable of on-column injections and an ECD or a halogen specific detector (HSD). Column 1: 1.8 meter by 4 mm with 1.5% SP-2250 / 1.95% SP-2401 on Supelcoport. Column 2: 1.8 meter by 4 mm with 3% OV-1 on Supelcoport.

RANGE: 8.5 to 400 µg/L

MDL: 0.004 µg/L (in reagent water)

PRACTICAL QUANTITATION LIMIT FACTORS FOR MULTIPLYING TIMES FID MDL VALUE

Matrix	Multiplication Factor
Groundwater	10
Low-level soil by sonication with GPC cleanup	670
High-level soil and sludge by sonication	10,000
Non-water miscible waste	100,000

PRECISION: 0.28X - 0.09 µg/L (overall precision)

ACCURACY: 0.85C + 0.14 µg/L (as recovery)

SAMPLING METHOD: Use 8 oz. widemouth glass bottles with Teflon lined caps for concentrated waste samples, soils, sediments and sludges. Use 1 or 2 1/2 gallon amber glass bottles with Teflon lined caps for liquid (water) samples.

STABILITY: Cool soil, sediment, sludge and liquid samples to 4°C. If residual chlorine is present in liquid samples add 3 mL of 10% sodium thiosulfate per gallon of sample and cool to 4°C.

M.H.T. = 14 days for concentrated waste, soil, sediment or sludge.

M.H.T. = 7 days for liquid samples.

All extracts must be analyzed within 40 days.

QUALITY CONTROL: A quality control check sample concentrate containing each analyte of interest is required. The QC check sample concentrate may be prepared from pure standard materials or purchased as certified solutions. Use appropriate trip, matrix, control site, method, reagent and solvent blanks. Internal, surrogate and five concentration level calibration standards are used. The quality control check sample concentrate should contain 4,4'-DDE at 2 μg/mL in acetone.

REFERENCE: Method 8080, SW-846, 3rd ed., Nov 1986.

PRIMARY NAME: 4,4'-DDT Method 8080

TITLE: Organochlorine Pesticides and PCBs

MATRIX: groundwater, soils, sludges water miscible liquid wastes, and non-water miscible wastes.

CAS # : 50-29-3

APPLICATION: This method is used for the analysis of 19 pesticides and 7 Aroclor (PCB) mixtures. Samples are extracted, concentrated and analyzed using direct injection of both neat and diluted organic liquid into a gas chromatograph (GC).

INTERFERENCES: Solvents, reagents and glassware may introduce artifacts. Other interferences may come from coextracted compounds from samples. Phthalate esters are common interferences when using an electron capture detector (ECD) so all plastics must be strictly avoided. Exhaustive cleanup of reagents and glassware may be required to eliminate phthalate contamination. Use of a halogen specific microcoulometric or electrolytic conductivity detector will eliminate phthalate interference. If the injection port or front of the GC column is dirty DDT may degrade to 4,4'-DDE and 4,4'-DDD. Sometimes toxaphene, chlordane, methoxychlor, BHC, PCBs and the ortho and para isomers of DDT are found as an interferences.

INSTRUMENTATION: GC capable of on-column injections and an ECD or a halogen specific detector (HSD). Column 1: 1.8 meter by 4 mm with 1.5% SP-2250 / 1.95% SP-2401 on Supelcoport. Column 2: 1.8 meter by 4 mm with 3% OV-1 on Supelcoport.

RANGE: 8.5 to 400 μg/L

MDL: 0.012 μg/L (in reagent water)

PRACTICAL QUANTITATION LIMIT FACTORS FOR MULTIPLYING TIMES FID MDL VALUE

Matrix	Multiplication Factor
Groundwater	10
Low-level soil by sonication with GPC cleanup	670
High-level soil and sludge by sonication	10,000
Non-water miscible waste	100,000

PRECISION: 0.31X - 0.21 μg/L (overall precision)

ACCURACY: 0.93C - 0.13 μg/L (as recovery)

SAMPLING METHOD: Use 8 oz. widemouth glass bottles with Teflon lined caps for concentrated waste samples, soils, sediments and sludges. Use 1 or 2 1/2 gallon amber glass bottles with Teflon lined caps for liquid (water) samples.

STABILITY: Cool soil, sediment, sludge and liquid samples to 4°C. If residual chlorine is present in liquid samples add 3 mL of 10% sodium thiosulfate per gallon of sample and cool to 4°C.

M.H.T. = 14 days for concentrated waste, soil, sediment or sludge.

M.H.T. = 7 days for liquid samples.

All extracts must be analyzed within 40 days.

QUALITY CONTROL: A quality control check sample concentrate containing each analyte of interest is required. The QC check sample concentrate may be prepared from pure standard materials or purchased as certified solutions. Use appropriate trip, matrix, control site, method, reagent and solvent blanks. Internal, surrogate and five concentration level calibration standards are used. The quality control check sample concentrate should contain 4,4'-DDT at 10 μg/mL in acetone.

REFERENCE: Method 8080, SW-846, 3rd ed., Nov 1986.

PRIMARY NAME: Demeton-O — Method 8140

TITLE: Organophosphorus Pesticides

MATRIX: groundwater, soils, sludges water miscible liquid wastes, and non-water miscible wastes.

CAS # : 298-03-3

APPLICATION: This method is used for the analysis of 21 organophosphorus pesticides. Samples are extracted, concentrated and analyzed using direct injection of both neat and diluted organic liquid into a gas chromatograph (GC).

INTERFERENCES: Solvents, reagents and glassware may introduce artifacts. Other interferences may come from coextracted compounds from samples. The use of Florisil cleanup materials may produce low recoveries. Elemental sulfur may interfere with some compounds when using a flame photometric detector. Sulfur cleanup (Method 3660) may alleviate sulfur interference.

INSTRUMENTATION: GC capable of on-column injections and a flame photometric detector (FPD) or a thermionic detector. Column 1: 1.8 meter by 2 mm with 5% SP-2401 on Supelcoport. Column 2: 1.8 meter by 2 mm with 3% SP-2401 on Supelcoport. Column 3: 50 cm by 1/8 inch Teflon with 15% SE-54 on Gas Chrom Q. The preferred column is Column Number 1.

RANGE: 11.9 to 314 μg/L

MDL: 0.25 μg/L (in reagent water)

PRACTICAL QUANTITATION LIMIT FACTORS FOR MULTIPLYING TIMES FID MDL VALUE

Matrix	Multiplication Factor
Groundwater	10
Low-level soil by sonication with GPC cleanup	670
High-level soil and sludge by sonication	10,000
Non-water miscible waste	100,000

PRECISION: 10.5% (single operator standard deviation)

ACCURACY: 67.4% (single operator average recovery)

SAMPLING METHOD: Use 8 oz. widemouth glass bottles with Teflon lined caps for concentrated waste samples, soils, sediments and sludges. Use 1 or 2 1/2 gallon amber glass bottles with Teflon lined caps for liquid (water) samples.

STABILITY: Cool soil, sediment, sludge and liquid samples to 4°C. If residual chlorine is present in liquid samples add 3 mL of 10% sodium thiosulfate per gallon of sample and cool to 4°C.

M.H.T. = 14 days for concentrated waste, soil, sediment or sludge.

M.H.T. = 7 days for liquid samples.

All extracts must be analyzed within 40 days.

QUALITY CONTROL: A quality control check sample concentrate containing this compound in acetone at a concentration 1,000 times more concentrated than the selected spike concentration is required. The QC check sample concentrate may be prepared from pure standard materials or purchased as certified solutions. Use appropriate trip, matrix, control site, method, reagent and solvent blanks. Internal, surrogate and five concentration level calibration standards are used.

REFERENCE: Method 8140, SW-846, 3rd ed., Sep 1986.

PRIMARY NAME: Demeton-S Method 8140

TITLE: Organophosphorus Pesticides

MATRIX: groundwater, soils, sludges water miscible liquid wastes, and non-water miscible wastes

CAS # : 126-75-0

APPLICATION: This method is used for the analysis of 21 organophosphorus pesticides. Samples are extracted, concentrated and analyzed using direct injection of both neat and diluted organic liquid into a gas chromatograph (GC).

INTERFERENCES: Solvents, reagents and glassware may introduce artifacts. Other interferences may come from coextracted compounds from samples.

The use of Florisil cleanup materials may produce low recoveries. Elemental sulfur may interfere with some compounds when using a flame photometric detector. Sulfur cleanup (Method 3660) may alleviate sulfur interference.

INSTRUMENTATION: GC capable of on-column injections and a flame photometric detector (FPD) or a thermionic detector. Column 1: 1.8 meter by 2 mm with 5% SP-2401 on Supelcoport. Column 2: 1.8 meter by 2 mm with 3% SP-2401 on Supelcoport. Column 3: 50 cm by 1/8 inch Teflon with 15% SE-54 on Gas Chrom Q. The preferred column is Column Number 1.

RANGE: 11.9 to 314 μg/L

MDL: 0.25 μg/L (in reagent water)

PRACTICAL QUANTITATION LIMIT FACTORS FOR MULTIPLYING TIMES FID MDL VALUE

Matrix	Multiplication Factor
Groundwater	10
Low-level soil by sonication with GPC cleanup	670
High-level soil and sludge by sonication	10,000
Non-water miscible waste	100,000

PRECISION: 10.5% (single operator standard deviation)

ACCURACY: 67.4% (single operator average recovery)

SAMPLING METHOD: Use 8 oz. widemouth glass bottles with Teflon lined caps for concentrated waste samples, soils, sediments and sludges. Use 1 or 2 1/2 gallon amber glass bottles with Teflon lined caps for liquid (water) samples.

STABILITY: Cool soil, sediment, sludge and liquid samples to 4°C. If residual chlorine is present in liquid samples add 3 mL of 10% sodium thiosulfate per gallon of sample and cool to 4°C.

M.H.T. = 14 days for concentrated waste, soil, sediment or sludge.

M.H.T. = 7 days for liquid samples.

All extracts must be analyzed within 40 days.

QUALITY CONTROL: A quality control check sample concentrate containing this compound in acetone at a concentration 1,000 times more concentrated than the selected spike concentration is required. The QC check sample concentrate may be prepared from pure standard materials or purchased as certified solutions. Use appropriate trip, matrix, control site, method, reagent and solvent blanks. Internal, surrogate and five concentration level calibration standards are used.

REFERENCE: Method 8140, SW-846, 3rd ed., Sep 1986.

PRIMARY NAME: Diazinon

Method 8140

TITLE: Organophosphorus Pesticides

MATRIX: groundwater, soils, sludges water miscible liquid wastes, and non-water miscible wastes.

CAS # : 333-41-5

APPLICATION: This method is used for the analysis of 21 organophosphorus pesticides. Samples are extracted, concentrated and analyzed using direct injection of both neat and diluted organic liquid into a gas chromatograph (GC).

INTERFERENCES: Solvents, reagents and glassware may introduce artifacts. Other interferences may come from coextracted compounds from samples. The use of Florisil cleanup materials may produce low recoveries. Elemental sulfur may interfere with some compounds when using a flame photometric detector. Sulfur cleanup (Method 3660) may alleviate sulfur interference.

INSTRUMENTATION: GC capable of on-column injections and a flame photometric detector (FPD) or a thermionic detector. Column 1: 1.8 meter by 2 mm with 5% SP-2401 on Supelcoport. Column 2: 1.8 meter by 2 mm with 3% SP-2401 on Supelcoport. Column 3: 50 cm by 1/8 inch Teflon with 15% SE-54 on Gas Chrom Q. The preferred column is Column Number 2.

RANGE: 5.6 µg/L only

MDL: 0.6 µg/L (in reagent water)

PRACTICAL QUANTITATION LIMIT FACTORS FOR MULTIPLYING TIMES FID MDL VALUE

Matrix	Multiplication Factor
Groundwater	10
Low-level soil by sonication with GPC cleanup	670
High-level soil and sludge by sonication	10,000
Non-water miscible waste	100,000

PRECISION: 6.0% (single operator standard deviation)

ACCURACY: 67.0% (single operator average recovery)

SAMPLING METHOD: Use 8 oz. widemouth glass bottles with Teflon lined caps for concentrated waste samples, soils, sediments and sludges. Use 1 or 2 1/2 gallon amber glass bottles with Teflon lined caps for liquid (water) samples.

STABILITY: Cool soil, sediment, sludge and liquid samples to 4°C. If residual chlorine is present in liquid samples add 3 mL of 10% sodium thiosulfate per gallon of sample and cool to 4°C.

M.H.T. = 14 days for concentrated waste, soil, sediment or sludge.

M.H.T. = 7 days for liquid samples.

All extracts must be analyzed within 40 days.

QUALITY CONTROL: A quality control check sample concentrate containing this compound in acetone at a concentration 1,000 times more concentrated than the selected spike concentration is required. The QC check sample concentrate may be prepared from pure standard materials or purchased as certified solutions. Use appropriate trip, matrix, control site, method, reagent and solvent blanks. Internal, surrogate and five concentration level calibration standards are used.

REFERENCE: Method 8140, SW-846, 3rd ed., Sep 1986.

PRIMARY NAME: Dicamba

Method 8150

TITLE: Chlorinated Herbicides

MATRIX: groundwater, soils, sludges water miscible liquid wastes, and non-water miscible wastes.

CAS # : 1918-00-9

APPLICATION: This method is used for the analysis of 10 chlorinated herbicides. Samples are extracted, hydrolyzed with potassium hydroxide, and extraneous organics are removed by a solvent wash. After acidification, the acids are extracted, concentrated and converted to their methyl esters using diazomethane. They are then analyzed using direct injection into a gas chromatograph (GC). Be very careful because diazomethane can explode under certain conditions and it is also a carcinogen.

INTERFERENCES: Organic acids and phenols (especially chlorinated acids and phenols) may cause interferences. Phthalate esters are not as significant an interference as with other GC-ECD methods if an electron capture detector is used. The herbicides may react readily with alkaline substances and be lost during analysis so all glassware and glass wool must be acid rinsed and sodium sulfate must be acidified with sulfuric acid prior to use. Sensitivity usually depends on the level of interferences rather than on instrumentation.

INSTRUMENTATION: GC capable of on-column injections and an electron capture detector (ECD)or a halogen specific detector. Column 1: 1.8 meter by 4 mm with 1.5% SP-2250 / 1.95% SP-2401 on Supelcoport. Column 2: 1.8 meter by 4 mm with 5% OV-210 on Gas Chrom Q. Column 3: 1.98 meter by 2 mm with 0.1% SP-1000 on Carbopack C. The preferred column is Column Number 1 or 2.

RANGE: Not listed

MDL: 0.27 μg/L (in reagent water; ECD)

PRACTICAL QUANTITATION LIMIT FACTORS FOR MULTIPLYING TIMES FID MDL VALUE

Matrix	Multiplication Factor
Groundwater	10
Low-level soil by sonication with GPC cleanup	670
High-level soil and sludge by sonication	10,000
Non-water miscible waste	100,000

PRECISION: (as standard deviation) 7% with 1.2 μg/L spike in drinking water; 9% with 1.1 μg/L in municipal water.

ACCURACY: (as mean recovery) 79% with 1.2 μg/L spike in drinking water; 86% with 1.1 μg/L in municipal water.

SAMPLING METHOD: Use 8 oz. widemouth glass bottles with Teflon lined caps for concentrated waste samples, soils, sediments and sludges. Use 1 or 2 1/2 gallon amber glass bottles with Teflon lined caps for liquid (water) samples.

STABILITY: Cool soil, sediment, sludge and liquid samples to 4°C. If residual chlorine is present in liquid samples add 3 mL of 10% sodium thiosulfate per gallon of sample and cool to 4°C.

M.H.T. = 14 days for concentrated waste, soil, sediment or sludge.

M.H.T. = 7 days for liquid samples.

All extracts must be analyzed within 40 days.

QUALITY CONTROL: A quality control check sample concentrate containing this compound in acetone at a concentration 1,000 times more concentrated than the selected spike concentration is required. The QC check sample concentrate may be prepared from pure standard materials or purchased as certified solutions. Use appropriate trip, matrix, control site, method, reagent and solvent blanks. Internal, surrogate and five concentration level calibration standards are used.

REFERENCE: Method 8150, SW-846, 3rd ed., Sep 1986.

PRIMARY NAME: Dichloroprop

Method 8150

TITLE: Chlorinated Herbicides

MATRIX: groundwater, soils, sludges water miscible liquid wastes, and non-water miscible wastes.

CAS # : 120-36-5

APPLICATION: This method is used for the analysis of 10 chlorinated herbicides. Samples are extracted, hydrolyzed with potassium hydroxide, and extraneous organics are removed by a solvent wash. After acidification, the acids are extracted, concentrated and converted to their methyl esters using diazomethane. They are then analyzed using direct injection into a gas chromatograph (GC). Be very careful because diazomethane can explode under certain conditions and it is also a carcinogen.

INTERFERENCES: Organic acids and phenols (especially chlorinated acids and phenols) may cause interferences. Phthalate esters are not as significant an interference as with other GC-ECD methods if an electron capture detector is used. The herbicides may react readily with alkaline substances and be lost during analysis so all glassware and glass wool must be acid rinsed and sodium sulfate must be acidified with sulfuric acid prior to use. Sensitivity usually depends on the level of interferences rather than on instrumentation.

INSTRUMENTATION: GC capable of on-column injections and an electron capture detector (ECD)or a halogen specific detector. Column 1: 1.8 meter by 4 mm with 1.5% SP-2250 / 1.95% SP-2401 on Supelcoport. Column 2: 1.8 meter by 4 mm with 5% OV-210 on Gas Chrom Q. Column 3: 1.98 meter by 2 mm with 0.1% SP-1000 on Carbopack C. The preferred column is Column Number 1.

RANGE: Not listed

MDL: 0.65 μg/L (in reagent water; ECD)

PRACTICAL QUANTITATION LIMIT FACTORS FOR MULTIPLYING TIMES FID MDL VALUE

Matrix	Multiplication Factor
Groundwater	10
Low-level soil by sonication with GPC cleanup	670
High-level soil and sludge by sonication	10,000
Non-water miscible waste	100,000

PRECISION: (as standard deviation) 2% with 10.7 μg/L spike in drinking water; 3% with 10.7 μg/L in municipal water.

ACCURACY: (as mean recovery) 97% with 10.7 μg/L spike in drinking water; 72% with 10.7 μg/L in municipal water.

SAMPLING METHOD: Use 8 oz. widemouth glass bottles with Teflon lined caps for concentrated waste samples, soils, sediments and sludges. Use 1 or 2 1/2 gallon amber glass bottles with Teflon lined caps for liquid (water) samples.

STABILITY: Cool soil, sediment, sludge and liquid samples to 4°C. If residual chlorine is present in liquid samples add 3 mL of 10% sodium thiosulfate per gallon of sample and cool to 4°C.

M.H.T. = 14 days for concentrated waste, soil, sediment or sludge.

M.H.T. = 7 days for liquid samples.

All extracts must be analyzed within 40 days.

QUALITY CONTROL: A quality control check sample concentrate containing this compound in acetone at a concentration 1,000 times more concentrated than the selected spike concentration is required. The QC check sample concentrate may be prepared from pure standard materials or purchased as certified solutions. Use appropriate trip, matrix, control site, method, reagent and solvent blanks. Internal, surrogate and five concentration level calibration standards are used.

REFERENCE: Method 8150, SW-846, 3rd ed., Sep 1986.

PRIMARY NAME: Dichlorovos

Method 8140

TITLE: Organophosphorus Pesticides

MATRIX: groundwater, soils, sludges water miscible liquid wastes, and non-water miscible wastes.

CAS # : 62-73-7

APPLICATION: This method is used for the analysis of 21 organophosphorus pesticides. Samples are extracted, concentrated and analyzed using direct injection of both neat and diluted organic liquid into a gas chromatograph (GC).

INTERFERENCES: Solvents, reagents and glassware may introduce artifacts. Other interferences may come from coextracted compounds from samples. The use of Florisil cleanup materials may produce low recoveries. Elemental sulfur may interfere with some compounds when using a flame photometric detector. Sulfur cleanup (Method 3660) may alleviate sulfur interference.

INSTRUMENTATION: GC capable of on-column injections and a flame photometric detector (FPD) or a thermionic detector. A halogen specific detector may also be used and may have the advantage of fewer interferences. Column 1: 1.8 meter by 2 mm with 5% SP-2401 on Supelcoport. Column 2: 1.8 meter by 2 mm with 3% SP-2401 on Supelcoport. Column 3: 50 cm by 1/8 inch Teflon with 15% SE-54 on Gas Chrom Q. The preferred column is Column Number 1 or 3.

RANGE: 15.6 to 517 μg/L

MDL: 0.1 μg/L (in reagent water)

PRACTICAL QUANTITATION LIMIT FACTORS FOR MULTIPLYING TIMES FID MDL VALUE

Matrix	Multiplication Factor
Groundwater	10
Low-level soil by sonication with GPC cleanup	670
High-level soil and sludge by sonication	10,000
Non-water miscible waste	100,000

PRECISION: 7.7% (single operator standard deviation)

ACCURACY: 72.1% (single operator average recovery)

SAMPLING METHOD: Use 8 oz. widemouth glass bottles with Teflon lined caps for concentrated waste samples, soils, sediments and sludges. Use 1 or 2 1/2 gallon amber glass bottles with Teflon lined caps for liquid (water) samples.

STABILITY: Cool soil, sediment, sludge and liquid samples to 4°C. If residual chlorine is present in liquid samples add 3 mL of 10% sodium thiosulfate per gallon of sample and cool to 4°C.

M.H.T. = 14 days for concentrated waste, soil, sediment or sludge.

M.H.T. = 7 days for liquid samples.

All extracts must be analyzed within 40 days.

QUALITY CONTROL: A quality control check sample concentrate containing this compound in acetone at a concentration 1,000 times more concentrated than the selected spike concentration is required. The QC check sample concentrate may be prepared from pure standard materials or purchased as certified solutions. Use appropriate trip, matrix, control site, method, reagent and solvent blanks. Internal, surrogate and five concentration level calibration standards are used.

REFERENCE: Method 8140, SW-846, 3rd ed., Sep 1986.

PRIMARY NAME: Dieldrin — Method 8080

TITLE: Organochlorine Pesticides and PCBs

MATRIX: groundwater, soils, sludges water miscible liquid wastes, and non-water miscible wastes.

CAS # : 60-57-1

APPLICATION: This method is used for the analysis of 19 pesticides and 7 Aroclor (PCB) mixtures. Samples are extracted, concentrated and analyzed using direct injection of both neat and diluted organic liquid into a gas chromatograph (GC).

INTERFERENCES: Solvents, reagents and glassware may introduce artifacts. Other interferences may come from coextracted compounds from samples. Phthalate esters are common interferences when using an electron capture detector (ECD) so all plastics must be strictly avoided. Exhaustive cleanup of reagents and glassware may be required to eliminate phthalate contamination. Use of a halogen specific microcoulometric or electrolytic conductivity detector will eliminate phthalate interference.

INSTRUMENTATION: GC capable of on-column injections and an ECD or a halogen specific detector (HSD). Column 1: 1.8 meter by 4 mm with 1.5% SP-2250 / 1.95% SP-2401 on Supelcoport. Column 2: 1.8 meter by 4 mm with 3% OV-1 on Supelcoport.

RANGE: 8.5 to 400 μg/L

MDL: 0.002 μg/L (in reagent water)

PRACTICAL QUANTITATION LIMIT FACTORS FOR MULTIPLYING TIMES FID MDL VALUE

Matrix	Multiplication Factor
Groundwater	10
Low-level soil by sonication with GPC cleanup	670
High-level soil and sludge by sonication	10,000
Non-water miscible waste	100,000

PRECISION: 0.16X + 0.16 μg/L (overall precision)

ACCURACY: 0.90C + 0.02 μg/L (as recovery)

SAMPLING METHOD: Use 8 oz. widemouth glass bottles with Teflon lined caps for concentrated waste samples, soils, sediments and sludges. Use 1 or 2 1/2 gallon amber glass bottles with Teflon lined caps for liquid (water) samples.

STABILITY: Cool soil, sediment, sludge and liquid samples to 4°C. If residual chlorine is present in liquid samples add 3 mL of 10% sodium thiosulfate per gallon of sample and cool to 4°C.

M.H.T. = 14 days for concentrated waste, soil, sediment or sludge.

M.H.T. = 7 days for liquid samples.

All extracts must be analyzed within 40 days.

QUALITY CONTROL: A quality control check sample concentrate containing each analyte of interest is required. The QC check sample concentrate may be prepared from pure standard materials or purchased as certified solutions. Use appropriate trip, matrix, control site, method, reagent and solvent blanks. Internal, surrogate and five concentration level calibration standards are used. The quality control check sample concentrate should contain dieldrin at 2 μg/mL in acetone.

REFERENCE: Method 8080, SW-846, 3rd ed., Nov 1986.

PRIMARY NAME: Dinoseb

Method 8150

TITLE: Chlorinated Herbicides

MATRIX: groundwater, soils, sludges water miscible liquid wastes, and non-water miscible wastes.

CAS # : 88-85-7

APPLICATION: This method is used for the analysis of 10 chlorinated herbicides. Samples are extracted, hydrolyzed with potassium hydroxide, and extraneous organics are removed by a solvent wash. After acidification, the acids are extracted, concentrated and converted to their methyl esters using diazomethane. They are then analyzed using direct injection into a gas chromatograph (GC). Be very careful because diazomethane can explode under certain conditions and it is also a carcinogen.

INTERFERENCES: Organic acids and phenols (especially chlorinated acids and phenols) may cause interferences. Phthalate esters are not as significant an interference as with other GC-ECD methods if an electron capture detector is used. The herbicides may react readily with alkaline substances and be lost during analysis so all glassware and glass wool must be acid rinsed and sodium sulfate must be acidified with sulfuric acid prior to use. Sensitivity usually depends on the level of interferences rather than on instrumentation.

INSTRUMENTATION: GC capable of on-column injections and an electron capture detector (ECD)or a halogen specific detector. Column 1: 1.8 meter by 4 mm with 1.5% SP-2250 / 1.95% SP-2401 on Supelcoport. Column 2: 1.8 meter by 4 mm with 5% OV-210 on Gas Chrom Q. Column 3: 1.98 meter by 2 mm with 0.1% SP-1000 on Carbopack C. The preferred column is Column Number 1.

RANGE: Not listed

MDL: 0.07 μg/L (in reagent water; ECD)

PRACTICAL QUANTITATION LIMIT FACTORS FOR MULTIPLYING TIMES FID MDL VALUE

Matrix	Multiplication Factor
Groundwater	10
Low-level soil by sonication with GPC cleanup	670
High-level soil and sludge by sonication	10,000
Non-water miscible waste	100,000

PRECISION: (as standard deviation) 4% with 0.5 μg/L spike in municipal water; 3% with 102 μg/L in municipal water.

ACCURACY: (as mean recovery) 86% with 0.5 μg/L spike in municipal water; 81% with 102 μg/L in municipal water.

SAMPLING METHOD: Use 8 oz. widemouth glass bottles with Teflon lined caps for concentrated waste samples, soils, sediments and sludges. Use 1 or 2 1/2 gallon amber glass bottles with Teflon lined caps for liquid (water) samples.

STABILITY: Cool soil, sediment, sludge and liquid samples to 4°C. If residual chlorine is present in liquid samples add 3 mL of 10% sodium thiosulfate per gallon of sample and cool to 4°C.

M.H.T. = 14 days for concentrated waste, soil, sediment or sludge.

M.H.T. = 7 days for liquid samples.

All extracts must be analyzed within 40 days.

QUALITY CONTROL: A quality control check sample concentrate containing this compound in acetone at a concentration 1,000 times more concentrated than the selected spike concentration is required. The QC check sample concentrate may be prepared from pure standard materials or purchased as certified solutions. Use appropriate trip, matrix, control site, method, reagent and solvent blanks. Internal, surrogate and five concentration level calibration standards are used.

REFERENCE: Method 8150, SW-846, 3rd ed., Sep 1986.

PRIMARY NAME: Disulfoton

Method 8140

TITLE: Organophosphorus Pesticides

MATRIX: groundwater, soils, sludges water miscible liquid wastes, and non-water miscible wastes.

CAS # : 298-04-4

APPLICATION: This method is used for the analysis of 21 organophosphorus pesticides. Samples are extracted, concentrated and analyzed using direct injection of both neat and diluted organic liquid into a gas chromatograph (GC).

INTERFERENCES: Solvents, reagents and glassware may introduce artifacts. Other interferences may come from coextracted compounds from samples. The use of Florisil cleanup materials may produce low recoveries. Elemental sulfur may interfere with some compounds when using a flame photometric detector. Sulfur cleanup (Method 3660) may alleviate sulfur interference.

INSTRUMENTATION: GC capable of on-column injections and a flame photometric detector (FPD) or a thermionic detector. Column 1: 1.8 meter by 2 mm with 5% SP-2401 on Supelcoport. Column 2: 1.8 meter by 2 mm with 3% SP-2401 on Supelcoport. Column 3: 50 cm by 1/8 inch Teflon with 15% SE-54 on Gas Chrom Q. The preferred column is Column Number 1.

RANGE: 5.2 to 92 μg/L

MDL: 0.20 μg/L (in reagent water)

PRACTICAL QUANTITATION LIMIT FACTORS FOR MULTIPLYING TIMES FID MDL VALUE

Matrix	Multiplication Factor
Groundwater	10
Low-level soil by sonication with GPC cleanup	670
High-level soil and sludge by sonication	10,000
Non-water miscible waste	100,000

PRECISION: 9.0% (single operator standard deviation)

ACCURACY: 81.9% (single operator average recovery)

SAMPLING METHOD: Use 8 oz. widemouth glass bottles with Teflon lined caps for concentrated waste samples, soils, sediments and sludges. Use 1 or 2 1/2 gallon amber glass bottles with Teflon lined caps for liquid (water) samples.

STABILITY: Cool soil, sediment, sludge and liquid samples to 4°C. If residual chlorine is present in liquid samples add 3 mL of 10% sodium thiosulfate per gallon of sample and cool to 4°C.

M.H.T. = 14 days for concentrated waste, soil, sediment or sludge.

M.H.T. = 7 days for liquid samples.

All extracts must be analyzed within 40 days.

QUALITY CONTROL: A quality control check sample concentrate containing this compound in acetone at a concentration 1,000 times more concentrated than the selected spike concentration is required. The QC check sample concentrate may be prepared from pure standard materials or purchased as certified solutions. Use appropriate trip, matrix, control site, method, reagent and solvent blanks. Internal, surrogate and five concentration level calibration standards are used.

REFERENCE: Method 8140, SW-846, 3rd ed., Sep 1986.

PRIMARY NAME: Endosulfan I | Method 8080

TITLE: Organochlorine Pesticides and PCBs

MATRIX: groundwater, soils, sludges water miscible liquid wastes, and non-water miscible wastes.

CAS # : 959-98-8

APPLICATION: This method is used for the analysis of 19 pesticides and 7 Aroclor (PCB) mixtures. Samples are extracted, concentrated and analyzed using direct injection of both neat and diluted organic liquid into a gas chromatograph (GC).

INTERFERENCES: Solvents, reagents and glassware may introduce artifacts. Other interferences may come from coextracted compounds from samples.

Phthalate esters are common interferences when using an electron capture detector (ECD) so all plastics must be strictly avoided. Exhaustive cleanup of reagents and glassware may be required to eliminate phthalate contamination. Use of a halogen specific microcoulometric or electrolytic conductivity detector will eliminate phthalate interference.

INSTRUMENTATION: GC capable of on-column injections and an ECD or a halogen specific detector (HSD). Column 1: 1.8 meter by 4 mm with 1.5% SP-2250 / 1.95% SP-2401 on Supelcoport. Column 2: 1.8 meter by 4 mm with 3% OV-1 on Supelcoport.

RANGE: 8.5 to 400 µg/L

MDL: 0.014 µg/L (in reagent water)

PRACTICAL QUANTITATION LIMIT FACTORS FOR MULTIPLYING TIMES FID MDL VALUE

Matrix	Multiplication Factor
Groundwater	10
Low-level soil by sonication with GPC cleanup	670
High-level soil and sludge by sonication	10,000
Non-water miscible waste	100,000

PRECISION: 0.18X + 0.08 µg/L (overall precision)

ACCURACY: 0.97C + 0.04 µg/L (as recovery)

SAMPLING METHOD: Use 8 oz. widemouth glass bottles with Teflon lined caps for concentrated waste samples, soils, sediments and sludges. Use 1 or 2 1/2 gallon amber glass bottles with Teflon lined caps for liquid (water) samples.

STABILITY: Cool soil, sediment, sludge and liquid samples to 4°C. If residual chlorine is present in liquid samples add 3 mL of 10% sodium thiosulfate per gallon of sample and cool to 4°C.

M.H.T. = 14 days for concentrated waste, soil, sediment or sludge.

M.H.T. = 7 days for liquid samples.

All extracts must be analyzed within 40 days.

QUALITY CONTROL: A quality control check sample concentrate containing each analyte of interest is required. The QC check sample concentrate may be prepared from pure standard materials or purchased as certified solutions. Use appropriate trip, matrix, control site, method, reagent and solvent blanks. Internal, surrogate and five concentration level calibration standards are used. The quality control check sample concentrate should contain endosulfan I at 2 μg/mL in acetone.

REFERENCE: Method 8080, SW-846, 3rd ed., Nov 1986.

PRIMARY NAME: Endosulfan II — Method 8080

TITLE: Organochlorine Pesticides and PCBs

MATRIX: groundwater, soils, sludges water miscible liquid wastes, and non-water miscible wastes.

CAS # : 33213-65-9

APPLICATION: This method is used for the analysis of 19 pesticides and 7 Aroclor (PCB) mixtures. Samples are extracted, concentrated and analyzed using direct injection of both neat and diluted organic liquid into a gas chromatograph (GC).

INTERFERENCES: Solvents, reagents and glassware may introduce artifacts. Other interferences may come from coextracted compounds from samples. Phthalate esters are common interferences when using an electron capture detector (ECD) so all plastics must be strictly avoided. Exhaustive cleanup of reagents and glassware may be required to eliminate phthalate contamination. Use of a halogen specific microcoulometric or electrolytic conductivity detector will eliminate phthalate interference.

INSTRUMENTATION: GC capable of on-column injections and an ECD or a halogen specific detector (HSD). Column 1: 1.8 meter by 4 mm with 1.5% SP-2250 / 1.95% SP-2401 on Supelcoport. Column 2: 1.8 meter by 4 mm with 3% OV-1 on Supelcoport.

RANGE: 8.5 to 400 μg/L

MDL: 0.004 μg/L (in reagent water)

PRACTICAL QUANTITATION LIMIT FACTORS FOR MULTIPLYING TIMES FID MDL VALUE

Matrix	Multiplication Factor
Groundwater	10
Low-level soil by sonication with GPC cleanup	670
High-level soil and sludge by sonication	10,000
Non-water miscible waste	100,000

PRECISION: 0.47X - 0.20 μg/L (overall precision)

ACCURACY: 0.93C + 0.34 μg/L (as recovery)

SAMPLING METHOD: Use 8 oz. widemouth glass bottles with Teflon lined caps for concentrated waste samples, soils, sediments and sludges. Use 1 or 2 1/2 gallon amber glass bottles with Teflon lined caps for liquid (water) samples.

STABILITY: Cool soil, sediment, sludge and liquid samples to 4°C. If residual chlorine is present in liquid samples add 3 mL of 10% sodium thiosulfate per gallon of sample and cool to 4°C.

M.H.T. = 14 days for concentrated waste, soil, sediment or sludge.

M.H.T. = 7 days for liquid samples.

All extracts must be analyzed within 40 days.

QUALITY CONTROL: A quality control check sample concentrate containing each analyte of interest is required. The QC check sample concentrate may be prepared from pure standard materials or purchased as certified solutions. Use appropriate trip, matrix, control site, method, reagent and solvent blanks. Internal, surrogate and five concentration level calibration standards are used. The quality control check sample concentrate should contain endosulfan II at 10 μg/mL in acetone.

REFERENCE: Method 8080, SW-846, 3rd ed., Nov 1986.

PRIMARY NAME: Endosulfan sulfate

Method 8080

TITLE: Organochlorine Pesticides and PCBs

MATRIX: groundwater, soils, sludges water miscible liquid wastes, and non-water miscible wastes.

CAS # : 1031-07-8

APPLICATION: This method is used for the analysis of 19 pesticides and 7 Aroclor (PCB) mixtures. Samples are extracted, concentrated and analyzed using direct injection of both neat and diluted organic liquid into a gas chromatograph (GC).

INTERFERENCES: Solvents, reagents and glassware may introduce artifacts. Other interferences may come from coextracted compounds from samples. Phthalate esters are common interferences when using an electron capture detector (ECD) so all plastics must be strictly avoided. Exhaustive cleanup of reagents and glassware may be required to eliminate phthalate contamination. Use of a halogen specific microcoulometric or electrolytic conductivity detector will eliminate phthalate interference.

INSTRUMENTATION: GC capable of on-column injections and an ECD or a halogen specific detector (HSD). Column 1: 1.8 meter by 4 mm with 1.5% SP-2250 / 1.95% SP-2401 on Supelcoport. Column 2: 1.8 meter by 4 mm with 3% OV-1 on Supelcoport.

RANGE: 8.5 to 400 μg/L

MDL: 0.066 μg/L (in reagent water)

PRACTICAL QUANTITATION LIMIT FACTORS FOR MULTIPLYING TIMES FID MDL VALUE

Matrix	Multiplication Factor
Groundwater	10
Low-level soil by sonication with GPC cleanup	670
High-level soil and sludge by sonication	10,000
Non-water miscible waste	100,000

PRECISION: 0.24X + 0.35 μg/L (overall precision)

ACCURACY: 0.89C - 0.37 μg/L (as recovery)

SAMPLING METHOD: Use 8 oz. widemouth glass bottles with Teflon lined caps for concentrated waste samples, soils, sediments and sludges. Use 1 or 2 1/2 gallon amber glass bottles with Teflon lined caps for liquid (water) samples.

STABILITY: Cool soil, sediment, sludge and liquid samples to 4°C. If residual chlorine is present in liquid samples add 3 mL of 10% sodium thiosulfate per gallon of sample and cool to 4°C.

M.H.T. = 14 days for concentrated waste, soil, sediment or sludge.

M.H.T. = 7 days for liquid samples.

All extracts must be analyzed within 40 days.

QUALITY CONTROL: A quality control check sample concentrate containing each analyte of interest is required. The QC check sample concentrate may be prepared from pure standard materials or purchased as certified solutions. Use appropriate trip, matrix, control site, method, reagent and solvent blanks. Internal, surrogate and five concentration level calibration standards are used. The quality control check sample concentrate should contain endosulfan sulfate at 10 μg/mL in acetone.

REFERENCE: Method 8080, SW-846, 3rd ed., Nov 1986.

PRIMARY NAME: Endrin

Method 8080

TITLE: Organochlorine Pesticides and PCBs

MATRIX: groundwater, soils, sludges water miscible liquid wastes, and non-water miscible wastes.

CAS # : 72-20-8

APPLICATION: This method is used for the analysis of 19 pesticides and 7 Aroclor (PCB) mixtures. Samples are extracted, concentrated and analyzed using direct injection of both neat and diluted organic liquid into a gas chromatograph (GC).

INTERFERENCES: Solvents, reagents and glassware may introduce artifacts. Other interferences may come from coextracted compounds from samples.

Phthalate esters are common interferences when using an electron capture detector (ECD) so all plastics must be strictly avoided. Exhaustive cleanup of reagents and glassware may be required to eliminate phthalate contamination. Use of a halogen specific microcoulometric or electrolytic conductivity detector will eliminate phthalate interference. If the injection port or front of the GC column is dirty endrin may degrade to endrin ketone and endrin aldehyde.

INSTRUMENTATION: GC capable of on-column injections and an ECD or a halogen specific detector (HSD). Column 1: 1.8 meter by 4 mm with 1.5% SP-2250 / 1.95% SP-2401 on Supelcoport. Column 2: 1.8 meter by 4 mm with 3% OV-1 on Supelcoport.

RANGE: 8.5 to 400 μg/L

MDL: 0.006 μg/L (in reagent water)

PRACTICAL QUANTITATION LIMIT FACTORS FOR MULTIPLYING TIMES FID MDL VALUE

Matrix	Multiplication Factor
Groundwater	10
Low-level soil by sonication with GPC cleanup	670
High-level soil and sludge by sonication	10,000
Non-water miscible waste	100,000

PRECISION: 0.24X + 0.25 μg/L (overall precision)

ACCURACY: 0.89C - 0.04 μg/L (as recovery)

SAMPLING METHOD: Use 8 oz. widemouth glass bottles with Teflon lined caps for concentrated waste samples, soils, sediments and sludges. Use 1 or 2 1/2 gallon amber glass bottles with Teflon lined caps for liquid (water) samples.

STABILITY: Cool soil, sediment, sludge and liquid samples to 4°C. If residual chlorine is present in liquid samples add 3 mL of 10% sodium thiosulfate per gallon of sample and cool to 4°C.

M.H.T. = 14 days for concentrated waste, soil, sediment or sludge.

M.H.T. = 7 days for liquid samples.

All extracts must be analyzed within 40 days.

QUALITY CONTROL: A quality control check sample concentrate containing each analyte of interest is required. The QC check sample concentrate may be prepared from pure standard materials or purchased as certified solutions. Use appropriate trip, matrix, control site, method, reagent and solvent blanks. Internal, surrogate and five concentration level calibration standards are used. The quality control check sample concentrate should contain endrin at 10 μg/mL in acetone.

REFERENCE: Method 8080, SW-846, 3rd ed., Nov 1986.

PRIMARY NAME: Endrin aldehyde Method 8080

TITLE: Organochlorine Pesticides and PCBs

MATRIX: groundwater, soils, sludges water miscible liquid wastes, and non-water miscible wastes.

CAS # : 7421-93-4

APPLICATION: This method is used for the analysis of 19 pesticides and 7 Aroclor (PCB) mixtures. Samples are extracted, concentrated and analyzed using direct injection of both neat and diluted organic liquid into a gas chromatograph (GC).

INTERFERENCES: Solvents, reagents and glassware may introduce artifacts. Other interferences may come from coextracted compounds from samples. Phthalate esters are common interferences when using an electron capture detector (ECD) so all plastics must be strictly avoided. Exhaustive cleanup of reagents and glassware may be required to eliminate phthalate contamination. Use of a halogen specific microcoulometric or electrolytic conductivity detector will eliminate phthalate interference.

INSTRUMENTATION: GC capable of on-column injections and an ECD or a halogen specific detector (HSD). Column 1: 1.8 meter by 4 mm with 1.5% SP-2250 / 1.95% SP-2401 on Supelcoport. Column 2: 1.8 meter by 4 mm with 3% OV-1 on Supelcoport.

RANGE: 8.5 to 400 μg/L

MDL: 0.023 μg/L (in reagent water)

PRACTICAL QUANTITATION LIMIT FACTORS FOR MULTIPLYING TIMES FID MDL VALUE

Matrix	Multiplication Factor
Groundwater	10
Low-level soil by sonication with GPC cleanup	670
High-level soil and sludge by sonication	10,000
Non-water miscible waste	100,000

PRECISION: Not determined

ACCURACY: Not determined

SAMPLING METHOD: Use 8 oz. widemouth glass bottles with Teflon lined caps for concentrated waste samples, soils, sediments and sludges. Use 1 or 2 1/2 gallon amber glass bottles with Teflon lined caps for liquid (water) samples.

STABILITY: Cool soil, sediment, sludge and liquid samples to 4°C. If residual chlorine is present in liquid samples add 3 mL of 10% sodium thiosulfate per gallon of sample and cool to 4°C.

M.H.T. = 14 days for concentrated waste, soil, sediment or sludge.

M.H.T. = 7 days for liquid samples.

All extracts must be analyzed within 40 days.

QUALITY CONTROL: A quality control check sample concentrate containing each analyte of interest is required. The QC check sample concentrate may be prepared from pure standard materials or purchased as certified solutions. Use appropriate trip, matrix, control site, method, reagent and solvent blanks. Internal, surrogate and five concentration level calibration standards are used. The quality control check sample concentrate should contain endrin aldehyde at 2 μg/mL in acetone.

REFERENCE: Method 8080, SW-846, 3rd ed., Nov 1986.

PRIMARY NAME: Ethoprop

Method 8140

TITLE: Organophosphorus Pesticides

MATRIX: groundwater, soils, sludges water miscible liquid wastes, and non-water miscible wastes.

CAS # : 13194-48-4

APPLICATION: This method is used for the analysis of 21 organophosphorus pesticides. Samples are extracted, concentrated and analyzed using direct injection of both neat and diluted organic liquid into a gas chromatograph (GC).

INTERFERENCES: Solvents, reagents and glassware may introduce artifacts. Other interferences may come from coextracted compounds from samples. The use of Florisil cleanup materials may produce low recoveries. Elemental sulfur may interfere with some compounds when using a flame photometric detector. Sulfur cleanup (Method 3660) may alleviate sulfur interference.

INSTRUMENTATION: GC capable of on-column injections and a flame photometric detector (FPD) or a thermionic detector. Column 1: 1.8 meter by 2 mm with 5% SP-2401 on Supelcoport. Column 2: 1.8 meter by 2 mm with 3% SP-2401 on Supelcoport. Column 3: 50 cm by 1/8 inch Teflon with 15% SE-54 on Gas Chrom Q. The preferred column is Column Number 2.

RANGE: 1.0 to 51.5 μg/L

MDL: 0.25 μg/L (in reagent water)

PRACTICAL QUANTITATION LIMIT FACTORS FOR MULTIPLYING TIMES FID MDL VALUE

Matrix	Multiplication Factor
Groundwater	10
Low-level soil by sonication with GPC cleanup	670
High-level soil and sludge by sonication	10,000
Non-water miscible waste	100,000

PRECISION: 4.1% (single operator standard deviation)

ACCURACY: 100.5% (single operator average recovery)

SAMPLING METHOD: Use 8 oz. widemouth glass bottles with Teflon lined caps for concentrated waste samples, soils, sediments and sludges. Use 1 or 2 1/2 gallon amber glass bottles with Teflon lined caps for liquid (water) samples.

STABILITY: Cool soil, sediment, sludge and liquid samples to 4°C. If residual chlorine is present in liquid samples add 3 mL of 10% sodium thiosulfate per gallon of sample and cool to 4°C.

M.H.T. = 14 days for concentrated waste, soil, sediment or sludge.

M.H.T. = 7 days for liquid samples.

All extracts must be analyzed within 40 days.

QUALITY CONTROL: A quality control check sample concentrate containing this compound in acetone at a concentration 1,000 times more concentrated than the selected spike concentration is required. The QC check sample concentrate may be prepared from pure standard materials or purchased as certified solutions. Use appropriate trip, matrix, control site, method, reagent and solvent blanks. Internal, surrogate and five concentration level calibration standards are used.

REFERENCE: Method 8140, SW-846, 3rd ed., Sep 1986.

PRIMARY NAME: Fensulfothion — Method 8140

TITLE: Organophosphorus Pesticides

MATRIX: groundwater, soils, sludges water miscible liquid wastes, and non-water miscible wastes.

CAS # : 115-90-2

APPLICATION: This method is used for the analysis of 21 organophosphorus pesticides. Samples are extracted, concentrated and analyzed using direct injection of both neat and diluted organic liquid into a gas chromatograph (GC).

INTERFERENCES: Solvents, reagents and glassware may introduce artifacts. Other interferences may come from coextracted compounds from samples. The use of Florisil cleanup materials may produce low recoveries. Elemental sulfur may interfere with some compounds when using a flame photometric detector. Sulfur cleanup (Method 3660) may alleviate sulfur interference.

INSTRUMENTATION: GC capable of on-column injections and a flame photometric detector (FPD) or a thermionic detector. Column 1: 1.8 meter by 2 mm with 5% SP-2401 on Supelcoport. Column 2: 1.8 meter by 2 mm with 3% SP-2401 on Supelcoport. Column 3: 50 cm by 1/8 inch Teflon with 15% SE-54 on Gas Chrom Q. The preferred column is Column Number 1.

RANGE: 23.9 to 110 µg/L

MDL: 1.5 µg/L (in reagent water)

PRACTICAL QUANTITATION LIMIT FACTORS FOR MULTIPLYING TIMES FID MDL VALUE

Matrix	Multiplication Factor
Groundwater	10
Low-level soil by sonication with GPC cleanup	670
High-level soil and sludge by sonication	10,000
Non-water miscible waste	100,000

PRECISION: 17.1% (single operator standard deviation)

ACCURACY: 94.1% (single operator average recovery)

SAMPLING METHOD: Use 8 oz. widemouth glass bottles with Teflon lined caps for concentrated waste samples, soils, sediments and sludges. Use 1 or 2 1/2 gallon amber glass bottles with Teflon lined caps for liquid (water) samples.

STABILITY: Cool soil, sediment, sludge and liquid samples to 4°C. If residual chlorine is present in liquid samples add 3 mL of 10% sodium thiosulfate per gallon of sample and cool to 4°C.

M.H.T. = 14 days for concentrated waste, soil, sediment or sludge.

M.H.T. = 7 days for liquid samples.

All extracts must be analyzed within 40 days.

QUALITY CONTROL: A quality control check sample concentrate containing this compound in acetone at a concentration 1,000 times more concentrated than the selected spike concentration is required. The QC check sample concentrate may be prepared from pure standard materials or purchased as certified solutions. Use appropriate trip, matrix, control site, method, reagent and solvent blanks. Internal, surrogate and five concentration level calibration standards are used.

REFERENCE: Method 8140, SW-846, 3rd ed., Sep 1986.

PRIMARY NAME: Fenthion

Method 8140

TITLE: Organophosphorus Pesticides

MATRIX: groundwater, soils, sludges water miscible liquid wastes, and non-water miscible wastes.

CAS # : 55-38-9

APPLICATION: This method is used for the analysis of 21 organophosphorus pesticides. Samples are extracted, concentrated and analyzed using direct injection of both neat and diluted organic liquid into a gas chromatograph (GC).

INTERFERENCES: Solvents, reagents and glassware may introduce artifacts. Other interferences may come from coextracted compounds from samples. The use of Florisil cleanup materials may produce low recoveries. Elemental sulfur may interfere with some compounds when using a flame photometric detector. Sulfur cleanup (Method 3660) may alleviate sulfur interference.

INSTRUMENTATION: GC capable of on-column injections and a flame photometric detector (FPD) or a thermionic detector. Column 1: 1.8 meter by 2 mm with 5% SP-2401 on Supelcoport. Column 2: 1.8 meter by 2 mm with 3% SP-2401 on Supelcoport. Column 3: 50 cm by 1/8 inch Teflon with 15% SE-54 on Gas Chrom Q. The preferred column is Column Number 1.

RANGE: 5.3 to 64 μg/L

MDL: 0.10 μg/L (in reagent water)

PRACTICAL QUANTITATION LIMIT FACTORS FOR MULTIPLYING TIMES FID MDL VALUE

Matrix	Multiplication Factor
Groundwater	10
Low-level soil by sonication with GPC cleanup	670
High-level soil and sludge by sonication	10,000
Non-water miscible waste	100,000

PRECISION: 19.9% (single operator standard deviation)

ACCURACY: 68.7% (single operator average recovery)

SAMPLING METHOD: Use 8 oz. widemouth glass bottles with Teflon lined caps for concentrated waste samples, soils, sediments and sludges. Use 1 or 2 1/2 gallon amber glass bottles with Teflon lined caps for liquid (water) samples.

STABILITY: Cool soil, sediment, sludge and liquid samples to 4°C. If residual chlorine is present in liquid samples add 3 mL of 10% sodium thiosulfate per gallon of sample and cool to 4°C.

M.H.T. = 14 days for concentrated waste, soil, sediment or sludge.

M.H.T. = 7 days for liquid samples.

All extracts must be analyzed within 40 days.

QUALITY CONTROL: A quality control check sample concentrate containing this compound in acetone at a concentration 1,000 times more concentrated than the selected spike concentration is required. The QC check sample concentrate may be prepared from pure standard materials or purchased as certified solutions. Use appropriate trip, matrix, control site, method, reagent and solvent blanks. Internal, surrogate and five concentration level calibration standards are used.

REFERENCE: Method 8140, SW-846, 3rd ed., Sep 1986.

PRIMARY NAME: Heptaclor

Method 8080

TITLE: Organochlorine Pesticides and PCBs

MATRIX: groundwater, soils, sludges water miscible liquid wastes, and non-water miscible wastes.

CAS # : 76-44-8

APPLICATION: This method is used for the analysis of 19 pesticides and 7 Aroclor (PCB) mixtures. Samples are extracted, concentrated and analyzed using direct injection of both neat and diluted organic liquid into a gas chromatograph (GC).

INTERFERENCES: Solvents, reagents and glassware may introduce artifacts. Other interferences may come from coextracted compounds from samples.

Phthalate esters are common interferences when using an electron capture detector (ECD) so all plastics must be strictly avoided. Exhaustive cleanup of reagents and glassware may be required to eliminate phthalate contamination. Use of a halogen specific microcoulometric or electrolytic conductivity detector will eliminate phthalate interference.

INSTRUMENTATION: GC capable of on-column injections and an ECD or a halogen specific detector (HSD). Column 1: 1.8 meter by 4 mm with 1.5% SP-2250 / 1.95% SP-2401 on Supelcoport. Column 2: 1.8 meter by 4 mm with 3% OV-1 on Supelcoport.

RANGE: 8.5 to 400 μg/L

MDL: 0.003 μg/L (in reagent water)

PRACTICAL QUANTITATION LIMIT FACTORS FOR MULTIPLYING TIMES FID MDL VALUE

Matrix	Multiplication Factor
Groundwater	10
Low-level soil by sonication with GPC cleanup	670
High-level soil and sludge by sonication	10,000
Non-water miscible waste	100,000

PRECISION: 0.16X + 0.08 μg/L (overall precision)

ACCURACY: 0.69C + 0.04 μg/L (as recovery)

SAMPLING METHOD: Use 8 oz. widemouth glass bottles with Teflon lined caps for concentrated waste samples, soils, sediments and sludges. Use 1 or 2 1/2 gallon amber glass bottles with Teflon lined caps for liquid (water) samples.

STABILITY: Cool soil, sediment, sludge and liquid samples to 4°C. If residual chlorine is present in liquid samples add 3 mL of 10% sodium thiosulfate per gallon of sample and cool to 4°C.

M.H.T. = 14 days for concentrated waste, soil, sediment or sludge.

M.H.T. = 7 days for liquid samples.

All extracts must be analyzed within 40 days.

QUALITY CONTROL: A quality control check sample concentrate containing each analyte of interest is required. The QC check sample concentrate may be prepared from pure standard materials or purchased as certified solutions. Use appropriate trip, matrix, control site, method, reagent and solvent blanks. Internal, surrogate and five concentration level calibration standards are used. The quality control check sample concentrate should contain heptachlor at 2 μg/mL in acetone.

REFERENCE: Method 8080, SW-846, 3rd ed., Nov 1986.

PRIMARY NAME: Heptaclor epoxide

Method 8080

TITLE: Organochlorine Pesticides and PCBs

MATRIX: groundwater, soils, sludges water miscible liquid wastes, and non-water miscible wastes.

CAS # : 1024-57-3

APPLICATION: This method is used for the analysis of 19 pesticides and 7 Aroclor (PCB) mixtures. Samples are extracted, concentrated and analyzed using direct injection of both neat and diluted organic liquid into a gas chromatograph (GC).

INTERFERENCES: Solvents, reagents and glassware may introduce artifacts. Other interferences may come from coextracted compounds from samples. Phthalate esters are common interferences when using an electron capture detector (ECD) so all plastics must be strictly avoided. Exhaustive cleanup of reagents and glassware may be required to eliminate phthalate contamination. Use of a halogen specific microcoulometric or electrolytic conductivity detector will eliminate phthalate interference.

INSTRUMENTATION: GC capable of on-column injections and an ECD or a halogen specific detector (HSD). Column 1: 1.8 meter by 4 mm with 1.5% SP-2250 / 1.95% SP-2401 on Supelcoport. Column 2: 1.8 meter by 4 mm with 3% OV-1 on Supelcoport.

RANGE: 8.5 to 400 μg/L

MDL: 0.083 μg/L (in reagent water)

PRACTICAL QUANTITATION LIMIT FACTORS FOR MULTIPLYING TIMES FID MDL VALUE

Matrix	Multiplication Factor
Groundwater	10
Low-level soil by sonication with GPC cleanup	670
High-level soil and sludge by sonication	10,000
Non-water miscible waste	100,000

PRECISION: 0.25X - 0.08 μg/L (overall precision)

ACCURACY: 0.89C + 0.10 μg/L (as recovery)

SAMPLING METHOD: Use 8 oz. widemouth glass bottles with Teflon lined caps for concentrated waste samples, soils, sediments and sludges. Use 1 or 2 1/2 gallon amber glass bottles with Teflon lined caps for liquid (water) samples.

STABILITY: Cool soil, sediment, sludge and liquid samples to 4°C. If residual chlorine is present in liquid samples add 3 mL of 10% sodium thiosulfate per gallon of sample and cool to 4°C.

M.H.T. = 14 days for concentrated waste, soil, sediment or sludge.

M.H.T. = 7 days for liquid samples.

All extracts must be analyzed within 40 days.

QUALITY CONTROL: A quality control check sample concentrate containing each analyte of interest is required. The QC check sample concentrate may be prepared from pure standard materials or purchased as certified solutions. Use appropriate trip, matrix, control site, method, reagent and solvent blanks. Internal, surrogate and five concentration level calibration standards are used. The quality control check sample concentrate should contain heptachlor epoxide at 2 μg/mL in acetone.

REFERENCE: Method 8080, SW-846, 3rd ed., Nov 1986.

PRIMARY NAME: 1,2,3,4,6,7,8-HpCDD

Method 8280

TITLE: Analysis of PCDDs and PCDFs

MATRIX: chemical wastes, fuel oils still bottoms, sludges, water, soil, fly ash, reactor residues.

CAS # : 35822-46-9

APPLICATION: This method is used for the analysis of tetra-, penta-, hexa-, hepta-, and octachlorinated dibenzo-p-dioxins (PCDDs) and dibenzofurans (PCDFs). The sensitivity of the method is dependent on the level of interferents within the matrix. Only experienced analysts should be used. Special safety precautions must be observed and an EPA-approved sample disposal plan must be used.

INTERFERENCES: Solvents, reagents and glassware may introduce artifacts. Other interferences may come from coextracted compounds from samples; PCBs and polychlorinated diphenyl ethers are common interferents.

INSTRUMENTATION: GC/MS with a fused silica capillary column. Also, solvent extraction and concentration glassware and either a gravity flow activated carbon AX-21 / silica gel Type 60 EM reagent column or a HPLC with a 10 mm by 7 cm silanized glass column with active carbon AX-21 and Spherisorb S10W silica for sample cleanup. One of three fused silica capillary GC columns may be used: column 1: 50 meter CP-Sil-88; Column 2: 30 meter by 0.25 mm DB-5; colunm 3: 30 meter SP-2250.

RANGE: 50 to 6,000 picograms

MDL: 2.77 ng/L (in reagent water) 1.87 µg/Kg (in Missouri soil) 1.41 µg/Kg (in fly ash) 4.65 µg/Kg (in industrial sludge) 4.59 µg/Kg (in still bottom) 8.14 µg/Kg (in fuel oil)
MDLs are for carbon-13 labeled analyte

PRECISION (as RSD): Not determined

ACCURACY (as Mean % Recovery): Not determined

SAMPLING METHOD: Use 1 liter (or quart) amber glass bottles with Teflon lined or solvent washed foil screw caps. Tape caps to bottle after sampling.

Compositing equipment must use glass containers and contain no Tygon or rubber tubing. Sample bottles must not be prewashed with the sample before its collection. Aqueous samples cannot be aliquoted from sample containers the entire sample must be used and the container is washed out with the extracting solvent.

STABILITY: Cool to 4°C and store at this temperature. When toluene is employed as the final solvent use a bonded phase GC column for separation. Otherwise, solvent exchange into tridecane is required for other liquid phases or the CP-Sil-88 GC column.

M.H.T. = 30 days;
samples must be completely analyzed within 45 days of collection.

QUALITY CONTROL: A method blank must be analyzed each time a set of samples is extracted or there is a change in reagents. A laboratory method blank must be run with each analytical batch of 20 or fewer samples. Field duplicates and field blanks should be analyzed periodically. GC column performance must be demonstrated initially and verified prior to analyzing any sample in a 12 hour period. A series of calibration standards must be processed through the procedure to validate elution patterns and absence of interferents from reagents. Both the alumina column and carbon column performance must be routinely checked for presence of the analyte. Performance evaluation samples and split samples with other laboratories are also expected to be periodically analyzed.

REFERENCE: Method 8280, SW-846, 3rd ed., Nov 1986.

PRIMARY NAME: 1,2,3,4,7,8-HxCDD

Method 8280

TITLE: Analysis of PCDDs and PCDFs

MATRIX: chemical wastes, fuel oils still bottoms, sludges, water, soil, fly ash, reactor residues.

CAS # : 57653-85-7

APPLICATION: This method is used for the analysis of tetra-, penta-, hexa-, hepta-, and octachlorinated dibenzo-p-dioxins (PCDDs) and dibenzofurans (PCDFs). The sensitivity of the method is dependent on the level of interferents within the matrix. Only experienced analysts should be used. Special safety precautions must be observed and an EPA-approved sample disposal plan must be used.

INTERFERENCES: Solvents, reagents and glassware may introduce artifacts. Other interferences may come from coextracted compounds from samples; PCBs and polychlorinated diphenyl ethers are common interferents.

INSTRUMENTATION: GC/MS with a fused silica capillary column. Also, solvent extraction and concentration glassware and either a gravity flow activated carbon AX-21 / silica gel Type 60 EM reagent column or a HPLC with a 10 mm by 7 cm silanized glass column with active carbon AX-21 and Spherisorb S10W silica for sample cleanup. One of three fused silica capillary GC columns may be used: column 1: 50 meter CP-Sil-88; Column 2: 30 meter by 0.25 mm DB-5; colunm 3: 30 meter SP-2250.

RANGE: 50 to 6,000 picograms MDL: Not determined

PRECISION (as RSD): 38% with 5 ng/g in clay; 8.8% with 25 ng/g soil; 3.4% with 125 ng/g in sludge.

ACCURACY (as Mean % Recovery): 46.8% with 5 ng/g in clay; 65.0% with 25 ng/g in soil; 81.9% with 125 ng/g in sludge; 125.4% with 46 ng/g in fly ash; 89.1% with 2500 ng/g in still bottom.

SAMPLING METHOD: Use 1 liter (or quart) amber glass bottles with Teflon lined or solvent washed foil screw caps. Tape caps to bottle after sampling. Compositing equipment must use glass containers and contain no Tygon or rubber tubing. Sample bottles must not be prewashed with the sample before its collection. Aqueous samples cannot be aliquoted from sample containers the entire sample must be used and the container is washed out with the extracting solvent.

STABILITY: Cool to 4°C and store at this temperature. When toluene is employed as the final solvent use a bonded phase GC column for separation. Otherwise, solvent exchange into tridecane is required for other liquid phases or the CP-Sil-88 GC column.

M.H.T. = 30 days;

samples must be completely analyzed within 45 days of collection.

QUALITY CONTROL: A method blank must be analyzed each time a set of samples is extracted or there is a change in reagents. A laboratory method blank must be run with each analytical batch of 20 or fewer samples. Field duplicates and field blanks should be analyzed periodically. GC column performance must be demonstrated initially and verified prior to analyzing any sample in a 12 hour period. A series of calibration standards must be processed through the procedure to validate elution patterns and absence of interferents from reagents. Both the alumina column and carbon column performance must be routinely checked for presence of the analyte. Performance evaluation samples and split samples with other laboratories are also expected to be periodically analyzed.

REFERENCE: Method 8280, SW-846, 3rd ed., Nov 1986.

PRIMARY NAME: 1,2,3,6,7,8-HxCDD

Method 8280

TITLE: Analysis of PCDDs and PCDFs

MATRIX: chemical wastes, fuel oils still bottoms, sludges, water, soil, fly ash, reactor residues.

CAS # : 34465-46-8

APPLICATION: This method is used for the analysis of tetra-, penta-, hexa-, hepta-, and octachlorinated dibenzo-p-dioxins (PCDDs) and dibenzofurans (PCDFs). The sensitivity of the method is dependent on the level of interferents within the matrix. Only experienced analysts should be used. Special safety precautions must be observed and an EPA-approved sample disposal plan must be used.

INTERFERENCES: Solvents, reagents and glassware may introduce artifacts. Other interferences may come from coextracted compounds from samples; PCBs and polychlorinated diphenyl ethers are common interferents.

INSTRUMENTATION: GC/MS with a fused silica capillary column. Also, solvent extraction and concentration glassware and either a gravity flow activated carbon AX-21 / silica gel Type 60 EM reagent column or a HPLC with a 10 mm by 7 cm silanized glass column with active carbon AX-21 and Spherisorb S10W silica for sample cleanup. One of three fused silica capillary GC columns may be used: column 1: 50 meter CP-Sil-88; Column 2: 30 meter by 0.25 mm DB-5; colunm 3: 30 meter SP-2250.

RANGE: 50 to 6,000 picograms

MDL: 2.21 ng/L (in reagent water) 1.25 μg/Kg (in Missouri soil) 0.55 μg/Kg (in fly ash) 2.30 μg/Kg (in industrial sludge) 6.21 μg/Kg (in still bottom) 5.02 μg/Kg (in fuel oil)
MDLs are for carbon-13 labeled analyte

PRECISION (as RSD): Not determined

ACCURACY (as Mean % Recovery): Not determined

SAMPLING METHOD: Use 1 liter (or quart) amber glass bottles with Teflon lined or solvent washed foil screw caps. Tape caps to bottle after sampling. Compositing equipment must use glass containers and contain no Tygon or rubber tubing. Sample bottles must not be prewashed with the sample before its collection. Aqueous samples cannot be aliquoted from sample containers the entire sample must be used and the container is washed out with the extracting solvent.

STABILITY: Cool to 4°C and store at this temperature. When toluene is employed as the final solvent use a bonded phase GC column for separation. Otherwise, solvent exchange into tridecane is required for other liquid phases or the CP-Sil-88 GC column.

M.H.T. = 30 days;

samples must be completely analyzed within 45 days of collection.

QUALITY CONTROL: A method blank must be analyzed each time a set of samples is extracted or there is a change in reagents. A laboratory method blank must be run with each analytical batch of 20 or fewer samples. Field duplicates and field blanks should be analyzed periodically. GC column performance must be demonstrated initially and verified prior to analyzing any sample in a 12 hour period. A series of calibration standards must be processed through the procedure to validate elution patterns and absence of interferents from reagents. Both the alumina column and carbon column performance must be routinely checked for presence of the analyte. Performance evaluation samples and split samples with other laboratories are also expected to be periodically analyzed.

REFERENCE: Method 8280, SW-846, 3rd ed., Nov 1986.

PRIMARY NAME: 1,2,3,4,7,8-HxCDF

Method 8280

TITLE: Analysis of PCDDs and PCDFs

MATRIX: chemical wastes, fuel oils still bottoms, sludges, water, soil, fly ash, reactor residues.

CAS # : 70648-26-9

APPLICATION: This method is used for the analysis of tetra-, penta-, hexa, hepta-, and octachlorinated dibenzo-p-dioxins (PCDDs) and dibenzofurans (PCDFs). The sensitivity of the method is dependent on the level of interferents within the matrix. Only experienced analysts should be used. Special safety precautions must be observed and an EPA-approved sample disposal plan must be used.

INTERFERENCES: Solvents, reagents and glassware may introduce artifacts. Other interferences may come from coextracted compounds from samples; PCBs and polychlorinated diphenyl ethers are common interferents.

INSTRUMENTATION: GC/MS with a fused silica capillary column. Also, solvent extraction and concentration glassware and either a gravity flow activated carbon AX-21 / silica gel Type 60 EM reagent column or a HPLC with a 10 mm by 7 cm silanized glass column with active carbon AX-21 and Spherisorb S10W silica for sample cleanup. One of three fused silica capillary GC columns may be used: column 1: 50 meter CP-Sil-88; Column 2: 30 meter by 0.25 mm DB-5; colunm 3: 30 meter SP-2250.

RANGE: 50 to 6,000 picograms

MDL: 2.53 ng/L (in reagent water) 0.83 µg/Kg (in Missouri soil) 0.30 µg/Kg (in fly ash) 2.17 µg/Kg (in industrial sludge) 2.27 µg/Kg (in still bottom) 2.09 µg/Kg (in fuel oil)
MDLs are for carbon-13 labeled analyte

PRECISION (as RSD): 26% with 5 ng/g in clay; 6.8% with 25 ng/g soil; 5.6% with 139 ng/g in sludge; 13.5% with 24.2 ng/g in fly ash.

ACCURACY (as Mean % Recovery): 54.2% with 5 ng/g in clay; 68.5% with 25 ng/g in soil; 82.2% with 125 ng/g in sludge; 91.0% with 46 ng/g in fly ash; 92.9% with 2500 ng/g in still bottom.

SAMPLING METHOD: Use 1 liter (or quart) amber glass bottles with Teflon lined or solvent washed foil screw caps. Tape caps to bottle after sampling. Compositing equipment must use glass containers and contain no Tygon or rubber tubing. Sample bottles must not be prewashed with the sample before its collection. Aqueous samples cannot be aliquoted from sample containers the entire sample must be used and the container is washed out with the extracting solvent.

STABILITY: Cool to 4°C and store at this temperature. When toluene is

employed as the final solvent use a bonded phase GC column for separation. Otherwise, solvent exchange into tridecane is required for other liquid phases or the CP-Sil-88 GC column.

M.H.T. = 30 days;
samples must be completely analyzed within 45 days of collection.

QUALITY CONTROL: A method blank must be analyzed each time a set of samples is extracted or there is a change in reagents. A laboratory method blank must be run with each analytical batch of 20 or fewer samples. Field duplicates and field blanks should be analyzed periodically. GC column performance must be demonstrated initially and verified prior to analyzing any sample in a 12 hour period. A series of calibration standards must be processed through the procedure to validate elution patterns and absence of interferents from reagents. Both the alumina column and carbon column performance must be routinely checked for presence of the analyte. Performance evaluation samples and split samples with other laboratories are also expected to be periodically analyzed.

REFERENCE: Method 8280, SW-846, 3rd ed., Nov 1986.

PRIMARY NAME: MCPA

Method 8150

TITLE: Chlorinated Herbicides

MATRIX: groundwater, soils, sludges water miscible liquid wastes, and non-water miscible wastes.

CAS # : 94-74-6

APPLICATION: This method is used for the analysis of 10 chlorinated herbicides. Samples are extracted, hydrolyzed with potassium hydroxide, and extraneous organics are removed by a solvent wash. After acidification, the acids are extracted, concentrated and converted to their methyl esters using diazomethane. They are then analyzed using direct injection into a gas chromatograph (GC). Be very careful because diazomethane can explode under certain conditions and it is also a carcinogen.

INTERFERENCES: Organic acids and phenols (especially chlorinated acids and phenols) may cause interferences. Phthalate esters are not as significant an interference as with other GC-ECD methods if an electron capture detector is used. The herbicides may react readily with alkaline substances and be lost during analysis so all glassware and glass wool must be acid rinsed and sodium sulfate must be acidified with sulfuric acid prior to use. Sensitivity usually depends on the level of interferences rather than on instrumentation.

INSTRUMENTATION: GC capable of on-column injections and an electron capture detector (ECD)or a halogen specific detector. Column 1: 1.8 meter by 4 mm with 1.5% SP-2250 / 1.95% SP-2401 on Supelcoport. Column 2: 1.8 meter by 4 mm with 5% OV-210 on Gas Chrom Q. Column 3: 1.98 meter by 2 mm with 0.1% SP-1000 on Carbopack C. The preferred column is Column Number 1.

RANGE: Not listed

MDL: 249 μg/L (in reagent water; ECD)

PRACTICAL QUANTITATION LIMIT FACTORS FOR MULTIPLYING TIMES FID MDL VALUE

Matrix	Multiplication Factor
Groundwater	10
Low-level soil by sonication with GPC cleanup	670
High-level soil and sludge by sonication	10,000
Non-water miscible waste	100,000

PRECISION: (as standard deviation) 4% with 2020 μg/L spike in drinking water; 3% with 2020 μg/L in municipal water.

ACCURACY: (as mean recovery) 98% with 2020 μg/L spike in drinking water; 73% with 2020 μg/L in municipal water.

SAMPLING METHOD: Use 8 oz. widemouth glass bottles with Teflon lined caps for concentrated waste samples, soils, sediments and sludges. Use 1 or 2 1/2 gallon amber glass bottles with Teflon lined caps for liquid (water) samples.

STABILITY: Cool soil, sediment, sludge and liquid samples to 4°C. If residual chlorine is present in liquid samples add 3 mL of 10% sodium thiosulfate per gallon of sample and cool to 4°C.

M.H.T. = 14 days for concentrated waste, soil, sediment or sludge.

M.H.T. = 7 days for liquid samples.

All extracts must be analyzed within 40 days.

QUALITY CONTROL: A quality control check sample concentrate containing this compound in acetone at a concentration 1,000 times more concentrated than the selected spike concentration is required. The QC check sample concentrate may be prepared from pure standard materials or purchased as certified solutions. Use appropriate trip, matrix, control site, method, reagent and solvent blanks. Internal, surrogate and five concentration level calibration standards are used.

REFERENCE: Method 8150, SW-846, 3rd ed., Sep 1986.

PRIMARY NAME: MCPP

Method 8150

TITLE: Chlorinated Herbicides

MATRIX: groundwater, soils, sludges water miscible liquid wastes, and non-water miscible wastes.

CAS # : 93-65-2

APPLICATION: This method is used for the analysis of 10 chlorinated herbicides. Samples are extracted, hydrolyzed with potassium hydroxide, and extraneous organics are removed by a solvent wash. After acidification, the acids are extracted, concentrated and converted to their methyl esters using diazomethane. They are then analyzed using direct injection into a gas chromatograph (GC). Be very careful because diazomethane can explode under certain conditions and it is also a carcinogen.

INTERFERENCES: Organic acids and phenols (especially chlorinated acids and phenols) may cause interferences. Phthalate esters are not as significant an interference as with other GC-ECD methods if an electron capture detector is used. The herbicides may react readily with alkaline substances and be lost during analysis so all glassware and glass wool must be acid rinsed and sodium sulfate must be acidified with sulfuric acid prior to use. Sensitivity usually depends on the level of interferences rather than on instrumentation.

INSTRUMENTATION: GC capable of on-column injections and an electron capture detector (ECD)or a halogen specific detector. Column 1: 1.8 meter by 4 mm with 1.5% SP-2250 / 1.95% SP-2401 on Supelcoport. Column 2: 1.8 meter by 4 mm with 5% OV-210 on Gas Chrom Q. Column 3: 1.98 meter by 2 mm with 0.1% SP-1000 on Carbopack C. The preferred column is Column Number 1.

RANGE: Not listed

MDL: 192 μg/L (in reagent water; ECD)

PRACTICAL QUANTITATION LIMIT FACTORS FOR MULTIPLYING TIMES FID MDL VALUE

Matrix	Multiplication Factor
Groundwater	10
Low-level soil by sonication with GPC cleanup	670
High-level soil and sludge by sonication	10,000
Non-water miscible waste	100,000

PRECISION: (as standard deviation) 4% with 2080 μg/L spike in drinking water; 3% with 2100 μg/L in municipal water.

ACCURACY: (as mean recovery) 94% with 2080 μg/L spike in drinking water; 97% with 2100 μg/L in municipal water.

SAMPLING METHOD: Use 8 oz. widemouth glass bottles with Teflon lined caps for concentrated waste samples, soils, sediments and sludges. Use 1 or 2 1/2 gallon amber glass bottles with Teflon lined caps for liquid (water) samples.

STABILITY: Cool soil, sediment, sludge and liquid samples to 4°C. If residual chlorine is present in liquid samples add 3 mL of 10% sodium thiosulfate per gallon of sample and cool to 4°C.

M.H.T. = 14 days for concentrated waste, soil, sediment or sludge.

M.H.T. = 7 days for liquid samples.

All extracts must be analyzed within 40 days.

QUALITY CONTROL: A quality control check sample concentrate containing this compound in acetone at a concentration 1,000 times more concentrated than the selected spike concentration is required. The QC check sample concentrate may be prepared from pure standard materials or purchased as certified solutions. Use appropriate trip, matrix, control site, method, reagent and solvent blanks. Internal, surrogate and five concentration level calibration standards are used.

REFERENCE: Method 8150, SW-846, 3rd ed., Sep 1986.

PRIMARY NAME: Merphos

Method 8140

TITLE: Organophosphorus Pesticides

MATRIX: groundwater, soils, sludges water miscible liquid wastes, and non-water miscible wastes.

CAS # : 150-50-5

APPLICATION: This method is used for the analysis of 21 organophosphorus pesticides. Samples are extracted, concentrated and analyzed using direct injection of both neat and diluted organic liquid into a gas chromatograph (GC).

INTERFERENCES: Solvents, reagents and glassware may introduce artifacts. Other interferences may come from coextracted compounds from samples. The use of Florisil cleanup materials may produce low recoveries. Elemental sulfur may interfere with some compounds when using a flame photometric detector. Sulfur cleanup (Method 3660) may alleviate sulfur interference.

INSTRUMENTATION: GC capable of on-column injections and a flame photometric detector (FPD) or a thermionic detector. Column 1: 1.8 meter by 2 mm with 5% SP-2401 on Supelcoport. Column 2: 1.8 meter by 2 mm with 3% SP-2401 on Supelcoport. Column 3: 50 cm by 1/8 inch Teflon with 15% SE-54 on Gas Chrom Q. The preferred column is Column Number 2.

RANGE: 1.0 to 50 μg/L

MDL: 0.25 μg/L (in reagent water)

PRACTICAL QUANTITATION LIMIT FACTORS FOR MULTIPLYING TIMES FID MDL VALUE

Matrix	Multiplication Factor
Groundwater	10
Low-level soil by sonication with GPC cleanup	670
High-level soil and sludge by sonication	10,000
Non-water miscible waste	100,000

PRECISION: 7.9% (single operator standard deviation)

ACCURACY: 120.7% (single operator average recovery)

SAMPLING METHOD: Use 8 oz. widemouth glass bottles with Teflon lined caps for concentrated waste samples, soils, sediments and sludges. Use 1 or 2 1/2 gallon amber glass bottles with Teflon lined caps for liquid (water) samples.

STABILITY: Cool soil, sediment, sludge and liquid samples to 4°C. If residual chlorine is present in liquid samples add 3 mL of 10% sodium thiosulfate per gallon of sample and cool to 4°C.

M.H.T. = 14 days for concentrated waste, soil, sediment or sludge.

M.H.T. = 7 days for liquid samples.

All extracts must be analyzed within 40 days.

QUALITY CONTROL: A quality control check sample concentrate containing this compound in acetone at a concentration 1,000 times more concentrated than the selected spike concentration is required. The QC check sample concentrate may be prepared from pure standard materials or purchased as certified solutions. Use appropriate trip, matrix, control site, method, reagent and solvent blanks. Internal, surrogate and five concentration level calibration standards are used.

REFERENCE: Method 8140, SW-846, 3rd ed., Sep 1986.

PRIMARY NAME: Methoxychlor

Method 8080

TITLE: Organochlorine Pesticides and PCBs

MATRIX: groundwater, soils, sludges water miscible liquid wastes, and non-water miscible wastes.

CAS # : 72-43-5

APPLICATION: This method is used for the analysis of 19 pesticides and 7 Aroclor (PCB) mixtures. Samples are extracted, concentrated and analyzed using direct injection of both neat and diluted organic liquid into a gas chromatograph (GC).

INTERFERENCES: Solvents, reagents and glassware may introduce artifacts. Other interferences may come from coextracted compounds from samples. Phthalate esters are common interferences when using an electron capture detector (ECD) so all plastics must be strictly avoided. Exhaustive cleanup of reagents and glassware may be required to eliminate phthalate contamination. Use of a halogen specific microcoulometric or electrolytic conductivity detector will eliminate phthalate interference. Toxaphene may sometimes be an interferent and coelute with methoxychlor.

INSTRUMENTATION: GC capable of on-column injections and an ECD or a halogen specific detector (HSD). Column 1: 1.8 meter by 4 mm with 1.5% SP-2250 / 1.95% SP-2401 on Supelcoport. Column 2: 1.8 meter by 4 mm with 3% OV-1 on Supelcoport.

RANGE: 8.5 to 400 μg/L

MDL: 0.176 μg/L (in reagent water)

PRACTICAL QUANTITATION LIMIT FACTORS FOR MULTIPLYING TIMES FID MDL VALUE

Matrix	Multiplication Factor
Groundwater	10
Low-level soil by sonication with GPC cleanup	670
High-level soil and sludge by sonication	10,000
Non-water miscible waste	100,000

PRECISION: Not determined

ACCURACY: Not determined

SAMPLING METHOD: Use 8 oz. widemouth glass bottles with Teflon lined caps for concentrated waste samples, soils, sediments and sludges. Use 1 or 2 1/2 gallon amber glass bottles with Teflon lined caps for liquid (water) samples.

STABILITY: Cool soil, sediment, sludge and liquid samples to 4°C. If residual chlorine is present in liquid samples add 3 mL of 10% sodium thiosulfate per gallon of sample and cool to 4°C.

M.H.T. = 14 days for concentrated waste, soil, sediment or sludge.

M.H.T. = 7 days for liquid samples.

All extracts must be analyzed within 40 days.

QUALITY CONTROL: A quality control check sample concentrate containing each analyte of interest is required. The QC check sample concentrate may be prepared from pure standard materials or purchased as certified solutions. Use appropriate trip, matrix, control site, method, reagent and solvent blanks. Internal, surrogate and five concentration level calibration standards are used. The quality control check sample concentrate should contain methoxychlor at 2 μg/mL in acetone.

REFERENCE: Method 8080, SW-846, 3rd ed., Nov 1986.

PRIMARY NAME: Methyl parathion — Method 8140

TITLE: Organophosphorus Pesticides

MATRIX: groundwater, soils, sludges water miscible liquid wastes, and non-water miscible wastes.

CAS # : 298-00-0

APPLICATION: This method is used for the analysis of 21 organophosphorus pesticides. Samples are extracted, concentrated and analyzed using direct injection of both neat and diluted organic liquid into a gas chromatograph (GC).

INTERFERENCES: Solvents, reagents and glassware may introduce artifacts. Other interferences may come from coextracted compounds from samples. The use of Florisil cleanup materials may produce low recoveries. Elemental sulfur may interfere with some compounds when using a flame photometric detector. Sulfur cleanup (Method 3660) may alleviate sulfur interference.

INSTRUMENTATION: GC capable of on-column injections and a flame photometric detector (FPD) or a thermionic detector. Column 1: 1.8 meter by 2 mm with 5% SP-2401 on Supelcoport. Column 2: 1.8 meter by 2 mm with 3% SP-2401 on Supelcoport. Column 3: 50 cm by 1/8 inch Teflon with 15% SE-54 on Gas Chrom Q. The preferred column is Column Number 2.

RANGE: 0.5 to 500 µg/L

MDL: 0.03 µg/L (in reagent water)

PRACTICAL QUANTITATION LIMIT FACTORS FOR MULTIPLYING TIMES FID MDL VALUE

Matrix	Multiplication Factor
Groundwater	10
Low-level soil by sonication with GPC cleanup	670
High-level soil and sludge by sonication	10,000
Non-water miscible waste	100,000

PRECISION: 5.3% (single operator standard deviation)

ACCURACY: 96.0% (single operator average recovery)

SAMPLING METHOD: Use 8 oz. widemouth glass bottles with Teflon lined caps for concentrated waste samples, soils, sediments and sludges. Use 1 or 2 1/2 gallon amber glass bottles with Teflon lined caps for liquid (water) samples.

STABILITY: Cool soil, sediment, sludge and liquid samples to 4°C. If residual chlorine is present in liquid samples add 3 mL of 10% sodium thiosulfate per gallon of sample and cool to 4°C.

M.H.T. = 14 days for concentrated waste, soil, sediment or sludge.

M.H.T. = 7 days for liquid samples.

All extracts must be analyzed within 40 days.

QUALITY CONTROL: A quality control check sample concentrate containing this compound in acetone at a concentration 1,000 times more concentrated than the selected spike concentration is required. The QC check sample concentrate may be prepared from pure standard materials or purchased as certified solutions. Use appropriate trip, matrix, control site, method, reagent and solvent blanks. Internal, surrogate and five concentration level calibration standards are used.

REFERENCE: Method 8140, SW-846, 3rd ed., Sep 1986.

PRIMARY NAME: Mevinphos

Method 8140

TITLE: Organophosphorus Pesticides

MATRIX: groundwater, soils, sludges water miscible liquid wastes, and non-water miscible wastes.

CAS # : 7786-34-7

APPLICATION: This method is used for the analysis of 21 organophosphorus pesticides. Samples are extracted, concentrated and analyzed using direct injection of both neat and diluted organic liquid into a gas chromatograph (GC).

INTERFERENCES: Solvents, reagents and glassware may introduce artifacts. Other interferences may come from coextracted compounds from samples. The use of Florisil cleanup materials may produce low recoveries. Elemental sulfur may interfere with some compounds when using a flame photometric detector. Sulfur cleanup (Method 3660) may alleviate sulfur interference.

INSTRUMENTATION: GC capable of on-column injections and a flame photometric detector (FPD) or a thermionic detector. Column 1: 1.8 meter by 2 mm with 5% SP-2401 on Supelcoport. Column 2: 1.8 meter by 2 mm with 3% SP-2401 on Supelcoport. Column 3: 50 cm by 1/8 inch Teflon with 15% SE-54 on Gas Chrom Q. The preferred column is Column Number 1.

RANGE: 15.5 to 520 μg/L

MDL: 0.3 μg/L (in reagent water)

PRACTICAL QUANTITATION LIMIT FACTORS FOR MULTIPLYING TIMES FID MDL VALUE

Matrix	Multiplication Factor
Groundwater	10
Low-level soil by sonication with GPC cleanup	670
High-level soil and sludge by sonication	10,000
Non-water miscible waste	100,000

PRECISION: 7.8% (single operator standard deviation)

ACCURACY: 56.5% (single operator average recovery)

SAMPLING METHOD: Use 8 oz. widemouth glass bottles with Teflon lined caps for concentrated waste samples, soils, sediments and sludges. Use 1 or 2 1/2 gallon amber glass bottles with Teflon lined caps for liquid (water) samples.

STABILITY: Cool soil, sediment, sludge and liquid samples to 4°C. If residual chlorine is present in liquid samples add 3 mL of 10% sodium thiosulfate per gallon of sample and cool to 4°C.

M.H.T. = 14 days for concentrated waste, soil, sediment or sludge.

M.H.T. = 7 days for liquid samples.

All extracts must be analyzed within 40 days.

QUALITY CONTROL: A quality control check sample concentrate containing this compound in acetone at a concentration 1,000 times more concentrated than the selected spike concentration is required. The QC check sample concentrate may be prepared from pure standard materials or purchased as certified solutions. Use appropriate trip, matrix, control site, method, reagent and solvent blanks. Internal, surrogate and five concentration level calibration standards are used.

REFERENCE: Method 8140, SW-846, 3rd ed., Sep 1986.

PRIMARY NAME: Naled

Method 8140

TITLE: Organophosphorus Pesticides

MATRIX: groundwater, soils, sludges water miscible liquid wastes, and non-water miscible wastes.

CAS # : 300-76-5

APPLICATION: This method is used for the analysis of 21 organophosphorus pesticides. Samples are extracted, concentrated and analyzed using direct injection of both neat and diluted organic liquid into a gas chromatograph (GC).

INTERFERENCES: Solvents, reagents and glassware may introduce artifacts. Other interferences may come from coextracted compounds from samples. The use of Florisil cleanup materials may produce low recoveries. Elemental sulfur may interfere with some compounds when using a flame photometric detector. Sulfur cleanup (Method 3660) may alleviate sulfur interference.

INSTRUMENTATION: GC capable of on-column injections and a flame photometric detector (FPD) or a thermionic detector. A halogen specific detector may also be used and may have the advantage of fewer interferences. Column 1: 1.8 meter by 2 mm with 5% SP-2401 on Supelcoport. Column 2: 1.8 meter by 2 mm with 3% SP-2401 on Supelcoport. Column 3: 50 cm by 1/8 inch Teflon with 15% SE-54 on Gas Chrom Q. The preferred column is Column Number 3.

RANGE: 25.8 to 294 μg/L

MDL: 0.1 μg/L (in reagent water)

PRACTICAL QUANTITATION LIMIT FACTORS FOR MULTIPLYING TIMES FID MDL VALUE

Matrix	Multiplication Factor
Groundwater	10
Low-level soil by sonication with GPC cleanup	670
High-level soil and sludge by sonication	10,000
Non-water miscible waste	100,000

PRECISION: 8.1% (single operator standard deviation)

ACCURACY: 78.0% (single operator average recovery)

SAMPLING METHOD: Use 8 oz. widemouth glass bottles with Teflon lined caps for concentrated waste samples, soils, sediments and sludges. Use 1 or 2 1/2 gallon amber glass bottles with Teflon lined caps for liquid (water) samples.

STABILITY: Cool soil, sediment, sludge and liquid samples to 4°C. If residual chlorine is present in liquid samples add 3 mL of 10% sodium thiosulfate per gallon of sample and cool to 4°C.

M.H.T. = 14 days for concentrated waste, soil, sediment or sludge.

M.H.T. = 7 days for liquid samples.

All extracts must be analyzed within 40 days.

QUALITY CONTROL: A quality control check sample concentrate containing this compound in acetone at a concentration 1,000 times more concentrated than the selected spike concentration is required. The QC check sample concentrate may be prepared from pure standard materials or purchased as certified solutions. Use appropriate trip, matrix, control site, method, reagent and solvent blanks. Internal, surrogate and five concentration level calibration standards are used.

REFERENCE: Method 8140, SW-846, 3rd ed., Sep 1986.

PRIMARY NAME: OCDD

Method 8280

TITLE: Analysis of PCDDs and PCDFs

MATRIX: chemical wastes, fuel oils still bottoms, sludges, water, soil, fly ash, reactor residues.

CAS # : 3268-87-9

APPLICATION: This method is used for the analysis of tetra-, penta-, hexa-, hepta-, and octachlorinated dibenzo-p-dioxins (PCDDs) and dibenzofurans (PCDFs). The sensitivity of the method is dependent on the level of interferents within the matrix. Only experienced analysts should be used. Special safety precautions must be observed and an EPA-approved sample disposal plan must be used.

INTERFERENCES: Solvents, reagents and glassware may introduce artifacts. Other interferences may come from coextracted compounds from samples; PCBs and polychlorinated diphenyl ethers are common interferents.

INSTRUMENTATION: GC/MS with a fused silica capillary column. Also, solvent extraction and concentration glassware and either a gravity flow activated carbon AX-21 / silica gel Type 60 EM reagent column or a HPLC with a 10 mm by 7 cm silanized glass column with active carbon AX-21 and Spherisorb S10W silica for sample cleanup. One of three fused silica capillary GC columns may be used: column 1: 50 meter CP-Sil-88; Column 2: 30 meter by 0.25 mm DB-5; colunm 3: 30 meter SP-2250.

RANGE: 50 to 6,000 picograms

MDL: 3.93 ng/L (in reagent water) 2.35 μg/Kg (in Missouri soil) 2.27 μg/Kg (in fly ash) 6.44 μg/Kg (in industrial sludge) 10.1 μg/Kg (in still bottom) 23.2 μg/Kg (in fuel oil)
MDLs are for carbon-13 labeled analyte

PRECISION (as RSD): Not determined

ACCURACY (as Mean % Recovery): Not determined

SAMPLING METHOD: Use 1 liter (or quart) amber glass bottles with Teflon lined or solvent washed foil screw caps. Tape caps to bottle after sampling. Compositing equipment must use glass containers and contain no Tygon or rubber tubing. Sample bottles must not be prewashed with the sample before its collection. Aqueous samples cannot be aliquoted from sample containers the entire sample must be used and the container is washed out with the extracting solvent.

STABILITY: Cool to 4°C and store at this temperature. When toluene is employed as the final solvent use a bonded phase GC column for separation. Otherwise, solvent exchange into tridecane is required for other liquid phases or the CP-Sil-88 GC column.

M.H.T. = 30 days;
samples must be completely analyzed within 45 days of collection.

QUALITY CONTROL: A method blank must be analyzed each time a set of samples is extracted or there is a change in reagents. A laboratory method blank must be run with each analytical batch of 20 or fewer samples. Field duplicates and field blanks should be analyzed periodically. GC column performance must be demonstrated initially and verified prior to analyzing any sample in a 12 hour period. A series of calibration standards must be processed through the procedure to validate elution patterns and absence of interferents from reagents. Both the alumina column and carbon column performance must be routinely checked for presence of the analyte. Performance evaluation samples and split samples with other laboratories are also expected to be periodically analyzed.

REFERENCE: Method 8280, SW-846, 3rd ed., Nov 1986.

PRIMARY NAME: OCDF

Method 8280

TITLE: Analysis of PCDDs and PCDFs

MATRIX: chemical wastes, fuel oils still bottoms, sludges, water, soil, fly ash, reactor residues.

CAS # : 39001-02-0

APPLICATION: This method is used for the analysis of tetra-, penta-, hexa-, hepta-, and octachlorinated dibenzo-p-dioxins (PCDDs) and dibenzofurans (PCDFs). The sensitivity of the method is dependent on the level of interferents within the matrix. Only experienced analysts should be used. Special safety precautions must be observed and an EPA-approved sample disposal plan must be used.

INTERFERENCES: Solvents, reagents and glassware may introduce artifacts. Other interferences may come from coextracted compounds from samples; PCBs and polychlorinated diphenyl ethers are common interferents.

INSTRUMENTATION: GC/MS with a fused silica capillary column. Also, solvent extraction and concentration glassware and either a gravity flow activated carbon AX-21 / silica gel Type 60 EM reagent column or a HPLC with a 10 mm by 7 cm silanized glass column with active carbon AX-21 and Spherisorb S10W silica for sample cleanup. One of three fused silica capillary GC columns may be used: column 1: 50 meter CP-Sil-88; Column 2: 30 meter by 0.25 mm DB-5; colunm 3: 30 meter SP-2250.

RANGE: 50 to 6,000 picograms MDL: Not determined

PRECISION (as RSD): 3.3% with 317 ng/g in sludge.

ACCURACY (as Mean % Recovery): 86.8% with 125 ng/g in sludge.

SAMPLING METHOD: Use 1 liter (or quart) amber glass bottles with Teflon lined or solvent washed foil screw caps. Tape caps to bottle after sampling. Compositing equipment must use glass containers and contain no Tygon or rubber tubing. Sample bottles must not be prewashed with the sample before its collection. Aqueous samples cannot be aliquoted from sample containers the entire sample must be used and the container is washed out with the extracting solvent.

STABILITY: Cool to 4°C and store at this temperature. When toluene is employed as the final solvent use a bonded phase GC column for separation. Otherwise, solvent exchange into tridecane is required for other liquid phases or the CP-Sil-88 GC column.

M.H.T. = 30 days;

samples must be completely analyzed within 45 days of collection.

QUALITY CONTROL: A method blank must be analyzed each time a set of samples is extracted or there is a change in reagents. A laboratory method blank must be run with each analytical batch of 20 or fewer samples. Field duplicates and field blanks should be analyzed periodically. GC column performance must be demonstrated initially and verified prior to analyzing any sample in a 12 hour period. A series of calibration standards must be processed through the procedure to validate elution patterns and absence of interferents from reagents. Both the alumina column and carbon column performance must be routinely checked for presence of the analyte. Performance evaluation samples and split samples with other laboratories are also expected to be periodically analyzed.

REFERENCE: Method 8280, SW-846, 3rd ed., Nov 1986.

PRIMARY NAME: 1,2,3,4,7-PeCDD Method 8280

TITLE: Analysis of PCDDs and PCDFs

MATRIX: chemical wastes, fuel oils still bottoms, sludges, water, soil, fly ash, reactor residues.

CAS # : 39227-61-7

APPLICATION: This method is used for the analysis of tetra-, penta-, hexa-, hepta-, and octachlorinated dibenzo-p-dioxins (PCDDs) and dibenzofurans (PCDFs). The sensitivity of the method is dependent on the level of interferents within the matrix. Only experienced analysts should be used. Special safety precautions must be observed and an EPA-approved sample disposal plan must be used.

INTERFERENCES: Solvents, reagents and glassware may introduce artifacts. Other interferences may come from coextracted compounds from samples; PCBs and polychlorinated diphenyl ethers are common interferents.

INSTRUMENTATION: GC/MS with a fused silica capillary column. Also, solvent extraction and concentration glassware and either a gravity flow activated carbon AX-21 / silica gel Type 60 EM reagent column or a HPLC with a 10 mm by 7 cm silanized glass column with active carbon AX-21 and Spherisorb S10W silica for sample cleanup. One of three fused silica capillary GC columns may be used: column 1: 50 meter CP-Sil-88; Column 2: 30 meter by 0.25 mm DB-5; colunm 3: 30 meter SP-2250.

RANGE: 50 to 6,000 picograms MDL: Not determined

PRECISION (as RSD): 10% with 5 ng/g in clay; 2.8% with 25 ng/g soil; 4.6% with 125 ng/g in sludge; 6.9% with 25.8 ng/g in fly ash.

ACCURACY (as Mean % Recovery): 58.4% with 5 ng/g in clay; 62.2% with 25 ng/g in soil; 79.2% with 125 ng/g in sludge; 102.4% with 46 ng/g in fly ash; 81.8% with 2500 ng/g in still bottom.

SAMPLING METHOD: Use 1 liter (or quart) amber glass bottles with Teflon lined or solvent washed foil screw caps. Tape caps to bottle after sampling. Compositing equipment must use glass containers and contain no Tygon or rubber tubing. Sample bottles must not be prewashed with the sample before its collection. Aqueous samples cannot be aliquoted from sample containers the entire sample must be used and the container is washed out with the extracting solvent.

STABILITY: Cool to 4°C and store at this temperature. When toluene is employed as the final solvent use a bonded phase GC column for separation. Otherwise, solvent exchange into tridecane is required for other liquid phases or the CP-Sil-88 GC column.

M.H.T. = 30 days; samples must be completely analyzed within 45 days of collection.

QUALITY CONTROL: A method blank must be analyzed each time a set of samples is extracted or there is a change in reagents. A laboratory method blank must be run with each analytical batch of 20 or fewer samples. Field duplicates and field blanks should be analyzed periodically. GC column performance must be demonstrated initially and verified prior to analyzing any sample in a 12 hour period. A series of calibration standards must be processed through the procedure to validate elution patterns and absence of interferents from reagents. Both the alumina column and carbon column performance must be routinely checked for presence of the analyte. Performance evaluation samples and split samples with other laboratories are also expected to be periodically analyzed.

REFERENCE: Method 8280, SW-846, 3rd ed., Nov 1986.

PRIMARY NAME: 1,2,3,7,8-PeCDD Method 8280

TITLE: Analysis of PCDDs and PCDFs

MATRIX: chemical wastes, fuel oils still bottoms, sludges, water, soil, fly ash, reactor residues.

CAS # : 40321-76-4

APPLICATION: This method is used for the analysis of tetra-, penta-, hexa-, hepta-, and octachlorinated dibenzo-p-dioxins (PCDDs) and dibenzofurans (PCDFs). The sensitivity of the method is dependent on the level of interferents within the matrix. Only experienced analysts should be used. Special safety precautions must be observed and an EPA-approved sample disposal plan must be used.

INTERFERENCES: Solvents, reagents and glassware may introduce artifacts. Other interferences may come from coextracted compounds from samples; PCBs and polychlorinated diphenyl ethers are common interferents.

INSTRUMENTATION: GC/MS with a fused silica capillary column. Also, solvent extraction and concentration glassware and either a gravity flow activated carbon AX-21 / silica gel Type 60 EM reagent column or a HPLC with a 10 mm by 7 cm silanized glass column with active carbon AX-21 and Spherisorb S10W silica for sample cleanup. One of three fused silica capillary GC columns may be used: column 1: 50 meter CP-Sil-88; Column 2: 30 meter by 0.25 mm DB-5; colunm 3: 30 meter SP-2250.

RANGE: 50 to 6,000 picograms

MDL: 1.27 ng/L (in reagent water) 0.70 μg/Kg (in Missouri soil) 0.25 μg/Kg (in fly ash) 1.34 μg/Kg (in industrial sludge) 2.46 μg/Kg (in still bottom) 2.09 μg/Kg (in fuel oil)
MDLs are for carbon-13 labeled analyte

PRECISION (as RSD): 25% with 5 ng/g in clay; 20% with 25 ng/g soil; 4.7% with 125 ng/g in sludge.

ACCURACY (as Mean % Recovery): 61.7% with 5 ng/g in clay; 68.4% with 25 ng/g in soil; 81.5% with 125 ng/g in sludge; 104.9% with 46 ng/g in fly ash; 84.0% with 2500 ng/g in still bottom.

SAMPLING METHOD: Use 1 liter (or quart) amber glass bottles with Teflon lined or solvent washed foil screw caps. Tape caps to bottle after sampling. Compositing equipment must use glass containers and contain no Tygon or rubber tubing. Sample bottles must not be prewashed with the sample before its collection. Aqueous samples cannot be aliquoted from sample containers the entire sample must be used and the container is washed out with the extracting solvent.

STABILITY: Cool to 4°C and store at this temperature. When toluene is

employed as the final solvent use a bonded phase GC column for separation. Otherwise, solvent exchange into tridecane is required for other liquid phases or the CP-Sil-88 GC column.

M.H.T. = 30 days; samples must be completely analyzed within 45 days of collection.

QUALITY CONTROL: A method blank must be analyzed each time a set of samples is extracted or there is a change in reagents. A laboratory method blank must be run with each analytical batch of 20 or fewer samples. Field duplicates and field blanks should be analyzed periodically. GC column performance must be demonstrated initially and verified prior to analyzing any sample in a 12 hour period. A series of calibration standards must be processed through the procedure to validate elution patterns and absence of interferents from reagents. Both the alumina column and carbon column performance must be routinely checked for presence of the analyte. Performance evaluation samples and split samples with other laboratories are also expected to be periodically analyzed.

REFERENCE: Method 8280, SW-846, 3rd ed., Nov 1986.

PRIMARY NAME: 1,2,3,7,8-PeCDF — Method 8280

TITLE: Analysis of PCDDs and PCDFs

MATRIX: chemical wastes, fuel oils still bottoms, sludges, water, soil, fly ash, reactor residues.

CAS # : 57117-41-6

APPLICATION: This method is used for the analysis of tetra-, penta-, hexa-, hepta-, and octachlorinated dibenzo-p-dioxins (PCDDs) and dibenzofurans (PCDFs). The sensitivity of the method is dependent on the level of interferents within the matrix. Only experienced analysts should be used. Special safety precautions must be observed and an EPA-approved sample disposal plan must be used.

INTERFERENCES: Solvents, reagents and glassware may introduce artifacts. Other interferences may come from coextracted compounds from samples; PCBs and polychlorinated diphenyl ethers are common interferents.

INSTRUMENTATION: GC/MS with a fused silica capillary column. Also, solvent extraction and concentration glassware and either a gravity flow activated carbon AX-21 / silica gel Type 60 EM reagent column or a HPLC with a 10 mm by 7 cm silanized glass column with active carbon AX-21 and Spherisorb S10W silica for sample cleanup. One of three fused silica capillary GC columns may be used: column 1: 50 meter CP-Sil-88; Column 2: 30 meter by 0.25 mm DB-5; colunm 3: 30 meter SP-2250.

RANGE: 50 to 6,000 picograms

MDL: 1.64 ng/L (in reagent water) 0.33 μg/Kg (in Missouri soil) 0.16 μg/Kg (in fly ash) 0.92 μg/Kg (in industrial sludge) 1.61 μg/Kg (in still bottom) 0.80 μg/Kg (in fuel oil)
MDLs are for carbon-13 labeled analyte

PRECISION (as RSD): 6.1% with 5 ng/g in clay; 5.0% with 25 ng/g soil; 4.8% with 125 ng/g in sludge.

ACCURACY (as Mean % Recovery): 57.4% with 5 ng/g in clay; 64.4% with 25 ng/g in soil; 84.8% with 125 ng/g in sludge; 105.8% with 46 ng/g in fly ash.

SAMPLING METHOD: Use 1 liter (or quart) amber glass bottles with Teflon lined or solvent washed foil screw caps. Tape caps to bottle after sampling. Compositing equipment must use glass containers and contain no Tygon or rubber tubing. Sample bottles must not be prewashed with the sample before its collection. Aqueous samples cannot be aliquoted from sample containers the entire sample must be used and the container is washed out with the extracting solvent.

STABILITY: Cool to 4°C and store at this temperature. When toluene is employed as the final solvent use a bonded phase GC column for separation. Otherwise, solvent exchange into tridecane is required for other liquid phases or the CP-Sil-88 GC column.

M.H.T. = 30 days;
samples must be completely analyzed within 45 days of collection.

QUALITY CONTROL: A method blank must be analyzed each time a set of samples is extracted or there is a change in reagents. A laboratory method blank must be run with each analytical batch of 20 or fewer samples. Field duplicates and field blanks should be analyzed periodically. GC column performance must be demonstrated initially and verified prior to analyzing any sample in a 12 hour period. A series of calibration standards must be processed through the procedure to validate elution patterns and absence of interferents from reagents. Both the alumina column and carbon column performance must be routinely checked for presence of the analyte. Performance evaluation samples and split samples with other laboratories are also expected to be periodically analyzed.

REFERENCE: Method 8280, SW-846, 3rd ed., Nov 1986.

PRIMARY NAME: Phorate

Method 8140

TITLE: Organophosphorus Pesticides

MATRIX: groundwater, soils, sludges water miscible liquid wastes, and non-water miscible wastes.

CAS # : 298-02-2

APPLICATION: This method is used for the analysis of 21 organophosphorus pesticides. Samples are extracted, concentrated and analyzed using direct injection of both neat and diluted organic liquid into a gas chromatograph (GC).

INTERFERENCES: Solvents, reagents and glassware may introduce artifacts. Other interferences may come from coextracted compounds from samples. The use of Florisil cleanup materials may produce low recoveries. Elemental sulfur may interfere with some compounds when using a flame photometric detector. Sulfur cleanup (Method 3660) may alleviate sulfur interference.

INSTRUMENTATION: GC capable of on-column injections and a flame photometric detector (FPD) or a thermionic detector. Column 1: 1.8 meter by 2 mm with 5% SP-2401 on Supelcoport. Column 2: 1.8 meter by 2 mm with 3% SP-2401 on Supelcoport. Column 3: 50 cm by 1/8 inch Teflon with 15% SE-54 on Gas Chrom Q. The preferred column is Column Number 1.

RANGE: 4.9 to 47 µg/L

MDL: 0.15 µg/L (in reagent water)

PRACTICAL QUANTITATION LIMIT FACTORS FOR MULTIPLYING TIMES FID MDL VALUE

Matrix	Multiplication Factor
Groundwater	10
Low-level soil by sonication with GPC cleanup	670
High-level soil and sludge by sonication	10,000
Non-water miscible waste	100,000

PRECISION: 8.9 % (single operator standard deviation)

ACCURACY: 62.7% (single operator average recovery)

SAMPLING METHOD: Use 8 oz. widemouth glass bottles with Teflon lined caps for concentrated waste samples, soils, sediments and sludges. Use 1 or 2 1/2 gallon amber glass bottles with Teflon lined caps for liquid (water) samples.

STABILITY: Cool soil, sediment, sludge and liquid samples to 4°C. If residual chlorine is present in liquid samples add 3 mL of 10% sodium thiosulfate per gallon of sample and cool to 4°C.

M.H.T. = 14 days for concentrated waste, soil, sediment or sludge.

M.H.T. = 7 days for liquid samples.

All extracts must be analyzed within 40 days.

QUALITY CONTROL: A quality control check sample concentrate containing this compound in acetone at a concentration 1,000 times more concentrated than the selected spike concentration is required. The QC check sample concentrate may be prepared from pure standard materials or purchased as certified solutions. Use appropriate trip, matrix, control site, method, reagent and solvent blanks. Internal, surrogate and five concentration level calibration standards are used.

REFERENCE: Method 8140, SW-846, 3rd ed., Sep 1986.

PRIMARY NAME: Ronnel

Method 8140

TITLE: Organophosphorus Pesticides

MATRIX: groundwater, soils, sludges water miscible liquid wastes, and non-water miscible wastes.

CAS # : 299-84-3

APPLICATION: This method is used for the analysis of 21 organophosphorus pesticides. Samples are extracted, concentrated and analyzed using direct injection of both neat and diluted organic liquid into a gas chromatograph (GC).

INTERFERENCES: Solvents, reagents and glassware may introduce artifacts. Other interferences may come from coextracted compounds from samples. The use of Florisil cleanup materials may produce low recoveries. Elemental sulfur may interfere with some compounds when using a flame photometric detector. Sulfur cleanup (Method 3660) may alleviate sulfur interference.

INSTRUMENTATION: GC capable of on-column injections and a flame photometric detector (FPD) or a thermionic detector. Column 1: 1.8 meter by 2 mm with 5% SP-2401 on Supelcoport. Column 2: 1.8 meter by 2 mm with 3% SP-2401 on Supelcoport. Column 3: 50 cm by 1/8 inch Teflon with 15% SE-54 on Gas Chrom Q. The preferred column is Column Number 2.

RANGE: 1.0 to 50 μg/L

MDL: 0.3 μg/L (in reagent water)

PRACTICAL QUANTITATION LIMIT FACTORS FOR MULTIPLYING TIMES FID MDL VALUE

Matrix	Multiplication Factor
Groundwater	10
Low-level soil by sonication with GPC cleanup	670
High-level soil and sludge by sonication	10,000
Non-water miscible waste	100,000

PRECISION: 5.6 % (single operator standard deviation)

ACCURACY: 99.2% (single operator average recovery)

SAMPLING METHOD: Use 8 oz. widemouth glass bottles with Teflon lined caps for concentrated waste samples, soils, sediments and sludges. Use 1 or 2 1/2 gallon amber glass bottles with Teflon lined caps for liquid (water) samples.

STABILITY: Cool soil, sediment, sludge and liquid samples to 4°C. If residual chlorine is present in liquid samples add 3 mL of 10% sodium thiosulfate per gallon of sample and cool to 4°C.

M.H.T. = 14 days for concentrated waste, soil, sediment or sludge.

M.H.T. = 7 days for liquid samples.

All extracts must be analyzed within 40 days.

QUALITY CONTROL: A quality control check sample concentrate containing this compound in acetone at a concentration 1,000 times more concentrated than the selected spike concentration is required. The QC check sample concentrate may be prepared from pure standard materials or purchased as certified solutions. Use appropriate trip, matrix, control site, method, reagent and solvent blanks. Internal, surrogate and five concentration level calibration standards are used.

REFERENCE: Method 8140, SW-846, 3rd ed., Sep 1986.

PRIMARY NAME: Stirophos (Tetrachlorvinphos) Method 8140

TITLE: Organophosphorus Pesticides

MATRIX: groundwater, soils, sludges water miscible liquid wastes, and non-water miscible wastes.

CAS # : 22248-79-9

APPLICATION: This method is used for the analysis of 21 organophosphorus pesticides. Samples are extracted, concentrated and analyzed using direct injection of both neat and diluted organic liquid into a gas chromatograph (GC).

INTERFERENCES: Solvents, reagents and glassware may introduce artifacts. Other interferences may come from coextracted compounds from samples. The use of Florisil cleanup materials may produce low recoveries. Elemental sulfur may interfere with some compounds when using a flame photometric detector. Sulfur cleanup (Method 3660) may alleviate sulfur interference.

INSTRUMENTATION: GC capable of on-column injections and a flame photometric detector (FPD) or a thermionic detector. A halogen specific detector may also be used and may have the advantage of fewer interferences. Column 1: 1.8 meter by 2 mm with 5% SP-2401 on Supelcoport. Column 2: 1.8 meter by 2 mm with 3% SP-2401 on Supelcoport. Column 3: 50 cm by 1/8 inch Teflon with 15% SE-54 on Gas Chrom Q. The preferred column is Column Number 1 or 3.

RANGE: 30.3 to 505 μg/L

MDL: 5.0 μg/L (in reagent water)

PRACTICAL QUANTITATION LIMIT FACTORS FOR MULTIPLYING TIMES FID MDL VALUE

Matrix	Multiplication Factor
Groundwater	10
Low-level soil by sonication with GPC cleanup	670
High-level soil and sludge by sonication	10,000
Non-water miscible waste	100,000

PRECISION: 5.9% (single operator standard deviation)

ACCURACY: 66.1% (single operator average recovery)

SAMPLING METHOD: Use 8 oz. widemouth glass bottles with Teflon lined caps for concentrated waste samples, soils, sediments and sludges. Use 1 or 2 1/2 gallon amber glass bottles with Teflon lined caps for liquid (water) samples.

STABILITY: Cool soil, sediment, sludge and liquid samples to 4°C. If residual chlorine is present in liquid samples add 3 mL of 10% sodium thiosulfate per gallon of sample and cool to 4°C.

M.H.T. = 14 days for concentrated waste, soil, sediment or sludge.

M.H.T. = 7 days for liquid samples.

All extracts must be analyzed within 40 days.

QUALITY CONTROL: A quality control check sample concentrate containing this compound in acetone at a concentration 1,000 times more concentrated than the selected spike concentration is required. The QC check sample concentrate may be prepared from pure standard materials or purchased as certified solutions. Use appropriate trip, matrix, control site, method, reagent and solvent blanks. Internal, surrogate and five concentration level calibration standards are used.

REFERENCE: Method 8140, SW-846, 3rd ed., Sep 1986.

PRIMARY NAME: 2,4,5-T Method 8150

TITLE: Chlorinated Herbicides

MATRIX: groundwater, soils, sludges water miscible liquid wastes, and non-water miscible wastes.

CAS # : 93-76-5

APPLICATION: This method is used for the analysis of 10 chlorinated herbicides. Samples are extracted, hydrolyzed with potassium hydroxide, and extraneous organics are removed by a solvent wash. After acidification, the acids are extracted, concentrated and converted to their methyl esters using diazomethane. They are then analyzed using direct injection into a gas chromatograph (GC). Be very careful because diazomethane can explode under certain conditions and it is also a carcinogen.

INTERFERENCES: Organic acids and phenols (especially chlorinated acids and phenols) may cause interferences. Phthalate esters are not as significant an interference as with other GC-ECD methods if an electron capture detector is used. The herbicides may react readily with alkaline substances and be lost during analysis so all glassware and glass wool must be acid rinsed and sodium sulfate must be acidified with sulfuric acid prior to use. Sensitivity usually depends on the level of interferences rather than on instrumentation.

INSTRUMENTATION: GC capable of on-column injections and an electron capture detector (ECD)or a halogen specific detector. Column 1: 1.8 meter by 4 mm with 1.5% SP-2250 / 1.95% SP-2401 on Supelcoport. Column 2: 1.8 meter by 4 mm with 5% OV-210 on Gas Chrom Q. Column 3: 1.98 meter by 2 mm with 0.1% SP-1000 on Carbopack C. The preferred column is Column Number 1 or 2.

RANGE: Not listed

MDL: 0.20 µg/L (in reagent water; ECD)

PRACTICAL QUANTITATION LIMIT FACTORS FOR MULTIPLYING TIMES FID MDL VALUE

Matrix	Multiplication Factor
Groundwater	10
Low-level soil by sonication with GPC cleanup	670
High-level soil and sludge by sonication	10,000
Non-water miscible waste	100,000

PRECISION: (as standard deviation) 6% with 1.1 µg/L spike in drinking water; 4% with 1.3 µg/L in municipal water.

ACCURACY: (as mean recovery) 85% with 1.1 µg/L spike in drinking water; 83% with 1.3 µg/L in municipal water.

SAMPLING METHOD: Use 8 oz. widemouth glass bottles with Teflon lined caps for concentrated waste samples, soils, sediments and sludges. Use 1 or 2 1/2 gallon amber glass bottles with Teflon lined caps for liquid (water) samples.

STABILITY: Cool soil, sediment, sludge and liquid samples to 4°C. If residual chlorine is present in liquid samples add 3 mL of 10% sodium thiosulfate per gallon of sample and cool to 4°C.

M.H.T. = 14 days for concentrated waste, soil, sediment or sludge.

M.H.T. = 7 days for liquid samples.

All extracts must be analyzed within 40 days.

QUALITY CONTROL: A quality control check sample concentrate containing this compound in acetone at a concentration 1,000 times more concentrated than the selected spike concentration is required. The QC check sample concentrate may be prepared from pure standard materials or purchased as certified solutions. Use appropriate trip, matrix, control site, method, reagent and solvent blanks. Internal, surrogate and five concentration level calibration standards are used.

REFERENCE: Method 8150, SW-846, 3rd ed., Sep 1986.

PRIMARY NAME: 1,2,3,4-TCDD Method 8280

TITLE: Analysis of PCDDs and PCDFs

MATRIX: chemical wastes, fuel oils still bottoms, sludges, water, soil, fly ash, reactor residues.

CAS # : 30746-58-8

APPLICATION: This method is used for the analysis of tetra-, penta-, hexa-, hepta-, and octachlorinated dibenzo-p-dioxins (PCDDs) and dibenzofurans (PCDFs). The sensitivity of the method is dependent on the level of interferents within the matrix. Only experienced analysts should be used. Special safety precautions must be observed and an EPA-approved sample disposal plan must be used.

INTERFERENCES: Solvents, reagents and glassware may introduce artifacts. Other interferences may come from coextracted compounds from samples; PCBs and polychlorinated diphenyl ethers are common interferents.

INSTRUMENTATION: GC/MS with a fused silica capillary column. Also, solvent extraction and concentration glassware and either a gravity flow activated carbon AX-21 / silica gel Type 60 EM reagent column or a HPLC with a 10 mm by 7 cm silanized glass column with active carbon AX-21 and Spherisorb S10W silica for sample cleanup. One of three fused silica capillary GC columns may be used: column 1: 50 meter CP-Sil-88; Column 2: 30 meter by 0.25 mm DB-5; colunm 3: 30 meter SP-2250.

RANGE: 50 to 6,000 picograms

MDL: Not determined

PRECISION (as RSD): 1.7% with 5 ng/g in clay; 1.1% with 25 ng/g soil; 9.0% with 125 ng/g in sludge; 7.9% with 38.5 ng/g in fly ash.

ACCURACY (as Mean % Recovery): 67.0% with 5 ng/g in clay; 60.3% with 25 ng/g in soil; 73.1% with 125 ng/g in sludge; 105.6% with 46 ng/g in fly ash; 93.8% with 2500 ng/g in still bottom.

SAMPLING METHOD: Use 1 liter (or quart) amber glass bottles with Teflon lined or solvent washed foil screw caps. Tape caps to bottle after sampling. Compositing equipment must use glass containers and contain no Tygon or rubber tubing. Sample bottles must not be prewashed with the sample before its collection. Aqueous samples cannot be aliquoted from sample containers the entire sample must be used and the container is washed out with the extracting solvent.

STABILITY: Cool to 4°C and store at this temperature. When toluene is employed as the final solvent use a bonded phase GC column for separation. Otherwise, solvent exchange into tridecane is required for other liquid phases or the CP-Sil-88 GC column.

M.H.T. = 30 days;

samples must be completely analyzed within 45 days of collection.

QUALITY CONTROL: A method blank must be analyzed each time a set of samples is extracted or there is a change in reagents. A laboratory method blank must be run with each analytical batch of 20 or fewer samples. Field duplicates and field blanks should be analyzed periodically. GC column performance must be demonstrated initially and verified prior to analyzing any sample in a 12 hour period. A series of calibration standards must be processed through the procedure to validate elution patterns and absence of interferents from reagents. Both the alumina column and carbon column performance must be routinely checked for presence of the analyte. Performance evaluation samples and split samples with other laboratories are also expected to be periodically analyzed.

REFERENCE: Method 8280, SW-846, 3rd ed., Nov 1986.

PRIMARY NAME: 1,2,7,8-TCDD

Method 8280

TITLE: Analysis of PCDDs and PCDFs

MATRIX: chemical wastes, fuel oils still bottoms, sludges, water, soil, fly ash, reactor residues.

CAS # : 34816-53-0

APPLICATION: This method is used for the analysis of tetra-, penta-, hexa-, hepta-, and octachlorinated dibenzo-p-dioxins (PCDDs) and dibenzofurans (PCDFs). The sensitivity of the method is dependent on the level of interferents within the matrix. Only experienced analysts should be used. Special safety precautions must be observed and an EPA-approved sample disposal plan must be used.

INTERFERENCES: Solvents, reagents and glassware may introduce artifacts. Other interferences may come from coextracted compounds from samples; PCBs and polychlorinated diphenyl ethers are common interferents.

INSTRUMENTATION: GC/MS with a fused silica capillary column. Also, solvent extraction and concentration glassware and either a gravity flow activated carbon AX-21 / silica gel Type 60 EM reagent column or a HPLC with a 10 mm by 7 cm silanized glass column with active carbon AX-21 and Spherisorb S10W silica for sample cleanup. One of three fused silica capillary GC columns may be used: column 1: 50 meter CP-Sil-88; Column 2: 30 meter by 0.25 mm DB-5; colunm 3: 30 meter SP-2250.

RANGE: 50 to 6,000 picograms MDL: Not determined

PRECISION (as RSD): 7.7% with 5 ng/g in clay; 9.0% with 25 ng/g soil; 7.7% with 125 ng/g in sludge; 23% with 2.6 ng/g in fly ash.

ACCURACY (as Mean % Recovery): 68.0% with 5 ng/g in clay; 75.3% with 25 ng/g in soil; 80.4% with 125 ng/g in sludge; 90.4% with 46 ng/g in fly ash; 88.4% with 2500 ng/g in still bottom.

SAMPLING METHOD: Use 1 liter (or quart) amber glass bottles with Teflon lined or solvent washed foil screw caps. Tape caps to bottle after sampling. Compositing equipment must use glass containers and contain no Tygon or rubber tubing. Sample bottles must not be prewashed with the sample before its collection. Aqueous samples cannot be aliquoted from sample containers the entire sample must be used and the container is washed out with the extracting solvent.

STABILITY: Cool to 4°C and store at this temperature. When toluene is employed as the final solvent use a bonded phase GC column for separation. Otherwise, solvent exchange into tridecane is required for other liquid phases or the CP-Sil-88 GC column.

M.H.T. = 30 days;

samples must be completely analyzed within 45 days of collection.

QUALITY CONTROL: A method blank must be analyzed each time a set of samples is extracted or there is a change in reagents. A laboratory method blank must be run with each analytical batch of 20 or fewer samples. Field duplicates and field blanks should be analyzed periodically. GC column performance must be demonstrated initially and verified prior to analyzing any sample in a 12 hour period. A series of calibration standards must be processed through the procedure to validate elution patterns and absence of interferents from reagents. Both the alumina column and carbon column performance must be routinely checked for presence of the analyte. Performance evaluation samples and split samples with other laboratories are also expected to be periodically analyzed.

REFERENCE: Method 8280, SW-846, 3rd ed., Nov 1986.

PRIMARY NAME: 1,2,8,9-TCDD

Method 8280

TITLE: Analysis of PCDDs and PCDFs

MATRIX: chemical wastes, fuel oils still bottoms, sludges, water, soil, fly ash, reactor residues.

CAS # : 62470-54-6

APPLICATION: This method is used for the analysis of tetra-, penta-, hexa-, hepta-, and octachlorinated dibenzo-p-dioxins (PCDDs) and dibenzofurans (PCDFs). The sensitivity of the method is dependent on the level of interferents within the matrix. Only experienced analysts should be used. Special safety precautions must be observed and an EPA-approved sample disposal plan must be used.

INTERFERENCES: Solvents, reagents and glassware may introduce artifacts. Other interferences may come from coextracted compounds from samples; PCBs and polychlorinated diphenyl ethers are common interferents.

INSTRUMENTATION: GC/MS with a fused silica capillary column. Also, solvent extraction and concentration glassware and either a gravity flow activated carbon AX-21 / silica gel Type 60 EM reagent column or a HPLC with a 10 mm by 7 cm silanized glass column with active carbon AX-21 and Spherisorb S10W silica for sample cleanup. One of three fused silica capillary GC columns may be used: column 1: 50 meter CP-Sil-88; Column 2: 30 meter by 0.25 mm DB-5; colunm 3: 30 meter SP-2250.

RANGE: 50 to 6,000 picograms

MDL: Not determined

PRECISION (as RSD): 10% with 5 ng/g in clay; 0.6% with 25 ng/g soil; 1.9% with 125 ng/g in sludge.

ACCURACY (as Mean % Recovery): 59.7% with 5 ng/g in clay; 60.3% with 25 ng/g in soil; 72.8% with 125 ng/g in sludge; 114.3% with 46 ng/g in fly ash; 81.2% with 2500 ng/g in still bottom.

SAMPLING METHOD: Use 1 liter (or quart) amber glass bottles with Teflon lined or solvent washed foil screw caps. Tape caps to bottle after sampling. Compositing equipment must use glass containers and contain no Tygon or rubber tubing. Sample bottles must not be prewashed with the sample before its collection. Aqueous samples cannot be aliquoted from sample containers the entire sample must be used and the container is washed out with the extracting solvent.

STABILITY: Cool to 4°C and store at this temperature. When toluene is employed as the final solvent use a bonded phase GC column for separation. Otherwise, solvent exchange into tridecane is required for other liquid phases or the CP-Sil-88 GC column.

M.H.T. = 30 days;

samples must be completely analyzed within 45 days of collection.

QUALITY CONTROL: A method blank must be analyzed each time a set of samples is extracted or there is a change in reagents. A laboratory method blank must be run with each analytical batch of 20 or fewer samples. Field duplicates and field blanks should be analyzed periodically. GC column performance must be demonstrated initially and verified prior to analyzing any sample in a 12 hour period. A series of calibration standards must be processed through the procedure to validate elution patterns and absence of interferents from reagents. Both the alumina column and carbon column performance must be routinely checked for presence of the analyte. Performance evaluation samples and split samples with other laboratories are also expected to be periodically analyzed.

REFERENCE: Method 8280, SW-846, 3rd ed., Nov 1986.

PRIMARY NAME: 1,3,6,8-TCDD

Method 8280

TITLE: Analysis of PCDDs and PCDFs

MATRIX: chemical wastes, fuel oils still bottoms, sludges, water, soil, fly ash, reactor residues.

CAS # : 33423-92-6

APPLICATION: This method is used for the analysis of tetra-, penta-, hexa-, hepta-, and octachlorinated dibenzo-p-dioxins (PCDDs) and dibenzofurans (PCDFs). The sensitivity of the method is dependent on the level of interferents within the matrix. Only experienced analysts should be used. Special safety precautions must be observed and an EPA-approved sample disposal plan must be used.

INTERFERENCES: Solvents, reagents and glassware may introduce artifacts. Other interferences may come from coextracted compounds from samples; PCBs and polychlorinated diphenyl ethers are common interferents.

INSTRUMENTATION: GC/MS with a fused silica capillary column. Also, solvent extraction and concentration glassware and either a gravity flow activated carbon AX-21 / silica gel Type 60 EM reagent column or a HPLC with a 10 mm by 7 cm silanized glass column with active carbon AX-21 and Spherisorb S10W silica for sample cleanup. One of three fused silica capillary GC columns may be used: column 1: 50 meter CP-Sil-88; Column 2: 30 meter by 0.25 mm DB-5; colunm 3: 30 meter SP-2250.

RANGE: 50 to 6,000 picograms MDL: Not determined

PRECISION (as RSD): 7.0% with 2.5 ng/g in clay; 5.1% with 25 ng/g soil; 3.1% with 125 ng/g in sludge.

ACCURACY (as Mean % Recovery): 39.4% with 2.5 ng/g in clay; 64% with 25 ng/g in soil; 64.5% with 125 ng/g in sludge; 127.5% with 46 ng/g in fly ash; 80.2% with 2500 ng/g in still bottom.

SAMPLING METHOD: Use 1 liter (or quart) amber glass bottles with Teflon lined or solvent washed foil screw caps. Tape caps to bottle after sampling. Compositing equipment must use glass containers and contain no Tygon or rubber tubing. Sample bottles must not be prewashed with the sample before its collection. Aqueous samples cannot be aliquoted from sample containers the entire sample must be used and the container is washed out with the extracting solvent.

STABILITY: Cool to 4°C and store at this temperature. When toluene is employed as the final solvent use a bonded phase GC column for separation. Otherwise, solvent exchange into tridecane is required for other liquid phases or the CP-Sil-88 GC column.

M.H.T. = 30 days;

samples must be completely analyzed within 45 days of collection.

QUALITY CONTROL: A method blank must be analyzed each time a set of samples is extracted or there is a change in reagents. A laboratory method blank must be run with each analytical batch of 20 or fewer samples. Field duplicates and field blanks should be analyzed periodically. GC column performance must be demonstrated initially and verified prior to analyzing any sample in a 12 hour period. A series of calibration standards must be processed through the procedure to validate elution patterns and absence of interferents from reagents. Both the alumina column and carbon column performance must be routinely checked for presence of the analyte. Performance evaluation samples and split samples with other laboratories are also expected to be periodically analyzed.

REFERENCE: Method 8280, SW-846, 3rd ed., Nov 1986.

PRIMARY NAME: 1,3,7,8-TCDD Method 8280

TITLE: Analysis of PCDDs and PCDFs

MATRIX: chemical wastes, fuel oils still bottoms, sludges, water, soil, fly ash, reactor residues.

CAS # : 50585-46-1

APPLICATION: This method is used for the analysis of tetra-, penta-, hexa-, hepta-, and octachlorinated dibenzo-p-dioxins (PCDDs) and dibenzofurans (PCDFs). The sensitivity of the method is dependent on the level of interferents within the matrix. Only experienced analysts should be used. Special safety precautions must be observed and an EPA-approved sample disposal plan must be used.

INTERFERENCES: Solvents, reagents and glassware may introduce artifacts. Other interferences may come from coextracted compounds from samples; PCBs and polychlorinated diphenyl ethers are common interferents.

INSTRUMENTATION: GC/MS with a fused silica capillary column. Also, solvent extraction and concentration glassware and either a gravity flow activated carbon AX-21 / silica gel Type 60 EM reagent column or a HPLC with a 10 mm by 7 cm silanized glass column with active carbon AX-21 and Spherisorb S10W silica for sample cleanup. One of three fused silica capillary GC columns may be used: column 1: 50 meter CP-Sil-88; Column 2: 30 meter by 0.25 mm DB-5; colunm 3: 30 meter SP-2250.

RANGE: 50 to 6,000 picograms MDL: Not determined

PRECISION (as RSD): 7.3% with 5 ng/g in clay; 1.3% with 25 ng/g soil; 5.8% with 125 ng/g in sludge; 3.5% with 16 ng/g in fly ash.

ACCURACY (as Mean % Recovery): 68.0% with 5 ng/g in clay; 79.3% with 25 ng/g in soil; 78.9% with 125 ng/g in sludge; 80.2% with 46 ng/g in fly ash; 90.5% with 2500 ng/g in still bottom.

SAMPLING METHOD: Use 1 liter (or quart) amber glass bottles with Teflon lined or solvent washed foil screw caps. Tape caps to bottle after sampling. Compositing equipment must use glass containers and contain no Tygon or rubber tubing. Sample bottles must not be prewashed with the sample before its collection. Aqueous samples cannot be aliquoted from sample containers the entire sample must be used and the container is washed out with the extracting solvent.

STABILITY: Cool to 4°C and store at this temperature. When toluene is employed as the final solvent use a bonded phase GC column for separation. Otherwise, solvent exchange into tridecane is required for other liquid phases or the CP-Sil-88 GC column.

M.H.T. = 30 days;

samples must be completely analyzed within 45 days of collection.

QUALITY CONTROL: A method blank must be analyzed each time a set of samples is extracted or there is a change in reagents. A laboratory method blank must be run with each analytical batch of 20 or fewer samples. Field duplicates and field blanks should be analyzed periodically. GC column performance must be demonstrated initially and verified prior to analyzing any sample in a 12 hour period. A series of calibration standards must be processed through the procedure to validate elution patterns and absence of interferents from reagents. Both the alumina column and carbon column performance must be routinely checked for presence of the analyte. Performance evaluation samples and split samples with other laboratories are also expected to be periodically analyzed.

REFERENCE: Method 8280, SW-846, 3rd ed., Nov 1986.

PRIMARY NAME: 1,3,7,9-TCDD

Method 8280

TITLE: Analysis of PCDDs and PCDFs

MATRIX: chemical wastes, fuel oils still bottoms, sludges, water, soil, fly ash, reactor residues.

CAS # : 62470-53-5

APPLICATION: This method is used for the analysis of tetra-, penta-, hexa-, hepta-, and octachlorinated dibenzo-p-dioxins (PCDDs) and dibenzofurans (PCDFs). The sensitivity of the method is dependent on the level of interferents within the matrix. Only experienced analysts should be used. Special safety precautions must be observed and an EPA-approved sample disposal plan must be used.

INTERFERENCES: Solvents, reagents and glassware may introduce artifacts. Other interferences may come from coextracted compounds from samples; PCBs and polychlorinated diphenyl ethers are common interferents.

INSTRUMENTATION: GC/MS with a fused silica capillary column. Also, solvent extraction and concentration glassware and either a gravity flow activated carbon AX-21 / silica gel Type 60 EM reagent column or a HPLC with a 10 mm by 7 cm silanized glass column with active carbon AX-21 and Spherisorb S10W silica for sample cleanup. One of three fused silica capillary GC columns may be used: column 1: 50 meter CP-Sil-88; Column 2: 30 meter by 0.25 mm DB-5; colunm 3: 30 meter SP-2250.

RANGE: 50 to 6,000 picograms MDL: Not determined

PRECISION (as RSD): 19% with 2.5 ng/g in clay; 2.3% with 25 ng/g soil; 6.5% with 125 ng/g in sludge.

ACCURACY (as Mean % Recovery): 68.5% with 2.5 ng/g in clay; 61.3% with 25 ng/g in soil; 78.4% with 125 ng/g in sludge; 85.0% with 46 ng/g in fly ash; 91.7% with 2500 ng/g in still bottom.

SAMPLING METHOD: Use 1 liter (or quart) amber glass bottles with Teflon lined or solvent washed foil screw caps. Tape caps to bottle after sampling. Compositing equipment must use glass containers and contain no Tygon or rubber tubing. Sample bottles must not be prewashed with the sample before its collection. Aqueous samples cannot be aliquoted from sample containers the entire sample must be used and the container is washed out with the extracting solvent.

STABILITY: Cool to 4°C and store at this temperature. When toluene is employed as the final solvent use a bonded phase GC column for separation. Otherwise, solvent exchange into tridecane is required for other liquid phases or the CP-Sil-88 GC column.

M.H.T. = 30 days;

samples must be completely analyzed within 45 days of collection.

QUALITY CONTROL: A method blank must be analyzed each time a set of samples is extracted or there is a change in reagents. A laboratory method blank must be run with each analytical batch of 20 or fewer samples. Field duplicates and field blanks should be analyzed periodically. GC column performance must be demonstrated initially and verified prior to analyzing any sample in a 12 hour period. A series of calibration standards must be processed through the procedure to validate elution patterns and absence of interferents from reagents. Both the alumina column and carbon column performance must be routinely checked for presence of the analyte. Performance evaluation samples and split samples with other laboratories are also expected to be periodically analyzed.

REFERENCE: Method 8280, SW-846, 3rd ed., Nov 1986.

PRIMARY NAME: 2,3,7,8-TCDD Method 8280

TITLE: Analysis of PCDDs and PCDFs

MATRIX: chemical wastes, fuel oils still bottoms, sludges, water, soil, fly ash, reactor residues.

CAS # : 1746-01-6

APPLICATION: This method is used for the analysis of tetra-, penta-, hexa-, hepta-, and octachlorinated dibenzo-p-dioxins (PCDDs) and dibenzofurans (PCDFs). The sensitivity of the method is dependent on the level of interferents within the matrix. Only experienced analysts should be used. Special safety precautions must be observed and an EPA-approved sample disposal plan must be used.

INTERFERENCES: Solvents, reagents and glassware may introduce artifacts. Other interferences may come from coextracted compounds from samples; PCBs and polychlorinated diphenyl ethers are common interferents.

INSTRUMENTATION: GC/MS with a fused silica capillary column. Also, solvent extraction and concentration glassware and either a gravity flow activated carbon AX-21 / silica gel Type 60 EM reagent column or a HPLC with a 10 mm by 7 cm silanized glass column with active carbon AX-21 and Spherisorb S10W silica for sample cleanup. One of three fused silica capillary

GC columns may be used: column 1: 50 meter CP-Sil-88; Column 2: 30 meter by 0.25 mm DB-5; colunm 3: 30 meter SP-2250.

RANGE: 50 to 6,000 picograms

MDL: 0.44 ng/L (in reagent water) 0.17 µg/Kg (in Missouri soil) 0.07 µg/Kg (in fly ash) 0.82 µg/Kg (in industrial sludge) 1.81 µg/Kg (in still bottom) 0.75 µg/Kg (in fuel oil)
MDLs are for carbon-13 labeled analyte

PRECISION (as RSD): 4.4% with 5 ng/g in clay; 2.8% with 378 ng/g soil; 4.8% with 125 ng/g in sludge; 24% with 487 ng/g in still bottom.

ACCURACY (as Mean % Recovery): 61.7% with 5 ng/g in clay; 90.0% with 125 ng/g in sludge; 90.0% with 46 ng/g in fly ash.

SAMPLING METHOD: Use 1 liter (or quart) amber glass bottles with Teflon lined or solvent washed foil screw caps. Tape caps to bottle after sampling. Compositing equipment must use glass containers and contain no Tygon or rubber tubing. Sample bottles must not be prewashed with the sample before its collection. Aqueous samples cannot be aliquoted from sample containers the entire sample must be used and the container is washed out with the extracting solvent.

STABILITY: Cool to 4°C and store at this temperature. When toluene is employed as the final solvent use a bonded phase GC column for separation. Otherwise, solvent exchange into tridecane is required for other liquid phases or the CP-Sil-88 GC column.

M.H.T. = 30 days; samples must be completely analyzed within 45 days of collection.

QUALITY CONTROL: A method blank must be analyzed each time a set of samples is extracted or there is a change in reagents. A laboratory method blank must be run with each analytical batch of 20 or fewer samples. Field duplicates and field blanks should be analyzed periodically. GC column performance must be demonstrated initially and verified prior to analyzing any sample in a 12 hour period. A series of calibration standards must be processed through the procedure to validate elution patterns and absence of interferents from reagents. Both the alumina column and carbon column performance must be routinely checked for presence of the analyte. Performance evaluation samples and split samples with other laboratories are also expected to be periodically analyzed.

REFERENCE: Method 8280, SW-846, 3rd ed., Nov 1986.

PRIMARY NAME: 1,2,7,8-TCDF Method 8280

TITLE: Analysis of PCDDs and PCDFs

MATRIX: chemical wastes, fuel oils still bottoms, sludges, water, soil, fly ash, reactor residues.

CAS # : 58802-20-3

APPLICATION: This method is used for the analysis of tetra-, penta-, hexa-, hepta-, and octachlorinated dibenzo-p-dioxins (PCDDs) and dibenzofurans (PCDFs). The sensitivity of the method is dependent on the level of interferents within the matrix. Only experienced analysts should be used. Special safety precautions must be observed and an EPA-approved sample disposal plan must be used.

INTERFERENCES: Solvents, reagents and glassware may introduce artifacts. Other interferences may come from coextracted compounds from samples; PCBs and polychlorinated diphenyl ethers are common interferents.

INSTRUMENTATION: GC/MS with a fused silica capillary column. Also, solvent extraction and concentration glassware and either a gravity flow activated carbon AX-21 / silica gel Type 60 EM reagent column or a HPLC with a 10 mm by 7 cm silanized glass column with active carbon AX-21 and Spherisorb S10W silica for sample cleanup. One of three fused silica capillary GC columns may be used: column 1: 50 meter CP-Sil-88; Column 2: 30 meter by 0.25 mm DB-5; colunm 3: 30 meter SP-2250.

RANGE: 50 to 6,000 picograms

MDL: Not determined

PRECISION (as RSD): 3.9% with 5 ng/g in clay; 1.0% with 25 ng/g soil; 7.2% with 125 ng/g in sludge; 7.6% with 7.4 ng/g in fly ash.

ACCURACY (as Mean % Recovery): 65.4% with 5 ng/g in clay; 71.1% with

25 ng/g in soil; 80.4% with 125 ng/g in sludge; 90.4% with 46 ng/g in fly ash; 104.5% with 2500 ng/g in still bottom.

SAMPLING METHOD: Use 1 liter (or quart) amber glass bottles with Teflon lined or solvent washed foil screw caps. Tape caps to bottle after sampling. Compositing equipment must use glass containers and contain no Tygon or rubber tubing. Sample bottles must not be prewashed with the sample before its collection. Aqueous samples cannot be aliquoted from sample containers the entire sample must be used and the container is washed out with the extracting solvent.

STABILITY: Cool to 4°C and store at this temperature. When toluene is employed as the final solvent use a bonded phase GC column for separation. Otherwise, solvent exchange into tridecane is required for other liquid phases or the CP-Sil-88 GC column.

M.H.T. = 30 days;

samples must be completely analyzed within 45 days of collection.

QUALITY CONTROL: A method blank must be analyzed each time a set of samples is extracted or there is a change in reagents. A laboratory method blank must be run with each analytical batch of 20 or fewer samples. Field duplicates and field blanks should be analyzed periodically. GC column performance must be demonstrated initially and verified prior to analyzing any sample in a 12 hour period. A series of calibration standards must be processed through the procedure to validate elution patterns and absence of interferents from reagents. Both the alumina column and carbon column performance must be routinely checked for presence of the analyte. Performance evaluation samples and split samples with other laboratories are also expected to be periodically analyzed.

REFERENCE: Method 8280, SW-846, 3rd ed., Nov 1986.

PRIMARY NAME: 2,3,7,8-TCDF

Method 8280

TITLE: Analysis of PCDDs and PCDFs

MATRIX: chemical wastes, fuel oils still bottoms, sludges, water, soil, fly ash, reactor residues.

CAS # : 51207-31-9

APPLICATION: This method is used for the analysis of tetra-, penta-, hexa-, hepta-, and octachlorinated dibenzo-p-dioxins (PCDDs) and dibenzofurans (PCDFs). The sensitivity of the method is dependent on the level of interferents within the matrix. Only experienced analysts should be used. Special safety precautions must be observed and an EPA-approved sample disposal plan must be used.

INTERFERENCES: Solvents, reagents and glassware may introduce artifacts. Other interferences may come from coextracted compounds from samples; PCBs and polychlorinated diphenyl ethers are common interferents.

INSTRUMENTATION: GC/MS with a fused silica capillary column. Also, solvent extraction and concentration glassware and either a gravity flow activated carbon AX-21 / silica gel Type 60 EM reagent column or a HPLC with a 10 mm by 7 cm silanized glass column with active carbon AX-21 and Spherisorb S10W silica for sample cleanup. One of three fused silica capillary GC columns may be used: column 1: 50 meter CP-Sil-88; Column 2: 30 meter by 0.25 mm DB-5; colunm 3: 30 meter SP-2250.

RANGE: 50 to 6,000 picograms

MDL: 0.63 ng/L (in reagent water) 0.11 µg/Kg (in Missouri soil) 0.06 µg/Kg (in fly ash) 0.46 µg/Kg (in industrial sludge) 0.26 µg/Kg (in still bottom) 0.48 µg/Kg (in fuel oil)
MDLs are for carbon-13 labeled analyte

PRECISION (as RSD): Not determined

ACCURACY (as Mean % Recovery): 64.9% with 5 ng/g in clay; 78.8% with 25 ng/g in soil; 78.6% with 125 ng/g in sludge; 88.6% with 46 ng/g in fly ash; 69.7% with 2500 ng/g in still bottom. The C-13 isomer was used.

SAMPLING METHOD: Use 1 liter (or quart) amber glass bottles with Teflon lined or solvent washed foil screw caps. Tape caps to bottle after sampling. Compositing equipment must use glass containers and contain no Tygon or rubber tubing. Sample bottles must not be prewashed with the sample before its collection. Aqueous samples cannot be aliquoted from sample containers the entire sample must be used and the container is washed out with the extracting solvent.

STABILITY: Cool to 4°C and store at this temperature. When toluene is

employed as the final solvent use a bonded phase GC column for separation. Otherwise, solvent exchange into tridecane is required for other liquid phases or the CP-Sil-88 GC column.

M.H.T. = 30 days;

samples must be completely analyzed within 45 days of collection.

QUALITY CONTROL: A method blank must be analyzed each time a set of samples is extracted or there is a change in reagents. A laboratory method blank must be run with each analytical batch of 20 or fewer samples. Field duplicates and field blanks should be analyzed periodically. GC column performance must be demonstrated initially and verified prior to analyzing any sample in a 12 hour period. A series of calibration standards must be processed through the procedure to validate elution patterns and absence of interferents from reagents. Both the alumina column and carbon column performance must be routinely checked for presence of the analyte. Performance evaluation samples and split samples with other laboratories are also expected to be periodically analyzed.

REFERENCE: Method 8280, SW-846, 3rd ed., Nov 1986.

PRIMARY NAME: Tokuthion (Prothiofos) Method 8140

TITLE: Organophosphorus Pesticides

MATRIX: groundwater, soils, sludges water miscible liquid wastes, and non-water miscible wastes.

CAS # : 34643-46-4

APPLICATION: This method is used for the analysis of 21 organophosphorus pesticides. Samples are extracted, concentrated and analyzed using direct injection of both neat and diluted organic liquid into a gas chromatograph (GC).

INTERFERENCES: Solvents, reagents and glassware may introduce artifacts. Other interferences may come from coextracted compounds from samples. The use of Florisil cleanup materials may produce low recoveries. Elemental sulfur may interfere with some compounds when using a flame photometric detector. Sulfur cleanup (Method 3660) may alleviate sulfur interference.

INSTRUMENTATION: GC capable of on-column injections and a flame

photometric detector (FPD) or a thermionic detector. Column 1: 1.8 meter by 2 mm with 5% SP-2401 on Supelcoport. Column 2: 1.8 meter by 2 mm with 3% SP-2401 on Supelcoport. Column 3: 50 cm by 1/8 inch Teflon with 15% SE-54 on Gas Chrom Q. The preferred column is Column Number 1.

RANGE: 5.3 to 64 µg/L

MDL: 0.5 µg/L (in reagent water)

PRACTICAL QUANTITATION LIMIT FACTORS FOR MULTIPLYING TIMES FID MDL VALUE

Matrix	Multiplication Factor
Groundwater	10
Low-level soil by sonication with GPC cleanup	670
High-level soil and sludge by sonication	10,000
Non-water miscible waste	100,000

PRECISION: 6.8% (single operator standard deviation)

ACCURACY: 64.6% (single operator average recovery)

SAMPLING METHOD: Use 8 oz. widemouth glass bottles with Teflon lined caps for concentrated waste samples, soils, sediments and sludges. Use 1 or 2 1/2 gallon amber glass bottles with Teflon lined caps for liquid (water) samples.

STABILITY: Cool soil, sediment, sludge and liquid samples to 4°C. If residual chlorine is present in liquid samples add 3 mL of 10% sodium thiosulfate per gallon of sample and cool to 4°C.

M.H.T. = 14 days for concentrated waste, soil, sediment or sludge.

M.H.T. = 7 days for liquid samples.

All extracts must be analyzed within 40 days.

QUALITY CONTROL: A quality control check sample concentrate containing this compound in acetone at a concentration 1,000 times more concentrated than the selected spike concentration is required. The QC check sample concentrate may be prepared from pure standard materials or purchased as certified solutions. Use appropriate trip, matrix, control site, method, reagent and solvent blanks. Internal, surrogate and five concentration level calibration standards are used.

REFERENCE: Method 8140, SW-846, 3rd ed., Sep 1986.

PRIMARY NAME: Toxaphene

Method 8080

TITLE: Organochlorine Pesticides and PCBs

MATRIX: groundwater, soils, sludges water miscible liquid wastes, and non-water miscible wastes.

CAS # : 8001-35-2

APPLICATION: This method is used for the analysis of 19 pesticides and 7 Aroclor (PCB) mixtures. Samples are extracted, concentrated and analyzed using direct injection of both neat and diluted organic liquid into a gas chromatograph (GC).

INTERFERENCES: Solvents, reagents and glassware may introduce artifacts. Other interferences may come from coextracted compounds from samples. Phthalate esters are common interferences when using an electron capture detector (ECD) so all plastics must be strictly avoided. Exhaustive cleanup of reagents and glassware may be required to eliminate phthalate contamination. Use of a halogen specific microcoulometric or electrolytic conductivity detector will eliminate phthalate interference. Sometimes 4,4'-DDT is an interferent that coelutes with toxaphene and methoxychlor may also be a common interferent.

INSTRUMENTATION: GC capable of on-column injections and an ECD or a halogen specific detector (HSD). Column 1: 1.8 meter by 4 mm with 1.5% SP-2250 / 1.95% SP-2401 on Supelcoport. Column 2: 1.8 meter by 4 mm with 3% OV-1 on Supelcoport.

RANGE: 8.5 to 400 μg/L

MDL: 0.24 μg/L (in reagent water)

PRACTICAL QUANTITATION LIMIT FACTORS FOR MULTIPLYING TIMES FID MDL VALUE

Matrix	Multiplication Factor
Groundwater	10
Low-level soil by sonication with GPC cleanup	670
High-level soil and sludge by sonication	10,000
Non-water miscible waste	100,000

PRECISION: 0.20X + 0.22 μg/L (overall precision)

ACCURACY: 0.80C + 1.74 μg/L (as recovery)

SAMPLING METHOD: Use 8 oz. widemouth glass bottles with Teflon lined caps for concentrated waste samples, soils, sediments and sludges. Use 1 or 2 1/2 gallon amber glass bottles with Teflon lined caps for liquid (water) samples.

STABILITY: Cool soil, sediment, sludge and liquid samples to 4°C. If residual chlorine is present in liquid samples add 3 mL of 10% sodium thiosulfate per gallon of sample and cool to 4°C.

M.H.T. = 14 days for concentrated waste, soil, sediment or sludge.

M.H.T. = 7 days for liquid samples.

All extracts must be analyzed within 40 days.

QUALITY CONTROL: A quality control check sample concentrate containing each analyte of interest is required. The QC check sample concentrate may be prepared from pure standard materials or purchased as certified solutions. Use appropriate trip, matrix, control site, method, reagent and solvent blanks. Internal, surrogate and five concentration level calibration standards are used. The quality control check sample concentrate should contain toxaphene at 50 μg/mL in acetone.

REFERENCE: Method 8080, SW-846, 3rd ed., Nov 1986.

PRIMARY NAME: 2,4,5-TP (Silvex) Method 8150

TITLE: Chlorinated Herbicides

MATRIX: groundwater, soils, sludges water miscible liquid wastes, and non-water miscible wastes.

CAS # : 93-72-1

APPLICATION: This method is used for the analysis of 10 chlorinated herbicides. Samples are extracted, hydrolyzed with potassium hydroxide, and extraneous organics are removed by a solvent wash. After acidification, the acids are extracted, concentrated and converted to their methyl esters using diazomethane. They are then analyzed using direct injection into a gas chromatograph (GC). Be very careful because diazomethane can explode under certain conditions and it is also a carcinogen.

INTERFERENCES: Organic acids and phenols (especially chlorinated acids and phenols) may cause interferences. Phthalate esters are not as significant an interference as with other GC-ECD methods if an electron capture detector is used. The herbicides may react readily with alkaline substances and be lost during analysis so all glassware and glass wool must be acid rinsed and sodium sulfate must be acidified with sulfuric acid prior to use. Sensitivity usually depends on the level of interferences rather than on instrumentation.

INSTRUMENTATION: GC capable of on-column injections and an electron capture detector (ECD)or a halogen specific detector. Column 1: 1.8 meter by 4 mm with 1.5% SP-2250 / 1.95% SP-2401 on Supelcoport. Column 2: 1.8 meter by 4 mm with 5% OV-210 on Gas Chrom Q. Column 3: 1.98 meter by 2 mm with 0.1% SP-1000 on Carbopack C. The preferred column is Column Number 1 or 2.

RANGE: Not listed

MDL: 0.17 µg/L (in reagent water; ECD)

PRACTICAL QUANTITATION LIMIT FACTORS FOR MULTIPLYING TIMES FID MDL VALUE

Matrix	Multiplication Factor
Groundwater	10
Low-level soil by sonication with GPC cleanup	670
High-level soil and sludge by sonication	10,000
Non-water miscible waste	100,000

PRECISION: (as standard deviation) 5% with 1.0 µg/L spike in drinking water; 4% with 1.3 µg/L in municipal water.

ACCURACY: (as mean recovery) 88% with 1.0 µg/L spike in drinking water; 88% with 1.3 µg/L in municipal water.

SAMPLING METHOD: Use 8 oz. widemouth glass bottles with Teflon lined caps for concentrated waste samples, soils, sediments and sludges. Use 1 or 2 1/2 gallon amber glass bottles with Teflon lined caps for liquid (water) samples.

STABILITY: Cool soil, sediment, sludge and liquid samples to 4°C. If residual chlorine is present in liquid samples add 3 mL of 10% sodium thiosulfate per gallon of sample and cool to 4°C.

M.H.T. = 14 days for concentrated waste, soil, sediment or sludge.

M.H.T. = 7 days for liquid samples.

All extracts must be analyzed within 40 days.

QUALITY CONTROL: A quality control check sample concentrate containing this compound in acetone at a concentration 1,000 times more concentrated than the selected spike concentration is required. The QC check sample concentrate may be prepared from pure standard materials or purchased as certified solutions. Use appropriate trip, matrix, control site, method, reagent and solvent blanks. Internal, surrogate and five concentration level calibration standards are used.

REFERENCE: Method 8150, SW-846, 3rd ed., Sep 1986.

PRIMARY NAME: Trichloronat — Method 8140

TITLE: Organophosphorus Pesticides

MATRIX: groundwater, soils, sludges water miscible liquid wastes, and non-water miscible wastes.

CAS # : 327-98-0

APPLICATION: This method is used for the analysis of 21 organophosphorus pesticides. Samples are extracted, concentrated and analyzed using direct injection of both neat and diluted organic liquid into a gas chromatograph (GC).

INTERFERENCES: Solvents, reagents and glassware may introduce artifacts. Other interferences may come from coextracted compounds from samples. The use of Florisil cleanup materials may produce low recoveries. Elemental sulfur may interfere with some compounds when using a flame photometric detector. Sulfur cleanup (Method 3660) may alleviate sulfur interference.

INSTRUMENTATION: GC capable of on-column injections and a flame photometric detector (FPD) or a thermionic detector. Column 1: 1.8 meter by 2 mm with 5% SP-2401 on Supelcoport. Column 2: 1.8 meter by 2 mm with

3% SP-2401 on Supelcoport. Column 3: 50 cm by 1/8 inch Teflon with 15% SE-54 on Gas Chrom Q. The preferred column is Column Number 1.

RANGE: 20 μg/L only

MDL: 0.15 μg/L (in reagent water)

PRACTICAL QUANTITATION LIMIT FACTORS FOR MULTIPLYING TIMES FID MDL VALUE

Matrix	Multiplication Factor
Groundwater	10
Low-level soil by sonication with GPC cleanup	670
High-level soil and sludge by sonication	10,000
Non-water miscible waste	100,000

PRECISION: 18.6% (single operator standard deviation)

ACCURACY: 105.0% (single operator average recovery)

SAMPLING METHOD: Use 8 oz. widemouth glass bottles with Teflon lined caps for concentrated waste samples, soils, sediments and sludges. Use 1 or 2 1/2 gallon amber glass bottles with Teflon lined caps for liquid (water) samples.

STABILITY: Cool soil, sediment, sludge and liquid samples to 4°C. If residual chlorine is present in liquid samples add 3 mL of 10% sodium thiosulfate per gallon of sample and cool to 4°C.

M.H.T. = 14 days for concentrated waste, soil, sediment or sludge.

M.H.T. = 7 days for liquid samples.

All extracts must be analyzed within 40 days.

QUALITY CONTROL: A quality control check sample concentrate containing this compound in acetone at a concentration 1,000 times more concentrated than the selected spike concentration is required. The QC check sample concentrate may be prepared from pure standard materials or purchased as certified solutions. Use appropriate trip, matrix, control site, method, reagent and solvent blanks. Internal, surrogate and five concentration level calibration standards are used.

REFERENCE: Method 8140, SW-846, 3rd ed., Sep 1986.

Chapter 6
Elements

PRIMARY NAME: Aluminum (Al)

Method 200.7

TITLE: Inductively Coupled Plasma (ICP)

MATRIX: Dissolved, suspended or total element in drinking and surface waters and in domestic and industrial wastewaters.

APPLICATION: The method covers the determination of 25 metals. Dissolved elements are determined in filtered and acidified samples after appropriate digestion (which increases dissolved solids). Its primary advantage is that ICP instruments allow simultaneous or rapid sequential determination of many elements in a short time. Samples are first nebulized and the aerosol is transported to a plasma torch in which element specific atomic line emission spectra are produced by a radio frequency inductively coupled plasma. Background correction is required for trace element detection except in the case of line broadening.

INTERFERENCES: There are spectral, physical and chemical interferences. The primary disadvantage of ICP instruments is background radiation from other elements and the plasma gases (spectral interferences). Changes in sample viscosity and surface tension with samples containing high dissolved solids (especially those exceeding 1500 mg/L) or high acid concentrations can cause physical interferences. Ionization effects, solute vaporization and molecular compound formation can cause chemical interferences. Manganese and vanadium can cause interference at the 100 mg/L level.

INSTRUMENTATION: Inductively Coupled Argon Plasma Emission Spectroscopy. 308.215 nm Wavelength

RANGE: Not listed MDL: 45 μg/L.

PRECISION: SD = 5.6% Mean @ true value 700 μg/L.

ACCURACY: Mean Recovery = 93% +/- 6% of spiked elements for all wastes.

SAMPLING METHOD: Wash sample container with detergent and tap water, rinse with 1+1 nitric acid and tap water, then rinse with 1+1 hydrochloric acid and tap water, then rinse with deionized, distilled water in that order. Perform any filtration or acid preservation steps when the sample is collected or as soon as possible thereafter.

STABILITY: Cool samples to 4°C. M.H.T. = 24 Hours.

QUALITY CONTROL: Mixed calibration standards, an instrument check standard and an interference check solution are used in addition to a quality control sample. The quality control sample should be prepared in the same acid matrix as the calibration standards at 10 times the instrumental detection limits and in accordance with the instructions provided by the supplier. Furthermore, two types of blanks are required: a calibration blank and a reagent blank.

REFERENCE: Method 200.7, U.S. EPA, EMSL-Cincinnati, OH, Nov. 1980

PRIMARY NAME: Aluminum (Al) Method 6010

TITLE: Inductively Coupled Plasma (ICP)

MATRIX: Applies to all matrices; ground water, EP extracts, soils, sludges, sediments, aqueous samples, solid and industrial wastes.

APPLICATION: The method covers the determination of 25 metals. Its primary advantage is that ICP instruments allow simultaneous or rapid sequential determination of many elements in a short time. Samples require digestion prior to analysis and are first nebulized and the aerosol is transported to a plasma torch in which element specific atomic line emission spectra are produced by a radio frequency inductively coupled plasma. Background correction is required for trace element detection except in the case of line broadning.

INTERFERENCES: There are spectral, physical and chemical interferences. The primary disadvantage of ICP instruments is background radiation from other elements and the plasma gases (spectral interferences). Changes in sample viscosity and surface tension with samples containing high dissolved solids or high acid concentrations can cause physical interferences. Ionization effects, solute vaporization and molecular compound formation can cause chemical interferences. Manganese and vanadium can cause interference at the 100 mg/L level.

INSTRUMENTATION: Inductively Coupled Argon Plasma Emission Spectroscopy. 308.215 nm Wavelength

RANGE: Not listed MDL: 45 μg/L.

PRECISION: SD = 5.6% Mean @ true value 700 μg/L.

ACCURACY: Mean Recovery = 93% +/- 6% of spiked elements for all wastes.

SAMPLING METHOD: Collect 400 mL of sample in plastic or glass containers.

STABILITY: Cool samples to 4°C. M.H.T. = 24 Hours.

QUALITY CONTROL: Mixed calibration standards, an instrument check standard and an interference check solution are used in addition to a quality control sample. The quality control sample should be prepared in the same acid matrix as the calibration standards at 10 times the instrumental detection limits and in accordance with the instructions provided by the supplier. Furthermore, two types of blanks are required: a calibration blank and a reagent blank.

REFERENCE: Method 6010, SW-846, 3rd ed., Nov 1986.

PRIMARY NAME: Aluminum (Al) Method 7020

TITLE: Atomic Absorption, (AA) Direct Aspiration

MATRIX: Drinking, Surface and Saline Waters. Wastewater

CAS # : 7429-90-5

APPLICATION: Sample is aspirated and atomized in a flame. A light beam from an aluminum hollow cathode lamp is directed through the flame into a monochromator and onto a detector. Since wavelength of light beam is specific for aluminum, light energy absorbed by detector is measure of aluminum concentration.

INTERFERENCES: The most troublesome type of interference is chemical, and caused by lack of absorption of atoms bound in molecular combination in the flame. High dissolved solids in a sample may result in nonatomic absorbance interference. Ionization and spectral interferences can occur.

INSTRUMENTATION: Atomic absorption spectrometer. Aluminum hollow cathode lamp or electrodeless discharge lamp. (309.3nm Wavelength).

RANGE: 5-50 mg/L

MDL: 0.1 mg/L

PRECISION: deviation = 299 µg/L @ 1205 µg/L (true value) 38 labs

ACCURACY: as bias = +6.3% @ 1205 µg/L (true value) 38 labs

SAMPLING METHOD: Use glass or plastic containers. Collect 200 g of solids and 600 mL of liquid samples.

STABILITY: Cool solid samples to 4°C. and analyze as soon as possible. Add nitric acid to liquid samples to pH < 2; M.H.T. = 6 months.

QUALITY CONTROL: At least one duplicate and one spike sample should be run every 20 samples or with each matrix type to verify precision of the method. For 20 or more samples per day, verify working standard curve. Run an additional standard at or near mid-range every 10 samples.

REFERENCE: Method 7020, SW-846, 3rd ed., Nov 1986.

PRIMARY NAME: Antimony (Sb) Method 200.7

TITLE: Inductively Coupled Plasma (ICP)

MATRIX: Dissolved, suspended or total element in drinking and surface waters and in domestic and industrial wastewaters.

APPLICATION: The method covers the determination of 25 metals. Dissolved elements are determined in filtered and acidified samples after appropriate digestion (which increases dissolved solids). Its primary advantage is that ICP instruments allow simultaneous or rapid sequential determination of many elements in a short time. Samples are first nebulized and the aerosol is transported to a plasma torch in which element specific atomic line emission spectra are produced by a radio frequency inductively coupled plasma. Background correction is required for trace element detection except in the case of line broadning.

INTERFERENCES: There are spectral, physical and chemical interferences. The primary disadvantage of ICP instruments is background radiation from other elements and the plasma gases (spectral interferences). Changes in sample viscosity and surface tension with samples containing high dissolved solids (especially those exceeding 1500 mg/L) or high acid concentrations can cause physical interferences. Ionization effects, solute vaporization and molecular compound formation can cause chemical interferences. Aluminum, chromium, iron, thallium, and vanadium can cause interferences at the 100 mg/L level.

INSTRUMENTATION: Inductively Coupled Argon Plasma Emission Spectroscopy. 206.833 nm Wavelength

RANGE: Not listed

MDL: 32 μg/L.

PRECISION: Not listed

ACCURACY: Mean Recovery = 93% +/- 6% of spiked elements for all wastes.

SAMPLING METHOD: Wash sample container with detergent and tap water, rinse with 1+1 nitric acid and tap water, then rinse with 1+1 hydrochloric acid and tap water, then rinse with deionized, distilled water in that order. Perform any filtration or acid preservation steps when the sample is collected or as soon as possible thereafter.

STABILITY: Cool samples to 4°C. M.H.T. = 24 Hours.

QUALITY CONTROL: Mixed calibration standards, an instrument check standard and an interference check solution are used in addition to a quality control sample. The quality control sample should be prepared in the same acid matrix as the calibration standards at 10 times the instrumental detection limits and in accordance with the instructions provided by the supplier. Furthermore, two types of blanks are required: a calibration blank and a reagent blank.

REFERENCE: Method 200.7, U.S. EPA, EMSL-Cincinnati, OH, Nov. 1980

PRIMARY NAME: Antimony (Sb) Method 6010

TITLE: Inductively Coupled Plasma (ICP)

MATRIX: Applies to all matrices; ground water, EP extracts, soils, sludges, sediments, aqueous samples, solid and industrial wastes.

APPLICATION: The method covers the determination of 25 metals. Its primary advantage is that ICP instruments allow simultaneous or rapid sequential determination of many elements in a short time. Samples require digestion prior to analysis and are first nebulized and the aerosol is transported to a plasma torch in which element specific atomic line emission spectra are produced by a radio frequency inductively coupled plasma. Background correction is required for trace element detection except in the case of line broadning.

INTERFERENCES: There are spectral, physical and chemical interferences. The primary disadvantage of ICP instruments is background radiation from other elements and the plasma gases (spectral interferences). Changes in sample viscosity and surface tension with samples containing high dissolved solids or high acid concentrations can cause physical interferences. Ionization effects, solute vaporization and molecular compound formation can cause chemical interferences. Aluminum, chromium, iron, thallium, and vanadium can cause interferences at the 100 mg/L level.

INSTRUMENTATION: Inductively Coupled Argon Plasma Emission Spectroscopy. 206.833 nm Wavelength

RANGE: Not listed MDL: 32 µg/L.

PRECISION: Not listed

ACCURACY: Mean Recovery = 93% +/- 6% of spiked elements for all wastes.

SAMPLING METHOD: Collect 400 mL of sample in plastic or glass containers.

STABILITY: Cool samples to 4°C. M.H.T. = 24 Hours.

QUALITY CONTROL: Mixed calibration standards, an instrument check standard and an interference check solution are used in addition to a quality control sample. The quality control sample should be prepared in the same acid matrix as the calibration standards at 10 times the instrumental detection limits and in accordance with the instructions provided by the supplier. Furthermore, two types of blanks are required: a calibration blank and a reagent blank.

REFERENCE: Method 6010, SW-846, 3rd ed., Nov 1986.

PRIMARY NAME: Antimony (Sb) Method 7040

TITLE: Atomic Absorption, (AA) Direct Aspiration

MATRIX: Drinking, Surface and Saline Waters. Wastewater

CAS # : 7440-36-0

APPLICATION: Sample is aspirated and atomized in a flame. A light beam from an Sb hollow cathode lamp is directed through the flame into a monochromator and onto a detector. Since wavelength of light beam is specific for Sb, light energy absorbed by detector is measure of Sb.

INTERFERENCES: The most troublesome type is chemical, caused by lack of absorption of atoms bound in molecular combination in the flame. High dissolved solids in sample may result in nonatomic absorbance interference. Lead, copper, nickel and excess acid can interfere.

INSTRUMENTATION: Atomic absorption spectrometer. Antimony hollow cathode lamp or electrodeless discharge lamp. (217.6nm Wavelength).

RANGE: 1-40 mg/L MDL: 0.2 mg/L

PRECISION: standard deviation = +/- 0.08 and 0.10% @ 5.0 and 15 mg Sb/L

ACCURACY: recoveries = 96% and 97% @ 5.0 mg and 15 mg Sb/L

SAMPLING METHOD: Use glass or plastic containers. Collect 200 g of solids and 600 mL of liquid samples.

STABILITY: Cool solid samples to 4°C. and analyze as soon as possible. Add nitric acid to liquid samples to pH < 2; M.H.T. = 6 months.

QUALITY CONTROL: At least one duplicate and one spike sample should be run every 20 samples or with each matrix type to verify precision of the method. For 20 or more samples per day, verify working standard curve. Run an additional standard at or near mid-range every 10 samples.

REFERENCE: Method 7040, SW-846, 3rd ed., Nov 1986.

PRIMARY NAME: Antimony (Sb) Method 7041

TITLE: Atomic Absorption, (AA) Furnace Technique

MATRIX: Wastes, mobility procedure extracts, soils, and groundwater

CAS # : 7440-36-0

APPLICATION: Aqueous samples, EP extracts, industrial wastes, soils, sludges, sediments and solid wastes require digestion before analysis. An aliquot of sample is placed in the graphite tube in the furnace and slowly evaporated, charred and atomized. Absorption of lamp radiation during atomization is proportional to (Sb) concentration.

INTERFERENCES: The furnace technique is subject to chemical interferences. Composition of sample matrix can have major effect on analysis. Modify matrix to remove interferences. High lead concentration may cause spectral interference @ 217.6 nm Line. (With interference, use 231.1 nm).

INSTRUMENTATION: Atomic absorption spectrometer. (Sb) hollow cathode lamp or electrodeless discharge lamp. Graphite furnace. Strip-chart recorder.

RANGE: 20-300 μg/L

MDL: 3 μg/L [217.6 nm line (primary)]

PRECISION: Not listed

ACCURACY: Not listed

SAMPLING METHOD: Use glass or plastic containers. Collect 200 g of solids and 600 mL of liquid samples.

STABILITY: Cool solid samples to 4°C. and analyze as soon as possible. Add nitric acid to liquid samples to pH < 2; their M.H.T. = 6 months.

QUALITY CONTROL: At least one duplicate and one spike sample should be run every 20 samples, or with each matrix type to verify method precision. If 20 or more samples are run a day, run a standard (at or near mid-range) every 10 samples.

REFERENCE: Method 7041, SW-846, 3rd ed., Nov 1986.

PRIMARY NAME: Arsenic (As)

Method 200.7

TITLE: Inductively Coupled Plasma (ICP)

MATRIX: Dissolved, suspended or total element in drinking and surface waters and in domestic and industrial wastewaters.

APPLICATION: The method covers the determination of 25 metals. Dissolved elements are determined in filtered and acidified samples after appropriate digestion (which increases dissolved solids). Its primary advantage is that ICP instruments allow simultaneous or rapid sequential determination of many elements in a short time. Samples are first nebulized and the aerosol is transported to a plasma torch in which element specific atomic line emission spectra are produced by a radio frequency inductively coupled plasma. Background correction is required for trace element detection except in the case of line broadning.

INTERFERENCES: There are spectral, physical and chemical interferences. The primary disadvantage of ICP instruments is background radiation from other elements and the plasma gases (spectral interferences). Changes in sample viscosity and surface tension with samples containing high dissolved solids (especially those exceeding 1500 mg/L) or high acid concentrations can cause physical interferences. Ionization effects, solute vaporization and molecular compound formation can cause chemical interferences. Aluminum, chromium and vanadium can cause interference at the 100 mg/L level.

INSTRUMENTATION: Inductively Coupled Argon Plasma Emission Spectroscopy. 193.696 nm Wavelength

RANGE: Not listed MDL: 53 μg/L.

PRECISION: SD = 7.5% Mean @ true value 200 μg/L.

ACCURACY: Mean Recovery = 93% +/- 6% of spiked elements for all wastes.

SAMPLING METHOD: Wash sample container with detergent and tap water, rinse with 1+1 nitric acid and tap water, then rinse with 1+1 hydrochloric acid and tap water, then rinse with deionized, distilled water in that order. Perform any filtration or acid preservation steps when the sample is collected or as soon as possible thereafter.

STABILITY: Cool samples to 4°C. M.H.T. = 24 Hours.

QUALITY CONTROL: Mixed calibration standards, an instrument check standard and an interference check solution are used in addition to a quality control sample. The quality control sample should be prepared in the same acid matrix as the calibration standards at 10 times the instrumental detection limits and in accordance with the instructions provided by the supplier. Furthermore, two types of blanks are required: a calibration blank and a reagent blank.

REFERENCE: Method 200.7, U.S. EPA, EMSL-Cincinnati, OH, Nov. 1980

PRIMARY NAME: Arsenic (As) Method 206.2

TITLE: Metals (Total, Dissolved, Suspended) AAS, Furnace Technique

MATRIX: drinking, surface and saline waters. Wastewater.

APPLICATION: Date issued 1978. A representative sample aliquot (containing nickel nitrate) is placed in graphite tube in furnace, evaporated to dryness, charred and atomized. Radiation from excited element is passed through vapor and decreases proportional to amount in vapor.

INTERFERENCES: Furnace technique subject to chemical and matrix interferences. Furnace gases may have molecular absorption bands enclosing analytical wavelength. Smoke-producing sample matrix can interfere. If As isn't volatilized and removed from furnace, memory effects occur.

INSTRUMENTATION: AAS. Arsenic(As) hollow cathode lamp or EDL. Graphite furnace. Pipets.

RANGE: 5-100 μg/L. MDL: 1 μg/L.

PRECISION: SD = +/- 1.6 @ 100 μg As/L.

ACCURACY: Recovery = 101% @ 100 μg As/L.

SAMPLING METHOD: Plastic or glass (pre-washed).

STABILITY: HNO3 to pH < 2. M.H.T. = 6 Months.

QUALITY CONTROL: A check standard should be run approximately after every 10 sample injections. Standards are run in part to monitor the life and performance of the graphite tube. Lack of reproducibility or significant change in the signal for the standard indicates tube should be replaced.

REFERENCE: Methods for the Chemical Analysis of Water and Wastes, EPA-600/4-79-020, USEPA, EMSL, 1979.

PRIMARY NAME: Arsenic (As) Method 6010

TITLE: Inductively Coupled Plasma (ICP)

MATRIX: Applies to all matrices; ground water, EP extracts, soils, sludges, sediments, aqueous samples, solid and industrial wastes.

APPLICATION: The method covers the determination of 25 metals. Its primary advantage is that ICP instruments allow simultaneous or rapid sequential determination of many elements in a short time. Samples require digestion prior to analysis and are first nebulized and the aerosol is transported to a plasma torch in which element specific atomic line emission spectra are produced by a radio frequency inductively coupled plasma. Background correction is required for trace element detection except in the case of line broadning.

INTERFERENCES: There are spectral, physical and chemical interferences. The primary disadvantage of ICP instruments is background radiation from other elements and the plasma gases (spectral interferences). Changes in sample viscosity and surface tension with samples containing high dissolved solids or high acid concentrations can cause physical interferences. Ionization effects, solute vaporization and molecular compound formation can cause chemical interferences. Aluminum, chromium and vanadium can cause interference at the 100 mg/L level.

INSTRUMENTATION: Inductively Coupled Argon Plasma Emission Spectroscopy. 193.696 nm Wavelength

RANGE: Not listed MDL: 53 µg/L.

PRECISION: SD = 7.5% Mean @ true value 200 µg/L.

ACCURACY: Mean Recovery = 93% +/- 6% of spiked elements for all wastes.

SAMPLING METHOD: Collect 400 mL of sample in plastic or glass containers.

STABILITY: Cool samples to 4°C. M.H.T. = 24 Hours.

QUALITY CONTROL: Mixed calibration standards, an instrument check standard and an interference check solution are used in addition to a quality control sample. The quality control sample should be prepared in the same acid matrix as the calibration standards at 10 times the instrumental detection limits and in accordance with the instructions provided by the supplier. Furthermore, two types of blanks are required: a calibration blank and a reagent blank.

REFERENCE: Method 6010, SW-846, 3rd ed., Nov 1986.

PRIMARY NAME: Arsenic (As) Method 7060

TITLE: Atomic Absorption, (AA) Furnace Technique

MATRIX: Wastes, mobility procedure extracts, soils and groundwater

APPLICATION: Sample preparation converts organic forms of As to inorganic forms. Sample preparation varies with matrix. Following appropriate dissolution of sample, a representative aliquot of digestate is spiked with nickel nitrate solution and placed in a graphite tube furnace.

INTERFERENCES: Elemental As and many of its compounds are volatilized. There may be losses in As during sample preparation. There can be severe nonspecific absorption and light scattering caused by matrix components during atomization. Memory effects occur if As isn't volatilized and removed from furnace. Aluminum is a severe positive interferent.

INSTRUMENTATION: Atomic absorption spectrometer, arsenic (As) hollow cathode lamp or electrodeless discharge lamp. Graphite furnace.

RANGE: 5-100 µg/L MDL: 1 µg/L

PRECISION: standard deviation = +/- 1.6 @ 100 µg As/L.

ACCURACY: recovery = 101% @ 100 µg As/L.

SAMPLING METHOD: Use plastic or glass containers (prewashed). Collect 100 mL of sample.

STABILITY: Add nitric acid to pH <2. M.H.T. = 6 Months.

QUALITY CONTROL: Run one spike duplicate sample for every 20 samples. Verify calibration with an independently prepared check standard every 15 samples. (Low wavelength, 193.7 nm, makes As analysis susceptible to problems)

REFERENCE: Method 7060, SW-846, 3rd ed., Nov 1986.

PRIMARY NAME: Arsenic (As) Method 7061

TITLE: Atomic Absorption, (AA) Gaseous hydride

MATRIX: Wastes, mobility procedure extracts, soils and groundwater

APPLICATION: Method approved only for sample matrices without high concentrations of Cr, Cu, Hg, Ni, Ag, Co and Mo after sample preparation with HNO3/H2SO4 digestion, Arsenic in digestate is reduced to trivalent form with SnCl2. As(III) is then converterd to a volatile hydride using hydrogen and is swept into an argon-hydrogen flame located in the optical path of an atomic absorption spectrometer. Absorption of the lamp radiation is proportional to the arsenic concentration.

INTERFERENCES: Traces of nitric acid left following sample work-up can result in analytical interferences. Elemental As and many of its compounds are volatilized, thus samples may be subject to losses during sample preparation. High concentrations of Cr, Cu, Hg, Ni, Ag, Co and Mo cause analytical interferences.

INSTRUMENTATION: Atomic absorption spectrometer. Arsenic (As) hollow cathode lamp or electrodeless discharge lamp. Burner (for argon-hydrogen flame). 193.7 nm Wavelength

RANGE: 2-20 μg/L MDL: 0.002 mg/L

PRECISION: standard deviation = +/- 0.9 @ 10 μg/L on o-arsenilic acid solution.

ACCURACY: Recovery = 93% @ 10 μg/L on o-arsenilic acid solution

SAMPLING METHOD: Use plastic or glass containers (prewashed). Collect 100 mL of sample.

STABILITY: Add nitric acid to pH < 2. M.H.T. = 6 Months.

QUALITY CONTROL: Run one spike duplicate sample for every 20 samples. Verify calibration with an independently prepared check standard every 15 samples.

REFERENCE: Method 7061, SW-846,3rd ed., Nov 1986.

PRIMARY NAME: Barium (Ba)

Method 200.7

TITLE: Inductively Coupled Plasma (ICP)

MATRIX: Dissolved, suspended or total element in drinking and surface waters and in domestic and industrial wastewaters.

APPLICATION: The method covers the determination of 25 metals. Dissolved elements are determined in filtered and acidified samples after appropriate digestion (which increases dissolved solids). Its primary advantage is that ICP instruments allow simultaneous or rapid sequential determination of many elements in a short time. Samples are first nebulized and the aerosol is transported to a plasma torch in which element specific atomic line emission spectra are produced by a radio frequency inductively coupled plasma. Background correction is required for trace element detection except in the case of line broadning.

INTERFERENCES: There are spectral, physical and chemical interferences. The primary disadvantage of ICP instruments is background radiation from other elements and the plasma gases (spectral interferences). Changes in sample viscosity and surface tension with samples containing high dissolved solids (especially those exceeding 1500 mg/L) or high acid concentrations can cause physical interferences. Ionization effects, solute vaporization and molecular compound formation can cause chemical interferences. Other metals do not cause interference at the 100 mg/L level.

INSTRUMENTATION: Inductively Coupled Argon Plasma Emission Spectroscopy. 455.403 nm Wavelength

RANGE: Not listed

MDL: 2 μg/L.

PRECISION: Not listed

ACCURACY: Mean Recovery = 93% +/- 6% of spiked elements for all wastes.

SAMPLING METHOD: Wash sample container with detergent and tap water, rinse with 1+1 nitric acid and tap water, then rinse with 1+1 hydrochloric acid and tap water, then rinse with deionized, distilled water in that order. Perform any filtration or acid preservation steps when the sample is collected or as soon as possible thereafter.

STABILITY: Cool samples to 4°C. M.H.T. = 24 Hours.

QUALITY CONTROL: Mixed calibration standards, an instrument check standard and an interference check solution are used in addition to a quality control sample. The quality control sample should be prepared in the same acid matrix as the calibration standards at 10 times the instrumental detection limits and in accordance with the instructions provided by the supplier. Furthermore, two types of blanks are required: a calibration blank and a reagent blank.

REFERENCE: Method 200.7, U.S. EPA, EMSL-Cincinnati, OH, Nov. 1980

PRIMARY NAME: Barium (Ba) Method 208.2

TITLE: Metals (Total, Dissolved, Suspended) AAS, Furnace Technique

MATRIX: drinking, surface and saline waters. Wastewater.

APPLICATION: Date issued 1978. A representative sample aliquot is placed in graphite tube in the furnace, evbaporated to dryness, charred and atomized. Radiation from excited element is passed through vapor and radiation intensity decreases proportional to amount of Ba in vapor.

INTERFERENCES: Furnace technique subject to chemical and matrix interferences. Furnace gases may have molecular absorption bands enclosing analytical wavelength. Smoke-producing sample matrix can interfere. If Ba isn't volitalized and removed from furnace, memory effects occur.

INSTRUMENTATION: AAS. Barium (Ba) hollow cathode lamp or EDL. Graphite furnace. Pipets.

RANGE: 10-200 µg/L. MDL: 2 µg/L.

PRECISION: SD = +/- 2.2 µg @ 1000 µg Ba/L.

ACCURACY: Recovery = 102% @ 1000 µg Ba/L.

SAMPLING METHOD: Plastic or glass (pre-washed).

STABILITY: HNO3 to pH < 2. M.H.T. = 6 Months.

QUALITY CONTROL: A check standard should be run approximately after every 10 sample injections. Standards are run in part to monitor the life and performance of the graphite tube. Lack of reproducibility or significant change in the signal for the standard indicates tube should be replaced.

REFERENCE: Methods for the chemical analysis of water and wastes, EPA-600/4-79-020, USEPA, EMSL, 1979.

PRIMARY NAME: Barium (Ba)

Method 6010

TITLE: Inductively Coupled Plasma (ICP)

MATRIX: Applies to all matrices; ground water, EP extracts, soils, sludges, sediments, aqueous samples, solid and industrial wastes.

APPLICATION: The method covers the determination of 25 metals. Its primary advantage is that ICP instruments allow simultaneous or rapid sequential determination of many elements in a short time. Samples require digestion prior to analysis and are first nebulized and the aerosol is transported to a plasma torch in which element specific atomic line emission spectra are produced by a radio frequency inductively coupled plasma. Background correction is required for trace element detection except in the case of line broadning.

INTERFERENCES: There are spectral, physical and chemical interferences. The primary disadvantage of ICP instruments is background radiation from other elements and the plasma gases (spectral interferences). Changes in sample viscosity and surface tension with samples containing high dissolved solids or high acid concentrations can cause physical interferences. Ionization effects, solute vaporization and molecular compound formation can cause chemical interferences. Other metals do not cause interference at the 100 mg/L level.

INSTRUMENTATION: Inductively Coupled Argon Plasma Emission Spectroscopy. 455.403 nm Wavelength

RANGE: Not listed

MDL: 2 μg/L.

PRECISION: Not listed

ACCURACY: Mean Recovery = 93% +/- 6% of spiked elements for all wastes.

SAMPLING METHOD: Collect 400 mL of sample in plastic or glass containers.

STABILITY: Cool samples to 4°C. M.H.T. = 24 Hours.

QUALITY CONTROL: Mixed calibration standards, an instrument check standard and an interference check solution are used in addition to a quality control sample. The quality control sample should be prepared in the same acid matrix as the calibration standards at 10 times the instrumental detection limits and in accordance with the instructions provided by the supplier. Furthermore, two types of blanks are required: a calibration blank and a reagent blank.

REFERENCE: Method 6010, SW-846, 3rd ed., Nov 1986.

PRIMARY NAME: Barium (Ba) Method 7080

TITLE: Atomic Absorption, (AA) Direct Aspiration

MATRIX: Drinking, Surface and Saline Waters. Wastewater

CAS # : 7440-39-3

APPLICATION: Sample is aspirated and atomized in a flame. A light beam from a Ba hollow cathode lamp is directed through the flame into a monochromator and onto a detector. Since wavelength of light beam is specific for barium, light energy absorbed by detector is a measure of barium.

INTERFERENCES: The most troublesome type is chemical, caused by lack of absorption of atoms bound in molecular combination in the flame. High dissolved solids in sample may result in nonatomic absorbance interference. Narrow spectral band pass must be used.

INSTRUMENTATION: Atomic absorption spectrometer. Barium hollow cathode lamp. (553.6 nm Wavelength; calcium also emits here).

RANGE: 1-20 mg/L MDL: 0.1 mg/L

PRECISION: standard deviation = +/- (0.043 and 0.13) @ 0.4 and 2 mg Ba/L

ACCURACY: recoveries = 94 and 113% @ 0.4 and 2 mg Ba/L

SAMPLING METHOD: Use glass or plastic containers. Collect 200 g of solids and 600 mL of liquid samples.

STABILITY: Cool solid samples to 4°C. and analyze as soon as possible. Add nitric acid to liquid samples to pH < 2; M.H.T. = 6 months.

QUALITY CONTROL: At least one duplicate and one spike sample should be run every 20 samples or with each matrix type to verify precision of the method. For 20 or more samples per day, verify working standard curve. Run an additional standard at or near mid-range every 10 samples.

REFERENCE: Method 7080, SW-846, 3rd ed., Nov 1986.

PRIMARY NAME: Barium (Ba) Method 7081

TITLE: Atomic Absorption, (AA) Furnace Technique

MATRIX: Wastes, mobility procedure extracts, soils and groundwater

CAS # : 7440-39-3

APPLICATION: Aqueous samples, EP extracts, industrial wastes, soils, sludges, sediments and solid wastes require digestion before analysis. An aliquot of sample is placed in the graphite tube in the furnace and slowly evaporated, charred and atomized. Absorption of lamp radiation during atomization is proportional to barium concentration.

INTERFERENCES: The furnace technique is subject to chemical interferences. Composition of sample matrix can have major effect on analysis. Modify matrix to remove interferences. Derived barium carbide causes loss of sensitivity and memory effects. Avoid halide acids.

INSTRUMENTATION: Atomic absorption spectrometer. Barium hollow cathode lamp or electrodeless discharge lamp. Graphite furnace. Strip-chart recorder

RANGE: 10-200 μg/L

MDL: 2 μg/L (553.6 nm Wavelength)

PRECISION: Not listed

ACCURACY: Not listed

SAMPLING METHOD: Use glass or plastic containers. Collect 200 g of solids and 600 mL of liquid samples.

STABILITY: Cool solid samples to 4°C. and analyze as soon as possible. Add nitric acid to liquid samples to pH < 2; their M.H.T. = 6 months.

QUALITY CONTROL: At least one duplicate and one spike sample should be run every 20 samples, or with each matrix type to verify method precision. If 20 or more samples are run a day, run a standard (at or near mid-range) every 10 samples.

REFERENCE: Method 7081, SW-846, 3rd ed., (Included as revision 0, December 1987)

PRIMARY NAME: Beryllium (Be)

Method 200.7

TITLE: Inductively Coupled Plasma (ICP)

MATRIX: Dissolved, suspended or total element in drinking and surface waters and in domestic and industrial wastewaters.

APPLICATION: The method covers the determination of 25 metals. Dissolved elements are determined in filtered and acidified samples after appropriate digestion (which increases dissolved solids). Its primary advantage is that ICP instruments allow simultaneous or rapid sequential determination of many elements in a short time. Samples are first nebulized and the aerosol is transported to a plasma torch in which element specific atomic line emission spectra are produced by a radio frequency inductively coupled plasma. Background correction is required for trace element detection except in the case of line broadning.

INTERFERENCES: There are spectral, physical and chemical interferences. The primary disadvantage of ICP instruments is background radiation from other elements and the plasma gases (spectral interferences). Changes in sample viscosity and surface tension with samples containing high dissolved solids (especially those exceeding 1500 mg/L) or high acid concentrations can cause physical interferences. Ionization effects, solute vaporization and molecular compound formation can cause chemical interferences. Thallium and vanadium can cause interference at the 100 mg/L level.

INSTRUMENTATION: Inductively Coupled Argon Plasma Emission Spectroscopy. 313.042 nm Wavelength

RANGE: Not listed MDL: 0.3 μg/L.

PRECISION: SD = 6.2% Mean @ true value 750 μg/L.

ACCURACY: Mean Recovery = 93% +/- 6% of spiked elements for all wastes.

SAMPLING METHOD: Wash sample container with detergent and tap water, rinse with 1+1 nitric acid and tap water, then rinse with 1+1 hydrochloric acid and tap water, then rinse with deionized, distilled water in that order. Perform any filtration or acid preservation steps when the sample is collected or as soon as possible thereafter.

STABILITY: Cool samples to 4°C. M.H.T. = 24 Hours.

QUALITY CONTROL: Mixed calibration standards, an instrument check standard and an interference check solution are used in addition to a quality control sample. The quality control sample should be prepared in the same acid matrix as the calibration standards at 10 times the instrumental detection limits and in accordance with the instructions provided by the supplier. Furthermore, two types of blanks are required: a calibration blank and a reagent blank.

REFERENCE: Method 200.7, U.S. EPA, EMSL-Cincinnati, OH, Nov. 1980

PRIMARY NAME: Beryllium (Be) Method 6010

TITLE: Inductively Coupled Plasma (ICP)

MATRIX: Applies to all matrices; ground water, EP extracts, soils, sludges, sediments, aqueous samples, solid and industrial wastes.

APPLICATION: The method covers the determination of 25 metals. Its primary advantage is that ICP instruments allow simultaneous or rapid sequential determination of many elements in a short time. Samples require digestion prior to analysis and are first nebulized and the aerosol is transported to a plasma torch in which element specific atomic line emission spectra are produced by a radio frequency inductively coupled plasma. Background correction is required for trace element detection except in the case of line broadning.

INTERFERENCES: There are spectral, physical and chemical interferences. The primary disadvantage of ICP instruments is background radiation from other elements and the plasma gases (spectral interferences). Changes in sample viscosity and surface tension with samples containing high dissolved solids or high acid concentrations can cause physical interferences. Ionization effects, solute vaporization and molecular compound formation can cause chemical interferences. Thallium and vanadium can cause interference at the 100 mg/L level.

INSTRUMENTATION: Inductively Coupled Argon Plasma Emission Spectroscopy. 313.042 nm Wavelength

RANGE: Not listed MDL: 0.3 µg/L.

PRECISION: SD = 6.2% Mean @ true value 750 µg/L.

ACCURACY: Mean Recovery = 93% +/- 6% of spiked elements for all wastes.

SAMPLING METHOD: Collect 400 mL of sample in plastic or glass containers.

STABILITY: Cool samples to 4°C. M.H.T. = 24 Hours.

QUALITY CONTROL: Mixed calibration standards, an instrument check standard and an interference check solution are used in addition to a quality control sample. The quality control sample should be prepared in the same acid matrix as the calibration standards at 10 times the instrumental detection limits and in accordance with the instructions provided by the supplier. Furthermore, two types of blanks are required: a calibration blank and a reagent blank.

REFERENCE: Method 6010, SW-846, 3rd ed., Nov 1986.

PRIMARY NAME: Beryllium (Be)

Method 7090

TITLE: Atomic Absorption, (AA) Direct Aspiration

MATRIX: Drinking, Surface and Saline Waters. Wastewater

CAS # : 7440-41-7

APPLICATION: Sample is aspirated and atomized in a flame. A light beam from a be hollow cathode lamp is directed through the flame into a monochromator and onto a detector. Since wavelength of light beam is specific for beryllium, light energy absorbed by detector is measure of be.

INTERFERENCES: The most troublesome type is chemical, caused by lack of absorption of atoms bound in molecular combination in the flame. High dissolved solids in sample may result in nonatomic absorbance interference. Aluminum concentrations greater than 500 ppm can interfere.

INSTRUMENTATION: Atomic absorption spectrometer. Beryllium hollow cathode lamp. (234.9 nm Wavelength).

RANGE: 0.05-2 mg/L

MDL: 0.005 mg/L

PRECISION: standard deviation = +/- 0.001 and 0.002 @ 0.01 and 0.25 mg Be/L

ACCURACY: recoveries = 100 and 97% @ 0.01 and 0.25 mg Be/L

SAMPLING METHOD: Use glass or plastic containers. Collect 200 g of solids and 600 mL of liquid samples.

STABILITY: Cool solid samples to 4°C. and analyze as soon as possible. Add nitric acid to liquid samples to pH < 2; M.H.T. = 6 months.

QUALITY CONTROL: At least one duplicate and one spike sample should be run every 20 samples or with each matrix type to verify precision of the method. For 20 or more samples per day, verify working standard curve. Run an additional standard at or near mid-range every 10 samples.

REFERENCE: Method 7090, SW-846, 3rd ed., Nov 1986.

PRIMARY NAME: Beryllium (Be) Method 7091

TITLE: Atomic Absorption, (AA) Furnace Technique

MATRIX: Wastes, mobility procedure extracts, soils and groundwater

CAS # : 7440-41-7

APPLICATION: Aqueous samples, EP extracts, industrial wastes, soils, sludges, sediments and solid wastes require digestion before analysis. An aliquot of sample is placed in the graphite tube in the furnace and slowly evaporated, charred and atomized. Absorption of lamp radiation during atomization is proportional to beryllium concentration.

INTERFERENCES: The furnace technique is subject to chemical interferences. Composition of sample matrix can have major effect on analysis. The long residence time and high concentrations of atomized sample in optical path can result in severe physical and chemical interferences.

INSTRUMENTATION: Atomic absorption spectrometer. Beryllium hollow cathode lamp or electrodeless discharge lamp. Graphite furnace. Strip-chart recorder.

RANGE: 1-30 μg/L (234.9 nm Wavelength) MDL: 0.2 μg/L

PRECISION: Not listed

ACCURACY: Not listed

SAMPLING METHOD: Use glass or plastic containers. Collect 200 g of solids and 600 mL of liquid samples.

STABILITY: Cool solid samples to 4°C. and analyze as soon as possible. Add nitric acid to liquid samples to pH < 2; their M.H.T. = 6 months.

QUALITY CONTROL: At least one duplicate and one spike sample should be run every 20 samples, or with each matrix type to verify method precision. If 20 or more samples are run a day, run a standard (at or near mid-range) every 10 samples.

REFERENCE: Method 7091, SW-846, 3rd ed., Nov 1986.

PRIMARY NAME: Boron (B) Method 200.7

TITLE: Inductively Coupled Plasma (ICP)

MATRIX: Dissolved, suspended or total element in drinking and surface waters and in domestic and industrial wastewaters.

APPLICATION: The method covers the determination of 25 metals. Dissolved elements are determined in filtered and acidified samples after appropriate digestion (which increases dissolved solids). Its primary advantage is that ICP instruments allow simultaneous or rapid sequential determination of many elements in a short time. Samples are first nebulized and the aerosol is transported to a plasma torch in which element specific atomic line emission spectra are produced by a radio frequency inductively coupled plasma. Background correction is required for trace element detection except in the case of line broadning.

INTERFERENCES: There are spectral, physical and chemical interferences. The primary disadvantage of ICP instruments is background radiation from other elements and the plasma gases (spectral interferences). Changes in sample viscosity and surface tension with samples containing high dissolved solids (especially those exceeding 1500 mg/L) or high acid concentrations can cause physical interferences. Ionization effects, solute vaporization and molecular compound formation can cause chemical interferences. Aluminum and iron can cause interference at the 100 mg/L level.

INSTRUMENTATION: Inductively Coupled Argon Plasma Emission Spectroscopy. 249.773 nm Wavelength

RANGE: Not listed

MDL: 5 μg/L.

PRECISION: Not listed

ACCURACY: Mean Recovery = 93% +/- 6% of spiked elements for all wastes.

SAMPLING METHOD: Wash sample container with detergent and tap water, rinse with 1+1 nitric acid and tap water, then rinse with 1+1 hydrochloric acid and tap water, then rinse with deionized, distilled water in that order. Perform any filtration or acid preservation steps when the sample is collected or as soon as possible thereafter.

STABILITY: Cool samples to 4°C. M.H.T. = 24 Hours.

QUALITY CONTROL: Mixed calibration standards, an instrument check standard and an interference check solution are used in addition to a quality control sample. The quality control sample should be prepared in the same acid matrix as the calibration standards at 10 times the instrumental detection limits and in accordance with the instructions provided by the supplier. Furthermore, two types of blanks are required: a calibration blank and a reagent blank.

REFERENCE: Method 200.7, U.S. EPA, EMSL-Cincinnati, OH, Nov. 1980

PRIMARY NAME: Boron (B) Method 6010

TITLE: Inductively Coupled Plasma (ICP)

MATRIX: Applies to all matrices; ground water, EP extracts, soils, sludges, sediments, aqueous samples, solid and industrial wastes.

APPLICATION: The method covers the determination of 25 metals. Its primary advantage is that ICP instruments allow simultaneous or rapid sequential determination of many elements in a short time. Samples require digestion prior to analysis and are first nebulized and the aerosol is transported to a plasma torch in which element specific atomic line emission spectra are produced by a radio frequency inductively coupled plasma. Background correction is required for trace element detection except in the case of line broadning.

INTERFERENCES: There are spectral, physical and chemical interferences. The primary disadvantage of ICP instruments is background radiation from other elements and the plasma gases (spectral interferences). Changes in sample viscosity and surface tension with samples containing high dissolved solids or high acid concentrations can cause physical interferences. Ionization effects, solute vaporization and molecular compound formation can cause chemical interferences. Aluminum and iron can cause interference at the 100 mg/L level.

INSTRUMENTATION: Inductively Coupled Argon Plasma Emission Spectroscopy. 249.773 nm Wavelength

RANGE: Not listed MDL: 5 µg/L.

PRECISION: Not listed

ACCURACY: Mean Recovery = 93% +/- 6% of spiked elements for all wastes.

SAMPLING METHOD: Collect 400 mL of sample in plastic or glass containers.

STABILITY: Cool samples to 4°C. M.H.T. = 24 Hours.

QUALITY CONTROL: Mixed calibration standards, an instrument check standard and an interference check solution are used in addition to a quality control sample. The quality control sample should be prepared in the same acid matrix as the calibration standards at 10 times the instrumental detection limits and in accordance with the instructions provided by the supplier. Furthermore, two types of blanks are required: a calibration blank and a reagent blank.

REFERENCE: Method 6010, SW-846, 3rd ed., Nov 1986.

PRIMARY NAME: Cadmium (Cd) Method 200.7

TITLE: Inductively Coupled Plasma (ICP)

MATRIX: Dissolved, suspended or total element in drinking and surface waters and in domestic and industrial wastewaters.

APPLICATION: The method covers the determination of 25 metals. Dissolved elements are determined in filtered and acidified samples after appropriate digestion (which increases dissolved solids). Its primary advantage is that ICP instruments allow simultaneous or rapid sequential determination of many elements in a short time. Samples are first nebulized and the aerosol is transported to a plasma torch in which element specific atomic line emission spectra are produced by a radio frequency inductively coupled plasma. Background correction is required for trace element detection except in the case of line broadning.

INTERFERENCES: There are spectral, physical and chemical interferences. The primary disadvantage of ICP instruments is background radiation from other elements and the plasma gases (spectral interferences). Changes in sample viscosity and surface tension with samples containing high dissolved solids (especially those exceeding 1500 mg/L) or high acid concentrations can cause physical interferences. Ionization effects, solute vaporization and molecular compound formation can cause chemical interferences. Iron and nickel can cause interference at the 100 mg/L level.

INSTRUMENTATION: Inductively Coupled Argon Plasma Emission Spectroscopy. 266.502 nm Wavelength

RANGE: Not listed MDL: 4 µg/L.

PRECISION: SD = 12% Mean @ true value 50 µg/L.

ACCURACY: Mean Recovery = 93% +/- 6% of spiked elements for all wastes.

SAMPLING METHOD: Wash sample container with detergent and tap water, rinse with 1+1 nitric acid and tap water, then rinse with 1+1 hydrochloric acid and tap water, then rinse with deionized, distilled water in that order. Perform any filtration or acid preservation steps when the sample is collected or as soon as possible thereafter.

STABILITY: Cool samples to 4°C. M.H.T. = 24 Hours.

QUALITY CONTROL: Mixed calibration standards, an instrument check standard and an interference check solution are used in addition to a quality control sample. The quality control sample should be prepared in the same acid matrix as the calibration standards at 10 times the instrumental detection limits and in accordance with the instructions provided by the supplier. Furthermore, two types of blanks are required: a calibration blank and a reagent blank.

REFERENCE: Method 200.7, U.S. EPA, EMSL-Cincinnati, OH, Nov. 1980

PRIMARY NAME: Cadmium (Cd) Method 6010

TITLE: Inductively Coupled Plasma (ICP)

MATRIX: Applies to all matrices; ground water, EP extracts, soils, sludges, sediments, aqueous samples, solid and industrial wastes.

APPLICATION: The method covers the determination of 25 metals. Its primary advantage is that ICP instruments allow simultaneous or rapid sequential determination of many elements in a short time. Samples require digestion prior to analysis and are first nebulized and the aerosol is transported to a plasma torch in which element specific atomic line emission spectra are produced by a radio frequency inductively coupled plasma. Background correction is required for trace element detection except in the case of line broadning.

INTERFERENCES: There are spectral, physical and chemical interferences. The primary disadvantage of ICP instruments is background radiation from other elements and the plasma gases (spectral interferences). Changes in sample viscosity and surface tension with samples containing high dissolved solids or high acid concentrations can cause physical interferences. Ionization effects, solute vaporization and molecular compound formation can cause chemical interferences. Iron and nickel can cause interference at the 100 mg/L level.

INSTRUMENTATION: Inductively Coupled Argon Plasma Emission Spectroscopy. 266.502 nm Wavelength

RANGE: Not listed MDL: 4 µg/L.

PRECISION: SD = 12% Mean @ true value 50 µg/L.

ACCURACY: Mean Recovery = 93% +/- 6% of spiked elements for all wastes.

SAMPLING METHOD: Collect 400 mL of sample in plastic or glass containers.

STABILITY: Cool samples to 4°C. M.H.T. = 24 Hours.

QUALITY CONTROL: Mixed calibration standards, an instrument check standard and an interference check solution are used in addition to a quality control sample. The quality control sample should be prepared in the same acid matrix as the calibration standards at 10 times the instrumental detection limits and in accordance with the instructions provided by the supplier. Furthermore, two types of blanks are required: a calibration blank and a reagent blank.

REFERENCE: Method 6010, SW-846, 3rd ed., Nov 1986.

PRIMARY NAME: cadmium (Cd) Method 7130

TITLE: Atomic Absorption, (AA) Direct Aspiration

MATRIX: Drinking, Surface and Saline Waters. Wastewater

CAS # : 7440-43-9

APPLICATION: Sample is aspirated and atomized in a flame. A light beam from a cadmium hollow cathode lamp is directed through the flame into a monochromator and onto a detector. Since wavelength of light beam is specific for cadmium,light energy absorbed by detector is measure of cadmium.

INTERFERENCES: The most troublesome type is chemical, caused by lack of absorption of atoms bound in molecular combination in the flame. High dissolved solids in sample may result in nonatomic absorbance interference. Non specific absorption and light scattering interfere.

INSTRUMENTATION: Atomic absorption spectrometer. Cadmium hollow cathode lamp (228.8 nm Wavelength).

RANGE: 0.05-2 mg/L MDL: 0.005 mg/L

PRECISION: standard deviation = 21 μg/L @ 71 μg/L (true value) 74 labs

ACCURACY: as bias = -2.2% @ 71 μg/L (true value) 74 labs

SAMPLING METHOD: Use glass or plastic containers. Collect 200 g of solids and 600 mL of liquid samples.

STABILITY: Cool solid samples to 4°C. and analyze as soon as possible. Add nitric acid to liquid samples to pH < 2; M.H.T. = 6 months.

QUALITY CONTROL: At least one duplicate and one spike sample should be run every 20 samples or with each matrix type to verify precision of the method. For 20 or more samples per day, verify working standard curve. Run an additional standard at or near mid-range every 10 samples.

REFERENCE: Method 7130, SW-846, 3rd ed., Nov 1986.

PRIMARY NAME: Cadmium (Cd) Method 7131

TITLE: Atomic Absorption, (AA) Furnace Technique

MATRIX: Wastes, mobility procedure extracts, soils and groundwater

CAS # : 7440-43-9

APPLICATION: Aqueous samples, EP extracts, industrial wastes, soils, sludges, sediments and solid wastes require digestion before analysis. An aliquot of sample is placed in the graphite tube in the furnace and slowly evaporated, charred and atomized. Absorption of lamp radiation during atomization is proportional to the cadmium concentration.

INTERFERENCES: The furnace technique is subject to chemical interferences. Composition of sample matrix can effect analysis. Cd analysis can suffer severe nonspecific absorption and light scattering; background correction is required. Use cadmium-free plastic pipet tips.

INSTRUMENTATION: Atomic absorption spectrometer. Cadmium hollow cathode lamp or electrodeless discharge lamp. Graphite furnace. Strip-chart recorder

RANGE: 0.5-10 μg/L

MDL: 0.1 μg/L (228.8 nm Wavelength)

PRECISION: standard deviation = +/- 0.10, 0.16, 0.33 @ 2.5, 5.0, 10.0 μg Cd/L

ACCURACY: recoveries = 96, 99, 98% @ 2.5, 5.0, 10.0 μg Cd/L

SAMPLING METHOD: Use glass or plastic containers. Collect 200 g of solids and 600 mL of liquid samples.

STABILITY: Cool solid samples to 4°C. and analyze as soon as possible. Add nitric acid to liquid samples to pH < 2; M.H.T. = 6 months.

QUALITY CONTROL: At least one duplicate and one spike sample should be run every 20 samples, or with each matrix type to verify method precision. If 20 or more samples are run a day, run a standard (at or near mid-range) every 10 samples.

REFERENCE: Method 7131, SW-846, 3rd ed., Nov 1986.

PRIMARY NAME: Calcium (Ca) Method 200.7

TITLE: Inductively Coupled Plasma (ICP)

MATRIX: Dissolved, suspended or total element in drinking and surface waters and in domestic and industrial wastewaters.

APPLICATION: The method covers the determination of 25 metals. Dissolved elements are determined in filtered and acidified samples after appropriate digestion (which increases dissolved solids). Its primary advantage is that ICP instruments allow simultaneous or rapid sequential determination of many elements in a short time. Samples are first nebulized and the aerosol is transported to a plasma torch in which element specific atomic line emission spectra are produced by a radio frequency inductively coupled plasma. Background correction is required for trace element detection except in the case of line broadning.

INTERFERENCES: There are spectral, physical and chemical interferences. The primary disadvantage of ICP instruments is background radiation from other elements and the plasma gases (spectral interferences). Changes in sample viscosity and surface tension with samples containing high dissolved solids (especially those exceeding 1500 mg/L) or high acid concentrations can cause physical interferences. Ionization effects, solute vaporization and molecular compound formation can cause chemical interferences. Chromium, iron, magnesium, manganese, thallium and vanadium can cause interference at the 100 mg/L level.

INSTRUMENTATION: Inductively Coupled Argon Plasma Emission Spectroscopy. 317.933 nm Wavelength

RANGE: Not listed MDL: 10 μg/L.

PRECISION: Not listed

ACCURACY: Mean Recovery = 93% +/- 6% of spiked elements for all wastes.

SAMPLING METHOD: Wash sample container with detergent and tap water, rinse with 1+1 nitric acid and tap water, then rinse with 1+1 hydrochloric acid and tap water, then rinse with deionized, distilled water in that order. Perform any filtration or acid preservation steps when the sample is collected or as soon as possible thereafter.

STABILITY: Cool samples to 4°C. M.H.T. = 24 Hours.

QUALITY CONTROL: Mixed calibration standards, an instrument check standard and an interference check solution are used in addition to a quality control sample. The quality control sample should be prepared in the same acid matrix as the calibration standards at 10 times the instrumental detection limits and in accordance with the instructions provided by the supplier. Furthermore, two types of blanks are required: a calibration blank and a reagent blank.

REFERENCE: Method 200.7, U.S. EPA, EMSL-Cincinnati, OH, Nov. 1980

PRIMARY NAME: Calcium (Ca) Method 215.1

TITLE: Metals (Total and Dissolved) AAS, Direct Aspiration

MATRIX: Drinking, Surface and Saline waters. Wastewater.

APPLICATION: Date issued 1971. Editorial revision 1974. Sample is aspirated and atomized in a flame. Light beam from hollow cathode (made of Ca) lamp is directed through flame into monochromator, then to detector which measures amount absorbed light. Energy absorbed is proportional to Ca.

INTERFERENCES: Phosphate, sulfate and aluminum interfere; are masked by lanthanum addition. Low Ca values, if sample pH > 7. (Prepare in dilute HCl solution). Low Ca values with magnesium conc >1000 mg/L. Control ionization interferences using large amounts of alkali for samples and standards.

INSTRUMENTATION: AAS. Calcium (Ca) hollow cathode lamp. Burner. Pipets. Strip chart recorder.

RANGE: 0.2-7 mg/L @ 422.7 nm Wavelength MDL: 0.001 mg/L.

PRECISION: SD = +/- (0.3 and 0.6) @ 9.0 and 36 mg Ca/L.

ACCURACY: Recoveries = 99% @ 9.0 and 36 mg Ca/L.

SAMPLING METHOD: Plastic or glass (pre-washed).

STABILITY: Add nitric acid to pH < 2. M.H.T. = 6 Months.

QUALITY CONTROL: After calibration curve composed of a minimum of a reagent blank and 3 standards has been prepared, subsequent calibration curves must be verified by use of at least a reagent blank and one standard near MCL. Must check within 10% of original curve. (For drinking water analysis)

REFERENCE: Methods for the Chemical Analysis of Water and Wastes, EPA-600/4-79-020, USEPA, EMSL, 1979.

PRIMARY NAME: Calcium (Ca) Method 6010

TITLE: Inductively Coupled Plasma (ICP)

MATRIX: Applies to all matrices; ground water, EP extracts, soils, sludges, sediments, aqueous samples, solid and industrial wastes.

APPLICATION: The method covers the determination of 25 metals. Its primary advantage is that ICP instruments allow simultaneous or rapid sequential determination of many elements in a short time. Samples require digestion prior to analysis and are first nebulized and the aerosol is transported to a plasma torch in which element specific atomic line emission spectra are produced by a radio frequency inductively coupled plasma.

Background correction is required for trace element detection except in the case of line broadning.

INTERFERENCES: There are spectral, physical and chemical interferences. The primary disadvantage of ICP instruments is background radiation from other elements and the plasma gases (spectral interferences). Changes in sample viscosity and surface tension with samples containing high dissolved solids or high acid concentrations can cause physical interferences. Ionization effects, solute vaporization and molecular compound formation can cause chemical interferences. Chromium, iron, magnesium, manganese, thallium and vanadium can cause interference at the 100 mg/L level.

INSTRUMENTATION: Inductively Coupled Argon Plasma Emission Spectroscopy. 317.933 nm Wavelength

RANGE: Not listed MDL: 10 µg/L.

PRECISION: Not listed

ACCURACY: Mean Recovery = 93% +/- 6% of spiked elements for all wastes.

SAMPLING METHOD: Collect 400 mL of sample in plastic or glass containers.

STABILITY: Cool samples to 4°C. M.H.T. = 24 Hours.

QUALITY CONTROL: Mixed calibration standards, an instrument check standard and an interference check solution are used in addition to a quality control sample. The quality control sample should be prepared in the same acid matrix as the calibration standards at 10 times the instrumental detection limits and in accordance with the instructions provided by the supplier. Furthermore, two types of blanks are required: a calibration blank and a reagent blank.

REFERENCE: Method 6010, SW-846, 3rd ed., Nov 1986.

PRIMARY NAME: Calcium (Ca) Method 7140

TITLE: Atomic Absorption, (AA) Direct Aspiration

MATRIX: Drinking, Surface and Saline Waters. Wastewater

CAS # : 7440-70-2

APPLICATION: Sample is aspirated and atomized in a flame. A light beam from a calcium hollow cathode lamp is directed through the flame into a monochromator and onto a detector. Since wavelength of light beam is specific for calcium, light energy absorbed by detector is measure of calcium.

INTERFERENCES: The most troublesome type is chemical, caused by lack of absorption of atoms bound in molecular combination in the flame. High dissolved solids in sample may result in nonatomic absorbance interference. Add lanthanum to prevent complexing problems.

INSTRUMENTATION: Atomic absorption spectrometer. Calcium hollow cathode lamp. (422.7 nm Wavelength).

RANGE: 0.2-7 mg/L MDL: 0.01 mg/L

PRECISION: standard deviation = +/- 0.3 and 0.6 @ 9.0 and 36 mg Ca/L

ACCURACY: recoveries = 99 and 99% @ 9.0 and 36 mg Ca/L

SAMPLING METHOD: Use glass or plastic containers. Collect 200 g of solids and 600 mL of liquid samples.

STABILITY: Cool solid samples to 4°C. and analyze as soon as possible. Add nitric acid to liquid samples to pH < 2; M.H.T. = 6 months.

QUALITY CONTROL: At least one duplicate and one spike sample should be run every 20 samples or with each matrix type to verify precision of the method. For 20 or more samples per day, verify working standard curve. Run an additional standard at or near mid-range every 10 samples.

REFERENCE: Method 7140, SW-846, 3rd ed., Nov 1986.

PRIMARY NAME: Chlorine (Cl_2) Method 330.1

TITLE: Inorganics, Non-Metallics MATRIX: Waters and Wastes
Total Residual

APPLICATION: Date issued 1974. Editorial revision 1978. (Titrimetric, amperometric). Amperometric Method applies to all types waters and wastes without substantial amount of organic matter. (Chlorine and chloramines stoichiometrically liberate iodine from KI at pH 4 or less)

INTERFERENCES: Stirring can lower chlorine values by volatilization. Copper and silver poison the electrode.

INSTRUMENTATION: Amperometer (microammeter with necessary electrical accessories); microburet.

RANGE: Not reported MDL: Not reported

PRECISION: SD = 24.8 and 12.5% @ 0.64 and 1.83 mg/L total chloride.

ACCURACY: Relative error = 8.5 and 8.8% @ 0.64 and 1.83 mg/L total Cl

SAMPLING METHOD: Plastic or glass (200 mL).

STABILITY: No preservation required. Analyze immediately.

QUALITY CONTROL: If dilution is necessary, it must be done with distilled water which is free of chlorine, chlorine-demand and ammonia. (Phenylarsine oxide or sodium thiosulfate is used as the standard reducing agent to titrate liberated iodine using amperometer to determine end point).

REFERENCE: Methods for the Chemical Analysis of Water and Wastes, EPA-600/4-79-020, USEPA, EMSL, 1979.

PRIMARY NAME: Chlorine (Cl_2) Method 330.2

TITLE: Inorganics, Non-Metallics Total Residual

MATRIX: All types waters, but especially Wastewaters.

APPLICATION: Date issued 1978. (Titrimetric, back, iodometric) (starch or amperometric endpoint). Iodometric backtitration is best for wastewaters but is applicable to all types waters. (Chlorine and chloramines stoichiometrically liberate iodine from KI at pH 4 or less)

INTERFERENCES: Manganese, iron and nitrite interference is minimized by buffering to pH 4 before adding KI. High concentrations of organics may cause uncertainty in the endpoint. Turbidity and color make endpoint difficult to detect. Practice runs with spikes may be necessary.

INSTRUMENTATION: Standard lab glassware. Microburet 0-2 mL or 0-10 mL is used. Amperometric titrater.

RANGE: Not reported MDL: Not reported

PRECISION: SD = +/- 0.12 mg/L Cl @ 3.51 mg/L Cl (river water)

ACCURACY: % recovery = 107.7% @ 0.84 mg/L Cl (river water)

SAMPLING METHOD: plastic or glass (200 mL).

STABILITY: No preservation required. Analyze immediately.

QUALITY CONTROL: Use chlorine free, chlorine-demand free distilled water for dilution. (Phenylarsine oxide is used as the standard reducing agent. Iodine quantitatively oxidizes reducing agent and excess is titrated with standard iodine titrant to starch-iodine or amperometric endpoint)

REFERENCE: Methods for the Chemical Analysis of Water and Wastes, EPA-600/4-79-020, USEPA, EMSL, 1979.

PRIMARY NAME: Chlorine (Cl_2) Method 330.3

TITLE: Inorganics, Non-Metallics Total Residual

MATRIX: Natural and Treated Waters.

APPLICATION: Date issued 1978. (Titrimetric, iodometric). Method applies to natural and treated waters at concentrations greater than 1 mg/L. Chlorine and chloramines liberate iodine from KI at pH 4 or less. (Iodine is titrated with a standard reducing agent using a starch indicator)

INTERFERENCES: Ferric, manganic and nitrite ions interfere, the neutral titration minimizes these interferences. Acetic acid is used for acid titration — never use HCl. Turbidity and color may make endpoint difficult to detect. Practice runs with spiked samples may be necessary.

INSTRUMENTATION: Standard lab glassware. Microburet 0-2 mL or 0-10 mL is used.

RANGE: Concentrations >1 mg/L Cl MDL: Not reported

PRECISION: SD = 27 and 23.6% @ 0.64 and 1.83 mg/L Cl.

ACCURACY: Relative error = 23.6 and 16.7% @ 0.64 and 1.83 mg/L Cl.

SAMPLING METHOD: plastic or glass (200 mL).

STABILITY: no preservation required. Analyze immediately.

QUALITY CONTROL: Phenylarsine oxide is used as the standardized reducing agent to titrate liberated iodine to starch iodine endpoint. (Titrate away from direct sunlight. Run blank titration).

REFERENCE: Methods for the Chemical Analysis of Water and Wastes, EPA-600/4-79-020, USEPA, EMSL, 1979.

PRIMARY NAME: Chlorine (Cl_2) Method 330.4

TITLE: Inorganics, Non-Metallics Total Residual

MATRIX: Natural and Treated Waters.

APPLICATION: Date issued 1978. (Titrimetric, DPD-FAS). The N,N-diethyl-p-phenylene diamine (DPD)-ferrous ammomium sulfate (FAS). Method applies to matrix listed at concentrations above 1 mg/L Cl. (Liberated iodine is titrated with FAS using DPD as indicator).

INTERFERENCES: Bromine, bromamine and iodine are interferences normally present in insignificant amounts. Oxidized manganese and copper interfere, but can be corrected for. Turbidity and color may make endpoint difficult to detect.

INSTRUMENTATION: Standard lab glassware. Microburet, 0-2 mL or 0-10 mL is used.

RANGE: Concentrations above 1 mg/L Cl MDL: Not listed

PRECISION: SD = 19.4 and 9.4% @ 0.64 and 1.83 mg/L Cl.

ACCURACY: Relative error = 8.1 and 4.3% @ 0.64 and 1.83 mg/L Cl.

SAMPLING METHOD: plastic or glass (200 mL).

STABILITY: No preservation required. Analyze immediately.

QUALITY CONTROL: This procedure gives a convenient direct reading (mL titrant = mg/L Cl) up to 4 mg/L. An aliquot should be diluted to 100 mL if higher concentrations are present. Use chlorine free distilled water to prepare indicator.

REFERENCE: Methods for the Chemical Analysis of Water and Wastes, EPA-600/4-79-020, USEPA, EMSL, 1979.

PRIMARY NAME: Chlorine (Cl_2) Method 330.5

TITLE: Inorganics, Non-Metallics Total Residual

MATRIX: Natural and Treated Waters.

APPLICATION: Date issued 1978. (Spectrophotometric, DPD). The N,N-diethyl-p-phenylene diamine (DPD) colorimetric method applies to matrix listed at concentrations from 0.2-4 mg/L Cl. (Liberated iodine reacts with DPD to produce a red colored solution read on spectrophotometer)

INTERFERENCES: Any oxidizing agents; these are usually present at insignificant concentrations compared to the residual chlorine concentrations. Turbidity and color will essentially prevent the colorimetric analysis. (Ferrous ammonium sulfate is titrant used on permanganate stds).

INSTRUMENTATION: Spectrophotometer. 515 nm. Cells of light path 1 cm or longer.

RANGE: 0.2-4.0 mg/L Cl

MDL: Not listed

PRECISION: SD = 27.6% @ 0.66 mg/L Cl.

ACCURACY: Relative error = 15.6% @ 0.66 mg/L Cl.

SAMPLING METHOD: plastic or glass (200 mL).

STABILITY: no preservation required. Analyze immediately.

QUALITY CONTROL: The solution is spectrophotometrically compared to a series of standards, using a graph or a regression analysis calculation. (Calculation is figured using absorbance and titrated concentrations of permanganate solutions (chlorine equivalent) and absorbance of sample).

REFERENCE: Methods for the Chemical Analysis of Water and Wastes, EPA-600/4-79-020, USEPA, EMSL, 1979.

PRIMARY NAME: Chromium (Cr) Method 200.7

TITLE: Inductively Coupled Plasma (ICP)

MATRIX: Dissolved, suspended or total element in drinking and surface waters and in domestic and industrial wastewaters.

APPLICATION: The method covers the determination of 25 metals. Dissolved elements are determined in filtered and acidified samples after appropriate digestion (which increases dissolved solids). Its primary advantage is that ICP instruments allow simultaneous or rapid sequential determination of many elements in a short time. Samples are first nebulized and the aerosol is transported to a plasma torch in which element specific atomic line emission spectra are produced by a radio frequency inductively coupled plasma. Background correction is required for trace element detection except in the case of line broadning.

INTERFERENCES: There are spectral, physical and chemical interferences. The primary disadvantage of ICP instruments is background radiation from other elements and the plasma gases (spectral interferences). Changes in sample viscosity and surface tension with samples containing high dissolved solids (especially those exceeding 1500 mg/L) or high acid concentrations can cause physical interferences. Ionization effects, solute vaporization and molecular compound formation can cause chemical interferences. Iron, manganese and vanadium can cause interference at the 100 mg/L level.

INSTRUMENTATION: Inductively Coupled Argon Plasma Emission Spectroscopy. 267.716 nm Wavelength

RANGE: Not listed

MDL: 7 μg/L.

PRECISION: SD = 3.8% Mean @ true value 150 μg/L.

ACCURACY: Mean Recovery = 93% +/- 6% of spiked elements for all wastes.

SAMPLING METHOD: Wash sample container with detergent and tap water, rinse with 1+1 nitric acid and tap water, then rinse with 1+1 hydrochloric acid and tap water, then rinse with deionized, distilled water in that order. Perform any filtration or acid preservation steps when the sample is collected or as soon as possible thereafter.

STABILITY: Cool samples to 4°C. M.H.T. = 24 Hours.

QUALITY CONTROL: Mixed calibration standards, an instrument check standard and an interference check solution are used in addition to a quality control sample. The quality control sample should be prepared in the same acid matrix as the calibration standards at 10 times the instrumental detection limits and in accordance with the instructions provided by the supplier. Furthermore, two types of blanks are required: a calibration blank and a reagent blank.

REFERENCE: Method 200.7, U.S. EPA, EMSL-Cincinnati, OH, Nov. 1980

PRIMARY NAME: Chromium (Cr) Method 6010

TITLE: Inductively Coupled Plasma (ICP)

MATRIX: Applies to all matrices; ground water, EP extracts, soils, sludges, sediments, aqueous samples, solid and industrial wastes.

APPLICATION: The method covers the determination of 25 metals. Its primary advantage is that ICP instruments allow simultaneous or rapid sequential determination of many elements in a short time. Samples require digestion prior to analysis and are first nebulized and the aerosol is transported to a plasma torch in which element specific atomic line emission spectra are produced by a radio frequency inductively coupled plasma. Background correction is required for trace element detection except in the case of line broadning.

INTERFERENCES: There are spectral, physical and chemical interferences. The primary disadvantage of ICP instruments is background radiation from other elements and the plasma gases (spectral interferences). Changes in sample viscosity and surface tension with samples containing high dissolved solids or high acid concentrations can cause physical interferences. Ionization effects, solute vaporization and molecular compound formation can cause chemical interferences. Iron, manganese and vanadium can cause interference at the 100 mg/L level.

INSTRUMENTATION: Inductively Coupled Argon Plasma Emission Spectroscopy. 267.716 nm Wavelength

RANGE: Not listed MDL: 7 μg/L.

PRECISION: SD = 3.8% Mean @ true value 150 μg/L.

ACCURACY: Mean Recovery = 93% +/- 6% of spiked elements for all wastes.

SAMPLING METHOD: Collect 400 mL of sample in plastic or glass containers.

STABILITY: Cool samples to 4°C. M.H.T. = 24 Hours.

QUALITY CONTROL: Mixed calibration standards, an instrument check standard and an interference check solution are used in addition to a quality control sample. The quality control sample should be prepared in the same acid matrix as the calibration standards at 10 times the instrumental detection limits and in accordance with the instructions provided by the supplier. Furthermore, two types of blanks are required: a calibration blank and a reagent blank.

REFERENCE: Method 6010, SW-846, 3rd ed., Nov 1986.

PRIMARY NAME: Chromium (Cr) Method 7190

TITLE: Atomic Absorption, (AA) Direct Aspiration

MATRIX: Drinking, Surface and Saline Waters. Wastewater

CAS # : 7440-47-3

APPLICATION: Sample is aspirated and atomized in a flame. A light beam from a chromium hollow cathode lamp is directed through the flame into a monochromator and onto a detector. Since wavelength of light beam is specific for chromium, light energy absorbed by detector is measure of chromium.

INTERFERENCES: The most troublesome type is chemical, caused by lack of absorption of atoms bound in molecular combination in the flame. High dissolved solids in sample may result in nonatomic absorbance interference. Very high alkali metal contents can interfere.

INSTRUMENTATION: Atomic absorption spectrometer. Chromium hollow cathode lamp. (357.9 nm Wavelength).

RANGE: 0.5-10 mg/L MDL: 0.05 mg/L

PRECISION: standard deviation = 105 µg/L @ 370 µg/L (true value) 74 labs

ACCURACY: as bias = -4.5% @ 370 µg/L (true value) 74 labs

SAMPLING METHOD: Use glass or plastic containers. Collect 200 g of solids and 600 mL of liquid samples.

STABILITY: Cool solid samples to 4°C. and analyze as soon as possible. Add nitric acid to liquid samples to pH < 2; M.H.T. = 6 months.

QUALITY CONTROL: At least one duplicate and one spike sample should be run every 20 samples or with each matrix type to verify precision of the method. For 20 or more samples per day, verify working standard curve. Run an additional standard at or near mid-range every 10 samples.

REFERENCE: Method 7190, SW-846, 3rd ed., Nov 1986.

PRIMARY NAME: Chromium (Cr) Method 7191

TITLE: Atomic Absorption, (AA) Furnace Technique

MATRIX: Wastes, mobility procedure extracts, soils and groundwater

CAS # : 7440-47-3

APPLICATION: Aqueous samples, EP extracts, industrial wastes, soils, sludges, sediments and solid wastes require digestion before analysis. An aliquot of sample is placed in the graphite tube in the furnace and slowly evaporated, charred and atomized. Absorption of lamp radiation during atomization is proportional to chromium concentration.

INTERFERENCES: The furnace technique is subject to chemical interferences. Composition of sample matrix can have major effect on analysis. Low concentrations of calcium and/or phosphate may interfere. Background correction may be required. Don't use nitrogen as purge gas.

INSTRUMENTATION: Atomic absorption spectrometer. Chromium hollow cathode lamp or electrodeless discharge lamp. Graphite furnace. Strip-chart recorder

RANGE: 5-100 μg/L

MDL: 1 μg/L (357.9 nm Wavelength)

PRECISION: standard deviation = +/- 0.10, 0.20, 0.80 @ 19, 48, 77 μg Cr/L

ACCURACY: recoveries = 97, 101, 102% @ 19, 48, 77 μg Cr/L

SAMPLING METHOD: Use glass or plastic containers. Collect 200 g of solids and 600 mL of liquid samples.

STABILITY: Cool solid samples to 4°C. and analyze as soon as possible. Add nitric acid to liquid samples to pH < 2; their M.H.T. = 6 months.

QUALITY CONTROL: At least one duplicate and one spike sample should be run every 20 samples, or with each matrix type to verify method precision. If 20 or more samples are run a day, run a standard (at or near mid-range) every 10 samples.

REFERENCE: Method 7191, SW-846, 3rd ed., Nov 1986.

PRIMARY NAME: Chromium (VI), Hexavalent (Cr)

Method 7195

TITLE: Coprecipitation

MATRIX: Ground Waters and Certain Wastewaters.

APPLICATION: Method is used to determine concentration of dissolved Cr(VI) in extraction procedure toxicity characteristic extracts and ground waters. Method may also apply to certain wastewaters,if no interfering substances are present. Cr(VI) is separated from solution by coprecipitation.

INTERFERENCES: Extracts containing either sulfate or chloride in concentrations above 1,000 mg/L should be diluted prior to analysis.

INSTRUMENTATION: Flame or furnace atomic absorption spectroscopy.

RANGE: Samples with >5 ∝g Cr(VI)/L. MDL: Not listed

PRECISION: Not listed

ACCURACY: Not listed

SAMPLING METHOD: Use plastic or glass containers.

STABILITY: Cool to 4°C. M.H.T. = 24 Hours.

QUALITY CONTROL: Dilute samples if they are more concentrated than the highest standard or if they fall on the plateau of a calibration curve. Run one spike duplicate sample for every 10 samples.

REFERENCE: Method 7195, SW-846, 3rd ed., Nov 1986.

PRIMARY NAME: Chromium (VI), Hexavalent (Cr) Method 7196

TITLE: Colorimetric

MATRIX: Ground Waters and Certain Wastewaters.

APPLICATION: Method is used to determine concentration of dissolved Cr(VI) in extraction procedure toxicity characteristic extracts and ground waters. Method may also apply to certain wastewaters, provided that no interfering substances are present. The reaction is very sensitive.

INTERFERENCES: The chromium reaction with diphenylcarbazide (DPC) is usually free of interferences. However, certain substances may interfere if the chromium concentration is relatively low. Hexavalent molybdenum, mercury, vanadium and iron in concentrations >1 mg/L can interfere.

INSTRUMENTATION: Spectro(or filter)photometer (with greenish yellow filter) 540 nm. Light path > = 1 cm.

RANGE: 0.5 to 50 mg Cr(VI) per liter. MDL: Not listed

PRECISION: Not listed

ACCURACY: Not listed

SAMPLING METHOD: Use plastic or glass containers.

STABILITY: Cool to 4°C. M.H.T. = 24 Hours.

QUALITY CONTROL: Dilute samples if they are more concentrated than the highest standard or if they fall on the plateau of a calibration curve. Run one spike duplicate sample for every 10 samples. A red-violet color is produced when DPC reacts with dissolved Cr(VI) in acid solution.

REFERENCE: Method 7196, SW-846, 3rd ed., Nov 1986.

PRIMARY NAME: Chromium (VI), Hexavalent (Cr) Method 7197

TITLE: Atomic Absorption, (AA) Chelation Extraction

MATRIX: EP (toxicity characteristic) extracts and ground water

CAS # : 7440-47-3

APPLICATION: Method also applies to certain wastewaters, provided that no interfering substances are present. Method is based on chelation of Cr(VI) with ammonium pyrolidine dithiocarbamate and extraction with methyl isobutyl ketone. The extract is aspirated and atomized in a flame of an atomic absorption spectrometer. A light beam from a chromium hollow cathode lamp is directed through the flame into a monochromator and onto a detector. Since wavelength is specific for chromium, the light energy absorbed is a measure of chromium.

INTERFERENCES: High concentrations of other metals, as may be found in wastewaters, may interfere.

INSTRUMENTATION: Atomic absorption spectrometer. Chromium hollow cathode lamp. Strip-chart recorder. (357.9 nm Wavelength)

RANGE: 1.0 to 25 μg Cr(VI)/L MDL: not listed

PRECISION: standard deviation = +/- 2.6 @ 50 μg Cr(VI)/L

ACCURACY: recovery = 96% @ 50 µg Cr(VI)/L

SAMPLING METHOD: Use plastic or glass containers. Sample as per chapter nine.

STABILITY: Cool to 4°C. Run as soon as possible. M.H.T. = 24 Hrs

QUALITY CONTROL: Run one spike duplicate sample for every 10 samples. Verify calibration with an independently prepared check standard every 15 samples.

REFERENCE: Method 7197, SW-846, 3rd ed., Nov 1986.

PRIMARY NAME: Chromium (VI), Hexavalent (Cr) Method 7198

TITLE: Differential Pulse Polarography MATRIX: EP Extracts, Natural and Wastewaters

CAS # : 7440-47-3

APPLICATION: The method measures the peak current produced from the reduction of Cr(VI) to Cr(III) at a dropping mercury electrode during a differential pulse voltage ramp. Method uses 0.125M ammonium hydroxide and 0.125M ammonium chloride as the supporting electrolyte. Cr(VI) reduction results in the peak current occurring at a peak potential of -0.25V versus a silver/silver chloride reference electrode.

INTERFERENCES: Copper ion at concentrations higher than Cr(VI) concentration is a potential interference due to peak overlap when using the 0.125M ammoniacal electrolyte. Reductants such as ferrous iron, sulfite, and sulfide will reduce Cr(VI) to Cr(III); thus it is imperative to analyze samples as soon as possible.

INSTRUMENTATION: Polarographic instrumentation with a dropping mercury electrode assembly. Reference electrode and strip chart recorder.

RANGE: 1.0-5.0 mg/L (higher by dilution) MDL: 10 µg/L

PRECISION: relative standard deviation = 0.69% @ 1.87 (Average value) 3 samples (leachate)

ACCURACY: Average recovery = 92.8% @ 5.0 mg/L spike. (8 samples; EP extracts)

SAMPLING METHOD: Use plastic or glass containers. Sample as per chapter nine.

STABILITY: Cool to 4°C. and run as soon as possible. M.H.T. = 24 Hrs.

QUALITY CONTROL: Quantitation must be performed by the method of standard additions. Verify calibration with an independently prepared check standard every 15 samples.

REFERENCE: Method 7198, SW-846, 3rd ed., Nov 1986.

PRIMARY NAME: Cobalt (Co) Method 200.7

TITLE: Inductively Coupled Plasma (ICP)

MATRIX: Dissolved, suspended or total element in drinking and surface waters and in domestic and industrial wastewaters.

APPLICATION: The method covers the determination of 25 metals. Dissolved elements are determined in filtered and acidified samples after appropriate digestion (which increases dissolved solids). Its primary advantage is that ICP instruments allow simultaneous or rapid sequential determination of many elements in a short time. Samples are first nebulized and the aerosol is transported to a plasma torch in which element specific atomic line emission spectra are produced by a radio frequency inductively coupled plasma. Background correction is required for trace element detection except in the case of line broadning.

INTERFERENCES: There are spectral, physical and chemical interferences. The primary disadvantage of ICP instruments is background radiation from other elements and the plasma gases (spectral interferences). Changes in sample viscosity and surface tension with samples containing high dissolved solids (especially those exceeding 1500 mg/L) or high acid concentrations can cause physical interferences. Ionization effects, solute vaporization and molecular compound formation can cause chemical interferences. Chromium, iron, nickel and thallium can cause interference at the 100 mg/L level.

INSTRUMENTATION: Inductively Coupled Argon Plasma Emission Spectroscopy. 228.616 nm Wavelength

RANGE: Not listed MDL: 7 µg/L.

PRECISION: SD = 10% Mean @ true value 700 µg/L.

ACCURACY: Mean Recovery = 93% +/- 6% of spiked elements for all wastes.

SAMPLING METHOD: Wash sample container with detergent and tap water, rinse with 1+1 nitric acid and tap water, then rinse with 1+1 hydrochloric acid and tap water, then rinse with deionized, distilled water in that order. Perform any filtration or acid preservation steps when the sample is collected or as soon as possible thereafter.

STABILITY: Cool samples to 4°C. M.H.T. = 24 Hours.

QUALITY CONTROL: Mixed calibration standards, an instrument check standard and an interference check solution are used in addition to a quality control sample. The quality control sample should be prepared in the same acid matrix as the calibration standards at 10 times the instrumental detection limits and in accordance with the instructions provided by the supplier. Furthermore, two types of blanks are required: a calibration blank and a reagent blank.

REFERENCE: Method 200.7, U.S. EPA, EMSL-Cincinnati, OH, Nov. 1980

PRIMARY NAME: Cobalt (Co) Method 6010

TITLE: Inductively Coupled Plasma (ICP)

MATRIX: Applies to all matrices; ground water, EP extracts, soils, sludges, sediments, aqueous samples, solid and industrial wastes.

APPLICATION: The method covers the determination of 25 metals. Its primary advantage is that ICP instruments allow simultaneous or rapid sequential determination of many elements in a short time. Samples require digestion prior to analysis and are first nebulized and the aerosol is transported to a plasma torch in which element specific atomic line emission spectra are produced by a radio frequency inductively coupled plasma.

Background correction is required for trace element detection except in the case of line broadning.

INTERFERENCES: There are spectral, physical and chemical interferences. The primary disadvantage of ICP instruments is background radiation from other elements and the plasma gases (spectral interferences). Changes in sample viscosity and surface tension with samples containing high dissolved solids or high acid concentrations can cause physical interferences. Ionization effects, solute vaporization and molecular compound formation can cause chemical interferences. Chromium, iron, nickel and thallium can cause interference at the 100 mg/L level.

INSTRUMENTATION: Inductively Coupled Argon Plasma Emission Spectroscopy. 228.616 nm Wavelength

RANGE: Not listed MDL: 7 µg/L.

PRECISION: SD = 10% Mean @ true value 700 µg/L.

ACCURACY: Mean Recovery = 93% +/- 6% of spiked elements for all wastes.

SAMPLING METHOD: Collect 400 mL of sample in plastic or glass containers.

STABILITY: Cool samples to 4°C. M.H.T. = 24 Hours.

QUALITY CONTROL: Mixed calibration standards, an instrument check standard and an interference check solution are used in addition to a quality control sample. The quality control sample should be prepared in the same acid matrix as the calibration standards at 10 times the instrumental detection limits and in accordance with the instructions provided by the supplier. Furthermore, two types of blanks are required: a calibration blank and a reagent blank.

REFERENCE: Method 6010, SW-846, 3rd ed., Nov 1986.

PRIMARY NAME: Cobalt (Co) Method 7200

TITLE: Atomic Absorption, (AA) Direct Aspiration

MATRIX: Drinking, Surface and Saline Waters. Wastewater

CAS # : 7440-48-4

APPLICATION: Sample is aspirated and atomized in a flame. A light beam from a Co hollow cathode lamp is directed through the flame into a monochromator and onto a detector. Since wavelength of light beam is specific for Co, light energy absorbed by detector is measure of cobalt.

INTERFERENCES: The most troublesomee type is chemical, caused by lack of absorption of atoms bound in molecular combination in the flame. High dissolved solids in sample may result in nonatomic absorbance interference. Excess of other transition metals may interfere.

INSTRUMENTATION: Atomic absorption spectrometer. Cobalt hollow cathode lamp. (240.7 nm Wavelength).

RANGE: 0.5-5 mg/L MDL: 0.05 mg/L

PRECISION: standard deviation = +/- 0.013, 0.01, 0.05 @ 0.20, 1.0, 5.0 mg Co/L

ACCURACY: recoveries = 98, 98, 97% @ 0.20, 1.0, 5.0 mg Co/L

SAMPLING METHOD: Use glass or plastic containers. Collect 200 g of solids and 600 mL of liquid samples.

STABILITY: Cool solid samples to 4°C. and analyze as soon as possible. Add nitric acid to liquid samples to pH < 2; M.H.T. = 6 months.

QUALITY CONTROL: At least one duplicate and one spike sample should be run every 20 samples or with each matrix type to verify precision of the method. For 20 or more samples per day, verify working standard curve. Run an additional standard at or near mid-range every 10 samples.

REFERENCE: Method 7200, SW-846, 3rd ed., Nov 1986.

PRIMARY NAME: Cobalt (Co) Method 7201

TITLE: Atomic Absorption, (AA) Furnace Technique

MATRIX: Wastes, mobility procedure extracts, soils and groundwater

CAS # : 7440-48-4

APPLICATION: Aqueous samples, EP extracts, industrial wastes, soils, sludges, sediments and solid wastes require digestion before analysis. An aliquot of sample is placed in the graphite tube in the furnace and slowly evaporated, charred and atomized. Absorption of lamp radiation during atomization is proportional to cobalt concentration.

INTERFERENCES: The furnace technique is subject to chemical interferences. Composition of sample matrix can effect analysis. Modify matrix to remove interferences or to stabilize the analyte. Background correction is required. Excess chloride may interfere.

INSTRUMENTATION: Atomic absorption spectrometer. Cobalt hollow cathode lamp or electrodeless discharge lamp. Graphite furnace. Strip-chart recorder

RANGE: 5-100 μg/L

MDL: 1 μg/L (240.7 nm Wavelength)

PRECISION: Not listed

ACCURACY: Not listed

SAMPLING METHOD: Use glass or plastic containers. Collect 200 g of solids and 600 mL of liquid samples.

STABILITY: Cool solid samples to 4°C. and analyze as soon as possible. Add nitric acid to liquid samples to pH < 2; M.H.T. = 6 months.

QUALITY CONTROL: At least one duplicate and one spike sample should be run every 20 samples, or with each matrix type to verify method precision. If 20 or more samples are run a day, run a standard (at or near mid-range) every 10 samples.

REFERENCE: Method 7201, SW-846, 3rd ed., Nov 1986.

PRIMARY NAME: Copper (Cu)

Method 200.7

TITLE: Inductively Coupled Plasma (ICP)

MATRIX: Dissolved, suspended or total element in drinking and surface waters and in domestic and industrial wastewaters.

APPLICATION: The method covers the determination of 25 metals. Dissolved elements are determined in filtered and acidified samples after appropriate digestion (which increases dissolved solids). Its primary advantage is that ICP instruments allow simultaneous or rapid sequential determination of many elements in a short time. Samples are first nebulized and the aerosol is transported to a plasma torch in which element specific atomic line emission spectra are produced by a radio frequency inductively coupled plasma. Background correction is required for trace element detection except in the case of line broadning.

INTERFERENCES: There are spectral, physical and chemical interferences. The primary disadvantage of ICP instruments is background radiation from other elements and the plasma gases (spectral interferences). Changes in sample viscosity and surface tension with samples containing high dissolved solids (especially those exceeding 1500 mg/L) or high acid concentrations can cause physical interferences. Ionization effects, solute vaporization and molecular compound formation can cause chemical interferences. Iron, thallium and vanadium can cause interference at the 100 mg/L level.

INSTRUMENTATION: Inductively Coupled Argon Plasma Emission Spectroscopy. 324.754 nm Wavelength

RANGE: Not listed MDL: 6 µg/L.

PRECISION: SD = 5.1% Mean @ true value 250 µg/L.

ACCURACY: Mean Recovery = 93% +/- 6% of spiked elements for all wastes.

SAMPLING METHOD: Wash sample container with detergent and tap water, rinse with 1+1 nitric acid and tap water, then rinse with 1+1 hydrochloric acid and tap water, then rinse with deionized, distilled water in that order. Perform any filtration or acid preservation steps when the sample is collected or as soon as possible thereafter.

STABILITY: Cool samples to 4°C. M.H.T. = 24 Hours.

QUALITY CONTROL: Mixed calibration standards, an instrument check standard and an interference check solution are used in addition to a quality control sample. The quality control sample should be prepared in the same acid matrix as the calibration standards at 10 times the instrumental detection limits and in accordance with the instructions provided by the supplier.

Furthermore, two types of blanks are required: a calibration blank and a reagent blank.

REFERENCE: Method 200.7, U.S. EPA, EMSL-Cincinnati, OH, Nov. 1980

PRIMARY NAME: Copper (Cu) Method 6010

TITLE: Inductively Coupled Plasma (ICP)

MATRIX: Applies to all matrices; ground water, EP extracts, soils, sludges, sediments, aqueous samples, solid and industrial wastes.

APPLICATION: The method covers the determination of 25 metals. Its primary advantage is that ICP instruments allow simultaneous or rapid sequential determination of many elements in a short time. Samples require digestion prior to analysis and are first nebulized and the aerosol is transported to a plasma torch in which element specific atomic line emission spectra are produced by a radio frequency inductively coupled plasma. Background correction is required for trace element detection except in the case of line broadning.

INTERFERENCES: There are spectral, physical and chemical interferences. The primary disadvantage of ICP instruments is background radiation from other elements and the plasma gases (spectral interferences). Changes in sample viscosity and surface tension with samples containing high dissolved solids or high acid concentrations can cause physical interferences. Ionization effects, solute vaporization and molecular compound formation can cause chemical interferences. Iron, thallium and vanadium can cause interference at the 100 mg/L level.

INSTRUMENTATION: Inductively Coupled Argon Plasma Emission Spectroscopy. 324.754 nm Wavelength

RANGE: Not listed

MDL: 6 μg/L.

PRECISION: SD = 5.1% Mean @ true value 250 μg/L.

ACCURACY: Mean Recovery = 93% +/- 6% of spiked elements for all wastes.

SAMPLING METHOD: Collect 400 mL of sample in plastic or glass containers.

STABILITY: Cool samples to 4°C. M.H.T. = 24 Hours.

QUALITY CONTROL: Mixed calibration standards, an instrument check standard and an interference check solution are used in addition to a quality control sample. The quality control sample should be prepared in the same acid matrix as the calibration standards at 10 times the instrumental detection limits and in accordance with the instructions provided by the supplier. Furthermore, two types of blanks are required: a calibration blank and a reagent blank.

REFERENCE: Method 6010, SW-846, 3rd ed., Nov 1986.

PRIMARY NAME: Copper (Cu) Method 7210

TITLE: Atomic Absorption, (AA) Direct Aspiration

MATRIX: Drinking, Surface and Saline Waters. Wastewater

CAS # : 7440-50-8

APPLICATION: Sample is aspirated and atomized in a flame. A light beam from a Cu hollow cathode lamp is directed through the flame into a monochromator and onto a detector. Since wavelength of light beam is specific for Cu, light energy absorbed by detector is measure of copper.

INTERFERENCES: The most troublesomee type is chemical, caused by lack of absorption of atoms bound in molecular combination in the flame. High dissolved solids in sample may result in nonatomic absorbance interference. Non specific absorption and scattering may interfere.

INSTRUMENTATION: Atomic absorption spectrometer. Copper hollow cathode lamp. (324.7 nm Wavelength).

RANGE: 0.2-5 mg/L MDL: 0.02 mg/L

PRECISION: standard deviation = 56 µg/L @ 332 µg/L (true value) 92 labs

ACCURACY: as bias = -2.4% @ 332 µg/L (true value) 92 labs

SAMPLING METHOD: Use glass or plastic containers. Collect 200 g of solids and 600 mL of liquid samples.

STABILITY: Cool solid samples to 4°C. and analyze as soon as possible. Add nitric acid to liquid samples to pH < 2; M.H.T. = 6 months.

QUALITY CONTROL: At least one duplicate and one spike sample should be run every 20 samples or with each matrix type to verify precision of the method. For 20 or more samples per day, verify working standard curve. Run an additional standard at or near mid-range every 10 samples.

REFERENCE: Method 7210, SW-846, 3rd ed., Nov 1986.

PRIMARY NAME: Copper (Cu) Method 7211

TITLE: Atomic Absorption, (AA) Furnace Technique

MATRIX: Wastes, mobility procedure extracts, soils and groundwater

CAS # : 7440-50-8

APPLICATION: Aqueous samples, EP extracts, industrial wastes, soils, sludges, sediments and solid wastes require digestion before analysis. An aliquot of sample is placed in the graphite tube in the furnace and slowly evaporated, charred and atomized. Absorption of lamp radiation during atomization is proportional to copper concentration.

INTERFERENCES: The furnace technique is subject to chemical interferences. Composition of sample matrix can have major effect on analysis. Modify matrix to remove interferences. Background correction may be required. Nonspecific absorption and scattering may be significant

INSTRUMENTATION: Atomic absorption spectrometer. Copper hollow cathode lamp or electrodeless discharge lamp. Graphite furnace. Strip-chart recorder

RANGE: 5-100 μg/L

MDL: 1 μg/L (324.7 nm Wavelength)

PRECISION: Not listed

ACCURACY: Not listed

SAMPLING METHOD: Use glass or plastic containers. Collect 200 g of solids and 600 mL of liquid samples.

STABILITY: Cool solid samples to 4°C. and analyze as soon as possible. Add nitric acid to liquid samples to pH < 2; M.H.T. = 6 months.

QUALITY CONTROL: At least one duplicate and one spike sample should be run every 20 samples, or with each matrix type to verify method precision. If 20 or more samples are run a day, run a standard (at or near mid-range) every 10 samples.

REFERENCE: Method 7211, SW-846, 3rd ed., (Included as revision 0, December 1987)

PRIMARY NAME: Iron (Fe) Method 200.7

TITLE: Inductively Coupled Plasma (ICP)

MATRIX: Dissolved, suspended or total element in drinking and surface waters and in domestic and industrial wastewaters.

APPLICATION: The method covers the determination of 25 metals. Dissolved elements are determined in filtered and acidified samples after appropriate digestion (which increases dissolved solids). Its primary advantage is that ICP instruments allow simultaneous or rapid sequential determination of many elements in a short time. Samples are first nebulized and the aerosol is transported to a plasma torch in which element specific atomic line emission spectra are produced by a radio frequency inductively coupled plasma. Background correction is required for trace element detection except in the case of line broadning.

INTERFERENCES: There are spectral, physical and chemical interferences. The primary disadvantage of ICP instruments is background radiation from other elements and the plasma gases (spectral interferences). Changes in sample viscosity and surface tension with samples containing high dissolved solids (especially those exceeding 1500 mg/L) or high acid concentrations can cause physical interferences. Ionization effects, solute vaporization and molecular compound formation can cause chemical interferences. Manganese can cause interference at the 100 mg/L level.

INSTRUMENTATION: Inductively Coupled Argon Plasma Emission Spectroscopy. 259.940 nm Wavelength

RANGE: Not listed MDL: 7 μg/L.

PRECISION: SD = 3.0% Mean @ true value 600 μg/L.

ACCURACY: Mean Recovery = 93% +/- 6% of spiked elements for all wastes.

SAMPLING METHOD: Wash sample container with detergent and tap water, rinse with 1+1 nitric acid and tap water, then rinse with 1+1 hydrochloric acid and tap water, then rinse with deionized, distilled water in that order. Perform any filtration or acid preservation steps when the sample is collected or as soon as possible thereafter.

STABILITY: Cool samples to 4°C. M.H.T. = 24 Hours.

QUALITY CONTROL: Mixed calibration standards, an instrument check standard and an interference check solution are used in addition to a quality control sample. The quality control sample should be prepared in the same acid matrix as the calibration standards at 10 times the instrumental detection limits and in accordance with the instructions provided by the supplier. Furthermore, two types of blanks are required: a calibration blank and a reagent blank.

REFERENCE: Method 200.7, U.S. EPA, EMSL-Cincinnati, OH, Nov. 1980

PRIMARY NAME: Iron (Fe) Method 236.2

TITLE: Metals (Total, Dissolved, Suspended) AA Furnace Technique

MATRIX: Drinking, Surface and Saline Waters. Wastewater

APPLICATION: Date issued 1978. A representative sample aliquot is placed in a graphite tube in furnace, evaporated to dryness, charred and atomized. Radiation from excited element is passed through vapor and radiation intensity decreases proportional to amount of Fe in vapor.

INTERFERENCES: Furnace technique subject to chemical and matrix interferences. Furnace gases may have molecular absorption bands enclosing analytical wavelength. Smoke-producing sample matrix can interfere. If Fe isn't volitalized and removed from furnace, memory effects occur.

INSTRUMENTATION: AAS. Iron (Fe) hollow cathode lamp or EDL. Graphite furnace. Pipets.

RANGE: 5-100 μg/L. MDL: 1 μg/L.

PRECISION: Not listed

ACCURACY: Not listed

SAMPLING METHOD: plastic or glass (pre-washed).

STABILITY: Add nitric acid to pH <2. M.H.T. = 6 Months.

QUALITY CONTROL: A check standard should be run approximately after every 10 sample injections. Standards are run in part to monitor the life and performance of the graphite tube. Lack of reproducibility or significant change in the signal for the standard indicates tube should be replaced.

REFERENCE: Methods for the Chemical Analysis of Water and Wastes, EPA-600/4-79-020, USEPA, EMSL, 1979.

PRIMARY NAME: Iron (Fe) Method 315 B

TITLE: Phenanthroline Method Total Dissolved Ferrous

MATRIX: Natural or Treated Waters

APPLICATION: Iron is brought into solution, reduced to ferrous state by boiling with acid and hydroxylamine and treated with 1,10-phenanthroline at pH 3.2 to 3.3. Three molecules of phenanthroline chelate each atom of ferrous iron to form an orange-red complex.

INTERFERENCES: Strong oxidizing agents, cyanide, nitrite, and phosphates (particularly polyphosphates), chromium, zinc, cobalt, copper, and nickel interfere. Bismuth, cadmium, mercury, molybdate and silver precipitate phenanthroline.

INSTRUMENTATION: Spectro (or filter) photometer with green filter. 510 nm. Light path > 1 cm.

RANGE: Not listed MDL: 10 μg/L.

PRECISION: SD = 25.5% @ 300 μg Fe/L (aqueous mixture of 8 metals)

ACCURACY: Relative error = 13.3% @ 300 μg Fe/L (aqueous mix of 8 metals)

SAMPLING METHOD: plastic or glass. Clean with acid. Rinse with distilled water.

STABILITY: Add nitric acid to pH <2. M.H.T. = 6 Months.

QUALITY CONTROL: Use reagents low in iron. Use iron-free distilled water in preparing standards and reagent solutions. Store reagents in glass stoppered bottles. Calculate ferric iron by subtracting ferrous iron from total iron. Don't expose phenanthroline solutions to sunlight.

REFERENCE: Standard Methods for the Examination of Water and Waste Water, 16th ed., Page 215, 1985.

PRIMARY NAME: Iron (Fe) Method 6010

TITLE: Inductively Coupled Plasma (ICP)

MATRIX: Applies to all matrices; ground water, EP extracts, soils, sludges, sediments, aqueous samples, solid and industrial wastes.

APPLICATION: The method covers the determination of 25 metals. Its primary advantage is that ICP instruments allow simultaneous or rapid sequential determination of many elements in a short time. Samples require digestion prior to analysis and are first nebulized and the aerosol is transported to a plasma torch in which element specific atomic line emission spectra are produced by a radio frequency inductively coupled plasma. Background correction is required for trace element detection except in the case of line broadning.

INTERFERENCES: There are spectral, physical and chemical interferences. The primary disadvantage of ICP instruments is background radiation from other elements and the plasma gases (spectral interferences). Changes in sample viscosity and surface tension with samples containing high dissolved solids or high acid concentrations can cause physical interferences. Ionization effects, solute vaporization and molecular compound formation can cause chemical interferences. Manganese can cause interference at the 100 mg/L level.

INSTRUMENTATION: Inductively Coupled Argon Plasma Emission Spectroscopy. 259.940 nm Wavelength

RANGE: Not listed MDL: 7 µg/L.

PRECISION: SD = 3.0% Mean @ true value 600 µg/L.

ACCURACY: Mean Recovery = 93% +/- 6% of spiked elements for all wastes.

SAMPLING METHOD: Collect 400 mL of sample in plastic or glass containers.

STABILITY: Cool samples to 4°C. M.H.T. = 24 Hours.

QUALITY CONTROL: Mixed calibration standards, an instrument check standard and an interference check solution are used in addition to a quality control sample. The quality control sample should be prepared in the same acid matrix as the calibration standards at 10 times the instrumental detection limits and in accordance with the instructions provided by the supplier. Furthermore, two types of blanks are required: a calibration blank and a reagent blank.

REFERENCE: Method 6010, SW-846, 3rd ed., Nov 1986.

PRIMARY NAME: Iron (Fe) Method 7380

TITLE: Atomic Absorption, (AA) Direct Aspiration

MATRIX: Drinking, Surface and Saline Waters. Wastewater

CAS # : 7439-89-6

APPLICATION: Sample is aspirated and atomized in a flame. A light beam from an Fe hollow cathode lamp is directed through the flame into a monochromator and onto a detector. Since wavelength of light beam is specific for Fe, light energy absorbed by detector is measure of iron.

INTERFERENCES: The most troublesomee type is chemical, caused by lack of absorption of atoms bound in molecular combination in the flame. High dissolved solids in sample may result in nonatomic absorbance interference. Iron is a universal contaminant; avoid contamination.

INSTRUMENTATION: Atomic absorption spectrometer. Iron hollow cathode lamp. [248.3 nm Wavelength(primary)]

RANGE: 0.3-5 mg/L

MDL: 0.03 mg/L

PRECISION: standard deviation = 173 µg/L @ 840 µg/L (true value) 82 labs

ACCURACY: as bias = +1.8% @ 840 µg/L (true value) 82 labs

SAMPLING METHOD: Use glass or plastic containers. Collect 200 g of solids and 600 mL of liquid samples.

STABILITY: Cool solid samples to 4°C. and analyze as soon as possible. Add nitric acid to liquid samples to pH < 2; M.H.T. = 6 months.

QUALITY CONTROL: At least one duplicate and one spike sample should be run every 20 samples or with each matrix type to verify precision of the method. For 20 or more samples per day, verify working standard curve. Run an additional standard at or near mid-range every 10 samples.

REFERENCE: Method 7380, SW-846, 3rd ed., Nov 1986.

PRIMARY NAME: Iron (Fe)

Method 7381

TITLE: Atomic Absorption, (AA) Furnace Technique

MATRIX: Wastes, mobility procedure extracts, soils and groundwater

CAS # : 7439-89-6

APPLICATION: Aqueous samples, EP extracts, industrial wastes, soils, sludges, sediments and solid wastes require digestion before analysis. An aliquot of sample is placed in the graphite tube in the furnace and slowly evaporated, charred and atomized. Absorption of lamp radiation during atomization is proportional to iron concentration.

INTERFERENCES: The furnace technique is subject to chemical interferences. Composition of sample matrix can have major effect on analysis. Modify matrix to remove interferences. Iron is a universal contaminant. Use great care to avoid contamination.

INSTRUMENTATION: Atomic absorption spectrometer. Iron hollow cathode lamp or electrodeless discharge lamp. Graphite furnace. Strip-chart recorder

RANGE: 5-100 µg/L

MDL: 1 µg/L (248.3 nm Wavelength)

PRECISION: Not listed

ACCURACY: Not listed

SAMPLING METHOD: Use glass or plastic containers. Collect 200 g of solids and 600 mL of liquid samples.

STABILITY: Cool solid samples to 4°C. and analyze as soon as possible. Add nitric acid to liquid samples to pH < 2; M.H.T. = 6 months.

QUALITY CONTROL: At least one duplicate and one spike sample should be run every 20 samples, or with each matrix type to verify method precision. If 20 or more samples are run a day, run a standard (at or near mid-range) every 10 samples.

REFERENCE: Method 7381, SW-846, 3rd ed., (Included as revision 0, December 1987)

PRIMARY NAME: Lead (Pb)

Method 200.7

TITLE: Inductively Coupled Plasma (ICP)

MATRIX: Dissolved, suspended or total element in drinking and surface waters and in domestic and industrial wastewaters.

APPLICATION: The method covers the determination of 25 metals. Dissolved elements are determined in filtered and acidified samples after appropriate digestion (which increases dissolved solids). Its primary advantage is that ICP instruments allow simultaneous or rapid sequential determination of many elements in a short time. Samples are first nebulized and the aerosol is transported to a plasma torch in which element specific atomic line emission spectra are produced by a radio frequency inductively coupled plasma. Background correction is required for trace element detection except in the case of line broadning.

INTERFERENCES: There are spectral, physical and chemical interferences. The primary disadvantage of ICP instruments is background radiation from other elements and the plasma gases (spectral interferences). Changes in sample viscosity and surface tension with samples containing high dissolved solids (especially those exceeding 1500 mg/L) or high acid concentrations can cause physical interferences. Ionization effects, solute vaporization and molecular compound formation can cause chemical interferences. Aluminum can cause interference at the 100 mg/L level.

INSTRUMENTATION: Inductively Coupled Argon Plasma Emission Spectroscopy. 220.353 nm Wavelength

RANGE: Not listed MDL: 42 µg/L.

PRECISION: SD = 16% Mean @ true value 250 µg/L.

ACCURACY: Mean Recovery = 93% +/- 6% of spiked elements for all wastes.

SAMPLING METHOD: Wash sample container with detergent and tap water, rinse with 1+1 nitric acid and tap water, then rinse with 1+1 hydrochloric acid and tap water, then rinse with deionized, distilled water in that order. Perform any filtration or acid preservation steps when the sample is collected or as soon as possible thereafter.

STABILITY: Cool samples to 4°C. M.H.T. = 24 Hours.

QUALITY CONTROL: Mixed calibration standards, an instrument check standard and an interference check solution are used in addition to a quality control sample. The quality control sample should be prepared in the same acid matrix as the calibration standards at 10 times the instrumental detection limits and in accordance with the instructions provided by the supplier. Furthermore, two types of blanks are required: a calibration blank and a reagent blank.

REFERENCE: Method 200.7, U.S. EPA, EMSL-Cincinnati, OH, Nov. 1980

PRIMARY NAME: Lead (Pb) Method 6010

TITLE: Inductively Coupled Plasma (ICP)

MATRIX: Applies to all matrices; ground water, EP extracts, soils, sludges, sediments, aqueous samples, solid and industrial wastes.

APPLICATION: The method covers the determination of 25 metals. Its primary advantage is that ICP instruments allow simultaneous or rapid sequential determination of many elements in a short time. Samples require digestion prior to analysis and are first nebulized and the aerosol is transported to a plasma torch in which element specific atomic line emission spectra are produced by a radio frequency inductively coupled plasma. Background correction is required for trace element detection except in the case of line broadning.

INTERFERENCES: There are spectral, physical and chemical interferences. The primary disadvantage of ICP instruments is background radiation from other elements and the plasma gases (spectral interferences). Changes in sample viscosity and surface tension with samples containing high dissolved solids or high acid concentrations can cause physical interferences. Ionization effects, solute vaporization and molecular compound formation can cause chemical interferences. Aluminum can cause interference at the 100 mg/L level.

INSTRUMENTATION: Inductively Coupled Argon Plasma Emission Spectroscopy. 220.353 nm Wavelength

RANGE: Not listed

MDL: 42 µg/L.

PRECISION: SD = 16% Mean @ true value 250 µg/L.

ACCURACY: Mean Recovery = 93% +/- 6% of spiked elements for all wastes.

SAMPLING METHOD: Collect 400 mL of sample in plastic or glass containers.

STABILITY: Cool samples to 4°C. M.H.T. = 24 Hours.

QUALITY CONTROL: Mixed calibration standards, an instrument check standard and an interference check solution are used in addition to a quality control sample. The quality control sample should be prepared in the same acid matrix as the calibration standards at 10 times the instrumental detection limits and in accordance with the instructions provided by the supplier. Furthermore, two types of blanks are required: a calibration blank and a reagent blank.

REFERENCE: Method 6010, SW-846, 3rd ed., Nov 1986.

PRIMARY NAME: Lead (Pb) Method 7420

TITLE: Atomic Absorption, (AA) Direct Aspiration

MATRIX: Drinking, Surface and Saline Waters. Wastewater

CAS # : 7439-92-1

APPLICATION: Sample is aspirated and atomized in a flame. A light beam from a Pb hollow cathode lamp is directed through the flame into a monochromator and onto a detector. Since wavelength of light beam is specific for Pb, light energy absorbed by detector is measure of lead.

INTERFERENCES: The most troublesomee type is chemical, caused by lack of absorption of atoms bound in molecular combination in the flame. High dissolved solids in sample may result in nonatomic absorbance interference. Background correction is required.

INSTRUMENTATION: Atomic absorption spectrometer. Lead hollow cathode lamp. [283.3 nm Wavelength(primary)]

RANGE: 1-20 mg/L MDL: 0.1 mg/L

PRECISION: standard deviation = 128 μg/L @ 367 μg/L (true value) 74 labs

ACCURACY: as bias = +2.9% @ 367 μg/L (true value) 74 labs

SAMPLING METHOD: Use glass or plastic containers. Collect 200 g of solids and 600 mL of liquid samples.

STABILITY: Cool solid samples to 4°C. and analyze as soon as possible. Add nitric acid to liquid samples to pH < 2; M.H.T. = 6 months.

QUALITY CONTROL: At least one duplicate and one spike sample should be run every 20 samples or with each matrix type to verify precision of the method. For 20 or more samples per day, verify working standard curve. Run an additional standard at or near mid-range every 10 samples.

REFERENCE: Method 7420, SW-846, 3rd ed., Nov 1986.

PRIMARY NAME: Lead (Pb) Method 7421

TITLE: Atomic Absorption, (AA) Furnace Technique

MATRIX: Drinking, Surface and Saline Waters. Wastewater.

APPLICATION: Pb in solution is readily determined by atomic absorption spectrometer, but detection limits, sensitivity and optimum range vary with the matrices and models of AA spectrophotometers. While drinking water may be analyzed directly, ground water, other aqueous samples, EP extracts, industrial wastes, soils, sludges, and sediments require digestion.

INTERFERENCES: "Chemical" interference is caused by lack of absorption of atoms bound in molecular combination in the flame. High dissolved solids in sample may cause interference from non atomic absorbance. Ionization and spectral interferences can occur.

INSTRUMENTATION: Atomic absorption spectrometer. Lead (Pb) hollow cathode lamp. Graphite furnace. 283 nm Wavelength.

RANGE: 5-100 µg/L MDL: 1 µg/L

PRECISION: standard deviation = +/- 3.7 @ 100 µg Pb/L.

ACCURACY: Recovery = 95% @ 100 µg Pb/L.

SAMPLING METHOD: Use glass or plastic containers. Collect 200 g of solids and 600 mL of liquid samples.

STABILITY: Cool solid samples to 4°C. and analyze as soon as possible. Add nitric acid to liquid samples to pH < 2; M.H.T. = 6 months.

QUALITY CONTROL: At least one duplicate and one spike sample should be run every 20 samples, or with each matrix type to verify method precision. If 20 or more samples are run, run a standard (at or near mid-range) every 10 samples.

REFERENCE: Method 7421, SW-846, 3rd ed., Nov 1986.

PRIMARY NAME: Magnesium (Mg) Method 200.7

TITLE: Inductively Coupled Plasma (ICP)

MATRIX: Dissolved, suspended or total element in drinking and surface waters and in domestic and industrial wastewaters.

APPLICATION: The method covers the determination of 25 metals. Dissolved elements are determined in filtered and acidified samples after appropriate digestion (which increases dissolved solids). Its primary advantage is that ICP instruments allow simultaneous or rapid sequential determination of many elements in a short time. Samples are first nebulized and the aerosol is transported to a plasma torch in which element specific atomic line emission spectra are produced by a radio frequency inductively coupled plasma. Background correction is required for trace element detection except in the case of line broadning.

INTERFERENCES: There are spectral, physical and chemical interferences. The primary disadvantage of ICP instruments is background radiation from other elements and the plasma gases (spectral interferences). Changes in sample viscosity and surface tension with samples containing high dissolved solids (especially those exceeding 1500 mg/L) or high acid concentrations can cause physical interferences. Ionization effects, solute vaporization and molecular compound formation can cause chemical interferences. Calcium, chromium, iron, manganese, thallium and vanadium can cause interference at the 100 mg/L level.

INSTRUMENTATION: Inductively Coupled Argon Plasma Emission Spectroscopy. 279.079 nm Wavelength

RANGE: Not listed

MDL: 30 µg/L.

PRECISION: Not listed

ACCURACY: Mean Recovery = 93% +/- 6% of spiked elements for all wastes.

SAMPLING METHOD: Wash sample container with detergent and tap water, rinse with 1+1 nitric acid and tap water, then rinse with 1+1 hydrochloric acid and tap water, then rinse with deionized, distilled water in that order. Perform any filtration or acid preservation steps when the sample is collected or as soon as possible thereafter.

STABILITY: Cool samples to 4°C. M.H.T. = 24 Hours.

QUALITY CONTROL: Mixed calibration standards, an instrument check standard and an interference check solution are used in addition to a quality control sample. The quality control sample should be prepared in the same acid matrix as the calibration standards at 10 times the instrumental detection limits and in accordance with the instructions provided by the supplier. Furthermore, two types of blanks are required: a calibration blank and a reagent blank.

REFERENCE: Method 200.7, U.S. EPA, EMSL-Cincinnati, OH, Nov. 1980

PRIMARY NAME: Magnesium (Mg) Method 242.1

TITLE: Metals (Total, Dissolved, Suspended) Direct Aspiration Atomic Absorption (AA)

MATRIX: Drinking, Surface and Saline Waters. Wastewater

APPLICATION: Date issued 1971. Editorial revision 1974 and 1978. Sample is aspirated and atomized in a flame. Light beam from hollow cathode(made of Mg) lamp is directed through the flame into monochromator, then to detector which measures amount absorbed light.

INTERFERENCES: Aluminum interference @ conc's > 2 mg/L masked by lanthanum addition. Phosphate interference is overcome by lanthanum addition. There can be interference from presence of high dissolved solids. Chemical and ionization interferences can occur.

INSTRUMENTATION: AAS. Magnesium (Mg) hollow cathode lamp. Burner. Pipets. Strip chart recorder.

RANGE: 0.02-0.5 mg/L @ 282.5 nm Wavelength MDL: 0.001 mg/L.

PRECISION: SD = +/- (0.1 and 0.2) @ 2.1 and 8.2 mg Mg/L.

ACCURACY: Recoveries = 100% @ 2.1 and 8.2 mg Mg/L.

SAMPLING METHOD: plastic or glass (pre-washed).

STABILITY: Add nitric acid to pH <2. M.H.T. = 6 Months.

QUALITY CONTROL: After calibration curve composed of a minimum of a reagent blank and 3 standards has been prepared, subsequent calibration curves must be verified by use of at least a reagent blank and one standard near MCL. Must check within 10% of original curve. (For drinking water analysis)

REFERENCE: Methods for the Chemical Analysis of Water and Wastes, EPA-600/4-79-020, USEPA, EMSL, 1979.

PRIMARY NAME: Magnesium (Mg) Method 6010

TITLE: Inductively Coupled Plasma (ICP)

MATRIX: Applies to all matrices; ground water, EP extracts, soils, sludges, sediments, aqueous samples, solid and industrial wastes.

APPLICATION: The method covers the determination of 25 metals. Its primary advantage is that ICP instruments allow simultaneous or rapid sequential determination of many elements in a short time. Samples require digestion prior to analysis and are first nebulized and the aerosol is transported to a plasma torch in which element specific atomic line emission spectra are produced by a radio frequency inductively coupled plasma. Background correction is required for trace element detection except in the case of line broadning.

INTERFERENCES: There are spectral, physical and chemical interferences. The primary disadvantage of ICP instruments is background radiation from other elements and the plasma gases (spectral interferences). Changes in sample viscosity and surface tension with samples containing high dissolved solids or high acid concentrations can cause physical interferences. Ionization effects, solute vaporization and molecular compound formation can cause chemical interferences. Calcium, chromium, iron, manganese, thallium and vanadium can cause interference at the 100 mg/L level.

INSTRUMENTATION: Inductively Coupled Argon Plasma Emission Spectroscopy. 279.079 nm Wavelength

RANGE: Not listed MDL: 30 μg/L.

PRECISION: Not listed

ACCURACY: Mean Recovery = 93% +/- 6% of spiked elements for all wastes.

SAMPLING METHOD: Collect 400 mL of sample in plastic or glass containers.

STABILITY: Cool samples to 4°C. M.H.T. = 24 Hours.

QUALITY CONTROL: Mixed calibration standards, an instrument check standard and an interference check solution are used in addition to a quality control sample. The quality control sample should be prepared in the same acid matrix as the calibration standards at 10 times the instrumental detection limits and in accordance with the instructions provided by the supplier. Furthermore, two types of blanks are required: a calibration blank and a reagent blank.

REFERENCE: Method 6010, SW-846, 3rd ed., Nov 1986.

PRIMARY NAME: Magnesium (Mg) Method 7450

TITLE: Atomic Absorption, (AA) Direct Aspiration

MATRIX: Drinking, Surface and Saline Waters. Wastewater

CAS # : 7439-95-4

APPLICATION: Sample is aspirated and atomized in a flame. A light beam from a Mg hollow cathode lamp is directed through the flame into a monochromator and onto a detector. Since wavelength of light beam is specific for Mg, light energy absorbed by detector is measure of magnesium.

INTERFERENCES: The most troublesomee type is chemical, caused by lack of absorption of atoms bound in molecular combination in the flame. High dissolved solids in sample may result in nonatomic absorbance interference. Add lanthanum to prevent complexing problems.

INSTRUMENTATION: Atomic absorption spectrometer. Magnesium hollow cathode lamp. (285.2 nm Wavelength)

RANGE: 0.02-0.05 mg/L MDL: 0.001 mg/L

PRECISION: standard deviation = +/- (0.1 and 0.2) @ (2.1 and 8.2)mg Mg/L

ACCURACY: recoveries = (100 and 100)% @ (2.1 and 8.2)mg Mg/L

SAMPLING METHOD: Use glass or plastic containers. Collect 200 g of solids and 600 mL of liquid samples.

STABILITY: Cool solid samples to 4°C. and analyze as soon as possible. Add nitric acid to liquid samples to pH < 2; M.H.T. = 6 months.

QUALITY CONTROL: At least one duplicate and one spike sample should be run every 20 samples or with each matrix type to verify precision of the method. For 20 or more samples per day, verify working standard curve. Run an additional standard at or near mid-range every 10 samples.

REFERENCE: Method 7450, SW-846, 3rd ed., Nov 1986.

PRIMARY NAME: Manganese (Mn) Method 200.7

TITLE: Inductively Coupled Plasma (ICP)

MATRIX: Dissolved, suspended or total element in drinking and surface waters and in domestic and industrial wastewaters.

APPLICATION: The method covers the determination of 25 metals. Dissolved elements are determined in filtered and acidified samples after appropriate digestion (which increases dissolved solids). Its primary advantage is that ICP instruments allow simultaneous or rapid sequential determination of many elements in a short time. Samples are first nebulized and the aerosol is transported to a plasma torch in which element specific atomic line emission spectra are produced by a radio frequency inductively coupled plasma. Background correction is required for trace element detection except in the case of line broadning.

INTERFERENCES: There are spectral, physical and chemical interferences. The primary disadvantage of ICP instruments is background radiation from other elements and the plasma gases (spectral interferences). Changes in sample viscosity and surface tension with samples containing high dissolved solids (especially those exceeding 1500 mg/L) or high acid concentrations can cause physical interferences. Ionization effects, solute vaporization and molecular compound formation can cause chemical interferences. Aluminum, chromium, iron, and magnesium can cause interference at the 100 mg/L level.

INSTRUMENTATION: Inductively Coupled Argon Plasma Emission Spectroscopy. 257.610 nm Wavelength

RANGE: Not listed MDL: 2 µg/L.

PRECISION: SD = 2.7% Mean @ true value 350 µg/L.

ACCURACY: Mean Recovery = 93% +/- 6% of spiked elements for all wastes.

SAMPLING METHOD: Wash sample container with detergent and tap water, rinse with 1+1 nitric acid and tap water, then rinse with 1+1 hydrochloric acid and tap water, then rinse with deionized, distilled water in that order. Perform any filtration or acid preservation steps when the sample is collected or as soon as possible thereafter.

STABILITY: Cool samples to 4°C. M.H.T. = 24 Hours.

QUALITY CONTROL: Mixed calibration standards, an instrument check standard and an interference check solution are used in addition to a quality control sample. The quality control sample should be prepared in the same acid matrix as the calibration standards at 10 times the instrumental detection limits and in accordance with the instructions provided by the supplier. Furthermore, two types of blanks are required: a calibration blank and a reagent blank.

REFERENCE: Method 200.7, U.S. EPA, EMSL-Cincinnati, OH, Nov. 1980

PRIMARY NAME: Manganese (Mn) Method 243.1

TITLE: Metals (Total, Dissolved, Suspended) Direct Aspiration Atomic Absorption (AA)

MATRIX: Drinking, Surface and Saline Waters. Wastewater.

APPLICATION: Date issued 1971. Editorial revision 1974 and 1978. Sample is aspirated and atomized in a flame. Light beam from hollow cathode (made of Mn) lamp is directed through the flame into monochromator, then to detector which measures amount absorbed light.

INTERFERENCES: Silica interference in determination of manganese is eliminated by adding calcium. There can be interference from presence of high dissolved solids. Chemical and ionization interferences can occur. (Use special procedure when Mn < 25 μg/L).

INSTRUMENTATION: AAS. Manganese (Mn) hollow cathode lamp. Burner. Pipets, strip chart recorder.

RANGE: 0.1-3 mg/L @ 279.5 nm Wavelength MDL: 0.01 mg/L.

PRECISION: SD = 31 mg/L @ 106 μg Mn/L. (70 Labs).

ACCURACY: As bias, -2.1 @ 106 μg Mn/L. (70 Labs).

SAMPLING METHOD: plastic or glass (pre-washed).

STABILITY: Add nitric acid to pH <2. M.H.T. = 6 Months.

QUALITY CONTROL: After calibration curve composed of a minimum of a reagent blank and 3 standards has been prepared, subsequent calibration curves must be verified by use of at least a reagent blank and one standard near MCL. Must check within 10% of original curve. (For drinking water analysis)

REFERENCE: Methods for the Chemical Analysis of Water and Wastes, EPA-600/4-79-020, USEPA, EMSL, 1979.

PRIMARY NAME: Manganese (Mn) Method 6010

TITLE: Inductively Coupled Plasma (ICP)

MATRIX: Applies to all matrices; ground water, EP extracts, soils, sludges, sediments, aqueous samples, solid and industrial wastes.

APPLICATION: The method covers the determination of 25 metals. Its primary advantage is that ICP instruments allow simultaneous or rapid sequential determination of many elements in a short time. Samples require digestion prior to analysis and are first nebulized and the aerosol is transported to a plasma torch in which element specific atomic line emission spectra are produced by a radio frequency inductively coupled plasma. Background correction is required for trace element detection except in the case of line broadning.

INTERFERENCES: There are spectral, physical and chemical interferences. The primary disadvantage of ICP instruments is background radiation from other elements and the plasma gases (spectral interferences). Changes in sample viscosity and surface tension with samples containing high dissolved solids or high acid concentrations can cause physical interferences. Ionization effects, solute vaporization and molecular compound formation can cause chemical interferences. Aluminum, chromium, iron, and magnesium can cause interference at the 100 mg/L level.

INSTRUMENTATION: Inductively Coupled Argon Plasma Emission Spectroscopy. 257.610 nm Wavelength

RANGE: Not listed MDL: 2 μg/L.

PRECISION: SD = 2.7% Mean @ true value 350 μg/L.

ACCURACY: Mean Recovery = 93% +/- 6% of spiked elements for all wastes.

SAMPLING METHOD: Collect 400 mL of sample in plastic or glass containers.

STABILITY: Cool samples to 4°C. M.H.T. = 24 Hours.

QUALITY CONTROL: Mixed calibration standards, an instrument check standard and an interference check solution are used in addition to a quality control sample. The quality control sample should be prepared in the same acid matrix as the calibration standards at 10 times the instrumental detection limits and in accordance with the instructions provided by the supplier. Furthermore, two types of blanks are required: a calibration blank and a reagent blank.

REFERENCE: Method 6010, SW-846, 3rd ed., Nov 1986.

PRIMARY NAME: Manganese (Mn) Method 7460

TITLE: Atomic Absorption, (AA) Direct Aspiration

MATRIX: Drinking, Surface and Saline Waters. Wastewater

CAS # : 7439-96-5

APPLICATION: Sample is aspirated and atomized in a flame. A light beam from a Mn hollow cathode lamp is directed through the flame into a monochromator and onto a detector. Since wavelength of light beam is specific for Mn, light energy absorbed by detector is measure of manganese.

INTERFERENCES: The most troublesomee type is chemical, caused by lack of absorption of atoms bound in molecular combination in the flame. High dissolved solids in sample may result in nonatomic absorbance interference. Background correction is required.

INSTRUMENTATION: Atomic absorption spectrometer. Manganese hollow cathode lamp. [279.5 nm (primary)]

RANGE: 0.1-3 mg/L MDL: 0.01 mg/L

PRECISION: standard deviation = 70 μg/L @ 426 μg/L (true value) 77 labs

ACCURACY: as bias = +1.5% @ 426 μg/L (true value) 77 labs

SAMPLING METHOD: Use glass or plastic containers. Collect 200 g of solids and 600 mL of liquid samples.

STABILITY: Cool solid samples to 4°C. and analyze as soon as possible. Add nitric acid to liquid samples to pH < 2; M.H.T. = 6 months.

QUALITY CONTROL: At least one duplicate and one spike sample should be run every 20 samples or with each matrix type to verify precision of the method. For 20 or more samples per day, verify working standard curve. Run an additional standard at or near mid-range every 10 samples.

REFERENCE: Method 7460, SW-846, 3rd ed., Nov 1986.

PRIMARY NAME: Manganese (Mn) Method 7461

TITLE: Atomic Absorption, (AA) Furnace Technique

MATRIX: Wastes, mobility procedure extracts, soils and groundwater

CAS # : 7439-96-5

APPLICATION: Aqueous samples, EP extracts, industrial wastes, soils, sludges, sediments and solid wastes require digestion before analysis. An aliquot of sample is placed in the graphite tube in the furnace and slowly evaporated, charred and atomized. Absorption of lamp radiation during atomization is proportional to (Mn) concentration.

INTERFERENCES: The furnace technique is subject to chemical interferences. Composition of sample matrix can have major effect on analysis. Modify matrix to remove interferences. Background correction must be used. Cross-contamination may be major error source.

INSTRUMENTATION: Atomic absorption spectrometer. Manganese hollow cathode lamp or electrodeless discharge lamp. Graphite furnace. Strip-chart recorder.

RANGE: 1-30 μg/L

MDL: 0.2 μg/L (279.5 nm Wavelength)

PRECISION: Not listed

ACCURACY: Not listed

SAMPLING METHOD: Use glass or plastic containers. Collect 200 g of solids and 600 mL of liquid samples.

STABILITY: Cool solid samples to 4°C. and analyze as soon as possible. Add nitric acid to liquid samples to pH < 2; M.H.T. = 6 months.

QUALITY CONTROL: At least one duplicate and one spike sample should be run every 20 samples, or with each matrix type to verify method precision. If 20 or more samples are run a day, run a standard (at or near mid-range) every 10 samples.

REFERENCE: Method 7461, SW-846, 3rd ed., (Included as revision 0, December 1987)

PRIMARY NAME: Mercury (Hg)

Method 7470

TITLE: AA, Manual Cold-Vapor Technique

MATRIX: Mobility procedure extracts, wastes, soils and groundwater

CAS # : 7439-97-6

APPLICATION: Before analysis, sample preparation of liquid waste samples is performed. Mercury is reduced to elemental state and aerated from solution in closed system. Mercury vapor passes through a cell positioned in light path of an atomic absorption spectrometer. Absorbance (peak height) is measured as a functioin of Hg concentration.

INTERFERENCES: Add potassium permanganate to eliminate sulfide problems. Concentrations >10 mg/L of copper can interfere. Matrices high in chlorides require extra permanganate since chlorides convert to free chlorine which absorbs at the mercury wavelength. Some volatile organics also absorb at this wavelength.

INSTRUMENTATION: Atomic absorption spectrometer. Mercury hollow cathode lamp or electrodeless discharge lamp. Air pump. Cold-vapor generator.

RANGE: varies with instrumentation.

MDL: 0.0002 mg/L (253.7 nm Wavelength)

PRECISION: standard deviation = 3.57 μg/L @ 9.6 μg/L (true value) 78 labs

ACCURACY: as bias = -5.2% @ 9.6 μg/L (true value) 78 labs

SAMPLING METHOD: Use glass or plastic containers (prewashed). Sample as per chapter 9.

STABILITY: Non-aqueous samples: cool to 4°C and run as soon as possible. Aqueous samples: add nitric acid to pH >2. M.H.T. = 28 days

QUALITY CONTROL: Run one spike duplicate sample for every 10 samples. Verify calibration with an independently prepared check standard every 15 samples.

REFERENCE: Method 7470, SW-846, 3rd ed., Nov 1986.

PRIMARY NAME: Mercury (Hg) Method 7471

TITLE: AA, Manual Cold-vapor Technique

MATRIX: Soils, Sediments, Sludges, Bottom Deposits

CAS # : 7439-97-6

APPLICATION: Before analysis, sample preparation of solid or semisolid waste samples is performed. Mercury is reduced to elemental state and aerated from solution in a closed system. Mercury vapor passes through a cell positioned in light path of an atomic absorption spectrometer. Absorbance (peak height) is measured as a function of Hg concentration.

INTERFERENCES: Add potassium permanganate to eliminate sulfide problems. Concentrations >10 mg/L of copper can interfere. Matrices high in chlorides require extra permanganate since chlorides convert to free chlorine which absorbs at the mercury wavelength. Some volatile organics also absorb at this wavelength. (Remove moisture from solid samples in a 60°C oven before analysis).

INSTRUMENTATION: Atomic absorption spectrometer. Mercury hollow cathode lamp or electrodeless discharge lamp. Air pump. Cold-vapor generator.

RANGE: 0.2-5 μg/g

MDL: 0.0002 μg/L (253.7 nm Wavelength)

PRECISION: standard deviation = +/- 0.02 and 0.03 @ O.29 and 0.82 μg Hg/g

ACCURACY: recoveries = 97 and 94% @ 0.29 and 0.82 μg Hg/g

SAMPLING METHOD: Use glass or plastic containers (prewashed). Sample as per Chapter 9.

STABILITY: Aqueous samples: add nitric acid to pH >2 M.H.T. = 28 days

QUALITY CONTROL: Run one spike duplicate sample for every 10 samples. Verify calibration with a singly prepared check standard every 15 samples.

REFERENCE: Method 7471, SW-846, 3rd ed., Nov 1986.

PRIMARY NAME: Molybdenum (Mo) Method 200.7

TITLE: Inductively Coupled Plasma (ICP)

MATRIX: Dissolved, suspended or total element in drinking and surface waters and in domestic and industrial wastewaters.

APPLICATION: The method covers the determination of 25 metals. Dissolved elements are determined in filtered and acidified samples after appropriate digestion (which increases dissolved solids). Its primary advantage is that ICP instruments allow simultaneous or rapid sequential determination of many elements in a short time. Samples are first nebulized and the aerosol is transported to a plasma torch in which element specific atomic line emission spectra are produced by a radio frequency inductively coupled plasma. Background correction is required for trace element detection except in the case of line broadning.

INTERFERENCES: There are spectral, physical and chemical interferences. The primary disadvantage of ICP instruments is background radiation from other elements and the plasma gases (spectral interferences). Changes in sample viscosity and surface tension with samples containing high dissolved solids (especially those exceeding 1500 mg/L) or high acid concentrations can cause physical interferences. Ionization effects, solute vaporization and molecular compound formation can cause chemical interferences. Aluminum and iron can cause interference at the 100 mg/L level.

INSTRUMENTATION: Inductively Coupled Argon Plasma Emission Spectroscopy. 202.030 nm Wavelength

RANGE: Not listed MDL: 8 µg/L.

PRECISION: Not listed

ACCURACY: Mean Recovery = 93% +/- 6% of spiked elements for all wastes.

SAMPLING METHOD: Wash sample container with detergent and tap water, rinse with 1+1 nitric acid and tap water, then rinse with 1+1 hydrochloric acid and tap water, then rinse with deionized, distilled water in that order. Perform any filtration or acid preservation steps when the sample is collected or as soon as possible thereafter.

STABILITY: Cool samples to 4°C. M.H.T. = 24 Hours.

QUALITY CONTROL: Mixed calibration standards, an instrument check standard and an interference check solution are used in addition to a quality control sample. The quality control sample should be prepared in the same acid matrix as the calibration standards at 10 times the instrumental detection limits and in accordance with the instructions provided by the supplier. Furthermore, two types of blanks are required: a calibration blank and a reagent blank.

REFERENCE: Method 200.7, U.S. EPA, EMSL-Cincinnati, OH, Nov. 1980

PRIMARY NAME: Molybdenum (Mo) Method 6010

TITLE: Inductively Coupled Plasma (ICP)

MATRIX: Applies to all matrices; ground water, EP extracts, soils, sludges, sediments, aqueous samples, solid and industrial wastes.

APPLICATION: The method covers the determination of 25 metals. Its primary advantage is that ICP instruments allow simultaneous or rapid sequential determination of many elements in a short time. Samples require digestion prior to analysis and are first nebulized and the aerosol is transported to a plasma torch in which element specific atomic line emission spectra are produced by a radio frequency inductively coupled plasma. Background correction is required for trace element detection except in the case of line broadning.

INTERFERENCES: There are spectral, physical and chemical interferences.

The primary disadvantage of ICP instruments is background radiation from other elements and the plasma gases (spectral interferences). Changes in sample viscosity and surface tension with samples containing high dissolved solids or high acid concentrations can cause physical interferences. Ionization effects, solute vaporization and molecular compound formation can cause chemical interferences. Aluminum and iron can cause interference at the 100 mg/L level.

INSTRUMENTATION: Inductively Coupled Argon Plasma Emission Spectroscopy. 202.030 nm Wavelength

RANGE: Not listed MDL: 8 μg/L.

PRECISION: Not listed

ACCURACY: Mean Recovery = 93% +/- 6% of spiked elements for all wastes.

SAMPLING METHOD: Collect 400 mL of sample in plastic or glass containers.

STABILITY: Cool samples to 4°C. M.H.T. = 24 Hours.

QUALITY CONTROL: Mixed calibration standards, an instrument check standard and an interference check solution are used in addition to a quality control sample. The quality control sample should be prepared in the same acid matrix as the calibration standards at 10 times the instrumental detection limits and in accordance with the instructions provided by the supplier. Furthermore, two types of blanks are required: a calibration blank and a reagent blank.

REFERENCE: Method 6010, SW-846, 3rd ed., Nov 1986.

PRIMARY NAME: Molybdenum (Mo) Method 7480

TITLE: Atomic Absorption, (AA) Direct Aspiration

MATRIX: Drinking, Surface and Saline Waters. Wastewater

CAS # : 7439-98-7

APPLICATION: Sample is aspirated and atomized in a flame. A light beam from a Mo hollow cathode lamp is directed through the flame into a monochromator and onto a detector. Since wavelength of light beam is specific for Mo, light energy absorbed by detector is measure of molybdenum.

INTERFERENCES: The most troublesomee type is chemical,caused by lack of absorption of atoms bound in molecular combination in the flame. High dissolved solids in sample may result in nonatomic absorbance interference. Addition of aluminum greatly reduces flame interference.

INSTRUMENTATION: Atomic absorption spectrometer. Molybdenum hollow cathode lamp. (313.3 nm Wavelength)

RANGE: 1-40 mg/L MDL: 0.1 mg/L

PRECISION: standard deviation = +/- 0.007, 0.02, 0.07 @ 0.30, 1.5, 7.5 mg Mo/L

ACCURACY: recoveries = 100, 96, 95% @ 0.30, 1.5, 7.5 mg Mo/L

SAMPLING METHOD: Use glass or plastic containers. Collect 200 g of solids and 600 mL of liquid samples.

STABILITY: Cool solid samples to 4°C. and analyze as soon as possible. Add nitric acid to liquid samples to pH < 2; M.H.T. = 6 months.

QUALITY CONTROL: At least one duplicate and one spike sample should be run every 20 samples or with each matrix type to verify precision of the method. For 20 or more samples per day, verify working standard curve. Run an additional standard at or near mid-range every 10 samples.

REFERENCE: Method 7480, SW-846, 3rd ed., Nov 1986.

PRIMARY NAME: Molybdenum (Mo) Method 7481

TITLE: Atomic Absorption, (AA) Furnace Technique

MATRIX: Wastes, mobility procedure extracts, soils and groundwater

CAS # : 7439-98-7

APPLICATION: Aqueous samples, EP extracts, industrial wastes, soils, sludges, sediments and solid wastes require digestion before analysis. An aliquot of sample is placed in the graphite tube in the furnace and slowly evaporated, charred and atomized. Absorption of lamp radiation during atomization is proportional to molybdenum concentration.

INTERFERENCES: The furnace technique is subject to chemical interferences. Composition of sample matrix can effect analysis. Molybdenum is prone to carbide formation. Use pyrolitically coated graphite tube. Memory effects are possible; clean furnace after concentration sample analysis.

INSTRUMENTATION: Atomic absorption spectrometer. Molybdenum hollow cathode lamp or electrodeless discharge lamp. Graphite furnace. Strip-chart recorder

RANGE: 3-60 μg/L

MDL: 1 μg/L (313.3 nm Wavelength)

PRECISION: Not listed

ACCURACY: Not listed

SAMPLING METHOD: Use glass or plastic containers. Collect 200 g of solids and 600 mL of liquid samples.

STABILITY: Cool solid samples to 4°C. and analyze as soon as possible. Add nitric acid to liquid samples to pH < 2; M.H.T. = 6 months.

QUALITY CONTROL: At least one duplicate and one spike sample should be run every 20 samples, or with each matrix type to verify method precision. If 20 or more samples are run a day, run a standard (at or near mid-range) every 10 samples.

REFERENCE: Method 7481, SW-846, 3rd ed., Nov 1986.

PRIMARY NAME: Nickel (Ni)

Method 200.7

TITLE: Inductively Coupled Plasma (ICP)

MATRIX: Dissolved, suspended or total element in drinking and surface waters and in domestic and industrial wastewaters.

APPLICATION: The method covers the determination of 25 metals. Dissolved elements are determined in filtered and acidified samples after appropriate digestion (which increases dissolved solids). Its primary advantage is that ICP instruments allow simultaneous or rapid sequential determination of many elements in a short time. Samples are first nebulized and the aerosol is transported to a plasma torch in which element specific atomic line emission spectra are produced by a radio frequency inductively coupled plasma. Background correction is required for trace element detection except in the case of line broadning.

INTERFERENCES: There are spectral, physical and chemical interferences. The primary disadvantage of ICP instruments is background radiation from other elements and the plasma gases (spectral interferences). Changes in sample viscosity and surface tension with samples containing high dissolved solids (especially those exceeding 1500 mg/L) or high acid concentrations can cause physical interferences. Ionization effects, solute vaporization and molecular compound formation can cause chemical interferences. No other elements cause interference at the 100 mg/L level.

INSTRUMENTATION: Inductively Coupled Argon Plasma Emission Spectroscopy. 231.604 nm Wavelength

RANGE: Not listed MDL: 15 µg/L.

PRECISION: SD = 5.8% Mean @ true value 250 µg/L.

ACCURACY: Mean Recovery = 93% +/- 6% of spiked elements for all wastes.

SAMPLING METHOD: Wash sample container with detergent and tap water, rinse with 1+1 nitric acid and tap water, then rinse with 1+1 hydrochloric acid and tap water, then rinse with deionized, distilled water in that order. Perform any filtration or acid preservation steps when the sample is collected or as soon as possible thereafter.

STABILITY: Cool samples to 4°C. M.H.T. = 24 Hours.

QUALITY CONTROL: Mixed calibration standards, an instrument check standard and an interference check solution are used in addition to a quality control sample. The quality control sample should be prepared in the same acid matrix as the calibration standards at 10 times the instrumental detection limits and in accordance with the instructions provided by the supplier.

Furthermore, two types of blanks are required: a calibration blank and a reagent blank.

REFERENCE: Method 200.7, U.S. EPA, EMSL-Cincinnati, OH, Nov. 1980

PRIMARY NAME: Nickel (Ni) Method 6010

TITLE: Inductively Coupled Plasma (ICP)

MATRIX: Applies to all matrices; ground water, EP extracts, soils, sludges, sediments, aqueous samples, solid and industrial wastes.

APPLICATION: The method covers the determination of 25 metals. Its primary advantage is that ICP instruments allow simultaneous or rapid sequential determination of many elements in a short time. Samples require digestion prior to analysis and are first nebulized and the aerosol is transported to a plasma torch in which element specific atomic line emission spectra are produced by a radio frequency inductively coupled plasma. Background correction is required for trace element detection except in the case of line broadning.

INTERFERENCES: There are spectral, physical and chemical interferences. The primary disadvantage of ICP instruments is background radiation from other elements and the plasma gases (spectral interferences). Changes in sample viscosity and surface tension with samples containing high dissolved solids or high acid concentrations can cause physical interferences. Ionization effects, solute vaporization and molecular compound formation can cause chemical interferences. No other elements cause interference at the 100 mg/L level.

INSTRUMENTATION: Inductively Coupled Argon Plasma Emission Spectroscopy. 231.604 nm Wavelength

RANGE: Not listed

MDL: 15 μg/L.

PRECISION: SD = 5.8% Mean @ true value 250 μg/L.

ACCURACY: Mean Recovery = 93% +/- 6% of spiked elements for all wastes.

SAMPLING METHOD: Collect 400 mL of sample in plastic or glass containers.

STABILITY: Cool samples to 4°C. M.H.T. = 24 Hours.

QUALITY CONTROL: Mixed calibration standards, an instrument check standard and an interference check solution are used in addition to a quality control sample. The quality control sample should be prepared in the same acid matrix as the calibration standards at 10 times the instrumental detection limits and in accordance with the instructions provided by the supplier. Furthermore, two types of blanks are required: a calibration blank and a reagent blank.

REFERENCE: Method 6010, SW-846, 3rd ed., Nov 1986.

PRIMARY NAME: Nickel (Ni) Method 7520

TITLE: Atomic Absorption, (AA) Direct Aspiration

MATRIX: Drinking, surface and saline saters. Wastewater

CAS # : 7440-02-0

APPLICATION: Sample is aspirated and atomized in a flame. A light beam from a Ni hollow cathode lamp is directed through the flame into a monochromator and onto a detector. Since wavelength of light beam is specific for Ni, light energy absorbed by detector is measure of nickel.

INTERFERENCES: The most troublesomee type is chemical, caused by lack of absorption of atoms bound in molecular combination in the flame. High dissolved solids in sample may result in nonatomic absorbance interference. High concentrations of iron, cobalt and chromium can interfere.

INSTRUMENTATION: Atomic absorption spectrometer. Nickel hollow cathode lamp. [232.0 nm (primary)].

RANGE: 0.3-5 mg/L MDL: 0.04 mg/L

PRECISION: standard deviation = +/- 0.011, 0.02, 0.04 @ 0.20, 1.5, 7.5 mg Mo/L

ACCURACY: recoveries = 100, 97, 93% @ 0.20, 1.5, 7.5 mg Mo/L

SAMPLING METHOD: Use glass or plastic containers. Collect 200 g of solids and 600 mL of liquid samples.

STABILITY: Cool solid samples to 4°C. and analyze as soon as possible. Add nitric acid to liquid samples to pH < 2; M.H.T. = 6 months.

QUALITY CONTROL: At least one duplicate and one spike sample should be run every 20 samples or with each matrix type to verify precision of the method. For 20 or more samples per day, verify working standard curve. Run an additional standard at or near mid-range every 10 samples.

REFERENCE: Method 7520, SW-846, 3rd ed., Nov 1986.

PRIMARY NAME: Osmium (Os) Method 7550

TITLE: Atomic Absorption, (AA) Direct Aspiration

MATRIX: Wastes, mobility procedure extracts, soils and groundwater

CAS # : 7440-04-2

APPLICATION: Prior to analysis, samples must be prepared for direct aspiration. Sample preparation method varies with the matrix. Following appropriate dissolution of the sample, an osmium lamp light beam is directed through an aspirated aliquot in a flame, into a monochromator and detector. Due to extreme toxicity of osmium and its compounds extreme care must be taken to handle it safely.

INTERFERENCES: Background correction is required. Monitor samples and standards for viscosity differences since viscosity may alter aspiration rate.

INSTRUMENTATION: Atomic absorption spectrometer. Osmium hollow cathode lamp. (290.0 nm Wavelength). Strip-chart recorder.

RANGE: 2-100 mg/L MDL: 0.3 mg/L

PRECISION: Not listed

ACCURACY: Not listed

SAMPLING METHOD: Use plastic or glass containers (prewashed). Sample as per chapter 9.

STABILITY: Non aqueous samples: cool to 4°C and analyze as soon as possible. Aqueous samples: add nitric acid to pH <2. M.H.T. = 6 Months

QUALITY CONTROL: Run one spike duplicate sample for every 10 samples. Verify calibration with an independently prepared check standard every 15 samples. Due to the very volatile nature of some osmium compounds, the applicable method must be verified by spiked samples or standard reference materials or both. Make osmium standards on daily basis, because of osmium volatility.

REFERENCE: Method 7550, SW-846, 3rd ed., Nov 1986.

PRIMARY NAME: Oxygen (O_2) Method 360.1

TITLE: Inorganics, Non-Metallics Dissolved Oxygen (DO)

MATRIX: Wastewaters and Streams.

APPLICATION: Date issued 1971. (Membrane electrode). This probe method is recommended for those samples containing materials which interfere with modified Winkler procedure. It is recommended for monitoring streams, lakes, outfalls, etc., with continuous record of DO.

INTERFERENCES: Probes with membranes respond to partial pressure of oxygen which in turn is a function of dissolved organic salts. Conversion factors for sea and brackish waters may be calculated. (Conversion factors for specific inorganic salts may be determined experimentally)

INSTRUMENTATION: Dissolved oxygen probe. Weston and Stack, YSI or Beckman.

RANGE: Not listed

MDL: Not listed

PRECISION: Manufacturer's specifications claim 0.1 mg/L repeatability.

ACCURACY: Manufacturer's specifications claim 0.1 mg/L repeatability.

SAMPLING METHOD: Glass container only (both bottle and top). Sample collection from shallow depths (less than 5 ft), use an apha type sampler. A Kemmerer type sampler is recommended for samples collected at depths > 5 ft. Fill 300 mL bottle to overflowing.

STABILITY: No preservation required. Analyze immediately.

QUALITY CONTROL: Record temperature at time of sampling.

REFERENCE: Methods for the Chemical Analysis of Water and Wastes, EPA-600/4-79-020, USEPA, EMSL, 1979.

PRIMARY NAME: Oxygen (O_2) Method 360.2

TITLE: Inorganics, Non-Metallics Dissolved Oxygen (DO)

MATRIX: Wastewater and Streams.

APPLICATION: Date issued 1971. (Modified Winkler, full bottle technique). This Method is applicable for use with most wastewaters and streams that contain nitrate nitrogen and not more than 1 mg/L ferrous iron. The DO probe technique gives comparable results.

INTERFERENCES: There are a number of interferences to the dissolved oxygen test, including oxidizing and reducing agents, nitrate ion, ferrous iron and organic matter. Most common interferences in the Winkler procedure may be overcome by use of the DO probe.

INSTRUMENTATION: A titration with 0.0375N sodium thiosulfate to a starch-iodine end point.

RANGE: Not listed

MDL: Not listed

PRECISION: exact data unavailable.

ACCURACY: exact data unavailable.

SAMPLING METHOD: glass container only (both bottle and top). Sample collection from shallow depths (less than 5 ft) and use an apha type sampler. A Kemmerer type sampler is recommended for samples collected at depths > 5ft. Fill a 300 mL bottle to overflowing.

STABILITY: Fix on site and store in dark. M.H.T. = 8 Hours.

QUALITY CONTROL: Record temperature at time of sampling.

REFERENCE: Methods for the Chemical Analysis of Water and Wastes, EPA-600/4-79-020, USEPA, EMSL, 1979.

PRIMARY NAME: Phosphorus (P) Method 365.1

TITLE: Inorganics, Non-Metallics

MATRIX: Drinking, Surface and Saline Waters. Wastewater.

APPLICATION: Date issued 1971. Editorial revision 1974 and 1978. (Colorimetric, automated, ascorbic acid). Applies to specified forms of phosphorus (P), based on reactions for the orthophosphate ion. Most analyses are run for phosphorus and dissolved P, orthophosphate, and dissolved orthophosphate.

INTERFERENCES: High iron concentrations cause precipitation of and loss of phosphorus. Sample turbidity must be removed by filtration prior to analysis for orthophosphate. Salt error for samples 5 to 20% salt, <1%. Arsenic concentrations > phosphorus concentration, may interfere.

INSTRUMENTATION: Technicon auto analyzer. 650-660 or 880 nm filter.

RANGE: 0.001-1.0 mg P/L range.

MDL: Not listed

PRECISION: 0.066 mg P/L @ 0.30 mg P/L (as orthophosphate)

ACCURACY: as bias, -0.04 mg P/L @ 0.30 mg P/L (as orthophosphate).

SAMPLING METHOD: plastic or glass. (50 mL).

STABILITY: Cool, 4°C. Add sulfuric acid to pH <2. M.H.T. = 28 Days.

QUALITY CONTROL: This method is based on reactions specific for the orthophosphate ion in which an antimony-phospho-molybdate complex is reduced to an intensely blue colored complex by ascorbic acid. Color is proportional to phosphorus concentration. Measure color on auto analyzer

REFERENCE: Methods for the Chemical Analysis of Water and Wastes, EPA-600/4-79-020, USEPA, EMSL, 1979.

PRIMARY NAME: Phosphorus (P) Method 365.2

TITLE: Inorganics, Non-Metallics MATRIX: Drinking, Surface and Saline Waters. Wastewater.

APPLICATION: Date issued 1971. (Colorimetric, ascorbic acid, single reagent). Applies to specified forms of phosphorus (P), based on reactions for orthophosphate ion. Most commonly measured forms are; phosphorus and dissolved P, orthophosphate and dissolved orthophosphate.

INTERFERENCES: High iron concentrations cause precipitation of and loss of phosphorus. Salt error for samples 5 to 20% salt, <1%. Arsenic concentration> phosphorus concentration, may interfere. (Only orthophosphate turns blue color in this test. Other (P) forms are converted to orthophosphate).

INSTRUMENTATION: Spectro(or filter)photometer. 650 or 880 nm. Light path of 1 cm or longer.

RANGE: 0.01-0.5 mg P/L. MDL: Not listed

PRECISION: SD = 0.018 mg P/L @ 0.335 mg P/L (as orthophosphate)

ACCURACY: As bias, -0.009 mg P/L L 0.335 mg P/L (as orthophosphate).

SAMPLING METHOD: plastic or glass (50 mL).

STABILITY: Cool, 4°C. Add sulfuric acid to pH <2. M.H.T. = 28 Days.

QUALITY CONTROL: This Method is based on reactions specific for the orthophosphate ion in which an antimony-phospho-molybdate complex is reduced to an intensely blue colored complex by ascorbic acid. Color is proportional to phosphorus concentration. Measure color on auto analyzer.

REFERENCE: Methods for the Chemical Analysis of Water and Wastes, EPA-600/4-79-020, USEPA, EMSL, 1979.

PRIMARY NAME: Phosphorus (P) Method 365.3

TITLE: Inorganics, Non-Metallics

MATRIX: Drinking, Surface and Saline Waters. Wastewater.

APPLICATION: Date issued 1978. (Colorimetric, ascorbic acid, two reagent). Applies to specified forms of phosphorus (P), based on reactions for the orthophosphate ion. Most commonly measured forms are; phosphorus and dissolved P, orthophosphate and dissolved orthophosphate.

INTERFERENCES: Arsenate interference is eliminated by reducing arsenic acid to arsenious acid using sodium bisulfite. High concentrations of iron cause low phosphorus recovery. The bisulfite treatment also eliminates this interference.

INSTRUMENTATION: Spectro(or filter)photometer. 660 or 880 nm. Light path of 1 cm or longer.

RANGE: 0.01-1.2 mg P/L.

MDL: Not listed

PRECISION: Not listed

ACCURACY: Recoveries = 99 and 100% @ 7.6 and 0.55 mg P/L (waste and sewage)

SAMPLING METHOD: plastic or glass (50 mL).

STABILITY: Cool, 4°C. Add sulfuric acid to pH < 2. M.H.T. = 28 Days.

QUALITY CONTROL: This Method is based on reactions specific for the orthophosphate ion in which an antimony-phospho-molybdate complex is reduced to an intensely blue colored complex by ascorbic acid. Color is proportional to phosphorus concentration. Measure color in auto analyzer.

REFERENCE: Methods for the Chemical Analysis of Water and Wastes, EPA-600/4-79-020, USEPA, EMSL, 1979.

PRIMARY NAME: Phosphorus (P) Method 365.4

TITLE: Inorganics, Non-Metallics MATRIX: Drinking, Surface and Waste Waters.

APPLICATION: Date issued 1974. (Colorimetric, automated, block digestor AAII). Sample is heated in presence of sulfuric acid, potassium sulfate and mercuric sulfate for 2 1/2 hours. Residue is cooled, diluted to 25 mL and placed on auto analyzer for phosphorus determination. Temperature of block digester during 2 1/2 hour digestion is 380°C.

INTERFERENCES: Only add 4-8 Teflon boiling chips during digestion. Too many boiling chips will cause sample to boil over.

INSTRUMENTATION: Block digestor BD-40. Technicon auto analyzer and method no.327-74W.

RANGE: 0.01-20 mg P/L. MDL: Not Listed

PRECISION: SD = +/- 0.06 @ 2.0 mg P/L (sewage sample)

ACCURACY: Recoveries = 95 and 98% @ 1.84 and 1.89 mg P/L (sewage samples)

SAMPLING METHOD: Plastic or glass (50 mL).

STABILITY: Cool, 4°C. Add sulfuric acid to pH < 2. M.H.T. = 28 Days.

QUALITY CONTROL: This Method covers the determination of total phosphorus. Prepare standard curve plotting peak heights of standards against concentration values. Compare sample peak heights with standard curve to compute concentration. (If TKN is determined, the sample should be diluted with ammonia-free water).

REFERENCE: Methods for the Chemical Analysis of Water and Wastes, EPA-600/4-79-020, USEPA, EMSL, 1979.

PRIMARY NAME: Potassium (K)

Method 200.7

TITLE: Inductively Coupled Plasma (ICP)

MATRIX: Dissolved, suspended or total element in drinking and surface waters and in domestic and industrial wastewaters.

APPLICATION: The method covers the determination of 25 metals. Dissolved elements are determined in filtered and acidified samples after appropriate digestion (which increases dissolved solids). Its primary advantage is that ICP instruments allow simultaneous or rapid sequential determination of many elements in a short time. Samples are first nebulized and the aerosol is transported to a plasma torch in which element specific atomic line emission spectra are produced by a radio frequency inductively coupled plasma. Background correction is required for trace element detection except in the case of line broadning.

INTERFERENCES: There are spectral, physical and chemical interferences. The primary disadvantage of ICP instruments is background radiation from other elements and the plasma gases (spectral interferences). Changes in sample viscosity and surface tension with samples containing high dissolved solids (especially those exceeding 1500 mg/L) or high acid concentrations can cause physical interferences. Ionization effects, solute vaporization and molecular compound formation can cause chemical interferences.

INSTRUMENTATION: Inductively Coupled Argon Plasma Emission Spectroscopy. 766.491 nm Wavelength

RANGE: Not listed

MDL: highly dependent on operating conditions and plasma position.

PRECISION: Not listed

ACCURACY: Mean Recovery = 93% +/- 6% of spiked elements for all wastes.

SAMPLING METHOD: Wash sample container with detergent and tap water, rinse with 1+1 nitric acid and tap water, then rinse with 1+1 hydrochloric acid and tap water, then rinse with deionized, distilled water in that order. Perform any filtration or acid preservation steps when the sample is collected or as soon as possible thereafter.

STABILITY: Cool samples to 4°C. M.H.T. = 24 Hours.

QUALITY CONTROL: Mixed calibration standards, an instrument check standard and an interference check solution are used in addition to a quality control sample. The quality control sample should be prepared in the same acid matrix as the calibration standards at 10 times the instrumental detection limits and in accordance with the instructions provided by the supplier. Furthermore, two types of blanks are required: a calibration blank and a reagent blank.

REFERENCE: Method 200.7, U.S. EPA, EMSL-Cincinnati, OH, Nov. 1980

PRIMARY NAME: Potassium (K) Method 6010

TITLE: Inductively Coupled Plasma (ICP)

MATRIX: Applies to all matrices; ground water, EP extracts, soils, sludges, sediments, aqueous samples, solid and industrial wastes.

APPLICATION: The method covers the determination of 25 metals. Its primary advantage is that ICP instruments allow simultaneous or rapid sequential determination of many elements in a short time. Samples require digestion prior to analysis and are first nebulized and the aerosol is transported to a plasma torch in which element specific atomic line emission spectra are produced by a radio frequency inductively coupled plasma. Background correction is required for trace element detection except in the case of line broadning.

INTERFERENCES: There are spectral, physical and chemical interferences. The primary disadvantage of ICP instruments is background radiation from other elements and the plasma gases (spectral interferences). Changes in sample viscosity and surface tension with samples containing high dissolved solids or high acid concentrations can cause physical interferences. Ionization effects, solute vaporization and molecular compound formation can cause chemical interferences.

INSTRUMENTATION: Inductively Coupled Argon Plasma Emission Spectroscopy. 766.491 nm Wavelength

RANGE: Not listed

MDL: highly dependent on operating conditions and plasma position.

PRECISION: Not listed

ACCURACY: Mean Recovery = 93% +/- 6% of spiked elements for all wastes.

SAMPLING METHOD: Collect 400 mL of sample in plastic or glass containers.

STABILITY: Cool samples to 4°C. M.H.T. = 24 Hours.

QUALITY CONTROL: Mixed calibration standards, an instrument check standard and an interference check solution are used in addition to a quality control sample. The quality control sample should be prepared in the same acid matrix as the calibration standards at 10 times the instrumental detection limits and in accordance with the instructions provided by the supplier. Furthermore, two types of blanks are required: a calibration blank and a reagent blank.

REFERENCE: Method 6010, SW-846, 3rd ed., Nov 1986.

PRIMARY NAME: Potassium (K) Method 7610

TITLE: Atomic Absorption, (AA) Direct Aspiration

MATRIX: Drinking, Surface and Saline Waters. Wastewater

CAS # : 7440-09-7

APPLICATION: Sample is aspirated and atomized in a flame. A light beam from a (K) hollow cathode lamp is directed through the flame into monochromator and onto detector. Since wavelength of light beam is specific for potassium, light energy absorbed by detector is measure of potassium.

INTERFERENCES: The most troublesomee type is chemical, caused by lack of absorption of atoms bound in molecular combination in the flame. High dissolved solids in sample may result in nonatomic absorbance interference. Other alkali salts in the sample and sodium can interfere.

INSTRUMENTATION: Atomic absorption spectrometer. Potassium hollow cathode lamp. (766.5 nm Wavelength)

RANGE: 0.1-2 mg/L MDL: 0.01 mg/L

PRECISION: standard deviation = +/- 0.20 and 0.50 @ 1.60 and 6.30 mg K/L

ACCURACY: recoveries = 103 and 102% @ 1.60 and 6.30 mg K/L

SAMPLING METHOD: Use glass or plastic containers. Collect 200 g of solids and 600 mL of liquid samples.

STABILITY: Cool solid samples to 4°C. and analyze as soon as possible. Add nitric acid to liquid samples to pH < 2; M.H.T. = 6 months.

QUALITY CONTROL: At least one duplicate and one spike sample should be run every 20 samples or with each matrix type to verify precision of the method. For 20 or more samples per day, verify working standard curve. Run an additional standard at or near mid-range every 10 samples.

REFERENCE: Method 7610, SW-846, 3rd ed., Nov 1986.

PRIMARY NAME: Selenium (Se) Method 200.7

TITLE: Inductively Coupled Plasma (ICP)

MATRIX: Dissolved, suspended or total element in drinking and surface waters and in domestic and industrial wastewaters.

APPLICATION: The method covers the determination of 25 metals. Dissolved elements are determined in filtered and acidified samples after appropriate digestion (which increases dissolved solids). Its primary advantage is that ICP instruments allow simultaneous or rapid sequential determination of many elements in a short time. Samples are first nebulized and the aerosol is transported to a plasma torch in which element specific atomic line emission spectra are produced by a radio frequency inductively coupled plasma. Background correction is required for trace element detection except in the case of line broadning.

INTERFERENCES: There are spectral, physical and chemical interferences. The primary disadvantage of ICP instruments is background radiation from other elements and the plasma gases (spectral interferences). Changes in sample viscosity and surface tension with samples containing high dissolved solids (especially those exceeding 1500 mg/L) or high acid concentrations can cause physical interferences. Ionization effects, solute vaporization and molecular compound formation can cause chemical interferences. Aluminum and iron can cause interference at the 100 mg/L level.

INSTRUMENTATION: Inductively Coupled Argon Plasma Emission Spectroscopy. 196.026 nm Wavelength

RANGE: Not listed MDL: 75 µg/L.

PRECISION: SD = 21.9% Mean @ true value 40 µg/L (results from only 2 labs).

ACCURACY: Mean Recovery = 93% +/- 6% of spiked elements for all wastes.

SAMPLING METHOD: Wash sample container with detergent and tap water, rinse with 1+1 nitric acid and tap water, then rinse with 1+1 hydrochloric acid and tap water, then rinse with deionized, distilled water in that order. Perform any filtration or acid preservation steps when the sample is collected or as soon as possible thereafter.

STABILITY: Cool samples to 4°C. M.H.T. = 24 Hours.

QUALITY CONTROL: Mixed calibration standards, an instrument check standard and an interference check solution are used in addition to a quality control sample. The quality control sample should be prepared in the same acid matrix as the calibration standards at 10 times the instrumental detection limits and in accordance with the instructions provided by the supplier. Furthermore, two types of blanks are required: a calibration blank and a reagent blank.

REFERENCE: Method 200.7, U.S. EPA, EMSL-Cincinnati, OH, Nov. 1980

PRIMARY NAME: Selenium (Se) Method 6010

TITLE: Inductively Coupled Plasma (ICP)

MATRIX: Applies to all matrices; ground water, EP extracts, soils, sludges, sediments, aqueous samples, solid and industrial wastes.

APPLICATION: The method covers the determination of 25 metals. Its primary advantage is that ICP instruments allow simultaneous or rapid sequential determination of many elements in a short time. Samples require digestion prior to analysis and are first nebulized and the aerosol is transported to a plasma torch in which element specific atomic line emission spectra are produced by a radio frequency inductively coupled plasma. Background correction is required for trace element detection except in the case of line broadning.

INTERFERENCES: There are spectral, physical and chemical interferences. The primary disadvantage of ICP instruments is background radiation from other elements and the plasma gases (spectral interferences). Changes in sample viscosity and surface tension with samples containing high dissolved solids or high acid concentrations can cause physical interferences. Ionization effects, solute vaporization and molecular compound formation can cause chemical interferences. Aluminum and iron can cause interference at the 100 mg/L level.

INSTRUMENTATION: Inductively Coupled Argon Plasma Emission Spectroscopy. 196.026 nm Wavelength

RANGE: Not listed MDL: 75 μg/L.

PRECISION: SD = 21.9% Mean @ true value 40 μg/L (results from only 2 labs).

ACCURACY: Mean Recovery = 93% +/- 6% of spiked elements for all wastes.

SAMPLING METHOD: Collect 400 mL of sample in plastic or glass containers.

STABILITY: Cool samples to 4°C. M.H.T. = 24 Hours.

QUALITY CONTROL: Mixed calibration standards, an instrument check standard and an interference check solution are used in addition to a quality control sample. The quality control sample should be prepared in the same acid matrix as the calibration standards at 10 times the instrumental detection limits and in accordance with the instructions provided by the supplier. Furthermore, two types of blanks are required: a calibration blank and a reagent blank.

REFERENCE: Method 6010, SW-846, 3rd ed., Nov 1986.

PRIMARY NAME: Selenium (Se) Method 7740

TITLE: Atomic Absorption, (AA) Furnace Technique

MATRIX: Wastes, mobility procedure extracts, soils and groundwater

CAS # : 7782-49-2

APPLICATION: Sample preparation converts organic Se to inorganic forms. Sample preparation varies with the matrix. An aliquot of digestate is placed in a graphite tube in the furnace and slowly evaporated, charred and atomized. Absorption of lamp radiation during atomization is proportional to selenium concentration.

INTERFERENCES: Elemental selenium and many of its compounds are volatile so there may be losses in selenium during sample preparation. There can be severe nonspecific absorption and light scattering caused by matrix components during atomization. Memory effects occur if selenium is not volatilized and removed from the furnace. Use of low wavelength (196.0 nm) makes selenium analysis susceptible to analytical problems.

INSTRUMENTATION: Atomic absorption spectrometer. Selenium hollow cathode lamp or electrodeless discharge lamp. Graphite furnace. Strip-chart recorder

RANGE: 5-100 µg/L

MDL: 2 µg/L

PRECISION: standard deviation = +/- 0.60, 0.40, 0.50 @ 5.0, 10, 20 µg Se/L

ACCURACY: recoveries = 92, 98, 100% @ 5.0, 10, 20 µg Se/L

SAMPLING METHOD: Use plastic or glass containers (prewashed). Sample as per chapter 9.

STABILITY: Non aqueous samples: cool to 4°C. Aqueous samples: add nitric acid to pH <2. M.H.T. = 6 Months

QUALITY CONTROL: Run one spike duplicate sample for every 10 samples. Verify calibration with an independently prepared check standard every 15 samples.

REFERENCE: Method 7740, SW-846, 3rd ed., Nov 1986.

PRIMARY NAME: Selenium (Se) Method 7741

TITLE: Atomic Absorption, (AA) Gaseous Hydride

MATRIX: Wastes, mobility procedure extracts, soils and groundwater

CAS # : 7782-49-2

APPLICATION: Method approved only for sample matrices without high concentrations of Cr, Cu, Hg, Ni, Ag, Co and Mo. After sample preparation with HNO_3/H_2SO_4 digestion, Se in digestate is reduced to Se(IV) with $SnCl_2$. Se(IV) is then converted to a volatile hydride using hydrogen and is swept into an argon-hydrogen flame located in the optical path of an atomic absorption spectrometer. Absorption of the lamp radiation is proportional to the selenium concentration.

INTERFERENCES: Traces of nitric acid left following sample work-up can result in analytical interferences. Elemental Se and many of its compounds are volatile; may be subject to losses during sample preparation. High concentrations of Cr, Cu, Hg, Ni, Ag, Co and Mo cause analytical interferences.

INSTRUMENTATION: Atomic absorption spectrometer. Selenium (Se) hollow cathode lamp or electrodeless discharge lamp. Burner (for argon-hydrogen flame)

RANGE: 2-20 µg/L MDL: 0.002 mg/L

PRECISION: standard deviation = +/- 1.1 @ 10 µg/L on selenium oxide solution.

ACCURACY: Recovery = 100% @ 10 µg/L on selenium oxide solution.

SAMPLING METHOD: Use plastic or glass containers (prewashed). Collect 100 mL of sample.

STABILITY: Add nitric acid to pH <2. M.H.T. = 6 Months.

QUALITY CONTROL: Run one spike duplicate sample for every 10 samples. Verify calibration with an independently prepared check standard every 15 samples.

REFERENCE: Method 7741, SW-846, 3rd ed., Nov 1986.

PRIMARY NAME: Silicon (Si) Method 200.7

TITLE: Inductively Coupled Plasma (ICP)

MATRIX: Dissolved, suspended or total element in drinking and surface waters and in domestic and industrial wastewaters.

APPLICATION: The method covers the determination of 25 metals. Dissolved elements are determined in filtered and acidified samples after appropriate digestion (which increases dissolved solids). Its primary advantage is that ICP instruments allow simultaneous or rapid sequential determination of many elements in a short time. Samples are first nebulized and the aerosol is transported to a plasma torch in which element specific atomic line emission spectra are produced by a radio frequency inductively coupled plasma. Background correction is required for trace element detection except in the case of line broadning.

INTERFERENCES: There are spectral, physical and chemical interferences. The primary disadvantage of ICP instruments is background radiation from other elements and the plasma gases (spectral interferences). Changes in sample viscosity and surface tension with samples containing high dissolved solids (especially those exceeding 1500 mg/L) or high acid concentrations can cause physical interferences. Ionization effects, solute vaporization and molecular compound formation can cause chemical interferences. Chromium and vanadium can cause interference at the 100 mg/L level.

INSTRUMENTATION: Inductively Coupled Argon Plasma Emission Spectroscopy. 288.158 nm Wavelength

RANGE: Not listed

MDL: 58 µg/L.

PRECISION: Not listed

ACCURACY: Mean Recovery = 93% +/- 6% of spiked elements for all wastes.

SAMPLING METHOD: Wash sample container with detergent and tap water, rinse with 1+1 nitric acid and tap water, then rinse with 1+1 hydrochloric acid and tap water, then rinse with deionized, distilled water in that order. Perform any filtration or acid preservation steps when the sample is collected or as soon as possible thereafter.

STABILITY: Cool samples to 4°C. M.H.T. = 24 Hours.

QUALITY CONTROL: Mixed calibration standards, an instrument check standard and an interference check solution are used in addition to a quality control sample. The quality control sample should be prepared in the same acid matrix as the calibration standards at 10 times the instrumental detection limits and in accordance with the instructions provided by the supplier. Furthermore, two types of blanks are required: a calibration blank and a reagent blank.

REFERENCE: Method 200.7, U.S. EPA, EMSL-Cincinnati, OH, Nov. 1980

PRIMARY NAME: Silicon (Si) Method 6010

TITLE: Inductively Coupled Plasma (ICP)

MATRIX: Applies to all matrices; ground water, EP extracts, soils, sludges, sediments, aqueous samples, solid and industrial wastes.

APPLICATION: The method covers the determination of 25 metals. Its primary advantage is that ICP instruments allow simultaneous or rapid sequential determination of many elements in a short time. Samples require digestion prior to analysis and are first nebulized and the aerosol is transported to a plasma torch in which element specific atomic line emission spectra are produced by a radio frequency inductively coupled plasma. Background correction is required for trace element detection except in the case of line broadning.

INTERFERENCES: There are spectral, physical and chemical interferences. The primary disadvantage of ICP instruments is background radiation from other elements and the plasma gases (spectral interferences). Changes in sample viscosity and surface tension with samples containing high dissolved solids or high acid concentrations can cause physical interferences. Ionization effects, solute vaporization and molecular compound formation can cause chemical interferences. Chromium and vanadium can cause interference at the 100 mg/L level.

INSTRUMENTATION: Inductively Coupled Argon Plasma Emission Spectroscopy. 288.158 nm Wavelength

RANGE: Not listed MDL: 58 µg/L.

PRECISION: Not listed

ACCURACY: Mean Recovery = 93% +/- 6% of spiked elements for all wastes.

SAMPLING METHOD: Collect 400 mL of sample in plastic or glass containers.

STABILITY: Cool samples to 4°C. M.H.T. = 24 Hours.

QUALITY CONTROL: Mixed calibration standards, an instrument check standard and an interference check solution are used in addition to a quality control sample. The quality control sample should be prepared in the same acid matrix as the calibration standards at 10 times the instrumental detection limits and in accordance with the instructions provided by the supplier. Furthermore, two types of blanks are required: a calibration blank and a reagent blank.

REFERENCE: Method 6010, SW-846, 3rd ed., Nov 1986.

PRIMARY NAME: Silver (Ag) Method 200.7

TITLE: Inductively Coupled Plasma (ICP)

MATRIX: Dissolved, suspended or total element in drinking and surface waters and in domestic and industrial wastewaters.

APPLICATION: The method covers the determination of 25 metals. Dissolved elements are determined in filtered and acidified samples after appropriate digestion (which increases dissolved solids). Its primary advantage is that ICP instruments allow simultaneous or rapid sequential determination of many elements in a short time. Samples are first nebulized and the aerosol is transported to a plasma torch in which element specific atomic line emission spectra are produced by a radio frequency inductively coupled plasma. Background correction is required for trace element detection except in the case of line broadning.

INTERFERENCES: There are spectral, physical and chemical interferences.

The primary disadvantage of ICP instruments is background radiation from other elements and the plasma gases (spectral interferences). Changes in sample viscosity and surface tension with samples containing high dissolved solids (especially those exceeding 1500 mg/L) or high acid concentrations can cause physical interferences. Ionization effects, solute vaporization and molecular compound formation can cause chemical interferences.

INSTRUMENTATION: Inductively Coupled Argon Plasma Emission Spectroscopy. 328.068 nm Wavelength

RANGE: Not listed MDL: 7 µg/L.

PRECISION: Not listed

ACCURACY: Mean Recovery = 93% +/- 6% of spiked elements for all wastes.

SAMPLING METHOD: Wash sample container with detergent and tap water, rinse with 1+1 nitric acid and tap water, then rinse with 1+1 hydrochloric acid and tap water, then rinse with deionized, distilled water in that order. Perform any filtration or acid preservation steps when the sample is collected or as soon as possible thereafter.

STABILITY: Cool samples to 4°C. M.H.T. = 24 Hours.

QUALITY CONTROL: Mixed calibration standards, an instrument check standard and an interference check solution are used in addition to a quality control sample. The quality control sample should be prepared in the same acid matrix as the calibration standards at 10 times the instrumental detection limits and in accordance with the instructions provided by the supplier. Furthermore, two types of blanks are required: a calibration blank and a reagent blank.

REFERENCE: Method 200.7, U.S. EPA, EMSL-Cincinnati, OH, Nov. 1980

PRIMARY NAME: Silver (Ag) Method 6010

TITLE: Inductively Coupled Plasma (ICP)

MATRIX: Applies to all matrices; ground water, EP extracts, soils, sludges, sediments, aqueous samples, solid and industrial wastes.

APPLICATION: The method covers the determination of 25 metals. Its primary advantage is that ICP instruments allow simultaneous or rapid sequential determination of many elements in a short time. Samples require digestion prior to analysis and are first nebulized and the aerosol is transported to a plasma torch in which element specific atomic line emission spectra are produced by a radio frequency inductively coupled plasma. Background correction is required for trace element detection except in the case of line broadning.

INTERFERENCES: There are spectral, physical and chemical interferences. The primary disadvantage of ICP instruments is background radiation from other elements and the plasma gases (spectral interferences). Changes in sample viscosity and surface tension with samples containing high dissolved solids or high acid concentrations can cause physical interferences. Ionization effects, solute vaporization and molecular compound formation can cause chemical interferences.

INSTRUMENTATION: Inductively Coupled Argon Plasma Emission Spectroscopy. 328.068 nm Wavelength

RANGE: Not listed MDL: 7 µg/L.

PRECISION: Not listed

ACCURACY: Mean Recovery = 93% +/- 6% of spiked elements for all wastes.

SAMPLING METHOD: Collect 400 mL of sample in plastic or glass containers.

STABILITY: Cool samples to 4°C. M.H.T. = 24 Hours.

QUALITY CONTROL: Mixed calibration standards, an instrument check standard and an interference check solution are used in addition to a quality control sample. The quality control sample should be prepared in the same acid matrix as the calibration standards at 10 times the instrumental detection limits and in accordance with the instructions provided by the supplier. Furthermore, two types of blanks are required: a calibration blank and a reagent blank.

REFERENCE: Method 6010, SW-846, 3rd ed., Nov 1986.

PRIMARY NAME: Silver (Ag) Method 7760

TITLE: Atomic Absorption, (AA) Direct Aspiration

MATRIX: Wastes, mobility procedure extracts, soils and groundwater

CAS # : 7440-22-4

APPLICATION: Prior to analysis, samples must be prepared for direct aspiration. Sample preparation method varies with the matrix. Following appropriate dissolution of the sample, a silver lamp light beam is directed through aspirated aliquot in the flame, into a monochromator and detector. The resulting absorption of hollow cathode radiation is proportional to the silver concentration.

INTERFERENCES: Background correction is required. Store ag standards and samples in brown bottles if possible. Silver chloride is insoluble, so avoid hydrochloric acid unless silver is in solution as a chloride complex. Monitor samples and standards for viscosity differences since this may alter the aspiration rate.

INSTRUMENTATION: Atomic absorption spectrometer. Silver hollow cathode lamp. (328.1 nm Wavelength). Strip-chart recorder.

RANGE: 0.1-4 mg/L MDL: 0.01 mg/L

PRECISION: standard deviation = +/- 8.8 @ 50 ∝g Ag/L (50 labs)

ACCURACY: relative error = 10.6% @ 50 ∝g Ag/L (50 labs)

SAMPLING METHOD: Use plastic or glass containers (prewashed). Sample as per chapter 9.

STABILITY: Non aqueous samples: cool to 4°C and analyze as soon as possible. Aqueous samples: add nitric acid to pH <2. M.H.T. = 6 Months

QUALITY CONTROL: Run one spike duplicate sample for every 10 samples. Verify calibration with an independently prepared check standard every 15 samples.

REFERENCE: Method 7760, SW-846, 3rd ed., Nov 1986.

PRIMARY NAME: Silver (Ag) Method 7761

TITLE: Atomic Absorption, (AA) Furnace Technique

MATRIX: Wastes, mobility procedure extracts, soils and groundwater

CAS # : 7440-22-4

APPLICATION: Sample preparation (digestion) of matrices is always necessary and varies with the matrix. An aliquot of digestate is placed in a graphite tube in the furnace and slowly evaporated, charred and atomized. Absorption of lamp radiation during atomization is proportional to silver concentration.

INTERFERENCES: The furnace technique is subject to chemical interferences. Composition of the sample matrix can effect analysis. Modify matrix to remove interferences. Avoid nonspecific absorption and scattering. Memory effects occur if silver is not volatilized and removed from the furnace.

INSTRUMENTATION: Atomic absorption spectrometer. Silver hollow cathode lamp or electrodeless discharge lamp. Graphite furnace. Strip-chart recorder. (328.1 nm wavelength).

RANGE: 1-25 µg/L MDL: 0.2 µg/L

PRECISION: standard deviation = +/- 0.40, 0.70, 0.90 @ 25, 50, 75 ∝g Ag/L

ACCURACY: recoveries = 94, 100, 104% @ 25, 50, 75 ∝g Ag/L

SAMPLING METHOD: Use plastic or glass containers (prewashed). Sample as per chapter 9.

STABILITY: Non aqueous samples: cool to 4°C. Aqueous samples: add nitric acid to pH <2. M.H.T. = 6 Months

QUALITY CONTROL: Run one spike replicate sample for every 10 samples or per analytical batch, whichever is more frequent. Verify calibration with an independently prepared check standard every 15 samples.

REFERENCE: Method 7761, SW-846, 3rd ed., (Included as revision 0, December 1987)

PRIMARY NAME: Sodium (Na)

Method 200.7

TITLE: Inductively Coupled Plasma (ICP)

MATRIX: Dissolved, suspended or total element in drinking and surface waters and in domestic and industrial wastewaters.

APPLICATION: The method covers the determination of 25 metals. Dissolved elements are determined in filtered and acidified samples after appropriate digestion (which increases dissolved solids). Its primary advantage is that ICP instruments allow simultaneous or rapid sequential determination of many elements in a short time. Samples are first nebulized and the aerosol is transported to a plasma torch in which element specific atomic line emission spectra are produced by a radio frequency inductively coupled plasma. Background correction is required for trace element detection except in the case of line broadning.

INTERFERENCES: There are spectral, physical and chemical interferences. The primary disadvantage of ICP instruments is background radiation from other elements and the plasma gases (spectral interferences). Changes in sample viscosity and surface tension with samples containing high dissolved solids (especially those exceeding 1500 mg/L) or high acid concentrations can cause physical interferences. Ionization effects, solute vaporization and molecular compound formation can cause chemical interferences. Thallium can cause interference at the 100 mg/L level.

INSTRUMENTATION: Inductively Coupled Argon Plasma Emission Spectroscopy. 588.995 nm Wavelength

RANGE: Not listed

MDL: 29 μg/L.

PRECISION: Not listed

ACCURACY: Mean Recovery = 93% +/- 6% of spiked elements for all wastes.

SAMPLING METHOD: Wash sample container with detergent and tap water, rinse with 1+1 nitric acid and tap water, then rinse with 1+1 hydrochloric acid and tap water, then rinse with deionized, distilled water in that order. Perform any filtration or acid preservation steps when the sample is collected or as soon as possible thereafter.

STABILITY: Cool samples to 4°C. M.H.T. = 24 Hours.

QUALITY CONTROL: Mixed calibration standards, an instrument check standard and an interference check solution are used in addition to a quality control sample. The quality control sample should be prepared in the same acid matrix as the calibration standards at 10 times the instrumental detection limits and in accordance with the instructions provided by the supplier. Furthermore, two types of blanks are required: a calibration blank and a reagent blank.

REFERENCE: Method 200.7, U.S. EPA, EMSL-Cincinnati, OH, Nov. 1980

PRIMARY NAME: Sodium (Na) Method 273.1

TITLE: Metals (Total, Dissolved, Suspended) Direct Aspiration (AA)

MATRIX: drinking, surface and saline Waters. Wastewater.

APPLICATION: Date issued 1971. Editorial revision 1974. Sample is aspirated and atomized in a flame. Light beam from hollow cathode (made of Na) lamp is directed through flame into monochromator, then to detector which measures amount absorbed light.

INTERFERENCES: There can be interference from presence of high dissolved solids in sample. Chemical and ionization interferences can occur. Ionization may be controlled by adding potassium (1000 mg/L) to both standards and samples.

INSTRUMENTATION: AAS. Sodium (Na) hollow cathode lamp. Burner. Pipets. Strip chart recorder.

RANGE: 0.03-1 mg/L @ 589.6 nm Wavelength MDL: 0.002 mg/L.

PRECISION: SD = +/- (0.1 and 0.8) @ 8.2 and 52 mg Na/L.

ACCURACY: Recoveries = (102 and 100)% @ 8.2 and 52 mg Na/L.

SAMPLING METHOD: plastic or glass (pre-washed).

STABILITY: Add nitric acid to pH <2. M.H.T. = 6 Months.

QUALITY CONTROL: After calibration curve composed of a minimum of a reagent blank and 3 standards has been prepared, subsequent calibration curves must be verified by use of at least a reagent blank and one standard near MCL. Must check within 10% of original curve. (For drinking water analysis)

REFERENCE: Methods for the Chemical Analysis of Water and Wastes. EPA-600/4-79-020, USEPA, EMSL, 1979.

PRIMARY NAME: Sodium (Na)

Method 6010

TITLE: Inductively Coupled Plasma (ICP)

MATRIX: Applies to all matrices; ground water, EP extracts, soils, sludges, sediments, aqueous samples, solid and industrial wastes.

APPLICATION: The method covers the determination of 25 metals. Its primary advantage is that ICP instruments allow simultaneous or rapid sequential determination of many elements in a short time. Samples require digestion prior to analysis and are first nebulized and the aerosol is transported to a plasma torch in which element specific atomic line emission spectra are produced by a radio frequency inductively coupled plasma. Background correction is required for trace element detection except in the case of line broadning.

INTERFERENCES: There are spectral, physical and chemical interferences. The primary disadvantage of ICP instruments is background radiation from other elements and the plasma gases (spectral interferences). Changes in sample viscosity and surface tension with samples containing high dissolved solids or high acid concentrations can cause physical interferences. Ionization effects, solute vaporization and molecular compound formation can cause chemical interferences. Thallium can cause interference at the 100 mg/L level.

INSTRUMENTATION: Inductively Coupled Argon Plasma Emission Spectroscopy. 588.995 nm Wavelength

RANGE: Not listed

MDL: 29 μg/L.

PRECISION: Not listed

ACCURACY: Mean Recovery = 93% +/- 6% of spiked elements for all wastes.

SAMPLING METHOD: Collect 400 mL of sample in plastic or glass containers.

STABILITY: Cool samples to 4°C. M.H.T. = 24 Hours.

QUALITY CONTROL: Mixed calibration standards, an instrument check standard and an interference check solution are used in addition to a quality control sample. The quality control sample should be prepared in the same acid matrix as the calibration standards at 10 times the instrumental detection limits and in accordance with the instructions provided by the supplier. Furthermore, two types of blanks are required: a calibration blank and a reagent blank.

REFERENCE: Method 6010, SW-846, 3rd ed., Nov 1986.

PRIMARY NAME: Sodium (Na) Method 7770

TITLE: Atomic Absorption, (AA) Direct Aspiration

MATRIX: Drinking, Surface and Saline Waters. Wastewater

CAS # : 7440-23-5

APPLICATION: Sample is aspirated and atomized in a flame. A light beam from a sodium hollow cathode lamp is directed through the flame into a monochromator and onto a detector. Since wavelength of light beam is specific for na, light energy absorbed by detector is measure of sodium.

INTERFERENCES: The most troublesomee type is chemical, caused by lack of absorption of atoms bound in molecular combination in the flame. High dissolved solids in sample may result in nonatomic absorbance interference. Sodium is a universal contaminant; use the method with great care.

INSTRUMENTATION: Atomic absorption spectrometer. Sodium hollow cathode lamp. (589.6 nm Wavelength)

RANGE: 0.03-1 mg/L MDL: 0.002 mg/L

PRECISION: standard deviation = +/- 0.10 and 0.80 @ 8.20 and 52 mg Na/L

ACCURACY: recoveries = 102 and 100% @ 8.20 and 52 mg Na/L

SAMPLING METHOD: Use glass or plastic containers. Collect 200 g of solids and 600 mL of liquid samples.

STABILITY: Cool solid samples to 4°C. and analyze as soon as possible. Add nitric acid to liquid samples to pH < 2; M.H.T. = 6 months.

QUALITY CONTROL: At least one duplicate and one spike sample should be run every 20 samples or with each matrix type to verify precision of the method. For 20 or more samples per day, verify working standard curve. Run an additional standard at or near mid-range every 10 samples.

REFERENCE: Method 7770, SW-846, 3rd ed., Nov 1986.

PRIMARY NAME: Thallium (Tl) Method 200.7

TITLE: Inductively Coupled Plasma (ICP)

MATRIX: Dissolved, suspended or total element in drinking and surface waters and in domestic and industrial wastewaters.

APPLICATION: The method covers the determination of 25 metals. Dissolved elements are determined in filtered and acidified samples after appropriate digestion (which increases dissolved solids). Its primary advantage is that ICP instruments allow simultaneous or rapid sequential determination of many elements in a short time. Samples are first nebulized and the aerosol is transported to a plasma torch in which element specific atomic line emission spectra are produced by a radio frequency inductively coupled plasma. Background correction is required for trace element detection except in the case of line broadning.

INTERFERENCES: There are spectral, physical and chemical interferences. The primary disadvantage of ICP instruments is background radiation from other elements and the plasma gases (spectral interferences). Changes in sample viscosity and surface tension with samples containing high dissolved solids (especially those exceeding 1500 mg/L) or high acid concentrations can cause physical interferences. Ionization effects, solute vaporization and molecular compound formation can cause chemical interferences. Aluminum can cause interference at the 100 mg/L level.

INSTRUMENTATION: Inductively Coupled Argon Plasma Emission Spectroscopy. 190.864 nm Wavelength

RANGE: Not listed MDL: 40 μg/L.

PRECISION: Not listed

ACCURACY: Mean Recovery = 93% +/- 6% of spiked elements for all wastes.

SAMPLING METHOD: Wash sample container with detergent and tap water, rinse with 1+1 nitric acid and tap water, then rinse with 1+1 hydrochloric acid and tap water, then rinse with deionized, distilled water in that order. Perform any filtration or acid preservation steps when the sample is collected or as soon as possible thereafter.

STABILITY: Cool samples to 4°C. M.H.T. = 24 Hours.

QUALITY CONTROL: Mixed calibration standards, an instrument check standard and an interference check solution are used in addition to a quality control sample. The quality control sample should be prepared in the same acid matrix as the calibration standards at 10 times the instrumental detection limits and in accordance with the instructions provided by the supplier. Furthermore, two types of blanks are required: a calibration blank and a reagent blank.

REFERENCE: Method 200.7, U.S. EPA, EMSL-Cincinnati, OH, Nov. 1980

PRIMARY NAME: Thallium (Tl) Method 6010

TITLE: Inductively Coupled Plasma (ICP)

MATRIX: Applies to all matrices; ground water, EP extracts, soils, sludges, sediments, aqueous samples, solid and industrial wastes.

APPLICATION: The method covers the determination of 25 metals. Its primary advantage is that ICP instruments allow simultaneous or rapid sequential determination of many elements in a short time. Samples require digestion prior to analysis and are first nebulized and the aerosol is transported to a plasma torch in which element specific atomic line emission spectra are produced by a radio frequency inductively coupled plasma. Background correction is required for trace element detection except in the case of line broadning.

INTERFERENCES: There are spectral, physical and chemical interferences. The primary disadvantage of ICP instruments is background radiation from other elements and the plasma gases (spectral interferences). Changes in sample viscosity and surface tension with samples containing high dissolved solids or high acid concentrations can cause physical interferences. Ionization effects, solute vaporization and molecular compound formation can cause chemical interferences. Aluminum can cause interference at the 100 mg/L level.

INSTRUMENTATION: Inductively Coupled Argon Plasma Emission Spectroscopy. 190.864 nm Wavelength

RANGE: Not listed MDL: 40 μg/L.

PRECISION: Not listed

ACCURACY: Mean Recovery = 93% +/- 6% of spiked elements for all wastes.

SAMPLING METHOD: Collect 400 mL of sample in plastic or glass containers.

STABILITY: Cool samples to 4°C. M.H.T. = 24 Hours.

QUALITY CONTROL: Mixed calibration standards, an instrument check standard and an interference check solution are used in addition to a quality control sample. The quality control sample should be prepared in the same acid matrix as the calibration standards at 10 times the instrumental detection limits and in accordance with the instructions provided by the supplier. Furthermore, two types of blanks are required: a calibration blank and a reagent blank.

REFERENCE: Method 6010, SW-846, 3rd ed., Nov 1986.

PRIMARY NAME: Thallium (Tl)

Method 7840

TITLE: Atomic Absorption, (AA) Direct Aspiration

MATRIX: Drinking, Surface and Saline Waters. Wastewater

CAS # : 7440-28-0

APPLICATION: Sample is aspirated and atomized in a flame. A light beam from a Tl hollow cathode lamp is directed through the flame into a monochromator and onto a detector. Since wavelength of light beam is specific for Tl, light energy absorbed by detector is measure of thallium.

INTERFERENCES: The most troublesomee type is chemical,caused by lack of absorption of atoms bound in molecular combination in the flame. High dissolved solids in sample may result in nonatomic absorbance interference. Background correction is required. Hydrochloric acid should not be used.

INSTRUMENTATION: Atomic absorption spectrometer. Thallium hollow cathode lamp. (276.8 nm Wavelength)

RANGE: 1-20 mg/L

MDL: 0.1 mg/L

PRECISION: standard deviation = +/- 0.018, 0.05, 0.20 @ 0.60, 3.0, 15 mg Tl/L

ACCURACY: recoveries = 100, 98, 98% @ 0.60, 3.0, 15 mg Tl/L

SAMPLING METHOD: Use glass or plastic containers. Collect 200 g of solids and 600 mL of liquid samples.

STABILITY: Cool solid samples to 4°C. and analyze as soon as possible. Add nitric acid to liquid samples to pH < 2; M.H.T. = 6 months.

QUALITY CONTROL: At least one duplicate and one spike sample should be run every 20 samples or with each matrix type to verify precision of the method. For 20 or more samples per day, verify working standard curve. Run an additional standard at or near mid-range every 10 samples.

REFERENCE: Method 7840, SW-846, 3rd ed., Nov 1986.

PRIMARY NAME: Thallium (Tl)

Method 7841

TITLE: Atomic Absorption, (AA) Furnace Technique

MATRIX: Wastes, mobility procedure extracts, soils and groundwater

CAS # : 7440-28-0

APPLICATION: Aqueous samples, EP extracts, industrial wastes, soils, sludges, sediments and solid wastes require digestion before analysis. An aliquot of sample is placed in the graphite tube in the furnace and slowly evaporated, charred and atomized. Absorption of lamp radiation during atomization is proportional to thallium concentration.

INTERFERENCES: The furnace technique is subject to chemical interferences. Composition of sample matrix can effect analysis. Hydrochloric acid or excessive chloride will cause thallium volatilization at low temperatures. Palladium is a suitable matrix modifier. Background correction is required.

INSTRUMENTATION: Atomic absorption spectrometer. Thallium hollow cathode lamp or electrodeless discharge lamp. Graphite furnace. Strip-chart recorder.

RANGE: 5-100 μg/L

MDL: 1 μg/L (276.8 nm Wavelength)

PRECISION: Not listed

ACCURACY: Not listed

SAMPLING METHOD: Use glass or plastic containers. Collect 200 g of solids and 600 mL of liquid samples.

STABILITY: Cool solid samples to 4°C. and analyze as soon as possible. Add nitric acid to liquid samples to pH < 2; M.H.T. = 6 months.

QUALITY CONTROL: At least one duplicate and one spike sample should be run every 20 samples, or with each matrix type to verify method precision. If 20 or more samples are run a day, run a standard (at or near mid-range) every 10 samples.

REFERENCE: Method 7841, SW-846, 3rd ed., Nov 1986.

PRIMARY NAME: Tin (Sn) Method 7870

TITLE: Atomic Absorption, (AA) Direct Aspiration

MATRIX: Drinking, Surface and Saline Waters. Wastewater

CAS # : 7440-31-5

APPLICATION: Sample is aspirated and atomized in a flame. A light beam from a Sn hollow cathode lamp is directed through the flame into a monochromator and onto a detector. Since wavelength of light beam is specific for Sn, light energy absorbed by detector is measure of tin.

INTERFERENCES: The most troublesomee type is chemical, caused by lack of absorption of atoms bound in molecular combination in the flame. High dissolved solids in sample may result in nonatomic absorbance interference. Ionization and spectral interferences can occur.

INSTRUMENTATION: Atomic absorption spectrometer. Tin hollow cathode lamp. (286.3 nm Wavelength)

RANGE: 10-300 mg/L MDL: 0.8 mg/L

PRECISION: standard deviation = +/- 0.25, 0.50, 0.50 @ 4.0, 20, 60 mg Sn/L

ACCURACY: recoveries = 96, 101, 101% @ 4.0, 20, 60 mg Sn/L

SAMPLING METHOD: Use glass or plastic containers. Collect 200 g of solids and 600 mL of liquid samples.

STABILITY: Cool solid samples to 4°C. and analyze as soon as possible. Add nitric acid to liquid samples to pH < 2; M.H.T. = 6 months.

QUALITY CONTROL: At least one duplicate and one spike sample should be run every 20 samples or with each matrix type to verify precision of the method. For 20 or more samples per day, verify working standard curve. Run an additional standard at or near mid-range every 10 samples.

REFERENCE: Method 7870, SW-846, 3rd ed., Nov 1986.

PRIMARY NAME: Vanadium (V)

Method 200.7

TITLE: Inductively Coupled Plasma (ICP)

MATRIX: Dissolved, suspended or total element in drinking and surface waters and in domestic and industrial wastewaters.

APPLICATION: The method covers the determination of 25 metals. Dissolved elements are determined in filtered and acidified samples after appropriate digestion (which increases dissolved solids). Its primary advantage is that ICP instruments allow simultaneous or rapid sequential determination of many elements in a short time. Samples are first nebulized and the aerosol is transported to a plasma torch in which element specific atomic line emission spectra are produced by a radio frequency inductively coupled plasma. Background correction is required for trace element detection except in the case of line broadning.

INTERFERENCES: There are spectral, physical and chemical interferences. The primary disadvantage of ICP instruments is background radiation from other elements and the plasma gases (spectral interferences). Changes in sample viscosity and surface tension with samples containing high dissolved solids (especially those exceeding 1500 mg/L) or high acid concentrations can cause physical interferences. Ionization effects, solute vaporization and molecular compound formation can cause chemical interferences. Chromium, iron and thallium can cause interference at the 100 mg/L level.

INSTRUMENTATION: Inductively Coupled Argon Plasma Emission Spectroscopy. 292.402 nm Wavelength

RANGE: Not listed

MDL: 8 μg/L.

PRECISION: SD = 1.8% Mean @ true value 750 μg/L.

ACCURACY: Mean Recovery = 93% +/- 6% of spiked elements for all wastes.

SAMPLING METHOD: Wash sample container with detergent and tap water, rinse with 1+1 nitric acid and tap water, then rinse with 1+1 hydrochloric acid and tap water, then rinse with deionized, distilled water in that order. Perform any filtration or acid preservation steps when the sample is collected or as soon as possible thereafter.

STABILITY: Cool samples to 4°C.

M.H.T. = 24 Hours.

QUALITY CONTROL: Mixed calibration standards, an instrument check standard and an interference check solution are used in addition to a quality control sample. The quality control sample should be prepared in the same acid matrix as the calibration standards at 10 times the instrumental detection limits and in accordance with the instructions provided by the supplier. Furthermore, two types of blanks are required: a calibration blank and a reagent blank.

REFERENCE: Method 200.7, U.S. EPA, EMSL-Cincinnati, OH, Nov. 1980

PRIMARY NAME: Vanadium (V)

Method 6010

TITLE: Inductively Coupled Plasma (ICP)

MATRIX: Applies to all matrices; ground water, EP extracts, soils, sludges, sediments, aqueous samples, solid and industrial wastes.

APPLICATION: The method covers the determination of 25 metals. Its primary advantage is that ICP instruments allow simultaneous or rapid sequential determination of many elements in a short time. Samples require digestion prior to analysis and are first nebulized and the aerosol is transported to a plasma torch in which element specific atomic line emission spectra are produced by a radio frequency inductively coupled plasma. Background correction is required for trace element detection except in the case of line broadning.

INTERFERENCES: There are spectral, physical and chemical interferences. The primary disadvantage of ICP instruments is background radiation from other elements and the plasma gases (spectral interferences). Changes in sample viscosity and surface tension with samples containing high dissolved solids or high acid concentrations can cause physical interferences. Ionization effects, solute vaporization and molecular compound formation can cause chemical interferences. Chromium, iron and thallium can cause interference at the 100 mg/L level.

INSTRUMENTATION: Inductively Coupled Argon Plasma Emission Spectroscopy. 292.402 nm Wavelength

RANGE: Not listed

MDL: 8 μg/L.

PRECISION: SD = 1.8% Mean @ true value 750 μg/L.

ACCURACY: Mean Recovery = 93% +/- 6% of spiked elements for all wastes.

SAMPLING METHOD: Collect 400 mL of sample in plastic or glass containers.

STABILITY: Cool samples to 4°C. M.H.T. = 24 Hours.

QUALITY CONTROL: Mixed calibration standards, an instrument check standard and an interference check solution are used in addition to a quality control sample. The quality control sample should be prepared in the same acid matrix as the calibration standards at 10 times the instrumental detection limits and in accordance with the instructions provided by the supplier. Furthermore, two types of blanks are required: a calibration blank and a reagent blank.

REFERENCE: Method 6010, SW-846, 3rd ed., Nov 1986.

PRIMARY NAME: Vanadium (V) Method 7910

TITLE: Atomic Absorption, (AA) Direct Aspiration

MATRIX: Drinking, Surface and Saline Waters. Wastewater

CAS # : 7440-62-2

APPLICATION: Sample is aspirated and atomized in a flame. A light beam from a vanadium hollow cathode lamp is directed through the flame into monochromator and onto detector. Since wavelength of light beam is specific for vanadium, light energy absorbed by detector is measure of vanadium.

INTERFERENCES: The most troublesomee type is chemical,caused by lack of absorption of atoms bound in molecular combination in the flame. High dissolved solids in sample may result in nonatomic absorbance interference. Adding aluminum to samples and standards controls interferences.

INSTRUMENTATION: Atomic absorption spectrometer. Vanadium hollow cathode lamp. (318.4 nm Wavelength)

RANGE: 2-100 mg/L MDL: 0.2 mg/L

PRECISION: standard deviation = +/- 0.10, 0.10, 0.20 @ 2.0, 10, 50 mg V/L

ACCURACY: recoveries = 100, 95, 97% @ 2.0, 10, 50 mg V/L

SAMPLING METHOD: Use glass or plastic containers. Collect 200 g of solids and 600 mL of liquid samples.

STABILITY: Cool solid samples to 4°C. and analyze as soon as possible. Add nitric acid to liquid samples to pH < 2; M.H.T. = 6 months.

QUALITY CONTROL: At least one duplicate and one spike sample should be run every 20 samples or with each matrix type to verify precision of the method. For 20 or more samples per day, verify working standard curve. Run an additional standard at or near mid-range every 10 samples.

REFERENCE: Method 7910, SW-846, 3rd ed., Nov 1986.

PRIMARY NAME: Vanadium (V) Method 7911

TITLE: Atomic Absorption, (AA) Furnace Technique

MATRIX: Wastes, mobility procedure extracts, soils and groundwater

CAS # : 7440-62-2

APPLICATION: Aqueous samples, EP extracts, industrial wastes, soils, sludges, sediments and solid wastes require digestion before analysis. An aliquot of sample is placed in the graphite tube in the furnace and slowly evaporated, charred and atomized. Absorption of lamp radiation during atomization is proportional to vanadium concentration.

INTERFERENCES: The furnace technique is subject to chemical interferences. Composition of sample matrix can effect analysis. Vanadium is refractory and prone to form carbides, so memory effects are common. Clean the furnace before and after analysis. Background correction is required.

INSTRUMENTATION: Atomic absorption spectrometer. Vanadium hollow cathode lamp or electrodeless discharge lamp. Graphite furnace. Strip-chart recorder.

RANGE: 10-200 μg/L

MDL: 4 μg/L (318.4 nm Wavelength)

PRECISION: Not listed

ACCURACY: Not listed

SAMPLING METHOD: Use glass or plastic containers. Collect 200 g of solids and 600 mL of liquid samples.

STABILITY: Cool solid samples to 4°C. and analyze as soon as possible. Add nitric acid to liquid samples to pH < 2; M.H.T. = 6 months.

QUALITY CONTROL: At least one duplicate and one spike sample should be run every 20 samples, or with each matrix type to verify method precision. If 20 or more samples are run a day, run a standard (at or near mid-range) every 10 samples.

REFERENCE: Method 7911, SW-846, 3rd ed., Nov 1986.

PRIMARY NAME: Zinc (Zn) Method 200.7

TITLE: Inductively Coupled Plasma (ICP)

MATRIX: Dissolved, suspended or total element in drinking and surface waters and in domestic and industrial wastewaters.

APPLICATION: The method covers the determination of 25 metals. Dissolved elements are determined in filtered and acidified samples after appropriate digestion (which increases dissolved solids). Its primary advantage is that ICP instruments allow simultaneous or rapid sequential determination of many elements in a short time. Samples are first nebulized and the aerosol is transported to a plasma torch in which element specific atomic line emission spectra are produced by a radio frequency inductively coupled plasma. Background correction is required for trace element detection except in the case of line broadning.

INTERFERENCES: There are spectral, physical and chemical interferences. The primary disadvantage of ICP instruments is background radiation from other elements and the plasma gases (spectral interferences). Changes in sample viscosity and surface tension with samples containing high dissolved solids (especially those exceeding 1500 mg/L) or high acid concentrations can cause physical interferences. Ionization effects, solute vaporization and molecular compound formation can cause chemical interferences. Copper and nickel can cause interference at the 100 mg/L level.

INSTRUMENTATION: Inductively Coupled Argon Plasma Emission Spectroscopy. 213.856 nm Wavelength

RANGE: Not listed MDL: 2 μg/L.

PRECISION: SD = 5.6% Mean @ true value 200 μg/L.

ACCURACY: Mean Recovery = 93% +/- 6% of spiked elements for all wastes.

SAMPLING METHOD: Wash sample container with detergent and tap water, rinse with 1+1 nitric acid and tap water, then rinse with 1+1 hydrochloric acid and tap water, then rinse with deionized, distilled water in that order. Perform any filtration or acid preservation steps when the sample is collected or as soon as possible thereafter.

STABILITY: Cool samples to 4°C. M.H.T. = 24 Hours.

QUALITY CONTROL: Mixed calibration standards, an instrument check standard and an interference check solution are used in addition to a quality control sample. The quality control sample should be prepared in the same acid matrix as the calibration standards at 10 times the instrumental detection limits and in accordance with the instructions provided by the supplier. Furthermore, two types of blanks are required: a calibration blank and a reagent blank.

REFERENCE: Method 200.7, U.S. EPA, EMSL-Cincinnati, OH, Nov. 1980

PRIMARY NAME: Zinc (Zn) Method 6010

TITLE: Inductively Coupled Plasma (ICP)

MATRIX: Applies to all matrices; ground water, EP extracts, soils, sludges, sediments, aqueous samples, solid and industrial wastes.

APPLICATION: The method covers the determination of 25 metals. Its primary advantage is that ICP instruments allow simultaneous or rapid sequential determination of many elements in a short time. Samples require digestion prior to analysis and are first nebulized and the aerosol is transported to a plasma torch in which element specific atomic line emission spectra are produced by a radio frequency inductively coupled plasma. Background correction is required for trace element detection except in the case of line broadning.

INTERFERENCES: There are spectral, physical and chemical interferences. The primary disadvantage of ICP instruments is background radiation from other elements and the plasma gases (spectral interferences). Changes in sample viscosity and surface tension with samples containing high dissolved solids or high acid concentrations can cause physical interferences. Ionization effects, solute vaporization and molecular compound formation can cause chemical interferences. Copper and nickel can cause interference at the 100 mg/L level.

INSTRUMENTATION: Inductively Coupled Argon Plasma Emission Spectroscopy. 213.856 nm Wavelength

RANGE: Not listed MDL: 2 μg/L.

PRECISION: SD = 5.6% Mean @ true value 200 μg/L.

ACCURACY: Mean Recovery = 93% +/- 6% of spiked elements for all wastes.

SAMPLING METHOD: Collect 400 mL of sample in plastic or glass containers.

STABILITY: Cool samples to 4°C. M.H.T. = 24 Hours.

QUALITY CONTROL: Mixed calibration standards, an instrument check standard and an interference check solution are used in addition to a quality control sample. The quality control sample should be prepared in the same acid matrix as the calibration standards at 10 times the instrumental detection limits and in accordance with the instructions provided by the supplier. Furthermore, two types of blanks are required: a calibration blank and a reagent blank.

REFERENCE: Method 6010, SW-846, 3rd ed., Nov 1986.

PRIMARY NAME: Zinc (Zn) Method 7950

TITLE: Atomic Absorption, (AA) Direct Aspiration

MATRIX: Drinking, Surface and Saline Waters. Wastewater

CAS # : 7440-66-6

APPLICATION: Sample is aspirated and atomized in a flame. A light beam from a Zn hollow cathode lamp is directed through the flame into a monochromator and onto a detector. Since wavelength of light beam is specific for Zn, light energy absorbed by detector is measure of zinc.

INTERFERENCES: The most troublesome type is chemical, caused by lack of absorption of atoms bound in molecular combination in the flame. High dissolved solids in sample may result in nonatomic absorbance interference. Zinc is a universal contaminent so use the method with great care.

INSTRUMENTATION: Atomic absorption spectrometer. Zinc hollow cathode lamp. (213.9 nm Wavelength)

RANGE: 0.05-1 mg/L MDL: 0.005 mg/L

PRECISION: standard deviation = 114 mg/L @ 310 μg/L (true value) 89 labs

ACCURACY: as bias = -0.7% @ 310 μg/L (true value) 89 labs

SAMPLING METHOD: Use glass or plastic containers. Collect 200 g of solids and 600 mL of liquid samples.

STABILITY: Cool solid samples to 4°C. and analyze as soon as possible. Add nitric acid to liquid samples to pH < 2; M.H.T. = 6 months.

QUALITY CONTROL: At least one duplicate and one spike sample should be run every 20 samples or with each matrix type to verify precision of the method. For 20 or more samples per day, verify working standard curve. Run an additional standard at or near mid-range every 10 samples.

REFERENCE: Method 7950, SW-846, 3rd ed., Nov 1986.

PRIMARY NAME: Zinc (Zn)

Method 7951

TITLE: Atomic Absorption, (AA) Furnace Technique

MATRIX: Wastes, mobility procedure extracts, soils and groundwater

CAS # : 7440-66-6

APPLICATION: Aqueous samples, EP extracts, industrial wastes, soils, sludges, sediments and solid wastes require digestion before analysis. An aliquot of sample is placed in the graphite tube in the furnace and slowly evaporated, charred and atomized. Absorption of lamp radiation during atomization is proportional to zinc concentration.

INTERFERENCES: The furnace technique is subject to chemical interferences. Composition of sample matrix can effect analysis. Modify matrix to remove interferences. Background correction correction must be used. Zinc is universal contaminant so use the method with great care.

INSTRUMENTATION: Atomic absorption spectrometer. Zinc hollow cathode lamp or electrodeless discharge lamp. Graphite furnace. Strip-chart recorder

RANGE: 0.2-4 μg/L

MDL: 0.05 μg/L (213.9 nm Wavelength)

PRECISION: Not listed

ACCURACY: Not listed

SAMPLING METHOD: Use glass or plastic containers. Collect 200 g of solids and 600 mL of liquid samples.

STABILITY: Cool solid samples to 4°C. and analyze as soon as possible. Add nitric acid to liquid samples to pH < 2; M.H.T. = 6 months.

QUALITY CONTROL: At least one duplicate and one spike sample should be run every 20 samples, or with each matrix type to verify method precision. If 20 or more samples are run a day, run a standard (at or near mid-range) every 10 samples.

REFERENCE: Method 7951, SW-846, 3rd ed., (Included as revision 0, December 1987)

Chapter 7
Water Quality Parameters

PRIMARY NAME: Acidity (Titrimetric) Method 305.1

TITLE: Inorganics, non-metallics MATRIX: Surface and waste waters

APPLICATION: Date issued 1971. Technical revision 1974. The pH of the sample is determined and a measured amount of standard acid is added, as needed, to lower pH to 4 or less. Hydrogen peroxide is added, the solution boiled for several minutes, cooled and tirated to pH 8.2. Method measures mineral acidity of a sample plus acidity from oxidation and hydrolysis of polyvalent cations, including salts of iron and aluminum.

INTERFERENCES: Suspended matter present in the sample, or precipitates formed during the titration may cause sluggish electrode response. (This is overcome by allowing 15-20 second pauses between titrant additions and drop by drop titrant additions near end point).

INSTRUMENTATION: pH meter suitable for electrometric titrations.

RANGE: 10-1000 mg/L as $CaCO_3$, on (50 mL) MDL: standard acid = (0.02N sulfuric)

PRECISION: +/- 10 mg/L on 4 sample concentrations (up to 2000 mg/L).

ACCURACY: Calculate acidity as mg/L $CaCO_3$ or as meq/L.

SAMPLING METHOD: Plastic or glass (100 mL).

STABILITY: Cool, 4°C. M.H.T. = 14 Days.

QUALITY CONTROL: Cool (boiled sample) to room temperature before titrating electrometrically with standard sodium hydroxide (0.02N) to pH 8.2

REFERENCE: Methods for the Chemical Analysis of Water and Wastes, EPA-600/4-79-020, USEPA, EMSL, 1979.

PRIMARY NAME: Alkalinity (Titrimetric, pH 4.5) Method 310.1

TITLE: Inorganics, Non-Metallics

MATRIX: Drinking, surface and saline waters, wastewaters.

APPLICATION: Date issued 1971. Editorial revision 1978. (Titrimetric, pH 4.5). An unaltered sample is titrated, using standard acid, to an electrometrically determined end point of pH 4.5. The sample must not be filtered, diluted, concentrated, or altered in any way.

INTERFERENCES: Substances such as salts of weak organic and inorganic acids present in large amounts, may cause interference in the electrometric pH measurements. Oil, and grease, by coating the pH electrode, may also interfere, causing sluggish response.

INSTRUMENTATION: pH meter or electrically operated titrator.

RANGE: for all alkalinity ranges.

MDL: keep titration volume <50 mL.

PRECISION: A standard deviation of 1 mg $CaCO_3$/L can be achieved.

ACCURACY: Not listed

SAMPLING METHOD: Plastic or glass. (100 mL).

STABILITY: Cool, 4°C. M.H.T. = 14 Days.

QUALITY CONTROL: Standardize and calibrate pH meter according to manufacturer's instructions. If automatic temperature compensation is not provided, make titration at 25 +/- 2°C. (For <1000 mg $CaCO_3$/L use 0.02 N titrant. For >1000 mg $CaCO_3$/L use 0.1 N titrant).

REFERENCE: Methods for the Chemical Analysis of Water and Wastes, EPA-600/4-79-020, USEPA, EMSL, 1979.

PRIMARY NAME: Alkalinity Method 310.2

TITLE: Inorganics, Non-Metallics

MATRIX: Drinking, Surface and Saline waters. Wastewater.

APPLICATION: Date issued 1971. Editorial revision 1974. (Colorimetric, automated, methyl orange). Methyl orange indicator is dissolved in a weak buffer @ pH 3.1, Just below equivalent point. Any alkali addition causes a loss of color directly proportional to amount of alkalinity

INTERFERENCES: Sample turbidity and color may interfere with this method. Turbidity must be removed by filtration prior to analysis. If sample is filtered, method is not approved for NPDES monitoring. Sample color absorbed in photometric range interferes.

INSTRUMENTATION: Technicon auto analyzer; 550 nm Filters; 15 mm tubular flow cells.

RANGE: 10-200 mg/L as $CaCO_3$. MDL: Not listed

PRECISION: SD = +/- 0.5 @ Conc's; 15, 57, 154, and 193 mg/L as $CaCO_3$

ACCURACY: Recoveries = 100 and 99% @ 31 and 149 mg/L as $CaCO_3$

SAMPLING METHOD: Plastic or glass (100 mL).

STABILITY: Cool, 4°C. M.H.T. = 14 Days.

QUALITY CONTROL: Place working standards in sampler in order of decreasing concentration. (Methyl orange is used as indicator because its pH range is in the same range as equivalent point for total alkalinity and has distinct color change that can be easily measured).

REFERENCE: Methods for the Chemical Analysis of Water and Wastes, EPA-600/4-79-020, USEPA, EMSL, 1979.

PRIMARY NAME: Biochemical Oxygen Demand (BOD) Method 405.1

TITLE: Organics

MATRIX: Municipal and industrial wastewaters

APPLICATION: Date issued 1971. Editorial revision 1974. The BOD test is an empirical bioassay-type procedure which measures the dissolved oxygen consumed by microbial life while assimilating and oxidizing the organic matter present.

INTERFERENCES: The actual environment conditions of temperature, biological population, water movement, sunlight, and oxygen concentration can not be accurately reproduced in the lab. Results obtained must take above factors into account when relating BOD results.

INSTRUMENTATION: Modified Winkler with full-bottle technique or probe method.

RANGE: Not listed

MDL: Not listed

PRECISION: At mean of 2.1 and 175 mg/L BOD, S.D. = 0.7 and 26 mg/L.

ACCURACY: No acceptable procedure to determine accuracy.

SAMPLING METHOD: Plastic or glass. (1000 mL).

STABILITY: Cool, 4°C. M.H.T. = 48 Hours.

QUALITY CONTROL: Sample of waste, or an appropriate dilution, is incubated for 5 days @ 20°C in the dark. The reduction in dissolved oxygen concentration during the incubation period yields a measure of the biochemical oxygen demand.

REFERENCE: Methods for the Chemical Analysis of Water and Wastes, EPA-600/4-79-020, USEPA, EMSL, 1979.

PRIMARY NAME: Bromide (Titrimetric) Method 320.1

TITLE: Inorganics, Non-metallics

MATRIX: Drinking, surface and saline waters, wastewaters.

APPLICATION: Date issued 1974. After pretreatment, sample is divided. One aliquot is run by converting iodide to iodate and titrating with phenylarsine oxide(pao)or sodium thiosulfate. Other aliquot is run for iodide+bromide by conversion to iodate and bromate and titrating.

INTERFERENCES: Iron, manganese and organic matter can interfere. (Calcium oxide pretreatment nullifies this interference). Color interferes with observation of indicator and bromine-water color changes. (Eliminate, using a pH meter and standard amounts of oxidant and oxidant-quencher).

INSTRUMENTATION: Laboratory iodometric titration equipment and glassware.

RANGE: 2-20 mg bromide/L. MDL: Not listed

PRECISION: SD = +/- 0.42 mg/L of bromide @ 20.3 mg/L of bromide

ACCURACY: Recovery = 99% @ 20.3 mg/L of bromide

SAMPLING METHOD: Plastic or glass (100 mL).

STABILITY: No preservation required. M.H.T. = 28 Days.

QUALITY CONTROL: When titrating either aliquot, run a distilled water blank with each sample set because of iodide, iodate, bromide and/or bromate in reagents. Calculate bromide by difference. [(iodide + bromide) - (iodide) = bromide].

REFERENCE: Methods for the Chemical Analysis of Water and Wastes, EPA-600/4-79-020, USEPA, EMSL, 1979.

PRIMARY NAME: Chemical Oxygen Demand (COD) Method 410.1

TITLE: Organics MATRIX: Wastewater

APPLICATION: Date issued 1971. Editorial revision 1978. (Titrimetric, mid level). Method determines the quantity of oxygen required to oxidize the organic matter in a waste sample, under specific conditions of oxidizing agent, temperature and time. (Use 0.25N $K_2Cr_2O_7$).

INTERFERENCES: Traces of organic material from glassware or atmosphere may cause gross positive error. Avoid inclusion of organic materials in distilled water for reagent preparation or sample dilution. Mercuric sulfate removes chloride interference. Cool flask during sulfuric acid addition.

INSTRUMENTATION: Reflux apparatus; 12 inch allihn condenser; ground glass joint connections.

RANGE: Wastewater (with TOC >50 mg/L) MDL: Not listed

PRECISION: SD = +/- 17.76 mg/L COD @ 270 mg/L COD.

ACCURACY: As bias, -4.7% @ 270 mg/L COD.

SAMPLING METHOD: Plastic or glass (50 mL).

STABILITY: Cool, 4°C. Add sulfuric acid to pH < 2. M.H.T. = 28 Days.

QUALITY CONTROL: Organic and oxidizable inorganic substances are oxidized by potassium dichromate in 50% sulfuric acid solution @ reflux temperature for 2 hrs. Silver sulfate is used as a catalyst. Excess dichromate is titrated with standard ferrous ammonium sulfate (0.25N) using an indicator.

REFERENCE: Methods for the Chemical Analysis of Water and Wastes, EPA-600/4-79-020, USEPA, EMSL, 1979.

PRIMARY NAME: Chemical Oxygen Demand (COD) Method 410.2

TITLE: Organics MATRIX: Surface water and wastewater.

APPLICATION: Date issued 1971. Editorial revision 1974 and 1978. (Titrimetric, low level). Method determines the quantity of oxygen required to oxidize the organic matter in a waste sample, under specific conditions of oxidizing agent, temp and time. (Use 0.025N $K_2Cr_2O_7$).

INTERFERENCES: Traces of organic material from glassware or atmosphere may cause gross positive error. Avoid inclusion of organic materials in distilled water for reagent preparation or sample dilution. Mercuric sulfate removes chloride interference. Cool flask during sulfuric acid addition.

INSTRUMENTATION: Reflux apparatus; 12 inch allihn condenser; ground glass joint connections.

RANGE: 5-50 mg/L COD MDL: Not listed

PRECISION: SD = +/- 4.15 mg/L COD @ 12.3 mg/L COD.

ACCURACY: As bias, 0.3% @ 12.3 mg/L COD.

SAMPLING METHOD: Plastic or glass (50 mL).

STABILITY: Cool, 4°C. Add sulfuric acid to pH < 2. M.H.T. = 28 Days.

QUALITY CONTROL: Organic and oxidizable inorganic substances are oxidized by potassium dichromate in 50% sulfuric acid solution @ reflux temperature for 2 hours. Excess dichromate is titrated with standard ferrous ammonium sulfate (0.025N) using orthophenanthroline ferrous complex as indicator.

REFERENCE: Methods for the Chemical Analysis of Water and Wastes, EPA-600/4-79-020, USEPA, EMSL, 1979.

PRIMARY NAME: Chemical Oxygen Demand (COD) Method 410.3

TITLE: Organics MATRIX: Saline waters

APPLICATION: Date issued 1971. Editorial revision 1978. (Titrimetric, high level for saline waters). Method determines the quantity of oxygen required to oxidize the organic matter in a waste sample, under specific conditions of oxidizing agent, temperature and time.

INTERFERENCES: Traces of organic material from glassware or atmosphere may cause gross positive error. Avoid inclusion of organic materials in distilled water for reagent preparation or sample dilution. Cool flask during sulfuric acid addition. Use chloride correction procedure as outlined.

INSTRUMENTATION: Reflux apparatus; 12 inch allihn condenser; ground glass joint connections.

RANGE: >250 mg/L COD @ chloride>1000 mg/L

MDL: Method has special calculation

PRECISION: Not listed

ACCURACY: Not listed

SAMPLING METHOD: Plastic or glass (50 mL).

STABILITY: Cool, 4°C. Add sulfuric acid to pH < 2. M.H.T. = 28 Days.

QUALITY CONTROL: Organic and oxidizable inorganic substances are oxidized by potassium dichromate in 50% sulfuric acid solution @ reflux temperature for 2 hours. Excess dichromate is titrated with standard ferrous ammonium sulfate (0.25N) using an indicator. (Use smaller sample, for COD >800 mg/L).

REFERENCE: Methods for the Chemical Analysis of Water and Wastes, EPA-600/4-79-020, USEPA, EMSL, 1979.

PRIMARY NAME: Chemical Oxygen Demand (COD) Method 410.4

TITLE: Organics

MATRIX: Surface waters and wastewaters

APPLICATION: Date issued 1978. (Colorimetric, automated, manual). Method determines the quantity of oxygen required to oxidize the organic matter in a waste sample, under specific conditions of oxidizing agent, temperature, and time.

INTERFERENCES: Chlorides are quantitatively oxidized by dichromate and represent a positive interference. Mercuric sulfate is added to the digestion tubes to complex the chlorides.

INSTRUMENTATION: Spectrophotometer (manually) or technicon auto analyzer (automated).

RANGE: 3-900 mg/L (automated). MDL: Not listed

PRECISION: Not listed

ACCURACY: Not listed

SAMPLING METHOD: Plastic or glass. (50 mL).

STABILITY: Cool, 4°C. Add sulfuric acid to pH < 2. M.H.T. = 28 Days.

QUALITY CONTROL: Process standards and blanks exactly as the samples. Samples, blanks and standards in sealed tubes are heated in an oven or block digester in presence of potassium dichromate (oxidizing agent) @ 150°C for 2 hours.

REFERENCE: Methods for the Chemical Analysis of Water and Wastes, EPA-600/4-79-020, USEPA, EMSL, 1979.

PRIMARY NAME: Chloride Method 325.1

TITLE: Inorganics, Non-metallics MATRIX: Drinking, surface and saline waters, wastewater.

APPLICATION: Date issued 1971. (Colorimetric, automated ferricyanide AAI). Thiocyanate ion (SCN) is liberated from mercuric thiocyanate through sequestration of mercury by chloride ion to form un-ionized mercuric chloride. SCN forms ferric thiocyanate with ferric ions. Ferric ammonium sulfate reagent provides ferric ion which forms highly colored ferric thyiocyanate in concentration proportional to the original chloride concentration.

INTERFERENCES: No significant interferences

INSTRUMENTATION: Technicon auto analyzer, 480 nm Filters, 15 mm tubular flow cell.

RANGE: 1 to 250 mg Cl/L. MDL: Not listed

PRECISION: SD = +/- 0.3 At conc. of 1, 100 and 250 mg Cl/L.

ACCURACY: At conc. of 10 and 100 mg Cl/L, 97% and 104% recoveries.

SAMPLING METHOD: Plastic or glass (50 mL).

STABILITY: No preservation required. M.H.T. = 28 Days.

QUALITY CONTROL: Place working standards in sampler in order of decreasing concentration.

REFERENCE: Methods for the Chemical Analysis of Water and Wastes, EPA-600/4-79-020, USEPA, EMSL, 1979.

PRIMARY NAME: Chloride Method 325.2

TITLE: Inorganics, Non-metallics

MATRIX: Drinking, surface and saline waters, wastewater.

APPLICATION: Date issued 1978. (Colorimetric, automated ferricyanide). Thiocyanate ion (SCN) is liberated from mercuric thiocyanate through sequestration of mercury by chloride ion to form un-ionized mercuric chloride. SCN forms ferric thiocyanate with ferric ions. Ferric nitrate reagent provides ferric ion which forms highly colored ferric thiocyanate in concentration proportional to original chloride concentration. The range may be extended with sample dilution.

INTERFERENCES: No significant interferences.

INSTRUMENTATION: Technicon auto analyzer, 480 nm Filters, 15 mm tubular flow cell.

RANGE: 1 to 200 mg Cl/L. MDL: Not listed

PRECISION: Not listed

ACCURACY: Not listed

SAMPLING METHOD: Plastic or glass (50 mL).

STABILITY: No preservation required; analyze immediately. M.H.T. = 28 Days.

QUALITY CONTROL: 1) where particulate matter is present, the sample must be filtered prior to the determination. Sample may be centrifuged in place of filtration, or use Technicon continuous filter. 2) Place working standards in sampler in order of decreasing concentration.

REFERENCE: Methods for the Chemical Analysis of Water and Wastes, EPA-600/4-79-020, USEPA, EMSL, 1979.

PRIMARY NAME: Chloride Method 325.3

TITLE: Inorganics, Non-metallics

MATRIX: Drinking, surface and saline waters, wastewater.

APPLICATION: Date issued 1971. Editorial revision 1978. (Titrimetric, mercuric nitrate). An acidified sample is titrated with mercuric nitrate in the presence of mixed diphenylcarbazone-bromphenol blue indicator. The end point is a blue-violet mercury DPC complex.

INTERFERENCES: Sulfite interference can be eliminated by oxidizing 50 ml sample solution with 0.5 To 1.0 mL H_2O_2. Special precautions are necessary when chromate and iron are present, especially ferric ion. (Automated titration may be used).

INSTRUMENTATION: Standard lab titrimetric equipment, including 1 or 5 mL microburet (0.01 mL Graduations)

RANGE: Suitable for all Cl ranges. MDL: Not listed

PRECISION: SD = 1.32 mg Cl/L at 18 mg Cl/L increment.

ACCURACY: As bias, +0.6 mg Cl/L at 18 mg Cl/L increment.

SAMPLING METHOD: Plastic or glass (50 mL).

STABILITY: No preservation required; analyze immediately M.H.T. = 28 Days.

QUALITY CONTROL: 1) check concentration of a 50 mL aliquot to determine strength of titrant and sample size to be used. 2) Make practice runs with indicator to become familiar with end point. (Alphazurine sharpens end point). 3) Store mercuric nitrate in dark bottle.

REFERENCE: Methods for the Chemical Analysis of Water and Wastes, EPA 600/4-79-020, USEPA, EMSL, 1979.

PRIMARY NAME: Chloride (Total) Method 300.0

TITLE: Inorganic Anions in Water MATRIX: Drinking, surface and mixed Wastewater

APPLICATION: A small volume of sample, typically 2 to 3 mL is introduced into an ion chromatograph. The anions of interest are separated and measured using a system comprised of a guard column, separator column, suppressor column and conductivity detector.

INTERFERENCES: Interferences can be caused by substances with retention times similar to and overlapping those of the ion of interest. Large amounts of an anion can interfere with peak resolution of adjacent anion. Method interference can be caused by reagent or equipment contamination.

INSTRUMENTATION: Ion chromatograph; analytical balance; guard, separator and suppressor columns.

RANGE: Not available MDL: 0.015 mg/L.

PRECISION: SD = 0.289 mg/L @ 10.0 mg/L chloride (drinking water).

ACCURACY: Recovery = 98.2% @ 10.0 mg/L chloride (drinking water)

SAMPLING METHOD: Plastic or glass.

STABILITY: No preservation required; analyze immediately M.H.T. = 28 Days.

QUALITY CONTROL: The laboratory should spike and analyze a minimum of 10% of all samples to monitor continuing lab performance. Field and laboratory duplicates should be analyzed. Measure retention times of standards.

REFERENCE: Test Method-The Determination of Inorganic Anions in Water by Ion Chromatography, (EPA-600/4-84-017).

PRIMARY NAME: Chlorine, Total Residual Method 330.1

TITLE: Inorganics, Non-Metallics MATRIX: Waters and wastes

APPLICATION: Date issued 1974. Editorial revision 1978. (Titrimetric, amperometric). Amperometric method applies to all types waters and wastes without substantial amount of organic matter. (Chlorine and chloramines stoichiometrically liberate iodine from KI at pH 4 or less)

INTERFERENCES: Stirring can lower chlorine values by volatilization. Copper and silver poison the electrode.

INSTRUMENTATION: Amperometer (microammeter with necessary electrical accessories); microburet.

RANGE: Not listed MDL: Not listed

PRECISION: SD = 24.8 and 12.5% @ 0.64 and 1.83 mg/L total chloride.

ACCURACY: Relative error = 8.5 and 8.8% @ 0.64 and 1.83 mg/L total Cl

SAMPLING METHOD: Plastic or glass (200 mL).

STABILITY: No preservation required; analyze immediately.

QUALITY CONTROL: If dilution is necessary, it must be done with distilled water which is free of chlorine, chlorine-demand and ammonia. (Phenylarsine oxide or sodium thiosulfate is used as the standard reducing agent to titrate liberated iodine using amperometer to determine end point).

REFERENCE: Methods for the Chemical Analysis of Water and Wastes, EPA-600/4-79-020, USEPA, EMSL, 1979.

PRIMARY NAME: Chlorine, Total Residual Method 330.2

TITLE: Inorganics, Non-Metallics

MATRIX: All types waters, but especially wastewaters.

APPLICATION: Date issued 1978. (Titrimetric, back, iodometric) (starch or amperometric endpoint). Iodometric backtitration is best for wastewaters but is applicable to all types waters. (Chlorine and chloramines stoichiometrically liberate iodine from KI at pH 4 or less)

INTERFERENCES: Manganese, iron and nitrite interference is minimized by buffering to pH 4 before adding KI. High concentrations of organics may cause uncertainty in the endpoint. Turbidity and color make endpoint difficult to detect. Practice runs with spikes may be necessary.

INSTRUMENTATION: Standard lab glassware; microburet 0-2 mL or 0-10 mL is used; amperometric titrater.

RANGE: Not listed

MDL: Not listed

PRECISION: SD = +/- 0.12 mg/L Cl @ 3.51 mg/L Cl (river water)

ACCURACY: % recovery = 107.7% @ 0.84 mg/L Cl (river water)

SAMPLING METHOD: plastic or glass (200 mL).

STABILITY: No preservation required; analyze immediately.

QUALITY CONTROL: Use chlorine free, chlorine-demand free distilled water for dilution. (Phenylarsine oxide is used as the standard reducing agent. Iodine quantitatively oxidizes reducing agent and excess is titrated with standard iodine titrant to starch-iodine or amperometric endpoint)

REFERENCE: Methods for the Chemical Analysis of Water and Wastes, EPA-600/4-79-020, USEPA, EMSL, 1979.

PRIMARY NAME: Chlorine, Total Residual Method 330.3

TITLE: Inorganics, Non-Metallics

MATRIX: Natural and treated waters.

APPLICATION: Date issued 1978. (Titrimetric, iodometric). Method applies to natural and treated waters at concentrations greater than 1 mg/L. Chlorine and chloramines liberate iodine from KI at pH 4 or less. (Iodine is titrated with a standard reducing agent using a starch indicator)

INTERFERENCES: Ferric, manganic and nitrite ions interfere, the neutral titration minimizes these interferences. Acetic acid is used for acid titration, never use HCl. Turbidity and color may make endpoint difficult to detect. Practice runs with spiked samples may be necessary.

INSTRUMENTATION: Standard lab glassware; microburet 0-2 mL or 0-10 mL is used.

RANGE: Concentrations >1 mg/L Cl MDL: Not listed

PRECISION: SD = 27 and 23.6% @ 0.64 and 1.83 mg/L Cl.

ACCURACY: Relative error = 23.6 and 16.7% @ 0.64 and 1.83 mg/L Cl.

SAMPLING METHOD: plastic or glass (200 mL).

STABILITY: No preservation required; analyze immediately.

QUALITY CONTROL: Phenylarsine oxide is used as the standardized reducing agent to titrate liberated iodine to starch iodine endpoint. (Titrate away from direct sunlight. Run blank titration).

REFERENCE: Methods for the Chemical Analysis of Water and Wastes, EPA-600/4-79-020, USEPA, EMSL, 1979.

PRIMARY NAME: Chlorine, Total Residual Method 330.4

TITLE: Inorganics, Non-Metallics MATRIX: Natural and treated waters.

APPLICATION: Date issued 1978. (Titrimetric, DPD-FAS). The N,N-diethyl-p-phenylene diamine (DPD)-ferrous ammomium sulfate (FAS). Method applies to matrix listed at concentrations above 1 mg/L Cl. (Liberated iodine is titrated with FAS using DPD as indicator).

INTERFERENCES: Bromine, bromamine and iodine are interferences normally present in insignificant amounts. Oxidized manganese and copper interfere, but can be corrected for. Turbidity and color may make endpoint difficult to detect.

INSTRUMENTATION: Standard lab glassware; microburet, 0-2 mL or 0-10 mL is used.

RANGE: Concentrations above 1 mg/L Cl MDL: Not listed

PRECISION: SD = 19.4 and 9.4% @ 0.64 and 1.83 mg/L Cl.

ACCURACY: Relative error = 8.1 and 4.3% @ 0.64 and 1.83 mg/L Cl.

SAMPLING METHOD: plastic or glass (200 mL).

STABILITY: No preservation required; analyze immediately.

QUALITY CONTROL: This procedure gives a convenient direct reading (mL titrant = mg/L Cl) up to 4 mg/L. An aliquot should be diluted to 100 mL if higher concentrations are present. Use chlorine free distilled water to prepare indicator.

REFERENCE: Methods for the Chemical Analysis of Water and Wastes, EPA-600/4-79-020, USEPA, EMSL, 1979.

PRIMARY NAME: Chlorine, Total Residual Method 330.5

TITLE: Inorganics, Non-Metallics MATRIX: Natural and treated waters.

APPLICATION: Date issued 1978. (Spectrophotometric, DPD). The N,N-diethyl-p-phenylene diamine (DPD) colorimetric method applies to matrix listed at concentrations from 0.2-4 mg/L Cl. (Liberated iodine reacts with DPD to produce a red colored solution read on spectrophotometer)

INTERFERENCES: Any oxidizing agents; these are usually present at insignificant concentrations compared to the residual chlorine concentrations. Turbidity and color will essentially prevent the colorimetric analysis. (Ferrous ammonium sulfate is titrant used on permanganate stds).

INSTRUMENTATION: Spectrophotometer, 515 nm; cells of light path 1 cm or longer.

RANGE: 0.2-4.0 mg/L Cl MDL: Not listed

PRECISION: SD = 27.6% @ 0.66 mg/L Cl.

ACCURACY: Relative error = 15.6% @ 0.66 mg/L Cl.

SAMPLING METHOD: plastic or glass (200 mL).

STABILITY: No preservation required; analyze immediately.

QUALITY CONTROL: The solution is spectrophotometrically compared to a series of standards, using a graph or a regression analysis calculation. (Calculation is figured using absorbance and titrated concentrations of permanganate solutions (chlorine equivalent) and absorbance of sample).

REFERENCE: Methods for the Chemical Analysis of Water and Wastes, EPA-600/4-79-020, USEPA, EMSL, 1979.

PRIMARY NAME: Conductance Method 120.1

TITLE: Physical Properties

MATRIX: Drinking, surface and saline waters, wastewater.

APPLICATION: Date issued 1971. (Specific conductance, umhos at 25°C). The specific conductance of a sample is measured by use of a self-contained conductivity meter. Field measurements with comparable instruments are reliable.

INTERFERENCES: Not listed

INSTRUMENTATION: Wheatstone type-bridge or equivalent.

RANGE: Not listed MDL: Not listed

PRECISION: SD = 7.55 @ specific conductance increment of 100.

ACCURACY: As bias, -2.0 umhos/cm @ specific conductance increment of 100.

SAMPLING METHOD: plastic or glass (100 mL).

STABILITY: Cool, 4°C. M.H.T. = 28 Days.

QUALITY CONTROL: Instrument must be standardized with KCl solution before daily use. Conductivity cell must be kept clean. Make temperature corrections, and report results @ 25°C, if sample not analyzed @ 25°C.

REFERENCE: Methods for the Chemical Analysis of Water and Wastes, EPA-600/4-79-020, USEPA, EMSL, 1979.

PRIMARY NAME: Dissolved Oxygen Uptake Rate Method 213 A

TITLE: Inorganics, Non-Metallics MATRIX: Activated sludge

APPLICATION: (DOUR; Also referred to as oxygen consumption rate). This test is used to determine the oxygen consumption rate of a biological suspension sample. A routine plant operation test, it will indicate changes in operating conditions at an early stage.

INTERFERENCES: Determination results are quite sensitive to temperature variations. The determination is sensitive to the time lag between sample collection and test initiation. Because test and sample site conditions can vary, observed measurements and actual DOUR may vary.

INSTRUMENTATION: Oxygen probe method and meter or manometric or respirometric device

RANGE: Not listed MDL: Not listed

PRECISION: Precision has not been determined.

ACCURACY: Accuracy is not applicable.

SAMPLING METHOD: Glass container only (300 mL). Bottle and top.

STABILITY: No preservation required; analyze immediately.

QUALITY CONTROL: Record appropriate DO data at time intervals of less than 1 min, depending on rate of consumption. Record data over 15 minute period or until DO becomes limiting, whichever occurs first.

REFERENCE: Standard Methods for the Examination of Water and Wastewater, 16th ed., Page 127, 1985.

PRIMARY NAME: Fluoride, Total

Method 340.1

TITLE: Inorganics, Non-Metallics

MATRIX: Drinking, surface and saline waters, wastewater.

APPLICATION: Date issued 1971. Editorial revision 1978. (Colorimetric, spadns with bellack distillation). After distillation to remove interferences, sample is treated with Spadns reagent. Loss of color resulting from fluoride reaction with z-s dye is function of fluoride concentration.

INTERFERENCES: A small error in reagent addition is most prominent source of error in this test. Care must be taken to avoid overheating flask above level of solution. (Maintain an even flame entirely under boiling flask). Extend range using fluoride ion selective method.

INSTRUMENTATION: Distillation equipment; spectrophotometer @ 570 nm or filterphotometer @ 550-580 nm.

RANGE: 0.1-2.5 mg/L F. (Can be extended)

MDL: Not listed

PRECISION: SD = +/- 0.089 mg/L F @ 0.83 mg/L F.

ACCURACY: Mean = 0.81 mg/L F @ 0.83 m/L F

SAMPLING METHOD: Use plastic containers.

STABILITY: No preservation required.

M.H.T. = 28 Days.

QUALITY CONTROL: Plot absorbance versus concentration. Prepare a new standard curve whenever fresh reagent is made. (Fluoride concentration is measured as the diffeence of absorbance in the blank and the sample).

REFERENCE: Methods for the Chemical Analysis of Water and Wastes, EPA-600/4-79-020, USEPA, EMSL, 1979.

PRIMARY NAME: Fluoride

Method 340.2

TITLE: Inorganics, Non-Metallics

MATRIX: Drinking, Surface and Saline Waters. Wastewater.

APPLICATION: Date issued 1971. Editorial revision 1974. (Potentiometric, ion selective electrode). Fluoride(F) is determined potentiometrically using a fluoride electrode in conjunction with a standard single junction sleeve type reference electrode and a pH meter.

INTERFERENCES: pH extremes interfere; sample pH should be between pH 5 and 9. Polyvalent cations of silicon, iron and aluminum interfere (form complexes with fluoride). Degree of interference depends on complexing cations, concentration of fluoride and pH of sample.

INSTRUMENTATION: Selective ion meter with direct concentration scale for (F) or pH meter with expanded mv scale

RANGE: 0.1-1000 mg/L F.

MDL: Not listed

PRECISION: SD = +/- 0.03 @ 0.85 mg/L F.

ACCURACY: Mean = 0.84 mg/L @ 0.85 mg/L F.

SAMPLING METHOD: Use plastic containers.

STABILITY: No preservation required.

M.H.T. = 28 Days.

QUALITY CONTROL: For industrial waste samples, regular amount of buffer may not be adequate. Analyst should check pH first. If highly basic (pH >9), add 1N HCl and adjust pH to 8.3. [Electrodes must remain in the solution at least 3 min or until reading has stabilized. (Up to 5 min)]

REFERENCE: Methods for the Chemical Analysis of Water and Wastes, EPA-600/4-79-020, USEPA, EMSL, 1979.

PRIMARY NAME: Fluoride | Method 340.3

TITLE: Inorganics, Non-Metallics

MATRIX: Drinking, surface and saline waters, wastewater.

APPLICATION: Date issued 1971. (Colorimetric, automated complexone). Fluoride (F) ion reacts with the red cerous chelate of alizarin complexone. There is a positive color developed in contrast to a bleaching action in other fluoride Methods. For total or total dissolved fluoride, the bellack distillation must be performed prior to complexone analysis.

INTERFERENCES: Aluminum forms an extremely stable fluoro compound, which is overcome by treatment with 8-hydroxyquinoline (complexes the aluminum) and extraction with chloroform. At aluminum levels < 0.2 mg/L, extraction procedure is not required.

INSTRUMENTATION: Technicon Auto Analyzer; 650 nm filters; 15 mm tubular flow cell.

RANGE: 0.05-1.5 mg F/L.

MDL: Not listed

PRECISION: SD = +/- 0.018 @ Concentrations of 0.06, 0.15 and 1.08 mg F/L.

ACCURACY: Recoveries = 89 and 102% @ concentrations of 0.14 and 1.25 mg F/L.

SAMPLING METHOD: Use plastic containers.

STABILITY: No preservation required.

M.H.T. = 28 Days.

QUALITY CONTROL: Arrange fluoride standards in sampler in order of decreasing concentration.

REFERENCE: Methods for the Chemical Analysis of Water and Wastes, EPA-600/4-79-020, USEPA, EMSL, 1979.

PRIMARY NAME: Hardness, Total (mg/L as CaCO3) Method 130.1

TITLE: Physical Properties

MATRIX: Drinking, surface and saline waters.

APPLICATION: Date issued 1971. (Colorimetric, automated EDTA). The magnesium EDTA exchanges magnesium on an equivalent basis for any calcium and/or other cations to form a more stable EDTA chelate than magnesium. The free Mg reacts with calmagite @ pH 10.

INTERFERENCES: No significant interferences. (The free magnesium (Mg) plus calgamite reaction gives a red-violet complex, and by measuring only magnesium concentration in the final reaction stream, total hardness can be measured accurately).

INSTRUMENTATION: Technicon Auto Analyzer, 520 nm Filters, 15 mm tubular flow cell.

RANGE: 10 to 400 mg/L as $CaCO_3$.

MDL: Not listed

PRECISION: SD = +/- 1.5 @ 19 and 120 mg/L as $CaCO_3$.

ACCURACY: At concentrations, 39 and 296 mg/L as $CaCO_3$, recoveries, 89 and 93%.

SAMPLING METHOD: Use plastic or glass containers and collect 100 mL samples.

STABILITY: Add nitric acid to pH <2. H2SO4 to pH <2.M.H.T. = 6 Months.

QUALITY CONTROL: For most wastewaters and highly polluted waters, the sample must be digested as in atomic absorption method. Arrange working standards in sampler in order of decreasing concentrations.

REFERENCE: Methods for the Chemical Analysis of Water and Wastes, EPA-600/4-79-020, USEPA, EMSL, 1979.

PRIMARY NAME: Hardness, Total (mg/L as $CaCO_3$) Method 130.2

TITLE: Physical Properties

MATRIX: Drinking, surface and saline waters, wastewater.

APPLICATION: Date issued 1971. Editorial revision 1978. (Titrimetric, EDTA). Calcium (Ca) and magnesium (Mg) ions are sequestered upon addition of disodium-EDTA. Reaction end point, using indicator, has red color in presence of Ca and Mg and a blue color when they're sequestered.

INTERFERENCES: Excessive amounts of heavy metals can interfere. This is usually overcome by complexing the metals with cyanide. Routine addition of sodium cyanide solution (caution:deadly poison) to prevent potential metallic interference is recommended.

INSTRUMENTATION: Standard laboratory titrimetric equipment.

RANGE: all concentration ranges of hardness MDL: Not listed

PRECISION: SD = 2.98 mg/L, $CaCO_3$ @ 194 mg/L, $CaCO_3$.

ACCURACY: As bias, -2.0 mg/L, $CaCO_3$ @ 194 mg/L, $CaCO_3$.

SAMPLING METHOD: plastic or glass (100 mL).

STABILITY: Add nitric acid or sulfuric acid to pH <2. M.H.T. = 6 Months.

QUALITY CONTROL: Use a sample aliquot containing not more than 25 mg calcium carbonate(CaCO3). Use inhibitors to reduce metal ion interference during titration. Automated titration may be used.

REFERENCE: Methods for the Chemical Analysis of Water and Wastes, EPA-600/4-79-020, USEPA, EMSL, 1979.

PRIMARY NAME: Iodide (Titrimetric) Method 345.1

TITLE: Inorganics, Non-Metallics

MATRIX: Drinking, surface and saline waters, wastewater.

APPLICATION: Date issued: 1974. After pretreatment to remove interferences, the sample is analyzed for iodide by converting the iodide to iodate with bromine water and titrating with phenylarsine oxide (PAO) or sodium thiosulfate.

INTERFERENCES: Iron, manganese and organic matter can interfere. (Calcium oxide pretreatment nullifies this interference). Color interferes with observation of indicator and bromine-water color changes. (Overcome by using a pH meter and standardized amounts of reagents).

INSTRUMENTATION: Laboratory iodometric titration equipment and glassware.

RANGE: 2-20 mg/L of iodide.

MDL: Not listed

PRECISION: SD = +/- 0.06 mg/L @ 11.6 mg/L of iodide.

ACCURACY: Recovery = 97% @ 11.6 mg/L of iodide.

SAMPLING METHOD: Use plastic or glass containers and collect 100 mL of sample.

STABILITY: No Federal Register rules apply.

QUALITY CONTROL: A distilled water blank must be run with each set of samples because of iodide in reagents.

REFERENCE: Methods for the Chemical Analysis of Water and Wastes, EPA-600/4-79-020, USEPA, EMSL, 1979.

PRIMARY NAME: Nitrate-N (Total) Method 300.0

TITLE: Inorganic Anions in Water

MATRIX: Drinking, surface and mixed wastewater

APPLICATION: A small volume of sample, typically 2 to 3 mL, is introduced into an ion chromatograph. The anions of interest are separated and measured using a system comprised of a guard column, separator column, suppressor column and conductivity detector.

INTERFERENCES: Interferences can be caused by substances with retention times similar to and overlapping those of ion of interest. Large amounts of an anion can interfere with peak resolution of adjacent anion. Method interference can be caused by reagent or equipment contamination.

INSTRUMENTATION: Ion chromatograph; analytical balance; guard, separator and suppressor columns.

RANGE: Not listed MDL: 0.013 mg/L.

PRECISION: SD = 0.356 mg/L @ 31.0 mg/L Nitrate-N (Drinking water).

ACCURACY: Mean recovery = 100.7% @ 31.0 mg/L Nitrate-N (Drinking water).

SAMPLING METHOD: Use plastic or glass containers to collect samples.

STABILITY: Cool, 4°C. M.H.T. = 48 Hours.

QUALITY CONTROL: The laboratory should spike and analyze a minimum of 10% of all samples to monitor continuing lab performance. Field and laboratory duplicates should be analyzed. Measure retention times of stds. (Nitrates exhibit great changes in retention times).

REFERENCE: Test Method - The Determination of Inorganic Anions in Water by Ion Chromatography, (EPA-600/4-84-017).

PRIMARY NAME: Nitrogen, Ammonia Method 350.1

TITLE: Inorganics, Non-Metallics MATRIX: Drinking, surface and saline waters, wastewater.

APPLICATION: Date issued 1974. Editorial revision 1978. (Colorimetric, automated phenate). Alkaline phenol and hypochlorite react with ammonia to form indophenol blue that is proportional to the ammonia concentration. (Blue color formed is intensified with sodium nitroprusside).

INTERFERENCES: Calcium and magnesium ions may be present in concentration sufficient to cause precipitation problems during analysis. Sample turbidity and color may interfere with this method.

INSTRUMENTATION: Technicon Auto Analyzer; 630-660 nm; 15 mm or 50 mm tubular flow cell.

RANGE: 0.01 To 2.0 mg/L NH_3 as N. MDL: Not listed

PRECISION: SD = +/- 0.005 At four concentrations (0.43-1.41 mg NH_3-N/L)

ACCURACY: At concentrations 0.16 and 1.44, recoveries were 107% and 99%.

SAMPLING METHOD: Use plastic or glass containers and collect 400 mL of sample.

STABILITY: Cool, 4°C. Add sulfuric acid to pH < 2. M.H.T. = 28 Days.

QUALITY CONTROL: Approximately 20-60 samples per hour can be analyzed. Arrange ammonia standards in sampler in order of decreasing concentration of nitrogen. All solutions must be made using ammonia-free water. When saline waters are analyzed, substitute ocean water is used to prepare standards.

REFERENCE: Methods for the Chemical Analysis of Water and Wastes, EPA-600/4-79-020, USEPA, EMSL, 1979.

PRIMARY NAME: Nitrogen, Ammonia Method 350.2

TITLE: Inorganics, Non-Metallics

MATRIX: Drinking, surface and saline waters, wastewater.

APPLICATION: Date issued 1971. Editorial revision 1974. (Colorimetric; titimetric; potentiometric distillation procedure). Method covers the range of 0.05 to 1.0 mg NH_3-N/L for colorimetric; 1.0 to 25 mg/L for titrimetric; and 0.05 to 1400 mg/L for electrode method.

INTERFERENCES: Cyanate hydrolyzes slightly, even at pH 9.5. Volatile alkaline compounds may cause an off-color upon nesslerization. (Some compounds may be eliminated by boiling off at low pH, 2-3). Residual chlorine must be removed by sample pretreatment with sodium thiosulfate.

INSTRUMENTATION: Spectrophotometer or filter photometer; 425 nm; light path of 1 cm or more.

RANGE: See Applicability. MDL: Not listed

PRECISION: SD = 0.244 mg NH_3-N/L @ 1.71 mg NH_3-N/L increment.

ACCURACY: As bias, +0.01 mg NH_3-N/L @ 1.71 mg NH_3-N/L increment.

SAMPLING METHOD: Use plastic or glasscontainers anc ollect 400 mL of sample.

STABILITY: Cool, 4°C. Add sulfuric acid to pH < 2. M.H.T. = 28 Days.

QUALITY CONTROL: 1) colorimetric determination is prepared in matched Nessler tubes and absorbance read @ 425 nm. 2) For titrimetric determination, the distillate in receiving flask is titrated with standard sulfuric acid. 3) For potential determination, consult Method 350.3: Selective ion electrode method.

REFERENCE: Methods for the Chemical Analysis of Water and Wastes, EPA-600/4-79-020, USEPA, EMSL, 1979.

PRIMARY NAME: Nitrogen, Ammonia Method 350.3

TITLE: Inorganics, Non-Metallics MATRIX: Drinking, surface and saline waters, wastewater.

APPLICATION: Date issued 1974. (Potentiometric, ion selective electrode). The ammonia is determined potentiometrically using an ion selective ammonia electrode. The NH_3 electrode uses a hydrophobic gas-permeable membrane to separate the sample from ammonium chloride internal soln.

INTERFERENCES: Volatile amines act as a positive interference. Mercury interferes by forming a complex with ammonia. Thus the sample can not be preserved with mercuric chloride.

INSTRUMENTATION: pH meter with expanded mv scale or specific ion meter.

RANGE: 0.03 To 1400 mg NH_3-N/L. MDL: Not listed

PRECISION: SD = +/- 0.038 @ 1.00 mg NH_3-N/L.

ACCURACY: Recoveries = 96 and 91% @ 0.19 and 0.13 mg NH3-N/L

SAMPLING METHOD: Use a plastic or glass container and collect 400 mL samples.

STABILITY: Cool, 4°C. Add sulfuric acid to pH < 2. M.H.T. = 28 Days.

QUALITY CONTROL: Distilled water must be ammonia free. When analyzing saline waters, standards must be made up in synthetic ocean water. See Method 350.1 for preparation directions.

REFERENCE: Methods for the Chemical Analysis of Water and Wastes, EPA-600/4-79-020, USEPA, EMSL, 1979.

PRIMARY NAME: Nitrogen, Kjeldahl, Total (TKN) Method 351.1

TITLE: Inorganics, Non-Metallics MATRIX: Surface and saline waters.

APPLICATION: Date issued 1971. Editorial revision 1974 and 1978. (Colorimetric, automated phenate). Total Kjeldahl nitrogen is defined as the sum of free-ammonia and of organic nitrogen compounds which are converted to ammonium sulfate under the conditions of digestion. Sample is automatically digested with a sulfuric acid solution containing catalyst. Organic nitrogen is converted to ammonium sulfate.

INTERFERENCES: Iron and chromium ions tend to catalyze while copper ions tend to inhibit the indophenol color reaction.

INSTRUMENTATION: Technicon Auto Analyzer; 630 nm Filters; 50 mm tubular flow cell.

RANGE: 0.05 to 2.0 mg N/L. MDL: Not listed

PRECISION: SD = 0.61 K-N mg N/L at 2.18 K-N mg N/L.

ACCURACY: As bias, -0.62 mg N/L at 2.18 K-N mg N/L.

SAMPLING METHOD: Use plastic or glass containers and collect 500 mL samples.

STABILITY: Cool, 4°C. Add sulfuric acid to pH < 2. M.H.T. = 28 Days.

QUALITY CONTROL: All solutions must be made using ammonia free water. Arrange standards in sampler cups in order of increasing concentration.

REFERENCE: Methods for the Chemical Analysis of Water and Wastes, EPA-600/4-79-020, USEPA, EMSL, 1979.

PRIMARY NAME: Nitrogen, Kjeldahl, Total (TKN) Method 351.2

TITLE: Inorganics, Non-Metallics MATRIX: Drinking, surface and wastewaters.

APPLICATION: Date issued 1978. (Colorimetric, semi-automated block digester, AAII). The sample is heated in the presence of sulfuric acid, potassium sulfate and mercuric sulfate for 2.5 Hrs. Residue is cooled, diluted to 25 mL and placed on autoanalyzer for NH_3 detmn. (The digested sample may also be used for phosphorus determination).

INTERFERENCES: The procedure converts nitrogen components of biological origin such as amino acids, proteins and peptides to ammonia, but may not convert the nitrogenous compounds of some industrial wastes such as amines, nitro compounds, hydrazones, semicarbazones and some amines.

INSTRUMENTATION: Block digester-40; Technicon manifold for ammonia.

RANGE: 0.1 to 20 mg/L TKN. MDL: Not listed

PRECISION: Not listed

ACCURACY: Not listed

SAMPLING METHOD: Use plastic or glass containers and collect 500 mL samples.

STABILITY: Cool, 4°C. Add sulfuric acid to pH < 2. M.H.T. = 28 Days.

QUALITY CONTROL: All solutions must be made using ammonia free water. Use Teflon boiling stones. The range may be extended with sample dilution.

REFERENCE: Methods for the Chemical Analysis of Water and Wastes, EPA-600/4-79-020, USEPA, EMSL, 1979.

PRIMARY NAME: Nitrogen, Kjeldahl, Total (TKN) Method 351.3

TITLE: Inorganics, Non-Metallics

MATRIX: Drinking, surface and saline waters, wastewater.

APPLICATION: Date issued 1971. Editorial revision 1974 and 1978. (Colorimetric; titrimetric; potentiometric). Three alternatives are listed for the determination of NH_3 after distillation. Titrimetric Method for concentrations above 1 mg N/L; colorimetric for concentrations <1 mg N/L; potentiometric for 0.05-1400 mg N/L.

INTERFERENCES: High nitrate concentrations (10 x or more than TKN level) result in low TKN values. The reaction between nitrate and ammonia (NH_3) can be prevented by the use of an ion exchange resin (chloride form) to remove the nitrate prior to the TKN analysis.

INSTRUMENTATION: Spectrophotometer; 425 nm; Light path 1 cm or longer.

RANGE: See Applicability. MDL: Not listed

PRECISION: SD = 1.056 mg N/L @ 4.10 K-N mg N/L.

ACCURACY: As Bias, +0.4 mg N/L @ 4.10 K-N mg N/L.

SAMPLING METHOD: Use plastic or glass containers and collect 500 mL samples.

STABILITY: Cool, 4°C. Add sulfuric acid to pH < 2. M.H.T. = 28 Days.

QUALITY CONTROL: Colorimetric determination is prepared in Nessler tubes and absorbance read @ 425 nm. For titrimetric determination, distillate in receiving flask is titrated with standard sulfuric acid. For potentiometric determination, consult Method 350.3. All solution must be made with ammonia free water.

REFERENCE: Methods for the Chemical Analysis of Water and Wastes, EPA-600/4-79-020, USEPA, EMSL, 1979.

PRIMARY NAME: Nitrogen, Kjeldahl, Total (TKN) Method 351.4

TITLE: Inorganics, Non-Metallics

MATRIX: Drinking, surface and wastewaters.

APPLICATION: Date issued 1978. (Potentiometric, ion selective electrode). Following digestion and cooling, distilled water is added to digestion flask and pH adjusted between 3 and 4.5 Using 10N sodium hydroxide. Sample is cooled and transferred to a 100 mL beaker. After inserting electrode into sample, sodium hydroxide-sodium iodide-EDTA solution is added and ammonia measured.

INTERFERENCES: Interference from metals are eliminated with addition of sodium iodide. High nitrate concentrations (10 x or more than TKN level) result in low TKN values. The nitrate and ammonia (NH_3) reaction is prevented by using an anion exchange resin to remove nitrate.

INSTRUMENTATION: pH meter (with expanded mv scale). NH_3 selective electrode. Digestion apparatus.

RANGE: 0.03-25 mg TKN/L.

MDL: Not listed

PRECISION: Not listed

ACCURACY: Not listed

SAMPLING METHOD: Use plastic or glass containers and collect 500 mL samples.

STABILITY: Cool, 4°C. H2SO4 to pH < 2. M.H.T. = 28 Days.

QUALITY CONTROL: Either macro or micro Kjeldahl system or block digestor can be used for digestion. All solutions must be made using ammonia-free water.

REFERENCE: Methods for the Chemical Analysis of Water and Wastes, EPA-600/4-79-020, USEPA, EMSL, 1979.

PRIMARY NAME: Nitrogen, Nitrate Method 352.1

TITLE: Inorganics, Non-Metallics

MATRIX: Drinking, surface and saline waters, wastewater.

APPLICATION: Date issued 1971. (Colorimetric, brucine). Method is based upon reaction of nitrate ion with brucine sulfate in a 13 N H_2SO_4 solution @ 100°C. The color of resulting complex is measured @ 410 nm. Temperature control of reaction is extremely critical.

INTERFERENCES: Dissolved organic matter causes an off color in 13 N sulfuric acid. Effects of salinity and residual chlorine interference can be eliminated. All strong oxidizing or reducing agents interfere. Ferrous and ferric iron and quadrivalent manganese interfere slightly.

INSTRUMENTATION: Spectro (or filter) photometer @ 410 nm. Water bath @ 100°C with stirring mechanism

RANGE: 0.1-2.0 mg NO_3-N/L. MDL: Not listed

PRECISION: SD = 0.214 mg N/L @ 1.24 mg N/L (as nitrogen, nitrate)

ACCURACY: as bias, +0.04 mg N/L @ 1.24 mg N/L (as NO_3-N)

SAMPLING METHOD: Use plastic or glass containers and collect 100 mL samples.

STABILITY: Cool, 4°C. M.H.T. = 48 Hours.

QUALITY CONTROL: Uneven heating of samples and standards during reaction time results in erratic values. Absolute control of temperature during critical color development period cannot be too strongly emphasized. Use distilled water free of nitrite and nitrate to prepare reagents and standards.

REFERENCE: Methods for the Chemical Analysis of Water and Wastes, EPA-600/4-79-020, USEPA, EMSL, 1979.

PRIMARY NAME: Nitrogen, Nitrate-Nitrite Method 353.1

TITLE: Inorganics, Non-Metallics MATRIX: Drinking, surface and wastewaters.

APPLICATION: Date issued 1971. Reissued with revision 1978. (Colorimetric, automated, hydrazine reduction). Nitrate is reduced to nitrite with hydrazine sulfate and the nitrite determined by diazotizing to form a highly colored dye which is measured colorimetrically.

INTERFERENCES: Sample color that absorbs in the photometric range used for analysis will interfere. The apparent nitrate (NO_3) and nitrite (NO_2) concentrations varied +/- 10% with concentrations of sulfide ion up to 10 mg/L.

INSTRUMENTATION: Colorimeter equipped with an 8, 15, or 50 mm flow cell; 529 nm filters.

RANGE: 0.01-10 mg/L NO_3-NO_2 nitrogen. MDL: Not listed

PRECISION: SD = +/- 0.03 @ 4.75 ug NO_3-N/L.

ACCURACY: recoveries were 99 and 101% @ 0.75 and 2.97 ug NO_3-N/L.

SAMPLING METHOD: Use plastic or glass containers and collect 100 mL samples.

STABILITY: Cool, 4°C. Add sulfuric acid to pH < 2. M.H.T. = 28 Days.

QUALITY CONTROL: Use continuous filter to remove precipitate. Place appropriate nitrate standards in sampler in order of decreasing concentration of nitrogen.

REFERENCE: Methods for the Chemical Analysis of Water and Wastes, EPA-600/4-79-020, USEPA, EMSL, 1979.

PRIMARY NAME: Nitrogen, Nitrate-Nitrite Method 353.2

TITLE: Inorganics, Non-Metallics MATRIX: Surface, saline and wastewater.

APPLICATION: Date issued 1971. Editorial revision 1974 and 1978. (Colorimetric, automated, cadmium reduction). Method pertains to determination of nitrite singly or nitrite and nitrate combined. A filtered sample is passed through a column containing granulated copper-cadmium.

INTERFERENCES: Build-up of suspended matter in reduction column restricts flow, low results may be found on samples with high concentrations of iron, copper or other metals, and samples with large concentrations of oil and grease will coat the surface of the cadmium.

INSTRUMENTATION: Technicon Auto Analyzer; 540 nm filters; 15 or 50 mm tubular flow cell.

RANGE: 0.05 To 10.0 mg/L NO_3-NO_2 N. MDL: Not listed

PRECISION: SD = 0.176 mg N/L @ 2.48 NO_3-N mg N/L.

ACCURACY: As bias, -0.067 mg N/L @ 2.48 NO_3-N mg N/L.

SAMPLING METHOD: Use plastic or glass containers and collect 100 mL samples.

STABILITY: Cool, 4°C. Add sulfuric acid to pH < 2. M.H.T. = 28 Days.

QUALITY CONTROL: Caution: samples for reduction column must not be preserved with mercuric chloride. When samples to be analyzed are saline waters, substitute ocean water should be used.(See Method 350.1). The range may be extended with sample dilution.

REFERENCE: Methods for the Chemical Analysis of Water and Wastes, EPA-600/4-79-020, USEPA, EMSL,

PRIMARY NAME: Nitrogen, Nitrate-Nitrite Method 353.3

TITLE: Inorganics, Non-Metallics

MATRIX: Drinking, surface and saline waters, wastewater.

APPLICATION: Date issued 1974. (Spectrophotometric, cadmium reduction). A filtered sample is passed through a column containing granulated copper-cadmium to reduce nitrate to nitrite. The nitrite is determined by diazotizing to form a highly colored azo dye.

INTERFERENCES: Build-up of suspended matter in the reduction column restricts sample flow. Low results may be obtained on samples with high concentrations of iron, copper or other metals. Samples with large amounts of oil and grease coat the surface of the cadmium.

INSTRUMENTATION: Spectrophotometer; 540 nm; light path of 1 cm or longer.

RANGE: 0.01 To 1.0 mg/L NO_3-NO_2 N. MDL: Not listed

PRECISION: SD = +/- 0.004 and 0.005 @ 0.24 and 0.55 mg NO_3+NO_2 N/L.

ACCURACY: Recoveries were, 100 and 102% @ 0.24 and 0.55 mg NO_3+NO_2 N/L

SAMPLING METHOD: Use plastic or glass containers and collect 100 mL samples.

STABILITY: Cool, 4°C. Add sulfuric acid to pH < 2. M.H.T. = 28 Days.

QUALITY CONTROL: Caution: samples for reduction must not be preserved with mercuric chloride. Carry out procedures for turbidity removal, oil and grease removal and add EDTA to eliminate high concentrations of metals interference. The range may be extended with sample dilution.

REFERENCE: Methods for the Chemical Analysis of Water and Wastes, EPA-600/4-79-020, USEPA,EMSL, 1979.

PRIMARY NAME: Nitrogen, Nitrite Method 354.1

TITLE: Inorganics, Non-Metallics

MATRIX: Drinking, surface and saline waters, wastewater.

APPLICATION: Date issued 1971. (Spectrophotometric). The diazonium compound formed by diazotation of sulfanilamide by nitrite in water under acid conditions is coupled with n-(1-naphthyl)ethylene diamine dihydrochloride to produce a reddish-purple color.

INTERFERENCES: There are very few known interferences at concentrations <1000 times that of nitrite. Strong oxidants or reductants readily affect nitrite concentrations. High alkalinity (>600 mg/L) give low results due to a pH shift.

INSTRUMENTATION: Spectrophotometer @ 540 nm; 1cm or larger cells.

RANGE: 0.01-1.0 mg NO_2 N/L.

MDL: Not listed

PRECISION: Not listed

ACCURACY: Not listed

SAMPLING METHOD: Use plastic or glass containers and collect 50 mL of sample.

STABILITY: Cool, 4°C. M.H.T. = 48 Hours.

QUALITY CONTROL: Use distilled water free of nitrite and nitrate to prepare all reagents and standards. If sample pH is > 10 or total alkalinity is >600 mg/L, adjust pH to 6 with 1:3 HCl. If necessary, filter sample through 0.45 ∝m filter using first portion of filtrate to rinse filter flask.

REFERENCE: Methods for the Chemical Analysis of Water and Wastes, EPA-600/4-79-020, USEPA, EMSL, 1979.

PRIMARY NAME: Organic Carbon, Total Method 415.1

TITLE: Organics - Combustion or oxidation products

MATRIX: Drinking, surface and saline waters, wastewater.

APPLICATION: Date issued 1971. Editorial revision 1974. Organic carbon is converted to carbon dioxide (CO_2) by catalytic combustion or wet chemical oxidation. CO_2 formed can be measured by infrared detector or converted to methane (CH_4) and measured by flame ionization.

INTERFERENCES: Carbonate and bicarbonate can interfere and must be removed or accounted for in calculation. This proceedure is applicable only to homogeneous samples which can be injected into apparatus reproducibly by syringe or pipet. Applies to TOC level above 1 mg/L.

INSTRUMENTATION: Apparatus for total and dissolved organic carbon; blender apparatus.

RANGE: Not listed

MDL: Not listed

PRECISION: SD = 8.32 TOC mg/L @ 107 TOC mg/L.

ACCURACY: As bias, +1.08 mg/L @ 107 TOC mg/L.

SAMPLING METHOD: Use plastic or glass containers and collect 25 mL samples.

STABILITY: Cool, 4°C. HCl or add sulfuric acid to pH<2.M.H.T. = 28 Days.

QUALITY CONTROL: Protect samples from sunlight and atmospheric oxygen. For instrument calibration, series of standards should encompass the expected concentration range of the samples. Instrument manufacturer's instructions should be followed.

REFERENCE: Methods for the Chemical Analysis of Water and Wastes, EPA-600/4-79-020, USEPA, EMSL, 1979.

PRIMARY NAME: Oxygen, Dissolved (DO) Method 360.1

TITLE: Inorganics, Non-Metallics MATRIX: Wastewaters and streams.

APPLICATION: Date issued 1971. (Membrane electrode). This probe method is recommended for those samples containing materials which interfere with modified Winkler procedure. It is recommended for monitoring streams, lakes, outfalls, etc., with continuous record of DO.

INTERFERENCES: Probes with membranes respond to partial pressure of oxygen which in turn is a function of dissolved organic salts. Conversion factors for sea and brackish waters may be calculated (conversion factors for specific inorganic salts may be determined experimentally).

INSTRUMENTATION: Dissolved oxygen probe; Weston and Stack, YSI or Beckman.

RANGE: Not listed MDL: Not listed

PRECISION: Manufacturer's specifications claim 0.1 mg/L repeatability.

ACCURACY: Manufacturer's specifications claim 0.1 mg/L repeatability.

SAMPLING METHOD: Glass container only (both bottle and top). For sample collection from shallow depths (less than 5 ft), use an apha type sampler. A Kemmerer type sampler is recommended for samples collected at depths > 5 ft. Fill a 300 mL bottle to overflowing.

STABILITY: No preservation required; analyze immediately.

QUALITY CONTROL: Record temperature at time of sampling.

REFERENCE: Methods for the Chemical Analysis of Water and Wastes, EPA-600/4-79-020, USEPA, EMSL, 1979.

PRIMARY NAME: Oxygen, Dissolved (DO) Method 360.2

TITLE: Inorganics, Non-Metallics MATRIX: Wastewater and streams.

APPLICATION: Date issued 1971. (Modified Winkler, full bottle technique). This method is applicable for use with most wastewaters and streams that contain nitrate nitrogen and not more than 1 mg/L ferrous iron. The DO probe technique gives comparable results.

INTERFERENCES: There are a number of interferences to the dissolved oxygen test, including oxidizing and reducing agents, nitrate ion, ferrous iron and organic matter. Most common interferences in the Winkler procedure may be overcome by use of the DO probe.

INSTRUMENTATION: A titration with 0.0375 N sodium thiosulfate to a starch-iodine end point.

RANGE: Not listed MDL: Not listed

PRECISION: Exact data unavailable.

ACCURACY: Exact data unavailable.

SAMPLING METHOD: Use a glass container only (both bottle and top). Sample collection from shallow depths (less than 5 ft), use an apha type sampler. A Kemmerer type sampler is recommended for samples collected at depths > 5 ft. Fill a 300 mL bottle to overflowing.

STABILITY: Fix on site and store in dark. M.H.T. = 8 Hours.

QUALITY CONTROL: Record temperature at time of sampling.

REFERENCE: Methods for the Chemical Analysis of Water and Wastes, EPA-600/4-79-020, USEPA, EMSL, 1979.

PRIMARY NAME: pH (electrometric) Method 150.1

TITLE: Physical Properties

MATRIX: Drinking, surface and saline waters, wastewater.

APPLICATION: Date issued 1971. Editorial revision 1978. At a given temperature the intensity of the acidic or basic nature of a solution is indicated by pH (hydrogen ion activity). Alkalinity and acidity are acid-and-base neutralizing abilities of water usually expressed as mg $CaCO_3$/L. pH is determined electrometrically using a glass electrode with a reference potential or a combination electrode.

INTERFERENCES: Coatings of oily or particulate matter, temperature effects, and sodium errors at pH levels >10 are interferences.

INSTRUMENTATION: pH meter, laboratory or field model; magnetic stirrer and Teflon stirring bar.

RANGE: pH meter range (0 to 14).

MDL: report pH to nearest 0.1 unit.

PRECISION: Not listed

ACCURACY: Limit of accuracy, +/- 0.1 pH unit.

SAMPLING METHOD: Use a plastic or glasscontainer and collect 25 mL of sample.

STABILITY: No preservation required; analyze immediately.

QUALITY CONTROL: Calibrate at minimum of two points that bracket expected pH of the samples and are approximately 3 pH units or more apart. Sample should be within 2°C of buffers, if automatic temperature compensation is not provided.

REFERENCE: Methods for the Chemical Analysis of Water and Wastes, EPA-600/4-79-020, USEPA, EMSL, 1979.

PRIMARY NAME: Phosphorus, All Forms

Method 365.1

TITLE: Inorganics, Non-Metallics

MATRIX: Drinking, surface and saline waters, wastewater.

APPLICATION: Date issued 1971. Editorial revision 1974 and 1978. (Colorimetric, automated, ascorbic acid). Applies to specified forms of phosphorus (P), based on reactions for the orthophosphate ion. Most samples are run for; phosphorus and dissolved P, orthophosphate, and dissolved orthophosphate.

INTERFERENCES: High iron concentrations cause precipitation of and loss of phosphorus. Sample turbidity must be removed by filtration prior to analysis for orthophosphate. Salt error for samples 5 to 20% salt is <1%. Arsenic concentrationswhich are greater than phosphorus concentration, may interfere.

INSTRUMENTATION: Technicon Auto Analyzer; 650-660 or 880 nm filter.

RANGE: 0.001-1.0 mg P/L. MDL: Not listed

PRECISION: SD = 0.066 mg P/L @ 0.30 mg P/L (as orthophosphate)

ACCURACY: As bias, -0.04 mg P/L @ 0.30 mg P/L (as orthophosphate).

SAMPLING METHOD: Use plastic or glass containers and collect 50 mL samples.

STABILITY: Cool, 4°C. Add sulfuric acid to pH <2. M.H.T. = 28 Days.

QUALITY CONTROL: This method is based on reactions specific for the orthophosphate ion in which an antimony-phospho-molybdate complex is reduced to an intensely blue colored complex by ascorbic acid. Color is proportional to phosphorus concentration. Measure color on an auto analyzer.

REFERENCE: Methods for the Chemical Analysis of Water and Wastes, EPA-600/4-79-020, USEPA, EMSL, 1979.

PRIMARY NAME: Phosphorus, All Forms Method 365.2

TITLE: Inorganics, Non-Metallics MATRIX: Drinking, surface and saline waters, wastewater.

APPLICATION: Date issued 1971. (Colorimetric, ascorbic acid, single reagent). Applies to specified forms of phosphorus (P), based on reactions for orthophosphate ion. Most commonly measured forms are; phosphorus and dissolved P, orthophosphate and dissolved orthophosphate.

INTERFERENCES: High iron concentrations cause precipitation of and loss of phosphorus. Salt error for samples 5 to 20% salt is <1%. Arsenic concentration which are greater than phosphorus concentration, may interfere. (Only orthophosphate turns blue color in this test. Other (P) forms are converted to orthophosphate).

INSTRUMENTATION: Spectro(or filter)photometer; 650 or 880 nm; Light path of 1cm or longer.

RANGE: 0.01-0.5 mg P/L. MDL: Not listed

PRECISION: SD = 0.018 mg P/L @ 0.335 mg P/L (as orthophosphate)

ACCURACY: As bias, -0.009 mg P/L L 0.335 mg P/L (as orthophosphate).

SAMPLING METHOD: Use plastic or glasscontainers and collect 50 mL samples.

STABILITY: Cool, 4°C. Add sulfuric acid to pH <2. M.H.T. = 28 Days.

QUALITY CONTROL: This method is based on reactions specific for the orthophosphate ion in which an antimony-phospho-molybdate complex is reduced to an intensely blue colored complex by ascorbic acid. Color is proportional to phosphorus concentration. Measure color on auto analyzer.

REFERENCE: Methods for the Chemical Analysis of Water and Wastes, EPA-600/4-79-020, USEPA, EMSL, 1979.

PRIMARY NAME: Phosphorus, All Forms Method 365.3

TITLE: Inorganics, Non-Metallics MATRIX: Drinking, surface and saline waters, wastewater.

APPLICATION: Date issued 1978. (Colorimetric, ascorbic acid, two reagent). Applies to specified forms of phosphorus (P), based on reactions for the orthophosphate ion. Most commonly measured forms are; phosphorus and dissolved P, orthophosphate and dissolved orthophosphate.

INTERFERENCES: Arsenate interference is eliminated by reducing arsenic acid to arsenious acid using sodium bisulfite. High concentrations of iron cause low phosphorus recovery. The bisulfite treatment also eliminates this interference.

INSTRUMENTATION: Spectro(or filter)photometer; 660 or 880 nm; Light path of 1cm or longer.

RANGE: 0.01-1.2 mg P/L. MDL: Not listed

PRECISION: Not listed

ACCURACY: Recoveries = 99 and 100% @ 7.6 and 0.55 mg P/L (waste and sewage)

SAMPLING METHOD: Use plastic or glass containers and collect 50 mL samples.

STABILITY: Cool, 4°C. Add sulfuric acid to pH < 2. M.H.T. = 28 Days.

QUALITY CONTROL: This method is based on reactions specific for the orthophosphate ion in which an antimony-phospho-molybdate complex is reduced to an intensely blue colored complex by ascorbic acid. Color is proportional to phosphorus concentration. Measure color in auto analyzer.

REFERENCE: Methods for the Chemical Analysis of Water and Wastes, EPA-600/4-79-020, USEPA, EMSL, 1979.

PRIMARY NAME: Phosphorus, Total Method 365.4

TITLE: Inorganics, Non-Metallics MATRIX: Drinking, surface and wastewaters.

APPLICATION: Date issued 1974. (Colorimetric, automated, block digestor AAII). Sample is heated in presence of sulfuric acid, potassium sulfate and mercuric sulfate for 2 1/2 hours. Residue is cooled, diluted to 25 mL and placed on auto analyzer for phosphorus determination. Temperature of block digester during 2 1/2 hour digestion is 380°C.

INTERFERENCES: Only add 4-8 Teflon boiling chips during digestion. Too many boiling chips will cause sample to boil over.

INSTRUMENTATION: Block digestor BD-40; Technicon Auto Analyzer and method No. 327-74W.

RANGE: 0.01-20 mg P/L. MDL: Not listed

PRECISION: SD = +/- 0.06 @ 2.0 mg P/L (sewage sample)

ACCURACY: Recoveries = 95 and 98% @ 1.84 and 1.89 mg P/L (sewage samples)

SAMPLING METHOD: Use plastic or glass containers and collect 50 mL samples.

STABILITY: Cool, 4°C. Add sulfuric acid to pH < 2. M.H.T. = 28 Days.

QUALITY CONTROL: This method covers the determination of total phosphorus. Prepare standard curve plotting peak heights of standards against concentration values. Compare sample peak heights with standard curve to compute concentration. (If TKN is determined, the sample should be diluted with ammonia-free water).

REFERENCE: Methods for the Chemical Analysis of Water and Wastes, EPA-600/4-79-020, USEPA, EMSL, 1979.

PRIMARY NAME: Residue, Filterable (TDS) Method 160.1

TITLE: Physical Properties

MATRIX: Drinking, surface and saline waters, wastewater.

APPLICATION: Date issued 1971. Also referred to as total dissolved solids. A well mixed sample is filtered through a standard glass fiber filter. The filtrate is evaporated and dried to constant weight @ 180°C. The increase in dish weight represents the total dissolved solids. (The filtrate from residue, non-filterable may be used).

INTERFERENCES: Highly mineralized waters with considerable calcium, magnesium, chloride, and/or sulfate content may be hygroscopic and will require prolonged drying, desiccation and rapid weighing. Samples with high concentrations of bicarbonates require prolonged drying.

INSTRUMENTATION: Glass fiber filter discs (Reeves Angel 934-AH, or equivalent; drying oven @180°C.

RANGE: 10 mg/L to 20,000 mg/L. MDL: Not listed

PRECISION: Not listed

ACCURACY: Not listed

SAMPLING METHOD: Use plastic or glasscontainers and collect 100 mL samples.

STABILITY: Cool, 4°C. M.H.T. = 48 Hours.

QUALITY CONTROL: Too much residue in evaporating dish will crust over and entrap water that will not be driven off during drying. Limit total residue to 200 mg.

REFERENCE: Methods for the Chemical Analysis of Water and Wastes, EPA-600/4-79-020, USEPA, EMSL, 1979.

PRIMARY NAME: Residue, Volatile (VSS) and (VS) Method 160.4

TITLE: Physical Properties

MATRIX: Sewage, sludge, waste, and sediments.

APPLICATION: Date issued 1971. Also referred to as volatile suspended solids and soluble solids. Residue from determination of total (a), filterable (b), or non-filterable (c) residue is ignited in a muffle furnace. (The loss of weight on ignition is reported as volatile solids). VSS = ignited solids from (c). VS = ignited solids from (a).

INTERFERENCES: Large source of error is failure to obtain a representative sample. The test is subject to errors due to loss of water of crystallization, loss of volatile organic matter prior to combustion, incomplete oxidation of certain complex organics and decomposition of mineral salts.

INSTRUMENTATION: Muffle furnace @ 550°C

RANGE: Not listed

MDL: Not listed

PRECISION: SD = +/- 11 mg/L at 170 mg/L volatile residue concentration.

ACCURACY: Not listed

SAMPLING METHOD: Use plastic or glass containers and collect 100 mL samples.

STABILITY: Cool, 4°C. M.H.T. = 7 Days.

QUALITY CONTROL: Ignite residue from (a), (b), or (c) (see Application) to constant weight in a muffle furnace @ 550°C. Usually 15 to 20 min ignition is required. Cool and weigh; repeat cycle of igniting, cooling, desiccating and weighing to constant weight.

REFERENCE: Methods for the Chemical Analysis of Water and Wastes, EPA-600/4-79-020, USEPA, EMSL, 1979.

PRIMARY NAME: Sulfate Method 375.1

TITLE: Inorganics, Non-Metallics MATRIX: Drinking, surface and wastewaters.

APPLICATION: Date issued 1971. (Colorimetric, automated, chloranilate). When solid barium chloranilate is added to a solution containing sulfate, barium sulfate is precipitated, releasing the highly colored acid chloranilate ion. Color intensity = amount of sulfate present.

INTERFERENCES: Cations, such as calcium, aluminum and iron interfere by precipitating the chloranilate. These ions are removed automatically by passage through an ion exchange column.

INSTRUMENTATION: Technicon Auto Analyzer; 520 nm filters; 15 mm tubular flow cells.

RANGE: 10-400 mg SO_4/L. MDL: Not listed

PRECISION: SD = +/- 0.8 @ 294 mg SO_4/L.

ACCURACY: Recoveries = 99 and 102% @ 82 and 295 mg SO_4/L.

SAMPLING METHOD: Use plastic or glass containers and collect 50 mL samples.

STABILITY: Cool, 4°C. M.H.T. = 28 Days.

QUALITY CONTROL: Place working standards in sampler in order of decreasing concentration. Approximately 15 samples per hour can be analyzed.

REFERENCE: Methods for the Chemical Analysis of Water and Wastes, EPA-600/4-79-020, USEPA, EMSL, 1979.

PRIMARY NAME: Sulfate Method 375.2

TITLE: Inorganics, Non-Metallics MATRIX: Drinking, surface and wastewaters.

APPLICATION: Date issued 1978. (Colorimetric, automated, methylthymol blue, AAII). After being passed through cation-exchange column, sample is reacted with alcohol solution of barium chloride and MTB @ pH 2.5-3.0 to form barium sulfate. This solution is raised to pH 12.5-13.0 so that excess barium reacts with MTB. Uncomplexed MTB = amount sulfate present]

INTERFERENCES: Multivalent cation interferences are eliminated by an ion exchange column. Samples with pH below 2 should be neutralized since high acid concentrations elute cations from ion exchange resin. Filter or centrifuge turbid samples.

INSTRUMENTATION: Technicon Auto Analyzer; 460 nm interference filters; 15 nm flow cell.

RANGE: 3-300 mg SO4/L (or) 0.5-30 mg SO4/LMDL: Not listed

PRECISION: SD = +/- 1.6 @ mean conc of 110 mg/L (26 samples).

ACCURACY: Mean recovery = 102% (on 24 surface and waste waters)

SAMPLING METHOD: Use plastic or glass containers and collect 50 mL samples.

STABILITY: Cool, 4°C. M.H.T. = 28 Days.

QUALITY CONTROL: Analyze all working standards in duplicate at beginning of a run to develop standard curve. Approximately 30 samples an hour can be analyzed.

REFERENCE: Methods for the Chemical Analysis of Water and Wastes, EPA-600/4-79-020, USEPA, EMSL, 1979.

PRIMARY NAME: Sulfate (Gravimetric) Method 375.3

TITLE: Inorganics, Non-Metallics

MATRIX: Drinking, surface and saline waters, wastewater.

APPLICATION: Date issued 1974. Editorial revision 1978. Sulfate is precipitated as barium sulfate ($BaSO_4$) in HCl medium by the addition of barium chloride. After digestion period, precipitate is filtered, washed with hot water until chloride free, ignited and weighed as $BaSO_4$.

INTERFERENCES: High results may be obtained for samples containing suspended matter, nitrate, sulfite and silica. Alkali metal sulfates frequently yield low results. This is especially true of alkali hydrogen sulfates. Heavy metals such as chromium and iron can interfere.

INSTRUMENTATION: Steam bath; drying oven; muffle furnace; analytical balance; filter paper (ashless)

RANGE: Not listed

MDL: Not listed

PRECISION: SD = 4.7% @ 259 mg/L sulfate (aqueous mix of 9 ions)

ACCURACY: Relative error = 1.9% @ 259 mg/L SO_4 (aqueous mix, 9 ions)

SAMPLING METHOD: Use plastic or glass containers and collect 50 mL samples.

STABILITY: Cool, 4°C.

M.H.T. = 28 Days.

QUALITY CONTROL: This is the most accurate method for sulfate concentrations above 10 mg/L. Use this method when greatest accuracy is required. Make sure precipitate is washed free of chloride. Do not let filter paper flame during ashing of precipitate.

REFERENCE: Methods for the Chemical Analysis of Water and Wastes, EPA-600/4-79-020, USEPA, EMSL, 1979.

PRIMARY NAME: Sulfate (Total) Method 300.0

TITLE: Inorganic Anions in Water MATRIX: Drinking, surface and mixed wastewater.

APPLICATION: A small volume of sample, typically 2 to 3 mL is introduced into an ion chromatograph. The anions of interest are separated and measured using a system comprised of a guard column, separator column, suppressor column and conductivity detector.

INTERFERENCES: Interferences can be caused by substances with retention times similar to and overlaping those of ion of interest. Large amounts of an anion can interfere with peak resolution of adjacent anion. Method interference can be caused by reagent or equipment contamination.

INSTRUMENTATION: Ion chromatograph; analytical balance; guard, separator and suppressor columns.

RANGE: (report results in mg/L). MDL: 0.206 mg/L.

PRECISION: SD = 1.475 mg/L @ 98.5 mg/L sulfate (drinking water).

ACCURACY: recovery = 104.3% @ 98.5 mg/L sulfate (drinking water)

SAMPLING METHOD: Use plastic or glass containers to collect samples.

STABILITY: Cool, 4°C. M.H.T. = 28 Days.

QUALITY CONTROL: The laboratory should spike and analyze a minimum of 10% of all samples to monitor continuing lab performance. Field and laboratory duplicates should be analyzed. Measure retention times of standards. (Sulfates exhibit great changes in retention times).

REFERENCE: Test Method-The Determination of Inorganic Anions in Water by Ion Chromatography, (EPA-600/4-84-017).

PRIMARY NAME: Sulfate (Turbidometric) Method 375.4

TITLE: Inorganics, Non-Metallics

MATRIX: Drinking, surface and wastewaters.

APPLICATION: Date issued 1971. Editorial revision 1978. Sulfate ion is converted to a barium sulfate suspension under controlled conditions. The resulting turbidity is determined using a photometer and compared to a curve prepared from standard sulfate solutions.

INTERFERENCES: Suspended matter and color interfere. Silica in concentrations over 500 mg/L will interfere.

INSTRUMENTATION: Nephelometer or [spectrophotometer @ 420 nm; light path of 4-5cm].

RANGE: Not listed

MDL: 1 mg/L sulfate

PRECISION: SD = 7.86 mg/L @ 110 mg SO_4/L.

ACCURACY: As bias, -3.3 mg/L @ 110 mg SO_4/L.

SAMPLING METHOD: Use plastic or glass containers and collect 50 mL samples.

STABILITY: Cool, 4°C.

M.H.T. = 28 Days.

QUALITY CONTROL: Correct for sample color and turbidity by running blanks from which barium chloride has been omitted. Suitable for all ranges of sulfate, but use sample aliquot with not more than 40 mg SO_4/L. Above 50 mg/L the accuracy decreases and suspensions lose stability.

REFERENCE: Methods for the Chemical Analysis of Water and Wastes, EPA-600/4-79-020, USEPA, EMSL, 1979.

PRIMARY NAME: Temperature (Thermometric) Method 170.1

TITLE: Physical Properties

MATRIX: Drinking, surface and saline waters, wastewater.

APPLICATION: Date issued 1971. Temperature readings are used in analyses calculations and general laboratory operations. In limnological studies, water temperatures as a function of depth often are required. Identification of deep wells is often possible by temperature measurements.

INTERFERENCES: For field operations, to prevent breakage, use a thermometer with a metal case. Thermometer should have a minimal thermal capacity to permit rapid equillibration.

INSTRUMENTATION: Thermometers (mercury-filled) or metallic (dial type) or a thermistor.

RANGE: Not listed

MDL: Not listed

PRECISION: Not determined.

ACCURACY: Not determined.

SAMPLING METHOD: Use plastic or glass containers and collect 100 mL samples.

STABILITY: No preservation required; analyze immediately.

QUALITY CONTROL: Thermometers should have a scale marked for every 0.1°C with markings etched on the capillary glass. Periodically check the thermometer against an NBS certified precision thermometer that is used with its certificate and collection chart.

REFERENCE: Methods for the Chemical Analysis of Water and Wastes, EPA-600/4-79-020, USEPA, EMSL, 1979.

PRIMARY NAME: Total Organic Halides (TOX)

Method 450.1

TITLE: Inorganics, Non-Metallics

MATRIX: Drinking and ground waters.

APPLICATION: Method determines TOX as Cl by carbon absorption. Organic halides are defined as all organic species containing chlorine, bromine, and iodine that are adsorbed by granular activated carbon under method conditions. Fluorine containing species are not determined by this method.

INTERFERENCES: Method interferences may be caused by contaminants, reagents, glassware and other sample processing hardware. Improperly handled (or stored) activated carbon (or carbon samples which register >1000 ng/40 mg) can interfere; also halogenated organic vapors can interfere.

INSTRUMENTATION: Dohrmann microcoulometric-titration and adsorption systems; strip-chart recorder

RANGE: Sensitivity limit = 5 mg/L. MDL: Not listed

PRECISION: SD = 4.3 @ 71 average halide ug Cl/L. (Tap water).

ACCURACY: Averagee recovery = 89% @ dose as 88 ug Cl/L (chloroform)

SAMPLING METHOD: Use a 250 mL amber glass bottle fitted with Teflon-lined cap.

STABILITY: Minimize volatile loss; store @ 4°C without headspace.

QUALITY CONTROL: The laboratory must develop and maintain a statement of method accuracy for their lab. Before analysis, analyst must demonstrate ability to generate acceptable precision and accuracy on appropriate QC check samples. Run all samples in duplicate.

REFERENCE: Total Organic Halide Method 405.1-Interim, EPA-600/4-81-056, USEPA, EMSL, 1980.

PRIMARY NAME: Total Suspended Solids (TSS) Method 160.2

TITLE: Physical Properties

MATRIX: Drinking, surface and saline waters, wastewater.

APPLICATION: Date issued 1971. Also referred to as residue, non filterable. A well mixed sample is filtered through a weighed glass fiber filter, and the residue on the filter is dried to constant weight at 103-105°C. Increase in weight of the filter represents TSS. The filtrate may be used for residue, filterable.

INTERFERENCES: Samples high in dissolved solids; saline waters, brines, and some wastes may be subject to a positive interference. Select filtering apparatus with care, so that washing of filter and dissolved solids in the filter minimizes this potential interference.

INSTRUMENTATION: Glass fiber filter discs (Reeves Angel 934-AH, or equivalent). Drying oven @ 103-105°C.

RANGE: 4 to 20,000 mg/L. MDL: Not listed

PRECISION: Not listed

ACCURACY: Not listed

SAMPLING METHOD: Use plastic or glass containers and collect 100 mL samples.

STABILITY: Cool, 4°C. M.H.T. = 7 Days.

QUALITY CONTROL: Non representative particulates such as leaves, sticks, fish and lumps of fecal matter should be excluded from the sample if it is determined that their inclusion is not desired in the final result.

REFERENCE: Methods for the Chemical Analysis of Water and Wastes, EPA-600/4-79-020, USEPA, EMSL, 1979.

PRIMARY NAME: Total Solids (TS) Method 160.3

TITLE: Physical Properties MATRIX: Drinking, surface and saline waters, wastewater.

APPLICATION: Date issued 1971. Also referred to as residue, total. A well mixed sample is evaporated in a weighed dish and dried to constant weight in an oven @ 103-105°C. Increase in weight over empty dish represents the total solids. Total solids is sum of homogenous suspended and dissolved materials in a sample.

INTERFERENCES: Non-representative particulates such as leaves, sticks, fish and lumps of fecal matter should be excluded from the sample if it is determined that their inclusion is not desired in the final result.

INSTRUMENTATION: Drying oven @ 103-105°C; porcelain evaporating dish (100 mL).

RANGE: 10 mg/L to 20,000 mg/L. MDL: Not listed

PRECISION: Not listed

ACCURACY: Not listed

SAMPLING METHOD: Use plastic or glass containers and collect 100 mL samples.

STABILITY: Cool, 4°C. M.H.T. = 7 Days.

QUALITY CONTROL: Floating oil and grease, if present, should be included in the sample and dispersed by a blender device before aliquoting.

REFERENCE: Methods for the Chemical Analysis of Water and Wastes, EPA-600/4-79-020, USEPA, EMSL, 1979.

PRIMARY NAME: Turbidity (Nephelometric) Method 180.1

TITLE: Physical Properties MATRIX: Drinking, surface and saline waters

APPLICATION: Date issued 1971. Editorial revisions 1974 and 1978. Method is based on comparison of light scattered by the sample under defined conditions with a standard reference suspension. The higher the intensity, the higher the turbidity.

INTERFERENCES: Presence of floating debris and coarse sediments which settle out rapidly will give low readings. Finely divided air bubbles will affect results in a positive manner. Presence of true color, dissolved substances which absorb light, cause low turbidities.

INSTRUMENTATION: Nephelometer (with light source) and one or more photo-electric detectors

RANGE: 0 to 40 NTU MDL: Not listed

PRECISION: SD = +/- 0.60 and 1.2 Units @ NTU levels of 26 and 75.

ACCURACY: Not listed

SAMPLING METHOD: Use plastic or glass containers and collect 100 mL samples.

STABILITY: Cool, 4°C. M.H.T. = 48 Hours.

QUALITY CONTROL: Use turbidity free water; sample tubes must be clear, colorless glass. Ntu = nephelometric turbidity units.

REFERENCE: Methods for the Chemical Analysis of Water and Wastes, EPA-600/4-79-020, USEPA, EMSL, 1979.

Chapter 8
Abbreviations

The following abbreviations are used with these summaries.

< — less than
> — greater than
@ — at
AA — Atomic absorption
AAS — atomic absorption spectrometer
BOD — biological oxygen demand
cm — centimeters
C — Centigrade (when used with deg.)
C — true value (when used with accuracy)
$CaCO_3$ — calcium carbonate
COD — carbon oxygen demand
DO — dissolved oxygen
EC — electron capture
ECD — electron capture detector
EP — extraction procedure
FLAA — flame atomic absorption
GC — gas chromatograph or gas chromatography
GC/MS — gas chromatograph/mass spectometer
GFAA — graphite furnace atomic absorption
H_2O_2 — hydrogen peroxide
HCl — hydrochloric acid
HNO_3 — nitric acid
H_2SO_4 — sulfuric acid
ICP — inductively coupled plasma emission spectrometry
kg — killogram
KI — potassium iodide

L	—	liter
MDL	—	method detection limit
mg	—	milligram
M.H.T.	—	maximum holding time for a sample
mL	—	milliliter
mm	—	millimeter
$Na_2S_2O_3$	—	sodium thiosulfate
NH_3	—	ammonia
NH_4OH	—	ammonium hydroxide
nm	—	nanometer
PID	—	photoionization detector
PQL	—	practical quantitation limit
RSD	—	relative standard deviation
SD	—	standard deviation
TDS	—	total dissolved solids
TKN	—	total Kjedahl nitrogen
TS	—	total suspended solids
µg	—	microgram
VOC	—	volatile organic compounds
X	—	mean recovery (used with accuracy)

Chapter 9
Definitions

Accuracy: the nearness of a result or the mean (X) of a set of results to the true value. Accuracy is assessed by means of reference samples and percent recoveries.

ACS Reagent Grade: See reagent grade.

Analytical Batch: the basic unit for analytical quality control is the analytical batch. The analytical batch is defined as samples which are analyzed together with the same method sequence and the same lots of reagents and with the manipulations common to each sample within the same time period or in continuous sequential time periods. Samples in each batch should be of similar composition.

Analytical Reagent Grade (AR): See reagent grade.

Background Samples: matrices minus the analytes of interest (matrix blanks) that are carried through all steps of the analytical procedure. All reagents, glassware, preparations, and instrumental analyses are included. These are necessary to account for the presence of spurious analytes, interferences, and background concentrations of the analyte of interest.

Blank: an artificial sample designed to monitor the introduction of artifacts into the process. For aqueous samples, reagent water is used as a blank matrix; however, a universal blank matrix does not exist for solid samples. The blank is taken through all the appropriate steps of the sample preparation and analysis process.

Calibration Check: verification of the ratio of instrument response to analyte amount. A calibration check is performed by analyzing for analyte standards in an appropriate solvent. Calibration check solutions are made from a stock solution which is different from the stock used to prepare standards.

Environmental Sample: An environmental sample or field sample is a representative sample of any material (aqueous, nonaqueous, or multimedia) collected from any source for which determination of composition or contamination is requested or required. For the purposes of this manual, environmental samples are classified as:

Surface Water — water from lakes, rivers, and streams.

Groundwater — water from wells.

Drinking Water — delivered (treated and untreated) water designed as potable water.

Water/Wastewater — raw source waters for public drinking water supplies, groundwaters, municipal influents and effluents, and industrial influents and effluents.

Sludge — municipal sludges and industrial sludges.

Waste — aqueous and nonaqueous liquid wastes, chemical solids, contaminated soils, and industrial liquid and solid wastes.

Equipment Blanks: samples of analyte-free media that have been used to rinse the sample equipment. They are used to document adequate decontamination of sampling equipment after its use. These blanks are collected after equipment decontamination and prior to using the equipment for sampling again.

Field Blanks: samples of analyte-free media similar to the sample matrix that are transferred from one vessel to another or exposed to the sampling environment at the sampling site. They are used to measure incidental or accidental contamination of a sample during the whole process (sampling, transport, sample preparation, and analysis). Capped and cleaned containers are taken to the sample collection site. Usually at least one field blank with a matrix comparable to the sample of interest is collected with each batch of samples.

Instrument Blanks: solvent or reagent blanks used to measure interference or contamination from an analytical instrument by cycling matrices containing materials that are normal to the analysis (but minus the analytes of interest) through the instrument.

Laboratory Blanks: See background samples.

Material Blanks: samples of construction materials such as those used in groundwater wells, pump and flow testing, etc. They are used to document decontamination (or measure artifacts) from use of these materials.

Matrix Samples: See background samples.

Matrix Spike: samples to which predetermined quantities of stock solutions of certain analytes are added prior to sample preparation (extraction, digestion) and analysis. Samples are split into duplicates, spiked and analyzed. Percent recoveries are calculated for each of the analytes detected. The relative percent difference between the samples is calculated and used to assess analytical accuracy in terms of recovery.

MDL: See method detection limit.

Method Blanks: See background samples.

Method Detection Limit: The minimum concentration of an analyte above true zero that can be detected and reported. The MDL does not take into account concentrations of the analyte in a background sample and may be equal to or even less than the Method Quantitation Limit (also called the Limit of Detection *"LOD"*)

Method Quantitation Limit: the minimum concentration of a substance that can be measured and reported.

MQL: See method quantitation limit.

Practical Quantitation Limit: the lowest level that can be reliably achieved within specified limits of precision and accuracy during routine laboratory operating conditions.

PQL: See Practical Quantitation Limit.

Precision: the measurement of agreement of a set of replicate results among themselves without assumption of any prior information as to the true result. Precision is assessed by means of duplicate or replicate sample analysis.

QC Check Sample: a blank which has been spiked with the analyte(s) from an independent source in order to monitor the execution of the analytical method. The level of the spike is at the regulatory action level when applicable. Otherwise, the spike is usually 5 times the estimate of the quantitation limit. The matrix used is phase matched with the samples and well characterized; for example, reagent grade water is appropriate for an aqueous sample.

RCRA: the Resource Conservation and Recovery Act.

Reagent Blank: an aliquot of analyte-free water or solvent analyzed with the analytical batch.

Reagent Grade: reagents which conform to the current specifications of the Committee on Analytical Reagents of the American Chemical Society.

Replicate Sample: a sample prepared by dividing a sample into two or more separate aliquots. Duplicte samples are considered to be two replicates. Replicate samples are used to measure the precision of the analytical methods used.

Rinsate Blanks: See equipment blanks.

Solvent Blanks: blanks consisting only of the solvent used to dilute or extract the sample. They are used to identify and/or correct for signals produced by the solvent or by impurities in the solvent.

Spiked Field Blanks: field blanks that have known amounts of the analytes of interest added to them. They are used to measure effects that the sample matrix may have on the analytical methods (usually analyte recovery).

Spiked Laboratory Blanks: laboratory blanks that have known amounts of the analytes of interest added to them. They are used to measure systematic bias from all laboratory sources.

Spiked Test Samples: samples used to measure effects that the sample matrix may have on the analytical methods (usually analyte recovery).

Standard Curve: a curve which plots concentrations of known analyte standard versus the instrument response to the analyte.

Surrogate: organic compounds which are similar to analytes of interest in chemical composition, extraction, and chromatography, but which are not normally found in environmental samples. These compunds are spiked into all blanks, standards, samples and spiked samples prior to analysis. Percent recoveries are calculated for each surrogate.

Trip Blanks: samples of analyte-free media taken from the laboratory to the sampling site and returned to the laboratory unopened. They are used to measure cross-contamination from the container and preservative during transport, field handling, and storage.